INTERNATIONAL ECONOMICS

International Economics
Ninth Edition

MC Vaish

Ex-Professor and Head
Department of Economics
University of Rajasthan, Jaipur

Sudama Singh

Ex-Professor
of Economics
Gorakhpur University, Gorakhpur

Oxford & IBH Publishing Co. Pvt. Ltd.
New Delhi
(*A Unit of* CBS Publishers & Distributors Pvt Ltd)

CBSPD

CBS Publishers & Distributors Pvt Ltd

New Delhi • Bengaluru • Chennai • Kochi • Kolkata • Lucknow • Mumbai
Hyderabad • Jharkhand • Nagpur • Patna • Pune • Uttarakhand

International Economics
Ninth Edition

ISBN: 978-81-204-1764-9

Ninth Edition 2013
 Reprint: 2017, 2018, **2024**

First Edition 1975
Second Edition 1977
Third Edition 1980
Fourth Edition 1983
Fifth Edition 1988
Sixth Edition 1993
Seventh Edition 1998
Eighth Edition 2006

OXFORD & IBH
New Delhi
(A Unit of CBS Publishers & Distributors Pvt Ltd)

Published by **Satish Kumar Jain** and produced by **Varun Jain** for

CBS Publishers & Distributors Pvt Ltd
4819/XI Prahlad Street, 24 Ansari Road, Daryaganj, New Delhi 110 002, India.
Ph: 011-23289259, 23266861   Website: www.cbspd.com
                                 e-mail: delhi@cbspd.com

Corporate Office: 204 FIE, Industrial Area, Patparganj, Delhi 110 092
Ph: 011-4934 4934   Fax: 011-4934 4935
                         e-mail: publishing@cbspd.com; publicity@cbspd.com

Branches

- **Bengaluru:** Seema House 2975, 17th Cross, KR Road, Banasankari 2nd Stage, Bengaluru 560 070, Karnataka, India
  Ph: +91-80-26771678/79   Fax: +91-80-26771680   e-mail: bangalore@cbspd.com
- **Chennai:** 7, Subbaraya Street, Shenoy Nagar, Chennai 600 030, Tamil Nadu, India
  Ph: +91-44-26680620, 26681266   e-mail: chennai@cbspd.com
- **Kochi:** 42/1325, 1326, Power House Road, Opp KSEB, Power House, Ernakulum Kochi 682 018, Kerala, India
  Ph: +91-484-4059061-65,67   Fax: +91-484-4059065   e-mail: kochi@cbspd.com
- **Kolkata:** 147, Hind Ceramics Compound, 1st Floor, Nilgunj Road, Belghoria, Kolkata-700056, West Bengal, India
  Ph: +033-25633055, 033-25633056   e-mail: kolkata@cbspd.com
- **Lucknow:** Basement, Khushnuma Complex, 7 Meerabai Marg (Behind Jawahar Bhawan), Lucknow-226001, UP, India
  Ph: +0522-4000032   e-mail: tiwari.lucknow@cbspd.com
- **Mumbai:** PWD Shed, Gala no 25/26, Ramchandra Bhatt Marg, Next to JJ Hospital Gate no. 2, Opp. Union Bank of India, Noorbaug, Mumbai-400009, Maharashtra, India
  Ph: 022-66661880/89   e-mail: mumbai@cbspd.com

Representatives

- Hyderabad   0-9885175004
- Patna   0-9334159340
- Jharkhand   0-9811541605
- Pune   0-9664372571
- Nagpur   0-8692091830
- Uttarakhand   0-9716462459

Printed at Chaman Enterprises, Daryaganj, Delhi, India

Preface to the Ninth Edition

The present edition of the book has been extensively revised and completely recast both in the form and substance by incorporating the latest literature on the subject which has left its mark on almost every page of the book. More than one-half of the figures in the book have been improved and redrawn. Three new chapters entitled Imperfect Competition and Intraindustry Trade and Monopolistic Competition and International Trade in part two of the book and one chapter entitled Internal and External Balance in part three of the book have been incorporated taking the total chapters to thirty sixs. All the chapters in Part Five have been updated by incorporating the fresh data from the latest annual reports of the IMF; IBRD; IDA and the ADB. Similarly, all the other chapters in the book have also been analytically improved and updated. Vast improvement has also been made in the language and the form of presentation of the matter in every chapter in the book. For the readers' convenience separate author index and subject index have also been given at the end of the book.

It is earnestly hoped that the teachers, students and others interested in the study of the subject of international economics and finance will find this thoroughly revised edition of the book more useful and engrossing than the preceding edition of the book and also compared with the other textbooks available on the subject. It is our constant endeavour to improve the quality of the book and with this end in view suggestions for improvement of the book from every quarter will be most welcomed and respected.

Jaipur:
September 1, 2012

M.C. VAISH
SUDAMA SINGH

Contents

PART THREE
FOREIGN EXCHANGE RATE AND THE BALANCE OF PAYMENTS

PART FOUR
TRADE POLICY

PART ONE

INTRODUCTION

CHAPTER 1

Nature and Importance of International Economics

Alfred Marshall[1] has very aptly defined Political Economy[2] or Economics as "a study of mankind in the ordinary business of life; it examines that part of individual and social action which is most closely connected with the attainment and with the use of the material requisites of wellbeing."[3] Thus, economics is concerned with the study of man's those manifold daily activities which are directly related to eking out a living, being in general concerned with the problems of consumption, production, exchange and distribution of the national product.

Meaning and Contents of International Economics

What is true of the general economics, is equally true of that specialised branch of economics which is called international economics. It may be narrowly defined as a set of problems, including monetary and physical problems, which arise from and in connection with the exchange of merchandise and commercial services between the nations. The basic concepts and analytical tools of international economics are precisely the same as those used in the principles of economics. The main difference, however, between the two lies only in the focus. International economics is mainly concerned with the study of economic relations between sovereign nations focusing on the international aspects of economic activities. Since the basis of all domestic and international economic relationships is *quid pro quo* involving mostly the exchange of physical goods and services between people, the core of international economic relationship comprises only that

1. Alfred Marshall (1842-1924) taught at the Cambridge University. He led the Cambridge School of Economists and developed a body of economic theory which served as the basis for the study of economics in the English-speaking countries for over two decades before 1914, although his influence persisted well after 1914. Among his well-known works, mention may only be made of his magnum opus *Principles of Economics* which was first published in 1890 and whose eighth edition came out in 1920. This edition of the book has since been reprinted several times. It is said that the publishers of the book Macmillan & Co. Ltd., sold this book successfully without offering any discount to the book trade. According to Mrs. Joan Robinson, who was Alfred Marshall's worthy student, this book was the Bible for the economics students at the Cambridge University. Marshall's other works include *Industry and Trade* published in 1919 and *Money, Credit and Commerce* published in 1923.

2. Before Alfred Marshall's book *Principles of Economics* which was published in 1890, all treatises on economics were published with the title of *Principles of Political Economy* because economics was not recognised as a separate discipline. See, for example, the titles of the works of Thomas Robert Malthus, David Ricardo, John Stuart Mill and others.

3. Alfred Marshall, *Principles of Economics*, Eighth Edition, 1920, Reset and Reprinted 1949, p. 1.

aspect of the entire range of international relations the basis of which is the exchange of physical goods and commercial services between the people belonging to different nations. Consequently, international economic relationship and international trade may be treated as synonymous for most purposes. International trade may be defined as the exchange of goods and services between different independent or sovereign states or countries in the world.

As a separate discipline, international economics has several tasks to perform. In the first place, it must explain and analyse the substantive contents of international economic relations. As already stated, the international exchange of physical goods and commercial services is quantitatively the single most important component of international economic relations. However, apart from the international exchange of goods and services, the second important component of the substantive contents of international economic relations relates to "factor" movements–movements of labour and capital–from one country to another. Historically, such international factor movements have played a significant role in the development of many countries' national economies. For example, in her early stages of economic development, the American economy was immensely helped by the massive immigration of cheap labour from Europe and large capital inflows from England. Similarly, Australia owes her present economic development to the massive foreign capital investments and large, scale immigration of labour to that country in the early part of the twentieth century. Today, the world's underdeveloped economies are being greatly assisted in their planned economic development by the massive equity and loan capital inflows from the capital-rich developed countries and international financial institutions such as the International Bank for Reconstruction and Development the International Development Association (IDA), the International Finance Corporation (IFC) etc. Consequently, such factor movements must be regarded as major components of international economics.

The "pure" theory of international economics answers two major and basic questions relating to international trade and factor movements. Firstly, it seeks to answer the question: Why do international specialisation and exchange and, parallel to this, international movements of labour and capital take place? What determines which goods and services will be exported and imported by each country, on what terms, and in what quantities? Secondly, it seeks to answer the question: What are the economic effects of free international trade and factor movements on the welfare of people residing in the individual countries and world as a whole? In other words, in more specific terms, it analyses the effects of international exchange of goods and services and factor movements on the productive efficiency and structure of the national economies and on the level and distribution of national income in the countries engaged in trading.

The pure theory of foreign trade describes the equilibrium conditions in real magnitudes and is a part of the general value and price theory. The pure theory of international trade has largely been developed in a static framework and discussions have taken place in terms of both the general and partial equilibrium analyses. The classical theory of comparative cost advantage and its elaborations are examples of the static general equilibrium theory while the study of the effects of an import duty levied on a commodity or a single industry (as against the economy as a whole) offers an example of the static partial equilibrium analysis.

Distinction is frequently made between the monetary and the pure (or equilibrium) theory of international trade. The monetary theory of international trade is mainly concerned with the study of the different methods of correcting the external balance of payments disequilibrium and with the determination of the short and long-term foreign exchange rates. The monetary theory of international trade is in part a dynamic theory and is closely related to the trade cycle and the income and employment determination theories. The pure theory of trade deals with the real variables which it assumes are determined independently of the monetary system, i.e., it assumes the neutrality of money. In other words, the real and the monetary variables of the economic system are determined independently of one another.

The third major part of international economics is related to the mechanism rather than the substance of economic relations between independent countries. Of special importance in this connection are the international monetary relations. Since each sovereign nation or country has her own independent monetary system, the study of international monetary relations is concerned with answering the various such questions as: What are the instruments for making payments from one country to another? From where and how does a country obtain the means of making the international payments? What are the causes and effects of disequilibrium in the external balance of payments of a country?

Apart from these important economic issues, international economics analyses the processes which bring different national economies into a set of balanced international relationships. Since each country is politically independent, being free to determine her own economic policies, there is a possibility of different national policies conflicting with one another. Consequently, the problem of maintaining equilibrium in international economic relationships and of restoring equilibrium in such relationships, if and when disturbed, is of prime importance.

It is also the business of international economics to study the major contemporary international economic problems with special reference to the problems of world's developing countries. Consequently, the study of special international economic institutions concerned with the trade and economic development of these countries such as the World Trade Organisation (WTO) which has replaced the General Agreement on Tariffs and Trade (GATT) from 1995, the United Nations Conference on Trade and Development (UNCTAD) and the International Development Association (IDA) assumes special importance in the context of the special trade and development problems of the developing countries.

The main propelling force behind all economic relationships is the individual self-interest. Just as in the case of domestic trade the act of exchange gratifies the self-interest of both the parties, similerly nations engage in trading because by doing so each nation serves her own self-interest by attaining the higher community welfare reflecting higher social wellbeing. This fact explains why people belonging to different ethnic groups living in different socio-political environment have carried on trading activities from very early times. In fact, in the early days the enterprising merchants from the Occident came to the Orient with goods laden on their frail boats after undertaking risky voyages across the vast and rough oceans in order to exchange their surplus goods against the spices and other goods with which they returned to their native lands. The early voyage undertaken by the Portuguese explorer Vasco da Gama who arrived in Calicut in 1498 and the trading activities of the East India Company, formed in 1600, to develop trade and commerce in South East Asia and India bear ample testimony to such commercial activities undertaken by the early enterprising merchants.

Literature on International Economics

Copious literature is available on international economics dealing with the theory and policy of international trade. A bulk of the analytical literature on the "pure" theory of international trade using sophisticated analytical tools is available in the form of academic articles written by eminent economists and published in the internationally-known academic journals and treatises. The theoretical literature on international economics has grown rapidly over time through the process of original contributions. Consequently, notions in the domain of international trade, which appear common-place today, were considered advanced theory fifty years back and were the result of original research undertaken another fifty years before that time.

The changing nature of the substantive contents of literature on the pure theory of international trade reflects the changing nature of the structure and framework within which international trade has been conducted over time. The literature of the Middle Ages and the mercantilist period is scattered mostly in the religious and political treatises and pamphlets of descriptive nature. In the mercantilist system, which dominated the scene during the sixteenth, seventeenth and the first-half of the eighteenth centuries, foreign trade was highly regulated and controlled by the omnipotent and omniscient state. Since the concept of nation was synonymous with that of state, every effort was made to make the state powerful. In this strategy, special stress was laid on the accumulation of specie. Consequently, no effort was spared to expand the export of manufactured goods in order to create a favourable external balance of payments to allow for the massive gold inflows which was considered essential for increase in the country's real wealth. The mercantilists advocated the restriction of imports and the encouragement of exports in order to induce an active balance of trade and an inflow of gold. Mercantilism, however, received its death-blow in 1752 with the publication of David Hume's *Political Discourses* which heralded the appearance in a systematic form of the classical theory of free trade. Reacting sharply against the faulty mercantilist approach, Adam Smith strongly denounced the mercantilist system in his magnum opus entitled *An Inquiry into the Nature and Causes of the Wealth of Nations* published in 1776. As against the mercantilists' emphasis on restricting imports and expanding exports to achieve a favourable external balance of trade, Adam Smith supported the system of free trade.

In the history of economic thought, the earliest systematic treatment of international trade is the classical theory cast against the background of the late eighteenth and nineteenth century economic liberalism amply reflected in the then-dominant economic philosophy of *laissez-faire laissez-passer*. The classical theory of international trade was first systematically propounded by Adam Smith—the father of political economy—and was subsequently elaborated and developed by his able followers, including Henry Thornton, David Ricardo, Nassau William Senior and John Stuart Mill. The classical theory of international trade was carefully built brick-by-brick at the hands of many economists. To the analytical framework prepared by the classical economists, the later-day economists, brought up in the classical tradition, including Alfred Marshall, John Elliott Cairnes, Charles F. Bastable and Frank William Taussig, to mention only a few, added substantially to make the classical theory of international trade more broad-based and comprehensive.

In the twentieth century, particularly during and after World War I, the nature of international trade underwent a substantial change necessitating modifications and refinements in the *naive* classical theory which was cast in the framework of static equilibrium analysis. Monopolies and international cartels appeared on the scene. The German transfer problem, post-war hyperinflation, abandonment of the international gold standard, great depression of the *thirties,* etc. caused chronic imbalances in the external balance of payments of many countries compelling them to abandon free trade and to take resort to restrictive trade measures, including foreign exchange controls, tariffs and physical quotas, thwarting the multilateral trading system.

All these changes in the nature of international trade were reflected in the growth of new literature on international economics. Frank Graham, Bertil Ohlin, Jacob Viner, Gottfried Von Haberler, Paul Einzig and others dealt with some of these problems in their excellent works. Jacob Viner's book entitled *Studies in the Theory of International Trade* (1937) presents a scholarly treatment of the history of doctrine and is unrivalled for its statement of the classical theory of comparative cost advantage and international values. As the nature of problems continued to change, new theoretical literature continued to flow in. Charles P. Kindleberger, Robert A. Mundell, Murray C. Kemp, Jaroslav Vanek, P.T. Ellsworth, Lorie Tarshis, Delbert A. Snider, Kelvin Lancaster, Richard E. Caves, James E. Meade, Bo Sodersten and H. Robert Heller, to name only a few, have written excellent treatises on the theory of international trade and students in most parts of the world use their books as recommended text books.[4] Since in the present-day world the problem of economic development of the underdeveloped regions of the world has assumed tremendous importance, an ever increasing number of economists have studied international trade in the context of the less developed regions of the world. Consequently, many analytical articles and excellent treatises have been published providing a comprehensive analysis of the interrelationship between foreign trade and economic development. Several writers, including Raul Prebisch, Jacob Viner, John R. Hicks, Charles P. Kindleberger, Trygve Haavelmo, Staffan Burenstam Linder, Hla Myint, Gottfried Von Haberler and Ragnar Nurkse have considered the various aspects of international trade from the viewpoint of the world's poor countries.

Economists have also presented several models for analysing those factors which affect the structure of comparative costs and movements in the terms of trade between the developing and developed countries. It was John R. Hicks who had first developed the model in his "Inaugural Lecture", published in *Oxford Economic Papers,* June 1953, and was later clarified by him in his *Essays in World Economics.* Note B. Harry G. Johnson's article entitled "Economic Expansion and International Trade", published in *Manchester School of Economic and Social Studies,* May 1955 and subsequently revised and extended in his *International Trade and Economic Growth* (1958) contains a comprehensive analysis of the problem. Jagdish Bhagwati's review article entitled "The Theory of International Trade", published in *The Indian Economic Journal,* July 1960, presents an excellent summary of Harry G. Johnson's analysis. Several other articles relating to Harry Johnson's model also clarify the effects of a country's development on its pattern of foreign trade and the terms of trade. Worthy of mention are W.M. Corden's article entitled "Economic Expansion and International Trade: A Geometric Approach", published in the June 1956 issue of *Oxford Economic Papers* and Jagdish Bhagwati's three illuminating articles.[5]

The effect of changes in the factor endowments on trade has been ably discussed by Rybczynski in his well-known article published in August 1959.[6] Findlay and Grubert have sorted out the relationship between technological progress and comparative costs.[7] Bensusan-Butt has

4. See Charles P. Kindleberger, *International Economics,* Fifth Edition, 1973, Richard D. Irwin Inc. Homewood, Illinois; Robert A. Mundell, *International Economics,* 1968, Macmillan Co., New York; Murray C. Kemp, *The Pure Theory of International Trade,* 1964, Englewood, Cliffs, Prentice Hall Inc.; Jaroslav Vanek, *International Trade: Theory and Economic Policy,* 1962, Richard D. Irwin Inc. Homewood, Illinois; P.T. Ellsworth and J. Clark Leith, *International Economy,* Fifth Edition, 1975, Macmillan Co., New York; Lorie Tarshis, *Introduction to International Trade and Finance,* Third Printing, 1962, John Wiley & Sons, Inc. New York; Delbert A. Snider, *Introduction to International Economics,* Sixth Edition, 1976, Richard D. Irwin Inc. Homewood, Illinois; Bo Sodersten, *International Economics,* Second Edition, Reprinted 1981, the Macmillan Press Ltd., London; H. Robert Heller, *International Trade,* Second Edition, 1973, Prentice Hall Inc. Englewood, Cliffs.
5. Jagdish Bhagwati, "Immiserizing Growth: A Geometrical Note", *The Review of Economic Studies,* June 1958: "International Trade and Economic Expansion". *The American Economic Review,* December 1958; "Growth, Terms of Trade and Comparative Advantage", *Economia Internazional,* August 1959.
6. T.M. Rybczynski, "Factor Endowment and Relative Commodity Prices", *Economica,* November 1955.
7. R. Findlay and H. Grubert, "Factor Intensity, Technological Progress, and the Terms of Trade", *Oxford Economic Papers,* February 1959.

made an interesting analysis of the effects of capital formation on trade.[8] Richard E. Baldwin, P.T. Ellsworth, Theodore Morgan, Gottfried Von Haberler, Jagdish Bhagwati and others have presented excellent papers published in different academic journals on the adverse secular trend in the commodity terms of trade of the developing countries.[9]

The balance of payments problems of the developing countries have been ably analysed by James E. Meade, Harry G. Johnson, Kenneth K. Kurihara, J.C. Ingram, W.M. Corden, and others.[10] Ragnar Nurkse has presented an excellent discussion of the balance of payments effects on capital formation.[11] Wolf and Sufrin's work presents a broad survey of the role of foreign investment in the less developed countries.[12] Sir Donald MacDougall's article entitled "The Benefits and Costs of Private Investment from Abroad: A Theoretical Approach", published in the March 1960 issue of *Economic Record* presents a rigorous analysis of the benefits and costs of foreign private investment. Equally penetrating are the discussions of Arndt, Knapp, Streeten and Cairncross.[13] A penetrating analysis of the problem of foreign debt servicing for the less developed countries has been presented by G.M. Alter in his essay "The Servicing of Foreign Capital Inflows by Underdeveloped Countries", published in Howard S. Elliss (ed.) work entitled *Economic Development for Latin America.*

On the issue of protectionist arguments in the context of underdeveloped countries, the study of literature by B.N. Ganguli, Gunnar Myrdal, Raul Prebisch, John Black, Everett E. Hagen, and several other economists can prove illuminating and rewarding.[14] Among the several studies made relating to customs unions or other preferential trading arrangements, those by Jacob Viner, James

8. D.M. Bensusan-Butt, "Model of Trade and Accumulation", *The American Economic Review*, September 1954.
9. Richard E. Baldwin, "Secular Movements in the Terms of Trade", *The American Economic Review. Papers and Proceedings*, May 1955; P.T. Ellsworth, "The Terms of Trade between Primary Producing and Industrial Countries", *Inter-American Economic Affairs*, Summer 1956; Theodore Morgan, "The Long-Run Terms of Trade between Agriculture and Manufacturing", *Economic Development and Cultural Change*, October 1959; Gottfried Von Haberler, "Terms of Trade and Economic Development", published in Howard S. Ellis (ed.), *Economic Developmenti for Latin America*, 1961; Jagdish Bhagwati, "A Skeptical Note on the Adverse Secular Trend in the Terms of Trade of Underdeveloped Countries", *Pakistan Economic Journal*, December 1960.
10. James E. Meade, *Balance of Payments*, Oxford University Press. 1951; Harry G. Johnson, *International Trade and Economic Growth*, 1958, Chapter VI; K.K. Kurihara, "Economic Development and the Balance of Payments", *Metroeconomica*, March 1958; J.C. Ingram, "Capital Imports and the Balance of Payments", *Southern Economic Journal*, Volume XXII, No. 4, and "Growth in Capacity and Canada's Balance of Payments", *The American Economic Review*, March 1957; W.M. Corden, "The Geometric Representation of Policies to Attain Internal and External Balance", *The Review of Economic Studies*, October 1960.
11. Ragnar Nurkse, "The Relation between Investment and External Balance in the Light of British Experience". *The Review of Economic Studies*, May 1956.
12. C. Wolf and S.C. Sufrin, *Capital Formation and Foreign Investment in Underdeveloped Areas*, 1955.
13. H.W. Arndt, "A Suggestion for Simplifying the Theory of International Capital Movements," *Economia Internazionale*, August 1954, and "Overseas Borrowing – The New Model", *Economic Record*, August 1957; J. Knapp, "Capital Exports and Growth", *Economic Journal*, September 1857; P.P. Streeten. *Economic Integration*, Chapter 4; A.K. Cairncross. "The Contribution of Foreign and Domestic Capital to Economic Development", *International Journal of Agrarian Affairs*, April 1961.
14. B.N. Ganguli, "Principles of Protection in the Context of Underdeveloped Countries", *The Indian Economic Review*, February 1952; Gunnar Myrdal, *An International Economy*, 1956, Chapter XIII; Raul Prebisch, "Commercial Policy in the Underdeveloped Countries", *The American Economic Review, Papers and Proceedings*. May 1959; John Black, "Arguments for Tariffs", *Oxford Economic Paper*, June 1959; Everett E. Hagen, "An Economic Justification for Protection", *The Quarterly Journal of Economics*, November 1958.

E. Meade, Richard G. Lipsey, Harry G. Johnson, Bela Balassa, R.L. Allen, Tibor Scitovsky and Paul Streeten deserve special mention for their perceptive analytical contents.[15]

There are those who strongly hold that underdevelopment of world's poor countries is due to the forces of the present international trading system. The most outspoken and comprehensive statement in this regard is that of Raul Prebisch[16] whose views on the matter are found in *Economic Development of Latin America and Its Principal Problems* (United Nations 1950) and *Economic Survey of Latin America, 1949* (United Nations 1950). On the problem of unequal distribution of the gains from trade, a careful reading of Hans W. Singer's stimulating article can prove rewarding.[17] For the criticisms of the traditional approach to economic growth through foreign trade, the student can gainfully glide through Gunnar Myrdal's two classics[18] and Hans W. Singer's illuminating article. The analysis of the problem presented by T. Balogh, Hla Myint and Paul A. Baran in their excellent articles is rigorous and thought-provoking.[19]

The above description of the literature is only a sketchy bird's eye-view of a cross-section of the contributions made in the realm of the theoretical literature on international economics. In fact, the vast canopy of the analytical literature cannot be described in a few pages, not even in a few chapters. The articles by Berrill and Hughes present an excellent discussion of the historical perspective on the development effects of export industries.[20]

Importance of the study of International Economics

Since time is a scarce resource having rival uses, it may be legitimately asked as to what is the utility of studying international economics? Why waste time on the fruitless study of international economics? In a way, the utility of the study of international economics is too obvious to be mentioned. The study of international economics has manifold advantages. Firstly, it broadens the mental outlook of the people by enabling them to think and act beyond the narrow confines of nationalism. It makes the rich nations realise the hard truth that their national prosperity cannot be sustained for long in a world surrounded by poor nations. It brings out the hard fact of

15. James E. Meade, *The Theory of Customs Unions*, 1955; Jacob Viner, *The Customs Union Issue*, 1950; Richard G. Lipsey, "The Theory of Customs Unions: A General Survey", *Economic Journal*, September 1960; Harry G. Johnson, "The Economic Theory of Customs Union", *Pakistan Economic Journal*, March 1960; Bela Balassa, *The Theory of Economic Integration*, 1961; R.L. Allen, "Integration in Less Developed Areas", *'Kyklo'* Volume XIV, No. 3, 1961; Tibor Scitovsky, "International Trade and Economic Integration as a Means of Overcoming the Disadvantages of a Small Nation" published in E.A.G. Robinson (ed.), *Economic Consequences of the Size of the Nations*, 1960; Paul Streeten, *Economic Integration*, 1961.

16. Raul Prebisch, who was the Secretary General of the United Nations Conference on Trade and Development (UNCTAD), received the coveted Jawaharlal Nehru Award for International Understanding for the year 1974. Dr. Prebisch has made a remarkable contribution towards promoting the cause of the Third World comprising the less developed countries. By sheer dint of his imagination and unusual skill he elevated the UNCTAD to the high position of world forum for the understanding and discussion of the development issues and of economic exploitation of the world's poor countries by the rich nations.

17. Hans W. Singer, "The Distribution of Gains between Investing and Borrowing Countries", *The American Economic Review, Papers and Proceedings*, May 1959.

18. Gunnar Myrdal, *An International Economy*, 1956; and *Rich Lands and Poor*, 1957.

19. T. Balogh, "Some Theoretical Problems of Post-War Foreign Investment", *Oxford Economic Papers*, March 1945; Hla Myint, "An Interpretation of Economic Backwardness", *Oxford Economic Papers*, June 1954; Paul A. Baran, "On the Political Economy of Backwardness", *Manchester School of Economic and Social Studies*, January 1952, and "The Political Economy of Growth", *Monthly Review Press*, 1957.

20. K.E. Berrill, "International Trade and the Rate of Economic Growth", *Economic History Review*, April 1960; J.R.T. Hughes, "Foreign Trade and Balanced Growth: The Historical Framework", *The American Economic Review, Papers and Proceedings*, May 1959.

complementarity between the different national economies and highlights the futility of autarky. Its judicious and unbiased study makes us realise the hard fact that a significant role has been played in the past in the development of national economies by foreign capital and labour. The developed economies of the world would not have been what they today are had it not been for the productive activities of massive international factor movements that took place across the national boundaries in the past. It is, therefore, obvious that if nations have to exist and prosper they have to act in the spirit of mutual give-and-take because no nation is self-sufficient in the matter of factor endowments.

It has been seen that slowing down in the growth of world trade slows down the economic growth of the national economies. For example, the slowdown in world trade from 8 per cent annual growth rate in 1985 to 3.5 per cent growth rate in 1986 was reflected in a slowdown of the average growth rate of output in the developing countries from 4.5 per cent in 1985 to 3.5 per cent in 1986. Similarly, the small increase of per cent in the volume of world merchandise trade in 1991 as against 5 per cent in 1990 and of only 1.5 per cent in terms of value as against a jump of 13.5 per cent in 1990 caused the recessionary trend in the world. During 2003, after a slow growth in the first half, world trade accelerated in the second half as a consequence of which the global trade in 2003 expanded by 4.5 per cent in real terms although it remained below the average rate of 6.5 per cent recorded in the 1990s. Similarly the recessionary trend in 2011 in the developed world also impacted the developing countries of the world rendering depression a global phenomenon. Consequently, a judicious study of international economics stresses the inevitability of mutual interdependence between the nations in the process of economic growth.

Secondly, the study of international economics causes development of international outlook among the men of vision, statesmen and economists. This outlook is essential for the tackling of many pressing current economic problems. The pooling of world's limited scarce economic resources is urgently needed to tackle the problem of mass poverty found almost ubiquitously in the world. Its study warns us that nations cannot for long afford to adopt and practise conflicting national economic policies. Its study points out the utter futility of adopting the beggar-my-neighbour policy by the rich countries of the world. The cogent lessons of international economics were forgotten and utterly disregarded by the leading world nations in the *thirties* much to their own distress and dismay.

Thirdly, its study tells us that many difficult problems which cannot be solved single-handed can be solved without much difficulty through international economic cooperation. The study of international economics lends strong support to the view that increasing international economic cooperation and increasing national wellbeing both go together. Moreover, a judicious study of international economics enables us to know about and understand the mechanism of operation of the different national monetary systems. It tells us in unambiguous terms that free trade, which safeguards the legitimate economic interests of every country, is the best guarantee against the outbreak of the global war. History bears ample testimony to the assertion that an era of free trade and multi-convertibility of currencies can never tolerate war. Peace prevailed in the world as long as the *laissez-faire* framework of free trade, reflected in the prewar international gold standard, functioned smoothly.

The phenomenal growth in the volume and value of world trade during the past five decades shows the great belief of world's statesmen in the processes of interdependence and globalisation as the sure means to economic prosperity. From a mere 376 billion US dollars in 1950, the world trade in merchandise goods and services stood at the new record level of US dollar 9.3 trillion[21] in

---

21. One trillion is equivalent to thousand billion.

2010 showing that in almost six decades it had grown by over twentyfour times. Since the average annual growth in world merchandise trade has exceeded the annual average growth in world merchandise output, there has been witnessed an increasing trend toward globalisation measured by the world's trade-output ratio. This trend toward increasing globalisation of world's national economies is particularly welcome as it facilitates the integration of world into a single entity by encouraging multilateral trading system.

A disquieting feature of this phenomenal growth of world trade in merchandise goods and commercial services has, however, been the glaring inequality in the share of this global trade as between the developed and the developing countries. While the share of the five developed countries represented by the United States of America (12.6%), Germany (9.4%), Japan (7.7%), France (5.3%) and the United Kingdom (5.2%) in the total world merchandise trade in 1999 was 40.2 per cent occupying the first five ranks, a large country like India accounted for a pittance of 0.7 per cent in world merchandise trade showing that the developing countries do not matter much in the global trade. The share of the nine developed countries in the world merchandise trade in 1999 was more than one-half being 55.4 per cent. Consequently, the developing countries are not able to extract concessions from the developed countries in their favour and improve their terms-of-trade. A stable world trading system should promise equal harnessing of the trading advantages for all the countries of the world.

The present world trading system in which the developing countries have been marginalised is not conducive to the lasting world peace and prosperity. It has the potential of generating conflicts capable of shaking its shaky foundation and consequently requires urgent reforms to the satisfaction of all the trading participants. Even the establishment of the World Trade Organisation (WTO) in 1995 with over 153 members acounting for over 90 per cent of world trade has so far not been successful in reforming the developed countries dominaing the old world trading system in which the developing countries are being continuously exploited by the developed countries.

Suggested Readings

Charles P. Kindleberger, *International Economics,* Fifth Edition, 1973, Chapter I.

Gerald M. Meier, *International Trade and Development* 1963, pp. 193-202.

Delbert A. Snider, *Introduction to International Economics,* Sixth Edition, 1976, Chapter I.

Questions

1. Discuss the nature and scope of international economics.
2. Why it is essential to study international economics? Discuss.
3. Comment on the mercantilist concepts of exports, imports and the balance of trade.

CHAPTER 2

Salient Features of International Trade

The concept of free international trade refers to that commercial system which does not discriminate between the domestic exchange and foreign exchange of commodities and services. Consequently, a free trade policy neither imposes additional burdens on the latter nor grants any special privileges to the former. The concept of free trade is opposed to every interference by the state with the free play of market economic forces. It is the external trade system of liberalism. It takes into consideration welfare of the whole world. The unrestricted international exchange of goods and services increases the real incomes of all the participating countries.

Appropriateness at Doctrine's Label

Several economists have argued against the use of the terms "international trade" or "foreign trade" on the ground that the theory deals with the trade between different *regions* irrespective of whether or not these regions are "nations" or "countries". Holding the view that the use of the term "international trade" was misleading, the well-known English economist Francis Ysidro Edgeworth had remarked that "international trade meaning in plain English trade between nations, it is not surprising that the term should mean something else in political economy."[1] It was recognised even by the early writers on the subject that the theory of international trade was not exclusively concerned with trade between the sovereign nations.

These writers, from David Hume onwards, who enunciated some theory in this field in terms of trade between different countries had also pointed out that the theory also applied to trade between different regions or provinces within a country. For example, John Stuart Mill had replied that although the statement was correct but he substituted "places" for "countries" in the statement when he was asked if Ricardo's statement that the same rule which regulates the relative values of the domestic goods does not regulate the values of the goods of different countries was correct. Charles F. Bastable had also favoured the replacement of "international" by "interregional". However, on having second thoughts he abandoned the idea of replacing the term international by the term interregional and concluded that "interregional would prove a troublesome word; it is better, therefore, to adhere to the old term."[2] It was the well-known Swedish economist Bertil Ohlin who adopted this troublesome word. However, in adopting the title "Interregional and International Trade" for his important book published in 1933, even he has given the impression that the term "international trade" does not fully embrace the term "interregional trade."

1. Francis Ysidro Edgeworth, *Collected Papers Relating to Political Economy,* Volume II, 1925, p. 5.
2. Charles F. Bastable, *Theory of International Trade,* Fourth Edition, 1903, p. 12, note.

Salient Features of and Justification for a Separate Theory of International Trade

There has taken place a good deal of debate on the issue whether international trade is fundamentally different from domestic or interregional trade. While holding that the theory of international trade should be regarded as a particular application of general economic theory, the classical economists had held that international trade had its own distinctive features which were absent in the case of trade taking place between different regions in the country. Bertil Ohlin, Gottfried Von Haberler and others have, however, argued that it is neither possible nor essential to draw any sharp distinction between international trade and domestic trade.

According to Bertil Ohlin, since there is no fundamental difference between the domestic and foreign trades, there is no need for a separate theory of international trade. According to Ohlin, different nations engage in trading for the same basic reasons for which different individuals or groups within the country trade with each other, instead of each one producing his own requirements. Individuals and, likewise, nations trade with one another because by trading they are enabled to exploit the substantial advantages of division of labour and specialisation to their mutual advantage. The economic gains accruing from practising the division of labour result from the increased scope for practising specialisation in production which makes the optimum use of 'varying ability' of each factor of production possible.

The fundamental cause of trade, be it domestic or foreign, is that all factor units are not alike. Some individuals have greater ability for performing certain tasks than others. Different natural and/ or acquired aptitudes make one person a better engineer, another an excellent doctor, third an efficient teacher, fourth an effective administrator, and so on. It is obvious that specialisation which follows as a corollary from the division of labour by making the adaptation of different tasks to different aptitudes possible is conducive to greater productive efficiency. Even if all individuals had exactly identical natural abilities, it would still pay to practise specialisation in one or more occupations because much greater skill could be acquired than if everyone produced everything required for himself. Moreover, specialisation would effect substantial economy in time which would otherwise be lost in frequently changing over from one occupation or process to another. In short, by causing substantial increase in skill and by saving time, specialisation enables society to enjoy the substantial advantages, in the form of increased production, of the economies of large-scale production. Adam Smith's classic example of pin-making factory brings home the fact of substantial gain resulting from practising the division of labour.

Different regions, like different individuals, are differently endowed with facilities for the production of different commodities because they are differently supplied with the productive factors. Nature's pattern of resource distribution over different parts of the globe is most non-egalitarian and it is amply reflected in the different factor endowments (natural resources) possessed by different countries. One region may abound in iron ore and coal reserves but may have no or little land suitable for growing cotton while another region may be richly endowed with land suitable for cotton cultivation but scantily supplied with iron ore and coal deposits. Obviously, according to the regional variations in resource endowments, the former region is better suited for the production of iron and steel and ill-suited for cotton growing than the latter. It is the differences in the relative proportions of factors in any particular region compared with the other regions which determine its suitability or otherwise for specific industries.

Strictly speaking, it is neither possible nor essential to draw a sharp line of distinction between the problems of domestic and foreign trade. If we examine the alleged peculiarities of the foreign trade, we find that the differences between the two trades turn out merely differences in degree rather than basic differences of a qualitative nature which will justify our drawing any short theoretical division between the two trades towing this line, according to Bertil Ohlin, there is no need for a separate theory of international trade. The mutual interdependence theory of pricing applicable to price determination in a single market could, with appropriate modifications, also study the determination of prices in several markets because the theory of international trade is only a part of the general localisation theory which takes into full account the space aspects of pricing. According to Ohlin, "international trade should be regarded as a special case within the general concept of interregional, or perhaps rather interlocal trade."[3] A close examination of the so-called much-publicised alleged special features of international trade leaves one in no doubt that these peculiarities of foreign trade turn out to be differences of degree rather than such basic differences of a qualitative nature as would justify drawing up of any sharp theoretical divisions leading to the need for a separate theory of international trade.

According to the classical economists, the most salient feature of international trade was the international immobility of factors of production. The belief that factors of production—labour and capital, if not land—although mobile within the country were immobile between the countries led the classical economists to argue nairely that the principles which determined the relative prices differed between the domestic (interregional) trade and foreign (international) trade. Critics have, however, questioned the validity of drawing any sharp distinction between the domestic and foreign trades on this basis alone. Firstly, perfect mobility of factors of production does not frequently exist in the domestic sphere either. Secondly, substantial movements of labour and capital often take place across the national boundaries. Immigration of labour has played an important role in the economic development of the United States of America, Australia, Israel and Argentina. In the post-war period, massive migration of workers took place from Spain, Portugal, Southern Italy, Greece, Yugoslavia and Turkey northward across the mountains to France, Germany and Switzerland. Today workers freely move between the European Economic Community. And what about the international flows of capital?

The economic development of the United States of America owes in no small measure to the massive inflows of British capital. Foreign equity capital and loans have contributed significantly in promoting the economic growth of the Third World's poor countries. It is, therefore, obvious that the difference between the domestic and international factor mobility is at best one of degree and not of kind. Consequently, this alone does not justify our drawing up of any sharp distinction between the interregional and international trade.

As a matter of fact, the classical economists were not oblivious of this fact. Adam smith had stressed the importance of emigration. John Stuart Mill and Charles F. Bastable had recognised that situations could exist in which factor mobility may exist between countries while in others factors may be immobile or at any rate not completely mobile within the country. This led them to infer that where factor immobility existed within a country the theory of international trade would apply whereas if labour and capital were mobile at the international level, a separate theory of international trade would be superfluous. To overcome the difficulty, John Eliatt Cairnes introduced the concept of "non-competing groups" of workers, i.e. groups of workers between which there

3. Bertil Ohlin, *Interregional and International Trade*, Harvard University Press, 1933, p. 589.

was no free movement. Frank William Taussig discussed the problem of differences in wage rates arising from inperfect mobility of labour with foreign trade.[4]

The absence of international factor mobility in the classical theory of international trade was a different kind of mobility from that which was assumed to be present in the case of domestic trade. What underlies the classical theory is the assumption of international *place* immobility of factors of production regardless of the occupation and the assumption of internal occupational mobility of the factors of production regardless of the location, and for a large part of their analysis only the former assumption is significant. Consequently, much of the criticism of the factor immobility assumption of the classical theory of international trade as being unrealistic is irrelevant as it is the result of failure on the part of critics to distinguish between the place and the occupational factor mobilities.

While granting the point of critics that difference between the international factor immobility and the internal factor mobility, even if valid, is valid only as a relative and not as an absolute difference, even so a relative difference in the factor mobility, provided it is of a substantial size, is sufficient to act as a foundation for a separate theory of international trade. "The differences in degrees of mobility of the factors of production, moreover, seem obviously to be great when *countries* are being considered and to be minor, or non-existent, or in the reverse direction, when neighbouring regions within a country are being considered, if the mobilities being compared are place mobility between areas, on the one hand, and occupational mobility within areas, on the other."[5]

Perfect long-run international factor mobility would exist only if sufficient place mobility exists to counter the persistent international differences in the monetary factor rewards in similar occupations. In this sense of the term, there is zero factor mobility so far as natural resources are concerned. As regards the international mobility of labour and capital, during the past seven decades or so, although their national mobility has considerably increased, international movements of labour and capital have been subjected to an ever increasing number of restrictions imposed by the governments in national interests. Who can deny the fact that the nationals of one country cannot enter another country without completing the formidable formalities of obtaining a valid passport and visa which are not always easily granted? Immigration laws severely restrict the entry of foreigners in the country. Lack of identity of language, customs and traditions between the countries act as formidable barriers to labour migration across the national borders. Capital is also more mobile within the country than between the countries because foreign capital cannot be invested in the country freely and in the enterprises of choice of the foreign investors. A vast canopy of national laws relating to company formation, taxation and investment has to be successfully crossed before foreigners can invest their capital in foreign countries. Foreign equity capital cannot be invested in any business and in any amount preferred by the foreign entrepreneurs. For example, in India until very recently before the advent of economic liberalisation, foreign equity investments were subject to Foreign Exchange Regulation Act (FERA) according to which foreign equity participation was restricted to less than one half of the total equity capital. Repatriation of foreign investment capital was also subject to several inhibiting restrictions.

4. This concept was introduced in 1875. See J.E. Cairnes's *Some Leading Principles of Political Economy*, 1875. The most detailed discussion of the question can be found in F.W. Taussig's *International Trade*, 1927, Chapter 6.
5. Jacob Viner, *Studies in the Theory of International Trade*, 1937, pp. 597–598.

These international immobilities of labour, natural resources and, to some extent, even of capital are all that is needed to justify the basis for a separate theory of international trade.[6] In short, in all fairness to the classical economists, it can be said that labour and capital are internationally much less mobile than these are mobile within a country and the problems associated with their mobility are of substantially different nature justifying our drawing of a sharp distinction between the domestic and foreign trades.

The second distinguishing feature of international trade is the existence of independent monetary systems in different countries giving rise to the need of exchanging one country's currency unit for another's at an agreed ratio. The purchase and sale of foreign currency is conducted in the foreign exchange market and the ratio at which different currencies are exchanged is known as *the foreign exchange rate*. Operations in foreign exchange often appear mysterious to the layman and, perhaps more than any other thing, except nationalism, these have convinced him that foreign trade is entirely distinct from the domestic trade. Although during the *laissez-faire* era of the prewar international gold standard, the presence of different national currencies did not matter much as the exchange ratio between them was determined on the basis of parity of their gold contents but in modern times differences in currencies which result in independent and different national monetary and credit policies are of far-reaching importance in so far as these influence the international movements of capital.

As a matter of fact, the presence of different independent national monetary systems need not necessarily change the character of international trade from its domestic counterpart. Under a system of free multi-convertibility of national currencies and stable foreign exchange rates, such as used to be the feature before World War I when a stable international monetary system existed, international payments are effected just as smoothly and easily as are the domestic payments. Under the prewar international gold standard, national currencies were linked to gold and were freely convertible. Consequently, an international payment was not fundamentally different from a domestic payment except that the former involved the residents of different countries.

Today, however, the situation is very different and international trade is sharply separated from domestic trade by the formidable restrictions imposed by the governments on foreign exchange transactions. Different national currencies are no longer freely convertible and foreign exchange rates, being no longer tied to gold, are unstable. When national currencies cease to be freely convertible, the character of the international trade is profoundly affected. An importer is unable to buy his merchandise from the country of his choice because he is unable to buy the necessary foreign exchange from his country's government. Instead, he may be compelled to import inferior or higher priced goods from a country to which the government allows him to make payments. This absence of freedom to buy (and sell) in the best market distinguishes contemporary international

---

6. According to John H. Williams, the classical assumption of international immobility of factors of production restricts the scope of the theory by preventing it to take into account the important economic consequences of substantial migrations of the factors of production which have occurred in the past and which have played an important role in the economic development of several countries in the nineteenth and early part of the twentieth centuries. Secondly, Williams has complained that as a theory of benefits from territorial division of labour, the conclusions of the classical theory of international trade contradict its premise of internal mobility of the factors of production. Trade means national specialisation for the international market. Specialisation is, therefore, the main characteristic feature and root idea of international trade. But specialisation, says Williams, is the anti-thesis of mobility, in this case of the domestic movement of the productive factors. John H. Williams, "The theory of International Trade Reconsidered", *The Economic Journal*, Volume 39, 1929, especially pp. 203–209.

trade from domestic trade. It is, however, important to note that it is the exercise of national sovereignty to restrict international payments than the mere presence of different independent monetary systems which is a distinguishing characteristic of international trade.

The third distinguishing feature of international trade arises from the fact that "the existence of political boundaries carries with it controls and regulations of international trade and payments, in the form of customs duties, quotas, exchange control, foreign trade monopolies, the more subtle measures of control referred to as "administrative protectionism" and so forth, which do not generally exist in the domestic trade area."[7] Governments draw distinction between domestic trade and foreign trade not on the basis of any objective economic criterion but merely on a judgment of value or a rule of law. Domestic trade means simply trade within that area whose prosperity interests a particular government or which is subject to its jurisdiction.

The fourth distinguishing feature of international trade lies in the existence of greater geographical distances and the consequent increases in the transport costs. According to Henry Sidgwick, the fact of distance which renders international exchange costly necessitates a special theory for the determination of international values. International trade, in most cases, involves the hauling of goods over longer distances involving long land, air and sea routes through different countries necessitating the use of different modes of transport. Consequently, transport costs become an important input in the case of international trade while these can in most cases be ignored in the case of domestic trade. Not only do the goods have to be transported over longer geographical distances but distinct problems of packing, insurance, banking and freight, which are generally absent in the case of domestic trade, require a good deal of attention in the case of international trade.

Fifthly, in the analysis of gains from trade, attention is focussed on particular boundaries enclosing areas of community of interest and these areas are generally different countries or nations. Henry Sidgwick correctly stated that "it is only in the case of the foreign trade that the investigation of the conditions of favourable interchange excite practical interest because it is only in this case that there has ever been a serious question of governmental interference with a view of making the interchange more favourable."[8] Moreover, in the theory of international trade in the inductive investigations the unit of investigation has almost invariably been a "country" partly because it has been an area of special interest to the investigator and partly because the greater interest of public and governments on country units has resulted in a relatively much greater supply of statistical information relating to such units than for "regions".

Sixthly, in the case of international trade, the sharing of the gains (terms of trade) from trade is an important macroeconomic issue because international trade involves the exchange of goods between the different people—it is a trade between 'us' and 'them' while interregional or domestic trade involves the exchange of goods between the same people in the sense that they are all members of the same nation—it is a trade between 'us'. Consequently, in the case of domestic trade although the commodity terms of trade may be important from the micro or individual viewpoint but from the macro or national point of view no problem of exploitation arises because a nation cannot exploit itself. In the case of international trade, however, the problem of exploitation of one nation by the other and consequently the issue of sharing the important gains from trade as determined by the terms of trade becomes extremely important.

7. Gottfried Von Haberler, *A Survey of International Trade Theory*, 1961, p. 2.
8. Henry Sidgwick, *The Principles of Political Economy*, Second Edition, 1887, p. 216.

Seventhly, international trade is carried on between severeign nations who, apart from distinct political identity, have distinct linguistic and cultural identities. It has, however, been argued by the critics that it is not a distinguishing feature of international trade as there are nations like India and Canada which have more than one official language and more than one cultural group in their populations. Moreover, several nations may have a common language and a common cultural heritage. Although there is some force in these arguments, it is nevertheless true that people are a like each other in more ways when they are members of the same nation than when they belong to different nations. Consequently, international trade—more than interregional trade—is a trade between people whose tastes, languages, customs, attitudes and other cultural traits differ.

Eighthly, in the case of international trade an adverse external balance of payments disequilibrium poses a special problem for the country while there is no such problem faced by a region or state in the country in the case of interregional or domestic trade because internal disequilibria are automatically financed. For example, there is no balance of trade equilibrium problem involved in the case of inter-state trade between the states of Haryana and Rajasthan. It is not, however, so if the trading regions are two independent sovereign nations like Pakistan and India whose balance of trade in the sense of the total value of exports and imports has to be ir equilibrium. It is partly due to the greater mobility of capital within the country than between the countries which means that the internal disequilibria are automatically financed. It is also partly the consequence of the fact that different regions or states in the country share with one another through the national budget, such sharing being virtually absent when the different sovereign countries are involved in a trading relationship.

Ninthly, the nature of the international markets and consequently of the internationally traded goods is different from the domestically traded goods. It is so because differences in the climate, tastes, customs, cultures, legal institutions, etc. are more marked when the people involved in trading belong to different nations than when they are citizens of the same country. Consequently, international trade is apt to be trade between the people whose languages, customs, attitudes and other cultural traits are different. For example, in India we drive our vehicles to our left hand side while in the United States of America people drive to their right-hand side. The consequence of these different administrative rules of the road is that in the United States of America the cars meant for domestic market are left-hand drive cars while the cars manufactured for sale in the export markets of India, Pakistan, Sri Lanka, Burma, Nepal etc., where following the British practice people move to their left on the roads, have to be right-hand drive cars with the steering fixed to the right-hand wheel side. Similarly, the electric gadgets—refrigerators, cooking range, toasters, mixers etc.—meant for sale in the domestic market in the United States of America must be operable on 110 volts while for sale in the export markets of India, Pakistan, Bangladesh etc. these gadgets should be operable on 220 volts. These examples show that the nature of the two goods and consequently of the two trades is different. In short, in a great many cases goods to be sold in the foreign countries must be specially designed to conform with the national traits and requirements of those countries.

Lastly, while the socio-political environment differs greatly between severeign nations, it is more uniform within a country. Households and business firms in the same country operate within the same legal framework, are subject to the same social institutions and are ruled by the same government. Similarity in the habits and business customs exist within the national boundaries rendering it easier for businessmen to strike dealings with the other economic units even if they are geographically far removed. All these conditions are only very rarely and imperfectly present in the case of trade taking place between different nations.

The following chart highlights the salient features of international trade bringing into sharp focus the principal differences between the domestic trade and foreign trade.

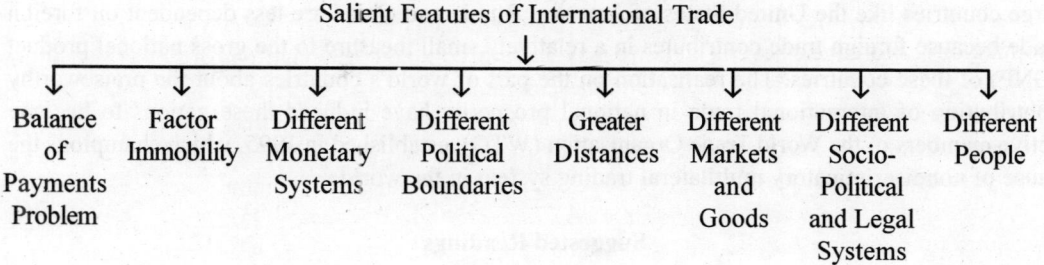

Salient Features of International Trade

| Balance of Payments Problem | Factor Immobility | Different Monetary Systems | Different Political Boundaries | Greater Distances | Different Markets and Goods | Different Socio-Political and Legal Systems | Different People |

Interdependence between International and Domestic Trades

No country in the world today is completely self-sufficient in the sense that it possesses all facilities for economical production of all the goods and services that are consumed by her people. Not even the United States of America and Russia—the richest nations in the world—are in such a happy situation and even they depend on foreign sources of supply for the greater part of their requirements of sugar, tea, coffee, natural rubber, silk, jute, tin, nickel and many other commodities. England imports large amounts of sugar, tea, wheat, cotton, meat, tobacco, silk, spices, metal alloys and many other goods. France imports coal in bulk since her own coal reserves are of inferior quality. Germany and Japan depend on the foreign sources of supply of iron ore and other goods.

Domestic and foreign trades have jointly contributed for many centuries toward meeting the needs of industry and the satisfaction of an ever expending consumer wants. Consequently, the two have become functionally connected. For hundreds of years, international trade has served as a vehicle for extension of improved techniques of production to an ever-increasing number of users. It has served as a medium for cross-fertilisation of different cultures and a connecting link between the activities of people all over the world. Consequently, restrictions on international trade imposed by some other country or group of countries tend to throw people in some other countries out of employment. As a result of this domestic unemployment caused by restrictions imposed on a country's foreign trade by foreign countries, the incomes of those rendered unemployed and consequently their purchases of domestic goods shrink. The unemployed persons in their search for employment seek new occupations modifying in the process domestic sales and purchases. In short, the effects of the imposition of trade restrictions by some one country spread to the other countries and from industry to industry in these other countries. The economy of every country is synchronised with the economy of every other country in the greater or less degree through the interdependence that has evolved gradually as the countries have themselves developed over the years.

In a highly interdependent modern world, all countries depend upon one another for the goods and factors, for financing and for the supply of raw materials. Although some nations could, no doubt, continue to exist without foreign trade but the domestic trade of no nation would continue to flow in its accustomed channels if its foreign trade ceased to exist. The economic aspect of interdependence between the domestic and foreign trades resolves itself largely into the consideration of (1) the gains in the productive efficiency that result from international division of labour, and (2) the extent to which existing population and national industrial economies have grown up with the international trade and have become dependent on it.

While the extent of dependence of different nations on international trade varies a great deal, yet all nations are dependent on it in some degree. While England, Japan, France, Germany, Denmark, etc. depend on foreign trade for their domestic economic prosperity to a large extent, large countries like the United States of America, Russia and China are less dependent on foreign trade because foreign trade contributes in a relatively small measure to the gross national product (GNP) of these countries. The realisation on the part of world's countries about the praiseworthy contribution of international trade in national prosperity have induced these nations to become active members of the World Trade Organisation (WTO) established in 1995, which champions the cause of nondiscriminatory multilateral trading system in the world.

Suggested Readings

Gottfried Von Haberler, *The Theory of International Trade,* 1950, pp. 3–8.
Gottfried Von Haberler, *A Survey of International Trade Theory,* 1961, Chapter 1, pp. 1–5.
Charles P. Kindleberger, *International Economics,* Fifth Edition, 1973, Chapter 1.
Jacob Viner, *Studies in the Theory of International Trade,* 1937, Appendix, pp. 549–601.

Questions

1. 'International trade should be regarded as a special case of interregional or perhaps rather interlocal trade.' (Bertil Ohlin)
   Discuss this statement fully. Give reasons to justify the need for a separate theory of international trade.
2. What are the basic reasons for trade? Justify the need for a separate theory of international trade.
3. Distinguish between international and interregional trade. Is it necessary to have a separate theory of international trade? Discuss fully.
4. Critically examine the following statements:
   (i) "In classical economics, the principles which determined relative values or prices differed between interregional and international trade because factors of production—labour and capital, if not land—were mobile within a country but immobile between countries." (Charles P. Kindleberger)
   (ii) "To the extent there are differences in factor mobility and equality of factor returns internationally as compared with interregionally, international trade will follow different laws." (Charles P. Kindleberger)

CHAPTER 3

Advantages of International Trade

The advantages of international trade[1] follow from the fact that it benefits the national economy by enabling a country to specialise in the production of those goods and services which she is best suited to produce looking to its endowments of natural resources, labour and capital. Some of the acknowledged advantages of international trade are as old as the science of political economy. Adam Smith, a staunch supporter of free trade, had launched his frontal attack on the mercantilist regulations of international trade. Stressing that international trade increases the wellbeing of all the trading nations and that any restriction on it was harmful, Adam Smith wrote: "... It is the maxim of every prudent master of a family never to attempt to make at home what it will cost him more to make than to buy. The tailor does not attempt to make his own shoes, but buys them from the shoemaker. The shoemaker does not attempt to make his own clothes, but employs a tailor. The farmer attempts to make neither the one nor the other, but employs those different artificers. All of them find it for their interest to employ their whole industry in a way in which they have some advantage over their neighbours, and to purchase with a part of its produce, or what is the same thing, with the price of a part of it, whatever else they have occasion for.

What is prudence in the conduct of every private family can scarcely be folly in that of a great kingdom. If a foreign country can supply us with a commodity cheaper than we ourselves can make it, better buy it from them with some part of the produce of our own industry employed in a way in which we have some advantage."[2]

Nothing can be more forceful than Adam Smith's above celebrated statement defending the principle of division of labour and international exchange of goods and services. Adam Smith stated forcefully that international trade enables each nation to increase her wealth and national wellbeing by making extended use of the principle of division of labour that makes specialisation in production possible. Adam Smith's following celebrated passage stating the advantages resulting from the division of labour and specialisation made possible by international trade bears repetition here.

"The natural advantages which one country has over another in producing particular commodities are sometimes so great that it is acknowledged by all the world to be in vain to struggle with them. By means of glasses, hotbeds, and hot walls, very good grapes can be raised in Scotland, and very good wine too can be made of them at about thirty times the expense for which at least

---

1. Refer also to Chapter 10.
2. Adam Smith, *An Inquiry into the Nature and Causes of the Wealth of Nations*, 1776, Volume One, Everyman's Library Edition, 1910, reprinted 1937, Book IV, Chapter II, p. 401.

equally good wine can be brought from foreign countries. Would it be a reasonable law to prohibit the importation of all foreign wines merely to encourage the making of claret and burgundy in Scotland? But if there would be a manifest absurdity in turning towards any employment thirty times more of the capital and industry of the country than would be necessary to purchase from foreign countries an equal quantity of the commodities wanted, there must be an absurdity, though not altogether so glaring, yet exactly of the same kind, in turning towards any such employment a thirtieth, or even a three-hundredth part more of either. Whether the advantages which one country has over another be natural or acquired is in this respect of no consequence. As long as the one country has those advantages, and the other wants them, it will always be more advantageous for the latter rather to buy of the former than to make. It is an acquired advantage only, which one artificer has over his neighbour, who exercises another trade; and yet they both find it more advantageous to buy of one another than to make what does not belong to their particular trades."[3]

It is difficult to come across a more forceful and clearer statement showing the advantages of international trade. After Adam Smith, David Ricardo brought home the same point by giving another forceful illustration in the following words.

"Two men can both make shoes and hats, and one is superior to the other in both employments, but in making hats he can only excel his competitor by one-fifth or 20 per cent, and in making shoes he can excel him by one-third or 33 per cent; will it not be for the interest of both that the superior man should employ himself exclusively in making shoes, and the inferior man in making hats?"[4]

Ricardo's statement, like Adam Smith's, shows the advantages of the division of labour and specialisation. It stresses that competence should specialise where competence counts most and incompetence should specialise where it counts least. In short, international trade enables the countries to exploit to their mutual advantage the substantial advantages of division of labour and specialisation. By ensuring that each country specialises in the production and exports of those goods and services which she is best suited to produce and imports those goods and services which she can obtain cheaper from abroad than what it costs to produce at home. International trade increases the real income and national wellbeing of all the participating countries.

By pushing forward specialisation by extending the scope of the division of labour, international trade lowers the prices of goods and services all over the world. Consequently, it stimulates their total consumption and demand which causes further specialisation making technological progress a reality.

By ensuring free competition, international trade reduces the monopolistic exploitation of consumers because goods and services are produced at the lowest per unit cost of production and the price is not higher than the average cost of production.

Being non-discriminatory in character, international trade ensures equal access to raw materials and world markets to all countries of the world. During the *thirties* some developed countries including Germany, Italy and Japan which did not possess raw material resources themselves raised objections in international economic conferences demanding free and equal access to the world's raw materials. It was so because in the *thirties* the multilateral trading system was throttled by various quantitative and foreign exchange controls. Under free trade, every country has equal access to scarce raw materials and other goods. Consequently, the sources of supply of essential raw materials cease to be the monopoly of the favoured few.

3. Adam Smith, *Op. cit.,* pp. 402–403.
4. David Ricardo, *The Principles of Political Economy and Taxation,* 1817, Everyman's Library Edition, p. 83, footnote.

International trade protects the economic interests of all the participating countries. During World War I and the interwar period, many countries were denied free access to world markets. Consequently, the problem of procuring the essential raw materials became so acute that it became almost impossible for certain countries like Japan, Germany and Italy to import raw materials from the raw materials producing countries which were mostly the colonies of England and France. Consequently, these countries agitated for the redistribution of colonies in the League of Nations and at other international forums. Japan attacked China and took away Manchuria which was the rich producer of coal, iron ore, soy bean etc.

International trade depends on the multilateral payments system which makes it possible to effect payments from the debtor to the creditor countries by enabling the former to create the necessary amount of export surplus in their balance of trade. International trade allows for the working of an international monetary system with free multilateral convertibility of currencies. Free trade is a prerequisite of international economic cooperation, brotherhood and lasting world peace.

International trade provides maximum scope for the optimum utilisation and allocation of world's scarce economics resources. Under the system of free trade, a country sells her products in those markets where she can get the best prices for her products and buys the essential raw materials and other consumer goods from the cheapest sources of supply. Consequently, a country enjoys the maximum advantages both as a consumer and as a producer of various goods and services. Exploitation of one country by another country is difficult since there are numerous buyers of her goods and she can also buy her requirements from various competing sources of supply. Thus, international trade is inimical to monopolistic and monopsonistic exploitation.

The contribution of international trade to national wellbeing is so great that few countries could become self-sufficient even with the greatest effort. The vast importance of free trade is evident from the phenomenal expansion in the volume of world trade. Between 1950 and 2008, world's merchandise exports in value terms expanded from mere 376 billion US dollars in 1950 to 9,300 billion US dollars in 2003, showing more than twenty five fold expansion in the world merchandise trade during little over five decades. In terms of volume, the index more than quadrupled during this period. Even allowing for the increase in the unit value, world trade expanded from 376 billion dollars in 1963 to over 2,200 billion dollars in 2008 at constant prices. The value of trade in commercial services had exceeded US $ 2,000 billion at the end of 2008.

Modem economies have been structured by international trade and specialisation of the past and their continued viability is dependent in no small measure on world economy. For many nations international trade is literally a matter of life and death. For example, it is physically impossible for the United Kingdom and Japan to feed, clothe and house their present population without imports from the other countries. Economic self-sufficiency for these two nations, which are among the world's few most developed nations, would mean virtual starvation and poverty unless mass-scale migration of population was possible. The survival of these countries essentially depends on the exports of their manufactures. Although the United Kingdom and Japan are examples of those countries which are highly dependent on international trade but even those countries which can supply their own people with the basic necessities of life from domestic production would be faced with substantial fall in their living standards if they were cut-off from international trade.[5] Australia

5. Much of the fall in the standards of living of people during war is due to the cessation of international trade as is illustrated by the experience of neutral countries such as Switzerland and Sweden on the one hand and the belligerent nations like Germany and Italy on the other in the Second World War. The Germans were deprived of the consumption of even the basic essential goods such as sugar as no imports of sugar and other essential consumption goods could be had from abroad during the war.

and New Zealand produce far more foodstuffs than is required to feed their sparse population. These countries trade-off this surplus for the manufactured goods from the industrial countries like the United Kingdom and Japan. Thus, for Australia and New Zealand although self-sufficiency would not mean starvation of their people, it would, nevertheless, deprive the residents of these two countries of the consumption of manufactured goods that are needed to sustain the modern high living standards. Since the contemporary New Zealand economy is itself the product of international specialisation and trade, a far different and poorer New Zealand economy would have evolved in the absence of world markets where she buys all that she needs but does not herself produce and sell all she produces but does not consume.

Even for a large country of the size of the United States of America which could afford to be self-sufficient, the cost of affording this self-sufficiency would be enormous and the average American would be reluctant to bear this cost reflected in the immediate perceptible fall in his living standard. A large range of foodstuffs which figure in the diet of every American of average means would no longer be available or would be available only at exorbitant prices in the absence of international trade. For example, the morning cup of coffee which has become an institution with the Americans would become a luxury for all but a few and most Americans would be forced to do without coffee or to use inferior substitutes. Similarly, for many countries sugar would have become an expensive item of consumption without international trade.

The unprecedented prosperity enjoyed by the OPEC–Organisation of the Petroleum Exporting Countries–nations would have been impossible without the ready world demand for their petrol and petroleum-based products made possible by international trade. Today the petrol-rich middle-east countries are booming with industrial activity owning a substantial chunk of equity and financial investments in international money and capital markets. Without international trade, their vast petrol reserves would have remained unexploited and these countries would have remained world's poorest desert countries. Thanks to international trade, their economic lot has been transformed and today these countries are among the world's rich nations.

More important than these advantages, are the advantages of trade which consist in the changes brought about by trade in the quality of labour and capital. "International trade changes the fundamental facts of economic life in trading nations, and cannot fail to affect in a thousand and one ways the factors governing the output of labour and capital. The far-reaching nature of the indirect influence is best realised if we ask what the world's population and capital equipment would be like if there had been no international trade, and how different it would be from the present situation. We can only say that the difference would be enormous, and that it cannot be adequately dealt with in quantitative terms. Trade changes the *quality* of the people, teaches them to consume new things and to use old things in new ways. Technical knowledge is largely the result of specialisation, which trade has made possible. The character, not only of so-called technical labour, but also of skilled and unskilled labour is affected."[6]

It is argued by the critics that international trade adversely affects wages, particularly when trade takes place between two countries in one of which wages are very low and in the other very high. For example, it is argued by the Americans that trade with India or China where wages are low will depress the wages of American workers. The fallacy of this argument was exposed by Frank William Taussig, a well-known American economist, when he stated that "perhaps the most familiar and most unfounded of all is the belief that complete freedom of trade would bring about an equalisation of money wages the world over .... There is no such tendency to equalisation. The

6. Bertil Ohlin, *Interregional and International Trade*, 1933, p. 125.

question of wages is at bottom one of productivity. The greater the productivity of industry at large, the higher will be the general level of wages."

Arguments against International Trade

Some of the familiar arguments against international trade include the following arguments:

1. A country which depends on the imports of goods and essential raw materials is in a vulnerable position during war. Where the country imports essential goods whose demand is inelastic she is placed in a disadvantageous position if the other country abruptly restricts or stops her exports of such goods during the war or national emergency.

2. International trade is a source of economic instability and interferes with national economic planning. It subscribes to the philosophy of *laissez faire* which conflicts with economic planning.

3. International trade inflicts losses on those home industries whose products are displaced by the imports.

While international trade brings the blessings of a higher standard of living for a nation, it also implies dependence on foreign markets as sources of supply of raw materials and also as outlets for domestic production. Some people look on this dependence as dangerous for national pride. According to them, the national interests demand that this dependence should be reduced or entirely eradicated. The most important enemy of international trade and specialisation is nationalism.

National Defence: It has been forcefully argued by some people that a nation which depends on foreign sources of supply is in a woeful situation during war. The harrowing experience of England during the two world wars is cited as a strong proof of this assertion. On two occasions the blockade of England by German submarines had brought England to her knees by completely cutting off imports of food and essential raw materials. This is, however, a political or military argument against free trade. Even Adam Smith, the great champion of free trade, had admitted that defence was better than opulence.

Instability and Economic Planning: International trade is also condemned as a source of economic instability. This argument gained currency in the 1930s when the great depression spread from one country to another by disrupting the international flow of goods, services and capital. Today this argument against international trade has been reinforced by government policies directed toward achieving full employment and economic growth. Advocates of economic planning regard international trade as positively harmful for a country's planned economic development. According to them, free trade is an anachronism and does not fit in with the requirements of a planned economy. It may, however, be pointed out that most nations are unable to achieve the objectives of full employment and growth except as members of an international trading system.

Protection: Traditionally, attacks on the international trade have been directed against imports. Several arguments, in addition to those already mentioned, have been advanced to justify the protection of domestic industry against foreign competition.[7] In fact, the arguments against international trade and specialisation are legion—national security, economic stability, full employment, planned economic development, protectionism and others of less importance. While all these arguments have a powerful emotional appeal for the layman, international trade has a tremendous vitality for growth and even the protectionists with all their arguments against the free

7. For greater details on protection see the chapter entitled Free Trade and Protection.

trade have not been able to dislodge it from its strong position. They have come to realise that it is a force to reckon with. The supreme solid advantages of international specialisation and trade make the alleged advantages of protection pale into insignifance.

The recent trend toward globalisation and the wide support extended to the nondiscriminatory multilateral trading system which is well reflected in the faith of over 150 world nations in the World Trade Organisation as its members show that the world economic prosperity can be translated into reality only through the nondiscreminatory multilateral trade and not through the unbridled protectionism.

Suggested Readings

Charles P. Kindleberger, *International Economics,* Fifth Edition, 1973, Chapter 5 and pp. 53–55.
Bertil Ohlin, *Interregional and International Trade,* 1933, pp. 123–126.

Questions

1. Discuss the advantages and disadvantages of international trade.
2. By pushing forward the scope for specialisation and division of labour, international trade has contributed to the high standards of living for the citizens of all nations. Do you agree with this statement? Discuss fully.

PART TWO

PURE THEORY OF INTERNATIONAL TRADE

CHAPTER 4

Theory of Comparative Cost Advantage

Introduction

Adam Smith had vigorously attacked the mercantilist regulated foreign trade and commercial policies because in the changed economic setting of the latter half of the eighteenth century these policies operated as impediments to national economic progress and wellbeing. Adam Smith was an eighteenth-century courier of the nineteenth-century economic liberalism enshrined in the well-known phrases *laissez-faire, laissez-passer*. Since England was the first country to outgrow the mercantilist policies, it is not surprising that the Smithian *laissez-faire* philosophy gained momentum in England earlier than elsewhere. By undermining the faulty logic of an outworn mercantilist system of trade restrictions and by glorifying the advantages of free commerce and division of labour, Adam Smith laid the firm foundation for the classical free trade doctrine applicable to both domestic and foreign trade. He championed the cause of free trade supporting the general principle of noninterference summed up in the physiocratic slogan of *laissez-faire, laissez-passer*.

Adam Smith had strongly opposed protection in all situations except for defence of the country. For him, free trade was the best source of opulence because by enlarging the size of the market[1] for goods, it makes it possible for the countries engaged in international exchange of goods and services to reap the utmost advantages of specialisation resulting from practising the division of labour. The theory of trade based on the application of the principle of division of labour is generally known as the classical theory of the comparative cost advantage.

Although the evolution of the doctrine of the comparative cost advantage dates back to David Hume and Adam Smith, the precise formulation of the theory was first given by David Ricardo[2]

1. Adam Smith (1723–1790), universally respected and regarded as the father of Economics, was professor of moral philosophy at the Glasgow University. He championed the cause of free trade and vehemently opposed mercantilism. He firmly believed in the blessings of the 'invisible hand'. He was the most influential economist who influenced the future generations of the classical economists.
2. David Ricardo (1772–1823) was the third son of a Dutch Jew who had immigrated to England and had acquired huge fortune in the British stock market. Ricardo possessed rare business talent. He had amassed huge fortune during the Napoleonic wars as a loading member of the London stock exchange through the sheer dint of his merit. At the early age of 30 years, he was a known wealthy man and had purchased a county estate. He also became a respectable member of the House of Commons and took keen interest in the controversial economic issues of his times. His economic reasoning was deeply influenced by the economic philosophy contained in Adam Smith's well-known book entitled *An Inquiry into the Nature and Causes of the Wealth of Nations first published in 1776*. He had elaborated his economic philosophy in his book entitled *The Principles of Political Economy and Taxation* published in 1817. His contributions ramify in many

followed by the great genius of John Stuart Mill. David Hume's contribution to the theory of free trade is contained in his well-known work entitled *Political Discourses* published in 1752 and is undoubtedly more significant and more original than Adam Smith's *magnum opus* entitled *An Inquiry into the Nature and Causes of the Wealth of Nations* published in 1776 notwithstanding the latter's more dominating influence on economic theory and practice.

According to Leser and Edwin R.A. Seligman, however, Robert Torrens was the first to propound the classical doctrine of free trade. The credit for propounding the classical theory of comparative cost advantage goes to David Ricardo mainly because it was from David Ricardo and not from Robert Torrens[3] that the latter-day economists learnt the doctrine. Grouped around these luminaries are the relatively less known but highly original writers, such as Thomas Robert Malthus, William Blake, John Wheatley, Mountifort Longfield and Nassau William Senior without mentioning whose contributions the discussion would remain incomplete. The theory of comparative cost advantage also called the "principle of comparative advantage" has, however, undergone substantial changes through improvements and refinements made by the latter-day economists, particularly by John Elliot Cairnes, Alfred Marshall, Charles F. Bastable, and others. The best known modern exponents of the theory are Frank William Taussig, Gottfried Von Haberler and Bertil Ohlin. The development of the theory of comparative cost advantage starting from Adam Smith has been discussed briefly in the following pages.

Basis of Trade

Trade, be it interregional or international, takes place when the domestic price ratios are different. In the case of international trade, a country will export that commodity which relative to the other commodity is cheaper at home than abroad and *vice versa*. In short, difference in the relative prices of goods between the different countries is the basis of international trade. The reason for the cheapness of one commodity and dearness of another relatively at home than abroad may be due to the differences either in the supply conditions or in the demand conditions of goods in the two countries. The classical economists, however, stressed only the differences in the supply conditions at home and abroad as the basis of international trade forgetting that situations may exist in which the direction of trade may be different from the one which would be expected on the considerations of comparative cost advantage alone. In other words, the theory of comparative cost advantage ignores the phenomenon of demand reversal because it focuses attention only on the supply side.

Adam Smith—Absolute Cost Advantage

The focal point of Adam Smith's argument for free trade rests on the advantages of the division of labour made possible by international trade. By enlarging the scope of the market, free trade makes possible the greater degree of specialisation of labour and consequently augments the gains from

directions ranging from practical issues of the effects of machines on workers, commercial laws and gold-bullion standard to the theories of value and distribution where his chief concern was to enunciate the principles which regulated the distribution of national product. The theory of comparative cost advantage has been expounded by David Ricardo in Chapter 7 entitled 'On Foreign Trade' in his above book.

3. Robert Torrens (1780–1864) was an officer in the British army during the Napoleonic wars. Subsequently, he turned to economics. In 1815 was published his pamphlet *Essay on the External Corn Trade* which contains the earliest formulation of the theory of comparative cost advantage.

territorial division of labour. Division of labour causes each country to specialise in the production of those commodities in which it has a comparative cost advantage over others. Adam Smith proves that the principle governing the international exchange of goods is basically the same which underlies the domestic exchange. He says: "It is the maxim of every prudent master of a family, never to attempt to make at home what it will cost him more to make than to buy. The tailor does not attempt to make his own shoes, but buys them from the shoemaker. The shoemaker does not attempt to make his own clothes, but employs a tailor ... What is prudence in the conduct of every private family can scarcely be folly in that of a great kingdom. If a foreign country can supply us with a commodity cheaper than we ourselves can make it, better buy it of them with some part of the produce of our own industry employed in a way in which we have some advantage.

The natural advantages which one country has over another in producing particular commodities are sometimes so great that it is acknowledged by all the world to be in vain to struggle with them. By means of glasses, hotbeds, and hot walls, very good grapes can be raised in Scotland, and very good wine too can be made of them at about thirty times the expense for which at least equally good wine can be brought from foreign countries. Would it be a reasonable law to prohibit the importation of all foreign wines merely to encourage the making of claret and burgundy in Scotland? But if there would be a manifest absurdity in turning towards any employment, thirty times more of the capital and industry of the country than would be necessary to purchase from foreign countries an equal quantity of the commodities wanted, there must be an absurdity, though not altogether so glaring, yet exactly of the same kind, in turning towards any such employment thirtieth, or even a three-hundredth part more of either. Whether the advantages which one country has over another be natural or acquired is in this respect of no consequence. As long as the one country has those advantages, and the other wants them, it will always be more advantageous for the latter rather to buy of the former than to make."[4]

Adam Smith developed the theory of international trade based on absolute difference in costs. Trade will emerge if one country has an absolute cost advantage in the production of one commodity and an absolute cost disadvantage in the production of the other. Each country will export that commodity in the production of which it has an absolute cost advantage and import that commodity in which it suffers from an absolute cost disadvantage. In other words, in order to export a commodity its price in the country should be low relatively to its price in the other country. It would be possible if the country possessed the comparative cost advantage in the production of that commodity and for this purpose the country must have an absolute cost advantage. To elucidate the point, let us assume that there are two countries India and Bangladesh producing two commodities cotton and jute. The unit cost of production measured in terms of the labour units of the two commodities in the two countries is as following:

Country	Unit Cost of Production (Man-hours)	
	Cotton	Jute
India	5	10
Bangladesh	10	5

4. Adam Smith, *An Inquiry into the Nature and Causes of the Wealth of Nations*, Volume I. 1776. Everyman's Library Edition, 1910, reprinted 1937, pp. 401–403.

It is evident that in India labour is more efficient in the production of cotton while in Bangladesh it is more efficient in the production of jute because one man-hour produces 0.2 unit of cotton or 0.1 unit of jute in India while in Bangladesh one man-hour produces 0.1 unit of cotton or 0.2 unit of jute. Consequently, India enjoys an absolute (and also comparative) cost advantage over Bangladesh in the production of cotton while Bangladesh enjoys an absolute (and also comparative) cost advantage over India in the production of jute. India's superiority in the production of cotton is evident from the fact that her relative factor productivity in cotton shown

by the ratio $\dfrac{0.2 \text{ unit of cotton in India}}{0.1 \text{ unit of cotton in Bangladesh}}$ is higher than her relative factor productivity in jute

expressed by the ratio $\dfrac{0.1 \text{ unit of jute in India}}{0.2 \text{ unit of jute in Bangladesh}}$.

Both India and Bangladesh will gain if India and exports cotton while Bangladesh exports jute because if India specialises in the production of cotton and Bangladesh specialises in the production of jute the total combined output of cotton and jute which can be obtained by incurring a cost of 15 man-hours in each country will be three units of each good. On the other hand, if both the countries produce both the commodities, the maximum combined output of cotton and jute involving the same total labour cost will be only two units of each good. Consequently, the gains of division of labour and specialisation are reflected in the additional output of one unit of each commodity which both countries can share. Specialisation based on the territorial division of labour made possible by free trade leads to the maximisation of total output resulting from using the given resources.

After specialisation, both the countries will gain from trade. India will gain from trade so long as she can obtain more than half unit of jute in exchange for one unit of cotton (this is the domestic exchange ratio in India) from Bangladesh. Bangladesh can give the maximum of two units of jute in exchange for one unit of cotton (this is the domestic exchange ratio in Bangladesh) from India. Obviously, the scope for exchange exists and international trade will take place.

Adam Smith's clear and lucid statement explains the basis of trade between the tropical zone countries and the temperate zone countries or between the industrial countries and the agricultural countries. Trade is based on the principle of comparative cost advantage emerging from absolute differences in costs. However, even so, it is not very convincing and does not go far enough. "It assumed without argument that international trade required a producer of exports to have an *absolute* advantage; that is, an exporting country must be able to produce, with a given amount of capital and labour, a larger output than any rival. But what if a country had no line of production in which it was clearly superior? Suppose a relatively backward country whose "capital and industry" in the broadest sense (compared with its more advanced neighbours) were inefficient, capable of producing less in all lines of production—a not too hypothetical case. Would it be forced to insulate itself against more efficient outside competition or see all its industry and agriculture subjected to ruinous competition?"[5] Adam Smith's analysis was too narrow and a more comprehensive and general theory capable of dealing with this kind of situation was needed. It was left to Robert Torrens and David Ricardo to attack the problem and to elaborate Adam Smith's statement to fit in a more general framework by formulating a more general theory of international trade.

5. P.T. Ellsworth and J. Clark Leith, *The International Economy,* Fifth Edition, 1975, p. 46.

David Ricardo—Comparative Cost Advantage

The theory of comparative cost advantage originated as an improvement and development of the eighteenth century criticisms of the mercantilist policy. Adam Smith ruthlessly criticised the mercantilist commercial system in his classic, *An Inquiry into the Nature and Causes of the Wealth of Nations published in 1776*. The doctrine of comparative cost advantage continues to command attention because it shows analytically the superiority of free trade over protection. The chief merit of the doctrine consists in removing the widely prevalent eighteenth century error that under free trade all commodities, abstracting from transport costs, would necessarily be produced in those countries where their absolute real costs of production were the lowest. Adam Smith's naive argument for free trade was based on this narrow view.

David Ricardo did not object to Adam Smith's analysis. Obviously, if one country commands an absolute advantage over the other country in the production of one commodity while the other country has an absolute advantage over the first country in the production of other commodity, both the countries can gain by trading. Bulk of trade, perhaps most trade, is subject to such absdute cost differences.

Recorde, however, discusses a different case, though loss favourable to fuetrade, which he considers typical. He assemes the one of the two countries can produce both goods with a smaller labour cost then the other country. what if one country is superior or more productive than the other country in producing all the goods. If country A produces all the goods at the lower that cost of production compared with country B, will both the countries still gain by trading? To this David Ricardo's answer was in affermative yes because so long as country B was not equally less productive in all branches of production, both countries will gain by trading.

It was David Ricardo who in his magnum opus entitled *The Principles of Political Economy and Taxation,* first published in 1817, formulated the pure theory of international trade. It may, however, be stated parenthetically that the theory was already formulated in 1815 by Robert Torrens although he was unaware of the implications of his idea.[6] At the centre of the Ricardian theory of international trade is the so-called "principle of comparative advantage". The idea that nations would gain by specialising in the production of those goods in which they possess special advantages has survived to the present day and David Ricardo's following statement has a force even today.

"It is.... important to the happiness of mankind that our enjoyments should be increased by the better distribution of labour, by each country producing those commodities for which by its situation, its climate and its other natural or artificial advantages, it is adapted, and by their exchanging them for the commodities of other countries."[7]

The classical doctrine of comparative cost advantage was developed by David Ricardo out of his celebrated labour theory of value. According to this theory, the value of any commodity is

6. Robert Torrens, *An Essay on the External Corn Trade,* 1815, pp. 264–266. J.H. Hollander has, however, strongly defended David Ricardo's claim to be the first to emphasise the theory by placing it in an appropriate setting and getting it accepted by the economists. According to him, much of the evidence presented by Seligman in support of Robert Torrens was either irrelevant or of questionable weight. According to Hollander, Torrens never realised the full significance of the theory of comparative cost advantage. The doctrine did not constitute an integral part of Torrens's thinking and he never made any explicit use of it. The mere fact that Torrens correctly stated the theory in outline in a single paragraph was, in Hollander's opinion, insufficient to pre-empt Ricardo's claim to priority. (For an excellent b r i e f discussion on this see Jacob Viner, *Studies in the Theory of International Trade,* 1937, pp. 442–443.)
7. David Ricardo, *The Principles, of Political Economy and Taxation, 1817* Everyman's Edition, Chapter VII.

determined by its labour cost of production. To express in Ricardo's words: "It is the comparative quantity of commodities which labour will produce that determines their present or past relative values."[8] The theory, however, breaks down as a determinant of value-in-exchange in international trade. To quote David Ricardo:

"The same rule which regulates the value of commodities in one country does not regulate the relative values of the commodities exchanged between two or more countries. The quantity of wine which Portugal shall give in exchange for the cloth of England is not determined by the respective quantities of labour devoted to the production of each, as it would be if both commodities were manufactured in England, or both in Portugal."[9]

But why the labour (cost) theory of value, which is valid in the case of domestic trade, cannot explain the exchange values in international trade? It is the immobility of labour as a factor of production between different countries which prohibits the application of this theory to trade taking place between different countries. If labour were perfectly mobile between the countries, commodities would exchange in the ratio of the quantities of labour embodied in them. Any divergence between their exchange ratio and their cost ratio would be eliminated by the market forces of demand and supply. If the product of a certain industry can be sold at more than the value of labour it contains, additional labour will be transferred to that industry from other occupations. Consequently, supply of the commodity will expand until the price falls to become equal to the value of labour embodied in it. Conversely, if the commodity sells for less than the value of its labour contents, labour will move away from that industry into other lines of production. Consequently, the supply of commodity will decrease and its price will rise until the price equals the labour cost of production of the commodity. Thus, the labour cost principle implies that there is a tendency of wages in different branches of production towards equality within a country so that prices of goods will be equal to the returns to labour in all the lines of production and regions within the country. This equilibrating mechanism does not, however, operate between the two countries due to the assumption of immobility of labour between the countries. Thus, if in the two countries the exchange ratio between the two commodities is different international trade will emerge.

The question, however, is: what determines the values of goods in international exchange? To explain this, David Ricardo used the doctrine of comparative cost advantage. According to this doctrine, a country will specialise in the production of that commodity for which its unit labour cost is comparatively lowest. In general terms, the doctrine emphasises the principle that competence should specialise where competence counts most and incompetence must specialise where incompetence counts least. It follows from this statement that a country would export that commodity in the production of which it has comparative cost advantage or superiority over others and import that commodity in which its cost advantage is least or in which it suffers from comparative cost disadvantage.

The principle of comparative cost advantage stresses that if trade is left free, each country in the long run tends to specialise in the production of and consequently tends to export those goods in the production of which it enjoys a comparative advantage in terms of real cost and to import those goods in the production of which it has comparative disadvantage in terms of real cost. Such specialisation is to the mutual advantage of both the countries participating in trade. This

8. David Ricardo, *Op. cit.,* p. 9.
9. David Ricardo, *Op. cit.,* pp. 81–82.

statement follows from the principle of division of labour which in general terms, applicable to both nations and individuals, stresses that competence should specialise where it counts most while incompetence should specialise where it counts least. David Ricardo's classic example of two men of differing efficiencies both of whom could make shoes and hats, amply illustrates the point.

Adam Smith's demonstration of the benefits of free trade was simple and powerful. It was not, however, very deep and subtle. Obviously, if one country enjoys an absolute cost advantage over the other country in the production of one good and the other country has an absolute advantage over the first country in the production of another good, both the countries could benefit by trading. Adam Smith did not, however, tackle the more difficult real case in which one country enjoys an absolute advantage over the other country in the production of both the goods. In other words, if country A can produce both the goods with smaller labour cost than country B, will the countries still benefit from trading? It was David Ricardo whose penetrating eyes saw the problem and answered in the affirmative yes. So long as country A is not equally more productive in both the lines of production, both the countries will benefit by trading.

In chapter VII of his *Principles of Political Economy and Taxation* David Ricardo explained the doctrine of comparative cost advantage through an arithmetical illustration by taking England and Portugal as the two countries and wine and cloth as the two goods whose production was assumed to be taking place under conditions of constant costs. Labour was the only factor of production with the result that the per unit labour cost of production was measured in terms of man-hours. Ricardo's arithmetical illustration is as follows:

Table 1 : Cost comparisons

Country	Man-hours required to produce one unit of	
	Cloth	Wine
Portugal	90	80
England	100	120

In the above illustration, Portugal commands an absolute cost advantage over England in the production of both the goods because the unit labour cost of production of both cloth and wine is lower in Portugal. It is evident from the fact that 90 < 100 and 80 < 120. She commands, however, comparative cost advantage over England only in the production of wine. It is evident from the fact that the ratio of the wine costs is less than ratio of the cloth costs, i.e., $\frac{80}{120} < \frac{90}{100}$. On the other hand, England suffers from an absolute cost disadvantage in the production of both the goods. It is evident by the fact that 100 > 90 and 120 > 80. She has, however, smaller comparative cost disadvantage in the production of cloth since the ratio of the cloth costs is less than the ratio of the wine costs, i.e., $\frac{100}{90} < \frac{120}{100}$. Under these circumstances, it would be to their mutual advantage if Portugal specialised in the production of wine and exchanged her surplus wine against cloth from England and if England specialised in the production of cloth exchanging her surplus cloth against wine from Portugal.

Haberler has explained the meaning of 'comparative advantage' by stating that "there must be at least two countries and two goods, and we have to compare *the ratio of the costs of production* of one good in both countries $\left(\dfrac{80}{120}\right)$ with the ratio of the costs of production of the other good in

both countries $\left(\dfrac{90}{100}\right)$. Expressed in words: Portugal has a comparative advantage over England in wine relatively to cloth. Conversely, the disadvantage of England is greater in wine than in cloth. Stated in another way, England has an *absolute* disadvantage in cloth, but at the same time she has

a *comparative* advantage in cloth. The above inequality $\left(\dfrac{80}{120} < \dfrac{90}{100}\right)$ states the position exactly..."[10]

Gain from trade is possible when and if there are comparative differences in costs between the countries concerned. Highlighting the fact of gain from trade to Portugal (and for the same reason to England) David Ricardo made the following statement.

"Though she (i.e., Portugal) could make the cloth with the labour of 90 men, she would import it from a country where it required the labour of 100 men to produce it, but it would be advantageous to her rather to employ her capital in the production of wine, for which she would obtain more cloth from England than she could produce by diverting a portion of her capital from the cultivation of wines to the manufacture of cloth."[11]

In order to demonstrate that both Portugal and England will gain by trading and specializing, the analysis in terms of the opportunity cost will prove helpful. The opportunity cost for wine or for that matter for cloth is the amount of cloth/wine which has to be given up in order to produce one (additional) unit of wine/cloth. The table below shows the opportunity costs of producing wine and cloth in Portugal and England.

Table 2: Opportunity Costs

	Opportunity costs for	
	Wine	Cloth
Portugal	80/90 = 8/9	90/80 = 9/8
England	120/100 = 12/10	100/120 = 10/12

It is evident from the table that Portugal has comparative advantage in producing wine since the opportunity cost of producing wine in Portugal is lower than in England while England has the lower opportunity cost in producing cloth. It shows that Portugal has a comparative advantage in the production of wine while England possesses a comparative advantage in the production of cloth.

To explain the meaning of the term 'comparative advantage' there must be at least two countries and two goods between which comparison has to be made. As long as the two countries' opportunity costs for one good differ, one country has comparative advantage in the production of one of the two goods while the other country has comparative advantage in the production of the

10. Gottfried Von Haberler, *The Theory of International Trade,* English Translation, Third Impression, 1950. pp. 128–129.
11. David Ricardo, *Op. cit.,* pp. 76–77.

other good. As long as this condition is satisfied, both the countries will gain from trade, regardless of the fact that one of the two countries has an absolute cost disadvantage in the production of both the goods.

The gain from trade reflected in the combined higher production of both wine and cloth if Portugal completely specialised in the production of wine and England completely specialised in the production of cloth compared with the combined total production of both the commodities in autarky is evident from the following illustration.

Assuming that the two commodities in the two countries are produced under constant costs and given the total factor endowment of 170 man-hours in Portugal and of 220 man-hours in England producing in autarky 1 unit of wine and 1 unit of cloth in each country, the combined output of both the goods in the two countries would be 2 units of wine and 2 units of cloth. Now, if Portugal specialised in the production of wine devoting all her 170 man-hours to wine production she could produce 2.125 units of wine and no cloth. Similarly, England could produce 2.2 units of cloth and no wine. Consequently, the combined output of the two commodities would be 2.125 units of wine and 2.2 units of cloth which is greater than the combined output of 2 units of wine and 2 units of cloth which is possible in autarky in the absence of specialisation and trade.

The principle of comparative cost advantage can be expressed algebraically. Let us assume that the unit labour cost of production of wine in Portugal is m_1 while in England it is m_2 and the unit labour cost of production of cloth in Portugal is m_3 while in England it is m_4. If the ratio of the unit cost of production of wine in Portugal and England is less than one $\left(\dfrac{m_1}{m_2} < 1\right)$ while that of the cloth is greater than one $\left(\dfrac{m_3}{m_4} > 1\right)$, i.e., if $\dfrac{m_1}{m_2} < 1 < \dfrac{m_3}{m_4}$, Portugal has an absolute and a "comparative cost advantage over England in the production of wine while England has an absolute and comparative cost advantage in the production of cloth. There is comparative difference in cost if the condition $\dfrac{m_1}{m_2} < \dfrac{m_3}{m_4} < 1$ holds. In our example, since $0.66 < 0.90 < 1.00$, it means that although Portugal has an absolute cost advantage over England in the production of both wine and cloth she has, however, a comparative cost advantage only in the production of wine. It is obvious that a comparative cost advantage is always accompanied by an absolute cost advantage. In other words, a country's comparative cost advantage is also her absolute cost advantage.

If the cost ratio $\dfrac{m_1}{m_3} = \dfrac{m_2}{m_4}$, there will be no trade since the comparative cost ratios in both the countries are equal. If the comparative cost ratios are equal in both the countries there will be no gain for either country by entering into trade. For example, if in Portugal one unit of wine costs 10 man-hours and one unit of cloth costs 5 man-hours while in England the production of one unit of wine costs 8 man-hours and one unit of cloth costs 4 man-hours, then the domestic cost (exchange) ratio in Portugal is $\dfrac{10}{5} = 2 : 1$ while in England it is $\dfrac{8}{4} = 2 : 1$. As the domestic cost (exchange) ratios are identical in both the countries there will be no gain from trade. Consequently, international trade will not take place.

In the case of comparative cost difference, there will be gain from trade. Returning to our example, the domestic exchange ratio in Portugal is one unit of wine = 0.88 unit of cloth. The domestic exchange ratio in England is one unit of wine = 1.20 units of cloth. If the international exchange ratio is one unit of wine for one unit of cloth and if Portugal exports one unit of wine to England she will get in exchange one unit of cloth from England. Consequently, her net gain from trade is 0.12 (= 1 − 0.88) unit of cloth because her domestic exchange ratio is 0.88 unit of cloth for one unit of wine. Similarly, England will gain 0.17 unit of wine by exporting one unit of cloth because her domestic exchange ratio is one unit of cloth for 0.83 unit of wine. The minimum amount of cloth which Portugal will accept from England for one unit of wine is 0.88 unit of cloth which equals her domestic exchange ratio while the maximum amount of cloth which England can give in exchange for one unit of wine is 1.20 units of cloth which is equal to her domestic exchange ratio.

If trade takes place at the domestic exchange ratio of Portugal (1 unit of wine = 0.88 unit of cloth) the entire benefit from trade will accrue to England. Conversely, if trade takes place at England's domestic exchange ratio (1 unit of wine = 1.2 units of cloth), Portugal will pocket the entire gain from trade. In practice, however, only in very rare situations international terms of trade will coincide with the domestic terms of trade of one or the other country because the *raison d' etre* for trade is that both the trading countries must gain from trade. Consequently, the actual international exchange ratio or the terms of trade will settle down anywhere between these upper and lower limits determined by the domestic exchange ratios of the two countries. But the important question is: what will be the actual exchange ratio or terms of trade between these two limits at which trade will take place between the two countries? In other words, in what proportion will the total gain from trade be divided between the two countries? Ricardo did not develop any analytical framework to study the division of the gain from trade. He merely stated that "England would give the produce of the labour of 100 man-hours (= 1 cloth) for the produce of the labour of 80 man-hours (= 1 wine)."[12] In other words, one unit of English cloth would exchange for one unit of Portuguese wine and the gain from trade would be divided approximately evenly between the two trading countries. Ricardo did not say anything beyond stating this in the matter and the question remained unanswered for more than three decades until John Stuart Mill arrived on the scene and grappled with the problem in 1848.

Assumptions

The classical theory of comparative cost advantage is based on the following assumptions.

1. Labour is the only factor of production. Consequently, the theory is based on the labour (cost) theory of value.

2. All units of labour are homogeneous.

3. Labour is perfectly mobile within the country and perfectly immobile between different countries.

4. Production functions are linearly homogeneous, i.e., production of goods obeys the law of constant returns to scale. In other words, the unit cost ratios of the two goods are constant regardless of the scale in production. Thus, the theory abstracts from considering the economies and diseconomies of scale of production.

5. The theory assumes trade taking place between only two countries and in only two goods produced by the single factor of production (labour). In other words, the naive classical theory is a 2 × 2 × 1 trade model.

12. David Ricardo, *Op. cit.*, pp. 76–77.

6. The theory abstracts from the transport costs, i.e, transport costs are absent.
7. The prices of goods are determined by their real (labour) cost of production.
8. The product and factor markets are perfectly competitive.

Diagrammatical Representation of the Theory

The classical theory of international trade enunciated by David Ricardo can be diagrammatically restated. Ricardo bases his theory on the labour theory of value which considers labour as the sole factor of production, the value and output being determined by the labour contents required in the production of unit of each good. Since both David Ricardo and John Stuart Mill assumed constant returns to scale, the production function is linearly homogeneous and the expansion path is a straight line. With these assumptions, the theory can be diagrammatically illustrated with the help of Figures 4.1, 4.2, 4.3, and 4.4.

Figure 4.1 has two parts. Part A of the figure shows the production of good X while part B shows the production of good Y in the country. The country produces two goods X and Y whose total output is measured on the Y-axis. The X-axis measures the total labour endowment of the country which is $0L_3$ fixed amount. If the entire labour is employed in the production of good X, the country can produce the maximum output of $0X_3$ amount. If the total labour is used in the production of good Y, the country can produce $0Y_3$ amount. If only $0L_2$ amount of labour is used in the production of good X, the country can produce only $0X_2$ output of good X which is less than the maximum output $0X_3$ which could be raised if $0L_3$ total labour force was employed in the production of good X. The remaining amount of labour $L_2 L_3$ (= $0L_2$ in Figure 4.1B) produces only $0Y_2$ amount of good Y which is less than $0Y_3$ amount which could be produced if the total labour amounting $0L_3$ was used in good Y's production. The two expansion rays 0R and 0T show the different slopes indicating that the production functions of good X and good Y although linearly homogeneous are not identical.

Figure 4.2 shows the country's production possibilities curve AB which depicts the different production possibilities sets of the two goods X and Y which can be produced with the given total

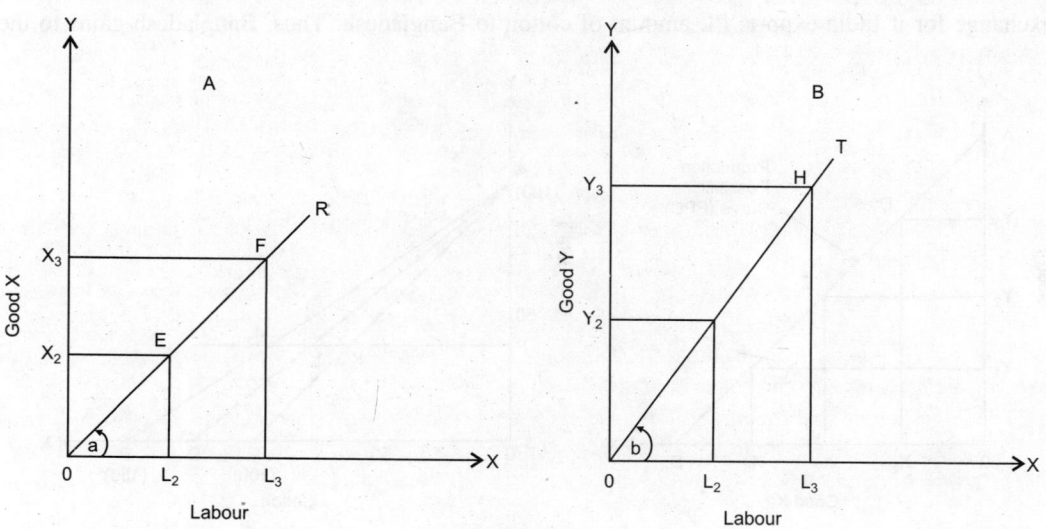

Figure 4.1

labour endowment of $0L_3$ amount. The maximum outputs of good X (point B) and good Y (point A) locate the extreme points of the production possibilities curve the slope of which is given by the ratio 0A/0B. The production possibilities curve of the other country for the two goods can be similarly derived.

In order to restate the theory of comparative cost advantage we may take the two countries India and Bangladesh and the two commodities cotton and jute. Assuming the constant returns to scale, perfect competition in the product and factor markets and the absence of transport costs, the cost structure of the two commodities in the two countries has been given below.

Country	Per Man-Year Output Units of	
	Cotton	Jute
India	150	100
Bangladesh	100	100

The above output units have been represented as intercepts on the jute (Y) and cotton (X) axes of the production possibilities curves of the two countries. In Figure 4.3, AB is the production possibilities curve of India while AC is the production possibilities curve of Bangladesh.

It is evident that the per unit jute production requires an equal amount of labour in both the countries while an equal amount of labour produces 50 per cent more units of cotton in India than in Bangladesh. It is, therefore, profitable for India to specialise in the production of cotton. In other words, India should produce and export cotton. The potential trading area in the figure has been shown by triangle ABC. If the actual international terms of trade line AD divides the potential trading triangular area into two equal parts, the total gain resulting from trade will be equally shared by both the countries. Nearer the international terms of trade line AD lies to the production possibilities curve of one or the other country, smaller will be the gain accruing from trade to the particular country and *vice versa*.

Turning to Figure 4.3, we assume that Bangladesh exports 0P amount of jute to India and in exchange for it India exports PL amount of cotton to Bangladesh. Thus, Bangladesh gains to the

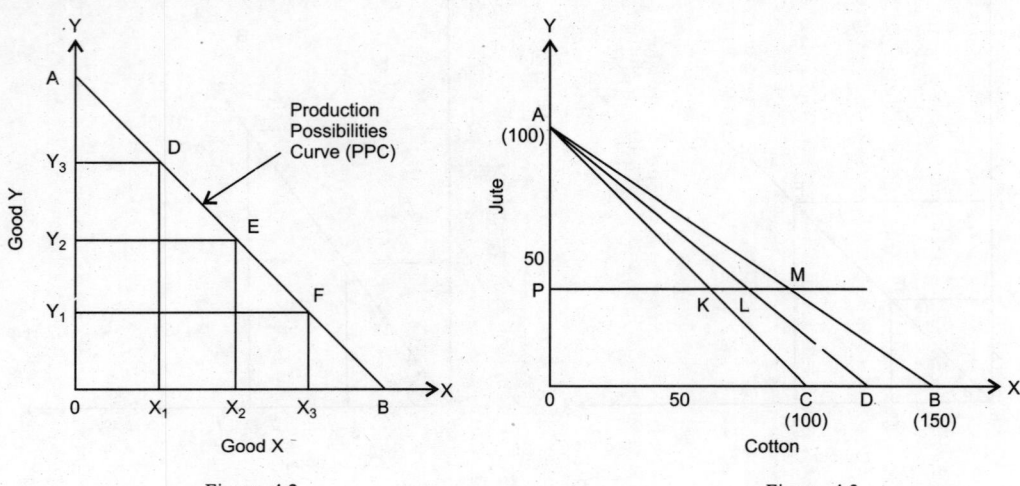

Figure 4.2   Figure 4.3

extent of KL amount of cotton because in autarky she could get only PK amount of cotton for the 0P amount of jute. Similarly, India saves LM amount of cotton because in her domestic market India had to offer PM amount of cotton for the 0P amount of jute. Evidently, both India and Bangladesh gain from trade.

The problem of determination of the actual terms of trade was not tackled by David Ricardo. It was left for John Stuart Mill, Alfred Marshall and Francis Ysidro Edgeworth who introduced the demand condition in the international trade in order to explain the process of the determination of the terms of trade. Thus, Ricardo's explanation of the comparative cost advantage theory was incomplete as it focussed attention only on the supply side and took the demand side for granted.

Classical Theory Expressed in Terms of the Reciprocal Demand Curves

John Stuart Mill examined the problem of determination of the actual terms of trade (exchange ratio) at which the countries exchanged the commodities. Mill's doctrine can be briefly stated below.

The actual exchange ratio at which goods are traded will depend on the strength and elasticity of each country's demand for the other country's product or on the reciprocal demand. The international exchange ratio will be stable when the value of each country's total exports is just enough to pay for her total imports.

In order to find out the actual terms of trade we construct a rectangle as shown in Figure 4.4 corresponding to the dimensions of Figure 4.3. We place India's production possibilities curve on the rectangle with its origin at point 0, flip Bangladesh's diagram over and place it on the rectangle with its origin at point 0'. With complete specialisation in each country, the production point of both India and Bangladesh will be at B with India producing only cotton and Bangladesh producing only jute. The outward limits to the terms of trade are determined by the slopes of the domestic exchange ratio lines AB and BC, i.e., 1 unit of jute : 1.5 units of cotton for India and 1 unit of jute : 1 unit of cotton for Bangladesh. The ABC triangle bound by the production possibilities

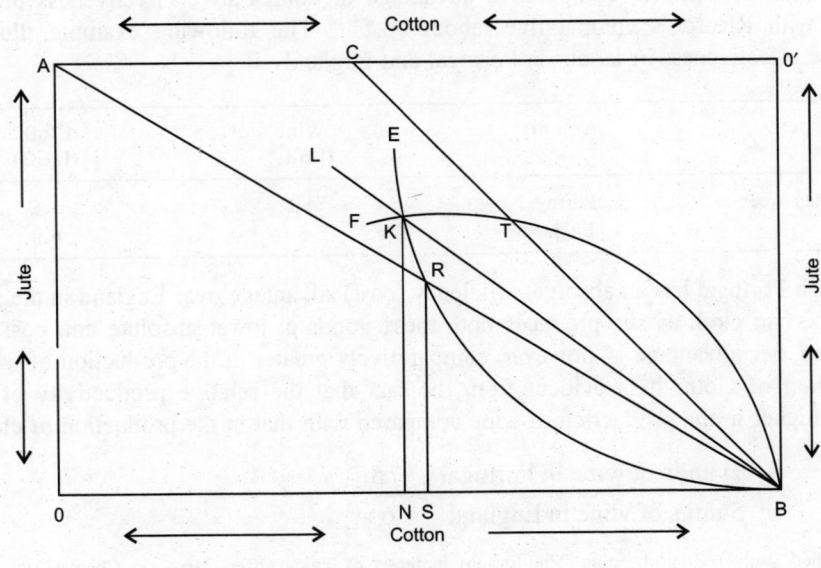

Figure 4.4

curves AB and BC represents the potential trading area. The actual exchange ratio or the international terms of trade will lie anywhere within this triangular area and will depend on the elasticities of demand of each country for the product of the other country.

India's and Bangladesh's international demand can be represented by their respective trade offer curves BRE and BTF which intersect at point K. Each trade offer curve shows each country's offer (supply) of exports in exchange for imports at the different terms of trade. If India is to consume a larger amount of Bangladesh's jute she must obtain a higher price for her cotton. In fact, the worst terms of trade acceptable for India will be her domestic terms of trade represented by the slope of the domestic exchange ratio line AB.

Suppose that at her domestic terms of trade India desires to consume only RS amount of jute and 0S amount of cotton. These quantities of jute and cotton will be produced in India in autarky since this is an attainable combination located at point R on India's production possibilities curve AB. Consequently, the BR part of India's trade offer curve is imaginary. As we move upward along the RK portion of India's trade offer curve, the terms of trade become more favourable for India. By similar reasoning, the BT part of Bangladesh's trade offer curve BTF is imaginary or non-operative. The two trade offer curves intersect each other at point K where the exports of each country are just sufficient to pay for her imports. Consequently, the slope of BL line which passes through point K shows the equilibrium international terms of trade. Any change in either country's demand for other country's good will alter the position of the terms of trade line. The more inelastic is the trade offer curve of the country, the more favourable will be the international terms of trade for her and *vice versa.*

Mill's analysis of the reciprocal demand was further translated into graphical terms by the noted Cambridge economists Alfred Marshall and Francis Ysidro Edgeworth. They developed and used the technique of trade offer curves to explain effectively the operation of Mill's reciprocal demand.[13]

The theory of comparative cost advantage was illustrated by John stuart Mill by assuming different outputs of the two goods for a given labour input in the two countries. "Thus his formulation ran in terms of comparative advantage or comparative effectiveness of labour, as contrasted with Ricardo's comparative labour cost."[14] The following example illustrates the comparative effectiveness of labour in Portugal and England.

Labour	Country	Wine (Units)	Cloth (Units)
One Man-Week	Portugal	10	7.5
	England	5	6.0

Although Portugal has an absolute efficiency (cost) advantage over England in the production of both wine and cloth as she produces both these goods at lower absolute unit cost compared with England, her advantage is, however, comparatively greater in the production of wine than in the production of cloth. It is evident from the fact that the relative productivity of labour in Portugal is higher in the production of wine compared with that in the production of cloth, i.e.,

$$\frac{10 \text{ units of wine in Portugal}}{5 \text{ units of wine in England}} > \frac{7.5}{6 \text{ι}}$$

13. For detailed study see sub-heading 'Equilibrium in terms of Trade Offer Curve' in Chapter 10.
14. P.T. Ellsworth and J.C. Leith, *The International Economy*, Fifth Edition, 1975, p. 49.

England suffers from absolute efficiency (cost) disadvantage in the production of both the goods. However, her comparative cost disadvantage is less in the production of cloth than in the production of wine. It is evident from the fact that the relative labour productivity in England is higher in the production of cloth than in the production of wine, i.e.,

$$\frac{6 \text{ units of cloth in England}}{7.5 \text{ units of cloth in Portugal}} > \frac{5}{10}$$

The principle of comparative costs may be stated in general terms in the following words.

If m_1 and m_2 are the quantities of wine and cloth produced by a given input in Portugal while m_3 and m_4 are the corresponding quantities of wine and cloth produced by the same given input in England, Portugal will export wine and import cloth if the ratio $\dfrac{m_1}{m_2}$ is greater than the ratio $\dfrac{m_3}{m_4}$, i.e., if $\dfrac{m_1}{m_2} > \dfrac{m_3}{m_4}$. It says that the production of wine is relatively more efficient in Portugal than in England as compared with the production of cloth.

Applying the general formula, Portugal will specialise in the production of wine which she will export to England while England will specialise in cloth and export it to Portugal. The range of the possible terms of trade is determined by the domestic exchange ratios established by the relative efficiency (productivity) of labour in each country. The ratio of labour cost of production of wine and cloth in Portugal—10 units of wine for 7.5 units of cloth—was one limit to the possible ratio of exchange between Portugal and England. The other limit to the possible exchange ratio between Portugal and England was 5 (or 10 units) of wine for 6 (or 12 units) of cloth determined by the ratio of labour cost of production of wine and cloth in England. Portugal would not pay more for the English cloth than what it costs her to produce it at home. Similarly, England would not pay more for the Portuguese wine than what it costs her to produce it. The upper and the lower limiting range of the possible barter terms of trade can be stated as:

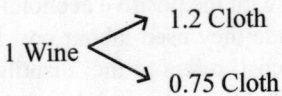

$$1 \text{ Wine} \begin{cases} 1.2 \text{ Cloth} \\ 0.75 \text{ Cloth} \end{cases}$$

If the barter terms of trade or the exchange ratio was either lower than 1 unit of wine to 0.75 unit of cloth or higher than 1 unit of wine to 1.2 units of cloth, there will be no trade between Portugal and England. Within the lower range of 1 unit of wine for 0.75 units of cloth and the upper range of 1 unit of wine for 1.2 units of cloth any exchange ratio may actually prevail. Within this range of 0.75 unit and 1.2 units of cloth to I unit of wine, the actual terms of trade will be determined by the intensity of each country's demand for the other country's product or by the intensity of reciprocal demand. By reciprocal demand Mill meant the quantity of exports which a country offers at the different terms of trade for the varying quantities of imports. The barter terms of trade or exchange ratio will be more favourable or advantageous for that country the foreign demand for whose product is inelastic while her own demand for the foreign country's product is highly elastic. At the equilibrium terms of trade, the total exports offered by each country will be just sufficient to pay for the total imports made from the other country. In other words, the balance of trade will be in equilibrium at the equilibrium terms of trade. To summarise:

1. the limiting range of the possible barter terms of trade is determined by the domestic or pre-trade (autarky) terms of trade which are set by the comparative efficiency in each country;

2. within this range, the actual barter terms of trade will be determined by each country's demand for the other country's product or good; and

3. only that barter terms of trade will be the equilibrium or stable barter terms of trade at which each country's total exports just pay for her total imports.

Criticisms

Until recently, the theory of comparative cost advantage formulated by David Ricardo and refined by John Stuart Mill was the most acceptable explanation of international trade. Praising the theory, Paul A. Samuelson stated that "if theories, like girls, could win beauty contests, comparative advantage would certainly rate high in that it is an elegantly logical structure. The theory of comparative advantage has in it a most important glimpse of truth .... A nation that neglects comparative advantage may have to pay a heavy price in terms of living standard, and potential rates of growth."[15] This, however, does not mean that the theory is free from criticisms. Its analytical structure is shaky due to its weak foundations resting on the several unrealistic assumptions. Consequently, it has been criticised by several economists, including Bertil Ohlin and Frank D. Graham.

1. An important criticism of the theory is the assumption of labour cost to explain the domestic exchange of goods. Critics have argued that labour is not the only factor of production used in the production of goods. Consequently, the total factor cost of production should also include the non-labour factor costs. According to the critics, the classical labour cost approach should be abandoned and international trade should be analysed in terms of the money costs since it is the money prices of the goods and services that determine which goods and services will be internationally traded and which country will produce them.

The supporters of the theory, however, argue that David Ricardo and John stuart Mill knew that in the production of goods labour was always used in combination with capital and/or land. They used only labour units to measure the value of commodities because their main concern was with the welfare economics and not with the positive economics. In order to measure the welfare in terms of the gain accruing from trade they used labour cost as a substitute of 'real cost'.

The classical concept of 'real cost' refers to the 'disutilities' or 'pain' (what Alfred Marshall calls "effort and sacrifice") involved in the employment of labour for the production of commodities. The classical economists assumed that disutility or pain involved in the production of different goods was proportional to the amount of labour embodied in their production. This is not, however, true in real life because disutilities are subjective costs which differ from hour to hour, person to person, place to place and according to the nature of work. For example, one hour's labour in a cold and bracing climate country like Canada or Germany would involve less exertion compared with the exertion involved in one hour's labour in the damp humid equatorial climate country like Congo or Brazil. Consequently, labour unit is an unapproachable and weak measure of the real unit cost of production of goods.

The various premises which form the basis of the orthodox value theory on which the classical doctrine of international trade is based are unrealistic. For example, the assumptions of a fixed relationship between wages of different categories of labour, an equal and fixed relationship between capital costs and costs of labour in all industries are unrealistic. There are highly labour-intensive industries in which wage costs are ten or even twenty times as high as capital costs while

---

15. Paul A. Samuelson, *Economics.* Fourth Edition, p. 658.

in others, with highly capital-intensive structure, the capital costs are considerably higher than the wage costs.

Moreover, the labour theory of value is faulty as it is based on the unrealistic assumptions of homogeneity of labour units, perfect mobility of labour within the country and perfect competition etc. It is, however, recognised that a country's labour force consists of different groups of labour, such as technical labour, skilled labour, semi-skilled labour, unskilled labour etc. called the 'non-competing groups'. These intergroups as well as the intergroup mobility of labour are far from being perfect even within the country. The competition both in the goods and factor markets is also imperfect.

Yet another unreal assumption of the labour theory of value is the assumption of a fixed relationship between the wages of different categories of labour which excludes the possibility of examining a fundamental problem of changes in the relative real wages of different categories of labour, e.g., the question: why have the real wages of office workers fallen while those of the manual labour have risen substantially during the past forty years? Evidently, such changes by affecting the relative prices of goods influence the size, composition and direction of trade. Consequently, the classical theory of international trade based on the labour theory of value is faulty.

Frank William Taussig has defended the classical theory of comparative cost advantage by pleading that the mere existence of such non-competing groups would not affect the classical theory of trade if in each country the relative wages were the same. The relative scale of wages will be the same in both the countries if the stratification of labour and the level of technological advancement in both the countries is the same. In the real world, different countries are at different levels of technological advancement which is reflected in the presence of technologically advanced countries like the United States of America, Japan and Germany and the backward Afro-Asian and the Latin American countries. Taussig's argument is not relevant in the real world and the classical labour theory of value should be discarded. If we discard the labour theory of value, the theory of comparative cost advantage is shattered to pieces.

2. The theory assumes that consumer tastes, production functions and factor endowments are fixed. The conclusions drawn from the static equilibrium analysis are irrelevant for a dynamic world where changes continuously occur in the technology, supply of factors of production, structure of industries and tastes of consumers. Consequently, it is difficult to calculate the comparative costs and the theory is unable to fit in any real circumstances.

3. The theory ignores the transport costs which is an important element in influencing both the direction and the volume of international trade. These costs are a special kind of input. The comparative cost advantage possessed by a country may be nullified by the transport costs and the commodity will not be traded unless the comparative cost difference exceeds the cost of transporting it from one country to another. The presence of transport costs leads to a third kind of commodities in addition to the export and import goods—domestic neutral goods—which are neither exported nor imported but are produced in each country. Some economists have attempted to overcome this difficulty by assuming that the country which exports a commodity also bears the transport costs and that the cost of production should also include the transportation costs.

The neglect of transport costs points out to another weakness of the comparative cost advantage theory. Numerous examples may be cited to show that one part of the country may import a commodity while the other part of the country may produce and even export it. For example, during the nineteenth century and early part of the twentieth century, German ports imported coal from England although Germany was herself an important producer and exporter of coal.

4. Another difficulty arises from the assumption of constant costs. According to the classical economists, the law of constant unit cost prevails in production so that the additional quantity of commodity can be produced at an unchanging per unit labour cost irrespective of the scale of production. In practice, however, the production of goods is subject to diminishing returns due to which after a certain level of output has been obtained the additional quantity of goods can be produced only at the increasing unit cost. In this situation, the cost ratios will change as the scale of output changes reducing the comparative cost advantage possessed by countries. Eventually, cost ratios in both the countries will become equal. Under equal cost ratios, there can be no trade since neither country would gain from trade. Thus, the law of diminishing returns implies another limitation of the theory of comparative cost advantage. It is, however, argued that the operation of diminishing returns does not invalidate the theory. In this situation, the only difference that arises is that specialisation is carried far less further than is possible under constant costs.

5. Frank D. Graham has proved that complete specialisation will be impossible even on the assumptions of the classical theory. Graham has cited examples to show that incomplete or partial specialisation will emerge if of the two trading countries one country is very small while the other is a very large (small and large in terms of total output and not in terms of geographical area) country. The small country will specialise completely in the production of one good while the large country will produce both the goods because (i) her large domestic demand will not be fully met by imports from the small country forcing her to produce the good at home in order to meet the unsatisfied domestic demand, and (ii) if she specialises completely in the production of the good in which she possesses comparative cost advantage, her surplus output will not be absorbed completely by the small importing country because the total domestic demand for the exports of the large country will be very small in the small country. In such a situation, the large country will continue to produce both the goods at home at her domestic exchange ratio. The small country is in an advantageous position of both selling her good and buying the good of the large country at the pre-trade domestic terms of trade (exchange ratio) of the large country. Here the small country is a price-taker and her smallness becomes a boon for the country since the entire gain from trade is appropriated by the small country.

Incomplete specialisation may also arise when the two goods traded are not of comparable values. When one good is a high-value good while the other good is a cheap low-value good, the country producing a high-value good will specialise completely while the country producing a low-value good will only partially specialise producing both the goods being exchanged at her domestic exchange ratio. Thus, where the two countries are of unequal size, the reciprocal demand principle will fail to operate and the small country may sell any quantity to the large country at the domestic exchange ratio of the latter reaping large gain from trade. The smallness of the country becomes her boon. Thus, unless trade takes place between the two countries of equal economic size or in the two commodities of equal value complete specialisation will not be possible. According to Graham, "the classical conclusion of complete specialisation between two countries can hold ground only when the dice is loaded by assuming trade in two commodities of approximately equal total consumption value and between two countries of approximately equal economic importance." These conditions are very rarely satisfied in the real world. Consequently, the classical theory of comparative cost advantage is unrealistic.

6. The theory of comparative cost advantage considers only the two countries and two commodities produced by a single input labour. In other words, the theory of comparative cost advantage is a $2 \times 2 \times 1$ trade model. According to the theory it is the comparative costs, measured in terms of labour units, which determine which good will be exported and which good will be imported by a country. Ricardo's classic example of trade between England and Portugal illustrates

this case. However, the moment we consider more than two commodities, we must also consider the demand conditions in order to determine not only the terms of trade but also the direction of trade. Even if we start from constant comparative costs, a change in the demand can cause such changes in the direction of international trade that goods which were exported are imported. In fact, international trade does not depend only on the comparative cost advantage or the supply conditions—it equally also depends on the conditions of demand.

7. The theory assumes that the entire production process of each commodity is carried on exclusively in the country, i.e., the process of producing a commodity is not divided between several countries. In other words, implicit in the theory is the assumption that all the inputs required for producing a commodity are available in the country. Difficulties are bound to arise in view of the fact that in many instances the process of producing a good is divided between several countries. For example, the cloth manufactured in India and forming her principal export good may require machinery, dyes and chemicals imported from Germany. In this case, the production of certain given quantity of cloth will require besides a certain number of Indian workers, a certain number of German workers also.

8. Yet another limitation of the principle of comparative cost advantage may arise as a result of a deliberate effort on the part of a country to produce a particular commodity although she may have no natural advantages in producing it and may import it at cheaper cost from other country. Nowadays, many countries follow the trade restrictionist policies for military and strategic reasons to avoid undue dependence on foreign countries. A good example is India's production of raw jute in Orissa and in some southern states at relatively high cost although cheaper jute can be imported from Bangladesh.

Some economists have shown that the theory of comparative cost advantage does not apply when a country imports one variety of a commodity while it produces and exports another variety of the commodity. It is not, however, a serious criticism of the theory. In economic analysis, each variety of the good is considered as a separate good. India may have a substantial comparative cost advantage in the production of coarse cloth while she may possess none in the production of superfine cloth. It is, therefore, natural for India to export the former and import the latter variety of cloth.

9. Bertil Ohlin has criticised the theory as being unduly cumbersome and unrealistic because it does not straightaway consider all the components of cost differences in different countries. It does not examine as to what extent the cheapness of production in one country is due to low wage, low interestrate, cheap transport and other facilities. The theory considers and compares only the output of wage-labour in the two countries leaving out all the other items of the unit cost of production. According to Bertil Ohlin, the classical approach is dangerous because while it considers specifically only the two countries and two commodities produced by only single input labour while it extends its conclusion uncritically and unhesitatingly to actual complex situations involving many countries and many commodities produced by several inputs. Moreover, the comparative costs theory is static. Consequently, it is ill-equipped to grapple with the dynamically changing subject-matter of international trade. Bertil Ohlin has criticised the theory in the following words.

"The doctrine of comparative costs as presented by Ricardo and Mill is unsatisfactory, not only because the scale of labour costs is built upon extreme simplifications, which cannot be abandoned without bringing down the whole fabric, but also because it neglects the influence of demand conditions on these scales themselves. The mutual interdependence is lost sight of. A simple description of certain conditions of production in terms of comparative cost schedules is put forward as determining the nature of international trade, while the play of reciprocal demand is given a secondary place as influencing only the extent of trade and the barter terms. As a matter of fact, the scale of comparative costs is not given *a priori,* but is affected by the play of reciprocal demand, as demonstrated already by Mongoldt."[16]

10. According to Gunnar Myrdal, the classical theory of international trade displays a strange isolation from the facts of life. Criticising the theory Gunnar Myrdal has stated that "on the whole, the literature is curiously devoid of attempts to relate the facts of international inequalities and the problems of underdevelopment and development to the theory of international trade."[17] Criticising the theory John H. Williams has stated that "the relation of international trade to the development of new resources and productive forces is a more significant part of the explanation of the present status of nations, of incomes, prices, well-being, than is the cross-section value analysis of the classical economists, with its assumption of given quantity of productive factors, already existent and employment."[18]

11. The classical theory is restricted to answering the narrow questions of what goods will be traded and what would be the gains from trade at any given moment. It does not tell us how the volume, composition and gains from trade undergo change through time. In other words, the theory does not analyse how the structure of comparative advantage changes over time. In order to extend this analysis to a developing economy it is necessary to consider how changes in factor supplies, technical progress, growth in productivity and changes in demand can alter the structure of comparative costs. For this reason it is necessary to introduce long period changes in the theory of comparative costs.

These criticisms together with the widespread interest displayed both at the theoretical and policy level in development make it obvious that the theory of comparative costs needs amendment and extension. Although the classicists had paid attention to the long-run growth of the domestic economy, their analysis of the international economy was essentially static. While the classical and neoclassical economists did not altogether ignore the effects of international trade on economic development nevertheless their treatment of the problem was sketchy. In order to examine the place of poor countries in the international economy, although the classical theory need not be completely supplanted by another theory but it must be supplemented in a broader frame of reference by removing it from the confines of full static equilibrium analysis.

Empirical Validity of the Theory

Several studies have been made to test the empirical validity of the comparative cost advantage theory. The most well-known among these studies are those made by G.D.A. MacDougal,[19] Robert Stern,[20] and Bela Balassa.[21] All these three economists worked with data relating to the United States of America and the United Kingdom in their studies. The data used by MacDougal pertained to 1937 while the data used by Robert Stern and Bela Balassa in their studies related to 1950 and

16. Bertil Ohlin, *Interregional and International Trade,* 1933, p. 23, footnote.

17. Gunnar Myrdal, *Rich Lands and Poor,* 1957, p. 153.

18. John H. Williams, "The Theory of International Trade Reconsidered", *The Economic Journal,* June, 1929, p. 196.

19. G.D.A. MacDougal, "British and American Exports: A Study Suggested by the Theory of Comparative Costs", *The Economic Journal,* Volume 61, December, 1951, pp. 697–724.

20. Robert Stern, "British and American Productivity and Comparative Costs in International Trade", *Oxford Economic Papers,* October, 1962.

21. Bela Balassa, "An Empirical Demonstration of Classical Comparative Cost Theory", *The Review of Economics and Statistics,* August, 1963.

1959, and 1950 respectively. The purpose of all these studies was to test the empirical validity of the labour theory of value, which was the foundation stone of the comparative cost advantage theory, as the main determinant of international trade.

According to the labour theory of value, differences in labour productivity which are reflected in differences in the cost of production of various commodities determine the pre-trade (autarky) prices of these commodities. If a country has a relatively low price of commodity A and a relatively high price of commodity B, it will export commodity A and import commodity B.

A major problem in testing this hypothesis arises from the difficulty of observing the pre-trade prices of commodities which would prevail under autarky, i.e., when countries did not trade with each other. In the real world, countries trade with each other as a consequence of which product prices are already equalised (barring differences due to tariffs and transport costs etc.) between different countries. In such a situation, it is extremely difficult to obtain any reliable information about the comparative cost advantage possessed by different countries in the production of different commodities. Consequently, the testing of the hypothesis is rendered nearly impossible. As an alternative solution, the hypothesis that a country which has a relatively high labour productivity in the production of a certain commodity will export that commodity was tested directly.

The second problem arises from the fact of existence of tariff and transport costs in the real world. Particularly in 1937, the year adopted by MacDougal for his investigation, both the United States of America and the United Kingdom had tariffs that were high enough to wipe out any comparative cost advantage which might have existed resulting in very little trade with each other. Consequently, the study's focus is on the export performance of the two countries in the third markets where the products of both the countries receive equal or same treatment with regard to tariffs and transport costs.

The third problem that arises in comparing the export performance of the two countries in the third markets, is the non-availability of export performance data in suitable form for comparison purposes. For certain commodities (industries), export data is available in physical quantity terms while for others it is available in terms of the total value of exports. This handicap is not, however, insurmountable since the value of exports is simply the physical quantity multiplied by the prices of exported commodities. As both the countries will obtain the same prices in the world market for the commodities, the two indices would give the same results.

The comparative cost advantage theory states that a country whose labour productivity in the production of a certain commodity is higher than the labour productivity of the other country will capture the whole world export market for this commodity. In other words, labour costs alone will be the determinant of trade. A pioneering attempt made at the empirical testing of this theory was made by Donald MacDougal and was published in 1951 and 1952. Contrary to the expectations that other factors like the capital costs, demand factors, political relationship, trade impediments etc., could have distorted the relationship between the labour productivity and export share, the theory explained the empirical pattern of trade.

For his study, MacDougal adopted the labour productivity ratios of the two countries and the export ratios of the two countries. He found that there was an approximately linear relationship between the labour productivity and export ratios showing that in those commodities in the production of which the US labour productivity (relative to the United Kingdom) was the highest, the United States of America would capture the largest share of the world export markets. As the

USA's relative advantage fell, her share in the export market shrank too. The process was continuous in the sense that a small comparative cost advantage did not enable any country to capture the whole export market. This was due to the (i) product differentiation and (ii) imperfectly competitive markets in the real world. The lack of homogeneity within commodity classes and the presence of oligopolistic or monopolistic industries enable the country to maintain some hold in the export markets even when cost advantage considerations speak otherwise.

MacDougal's study revealed that in order to capture a larger share of the export market the US labour productivity must be more than twice the British labour productivity. A plausible explanation for this was that in 1937 wages in the United States of America were twice of those in the United Kingdom. Consequently, with twice the labour productivity in comparison to that of the UK worker, the American goods would command no comparative cost advantage in the export markets. Moreover, the amount of indirect labour (labour spent in transporting, distributing and servicing) spent on the commodity was greater for the United States of America than for the United Kingdom and this amount was not included in the productivity data. Consequently, to provide for this, the labour productivity in the United States had to be greater than twice of that in the United Kingdom to make it equal. In other words, even with a labour productivity higher than double of that in the United Kingdom, the American goods would command no advantage over the British goods in the third markets. Finally, the Commonwealth preference under which countries members of the British Commonwealth granted preferential treatment to the British goods tended to reduce the comparative cost advantage possessed by America in the production of commodities. In other words, the system of imperial preference extended to the British goods in the form of either allowing the imports of British goods duty-free or at concessional import duty[22] discriminated against the United States of America requiring her to possess a still much higher labour productivity in comparison to that in the United Kingdom in order to enable her to capture the larger share of the export market.

MacDougal's study supported the Ricardian theory of comparative cost advantage. Robert Stern's and Bela Balassa's studies made for the post-war period confirmed and amplified the conclusion of MacDougal's pioneering study. The conclusion reached in both these studies was that other factors such as the unit capital cost did not significantly influence the export performance of countries. Both the studies confirmed MacDougal's findings showing a high correlation between the productivity of labour and country's share in the export market.

Jagdish Bhagwati[23] has, however, doubted the conclusion of these studies. Employing a more sophisticated technique, his findings show that linear regressions of export price ratios (USA/UK) on the labour productivity ratios did not yield any significant regression coefficients. So also regressions of unit labour cost ratios on export price ratios for these two countries yielded no significant regression coefficients. Consequently, in the light of Jagdish Bhagwati's findings the strong positive findings of Donald MacDougal, Robert Stern, and Bela Balassa about the empirical validity of the comparative cost advantage doctrine should not be taken as a final word on the subject until more conclusive evidence supporting these studies becomes available as a result of further analytical studies.

22. Under the policy of imperial preference and discriminating protection, British goods were allowed duty-free concessional imports in India while imports from the Continent and Japan were subjected to higher import duty.

23. Jagdish Bhagwati, "The Pure Theory of International Trade: A Survey", The Economic Journal, March 1964, Section I, Theorems in Statics: The Pattern of Trade.

Suggested Readings

P.T. Ellsworth and J. Clark Leith, *The International Economy,* Fifth Edition, 1975, Chapter 3.

Gottfried Von Haberler, *The Theory of International Trade,* 1950, Chapters 9–11.

Gottfried Von Haberler, *A Survey of International Trade Theory,* 1961, Chapter 1.

Bertil Ohlin, *Interregional and International Trade,* 1933, Appendix III.

Bo Sodersten, *International Economics,* Second Edition, Reprinted 1981, Part 1, Chapter 1.

Frank William Taussig, *International Trade,* 1927, Chapters 1–10.

Jacob Viner, *Studies in the Theory of International Trade,* 1937, Chapter VIII.

Questions

1. Discuss with example the classical comparative cost advantage theory of international trade. What are its main criticisms?

2. The classical trade theory emphasises differences in the comparative costs to explain the trade patterns. What other factors need to be taken into account if actual trade patterns are to be fully explained?

3. Critically examine the Ricardian comparative cost advantage theory of international trade. To what extent is this theory supported by empirical evidence?

4. Explain the Ricardian theory of comparative cost advantage. How will the theory be modified when the assumptions of (a) labour as the only factor of production, and (b) the law of constant cost are dropped?

CHAPTER 5

Development of Theory of Comparative Cost Advantage

The theory of comparative cost advantage rests on several assumptions. In order to make the doctrine a more accurate presentation of the reality, John Elliot Cairnes, Alfred Marshall, Frank William Taussig, Gottfried Von Haberler, Charles F. Bastable and others further developed and extended the ideas of Adam Smith, David Ricardo and John Stuart Mill. The doctrine of free trade was applied to the multi-country and multi-commodity world by Francis Ysidro Edgeworth, Bastable and Frank D Graham. It was further refined with the aid of linear programming technique and was deepened by a model introduced by Eli Heckscher which explained the differences in productivity which had been merely postulated by David Ricardo and his followers. These modifications do not, however, discard the results of the classical theory of trade, rather these help in clarifying the exposition of the theory without affecting its essential features. Some of the modifications made in the comparative cost advantage theory have been discussed below.

Theory Expressed in Money Terms

The Ricardian theory of comparative cost advantage was explained in terms of the labour cost of production. The modern economies are money economies where the volume and direction of international trade is not determined by the comparative differences in the labour costs of production but by differences in the money prices of goods. This does not, however, present much difficulty because the comparative cost differences in the labour costs of producing commodities can be translated into the absolute differences in money prices. Such translation will not alter the exchange relationships between commodities which lie behind the differences in money prices. We can now explain the translation of comparative cost differences into absolute money cost differences with the help of the following example.

In India 10 man-days' labour produces 100 units of cotton.

In India 10 man-days' labour produces 100 units of jute.

In Bangladesh 10 man-days' labour produces 40 units of cotton.

In Bangladesh 10 man-days' labour produces 80 units of jute.

In this example, although India has an absolute advantage over Bangladesh in the production of both the goods, she has comparative advantage over Bangladesh only in the production of cotton. On the other hand, although Bangladesh is inefficient in the production of both the commodities she is however comparatively less inefficient in the production of jute. If international trade takes place, India will specialise in the production of cotton and Bangladesh will specialise in the production of jute. Let us convert this real cost example into money cost by assuming that the daily money wages in India and Bangladesh are Rs. 3 and Rs. 2[1] respectively.

It is obvious that the per unit money cost of production of cotton is lower in India. In India it is 0.30 rupee per unit while in Bangladesh it is 0.50 rupee per unit. The money cost of production of jute is, however, lower in Bangladesh being 0.30 rupee per unit in India and only 0.25 rupee in Bangladesh. India will, therefore, specialise in the production of cotton and Bangladesh will specialise in the production of jute. This result is in harmony with the classical theory of comparative cost advantage. It can, however, be argued by the critics that we have arbitrarily chosen the money wage rates in the two countries. Since the same amount of labour (10 man-days) in the two countries produces different quantities of cotton and jute, the upper and the lower limits of money wage differences between the two countries are shown by the differences in the productivity (output). The maximum advantage which India enjoys over Bangladesh in the

production of cotton is equal to 2.5 times $\left(\dfrac{100 \text{ units of cotton in India}}{40 \text{ units of cotton in Bangladesh}} \right)$ determined by the

relative productive efficiency of labour in India in the production of cotton.

Comparative Cost Advantage Expressed In Money

Country	Daily wages (in rupees)	Total wage-bill (in rupees)	Total output	Money cost or supply price per unit (in rupees)
India	3	30	100 units of cotton	0.30
India	3	30	100 units of jute	0.30
Bangladesh	2	20	40 units of cotton	0.50
Bangladesh	2	20	80 units of jute	0.25

If the money wage in Bangladesh is assumed two rupees, the maximum money wage in India can be 2.5 times of the money wage in Bangladesh, i.e., five rupees. The lower limit of money wage in India is indicated by the maximum advantage which India has over Bangladesh in the production

of jute, i.e., 1.25 times $\left(\dfrac{100 \text{ units of jute in India}}{80 \text{ units of cotton in Bangladesh}} \right)$. The lower limit of money wage in India,

therefore, is Rs. 2.50. If we assume that money wage in Bangladesh is Rs. 2, money wage in India will be any amount between Rs. 2.50 and Rs. 5. It is only within these two upper and lower limits that we have chosen the money wage of Rs. 3 in India.

It can be shown that money wage differences cannot cross these two limits set by the differences in labour productivity in the two countries in the production of cotton and jute. If money wage in India rises to Rs. 5, the price per unit of cotton as well as of jute will be 50 paisa. Consequently, cotton will not be exported from India to Bangladesh since in Bangladesh the per unit cost of growing cotton is also 50 paisa. Jute will, however, be imported by India from Bangladesh causing deficit in India's balance of trade followed by an outflow (depletion) of gold from India. According to the theory of the gold movements and the quantity theory of money, the

1. For simplicity's sake Bangladesh's currency unit taka has been treated equivalent to India's currency unit rupee.

loss of gold will result in the contraction of the total money supply in India. Consequently, prices will fall in India and rise in Bangladesh since gold will be imported into the country. Conversely, if money wage in India is below the lower limit of Rs. 2.50, the unit cost of production in India will decrease. The unit cost of production of jute in India will be lower compared with that in Bangladesh. The result will be that India will continue to export cotton but will stop importing jute from Bangladesh. This will result in the surplus balance of trade for India resulting in the inflow of gold. Consequently, the supply of money will expand causing prices to rise in India and fall in Bangladesh since gold will be exported from Bangladesh. Thus, the ratio of money wages in the two countries must lie somewhere between the upper and the lower limit points.

We cannot, however, state from the cost data alone as to where exactly within these two upper and lower limits the actual ratio of money wages in the two countries and, therefore, the international terms of trade for the two commodities will settle. It will depend upon the conditions of the reciprocal demand of each country for the other's goods. The ratio of exchange will be favourable to India if Bangladesh's demand for Indian cotton is more intense than is India's demand for Bangladesh's jute. John Stuart Mill had pointed out that the actual exchange ratio would be determined, given the conditions of demand, by the fact that the total value of each country's exports must be equal to the total value of her imports.

Theory Applied to More than Two Goods

The comparative cost advantage theory of international trade developed by David Ricardo was confined to trade between only two countries and in only two commodities. According to the doctrine, a country will specialise in the production and export of that good in which it commands a comparative cost advantage over the other country and import that good in the production of which it suffers from comparative disadvantage. In the real world, however, many commodities are simultaneously exported and imported between any two trading countries. When the Ricardian theory is, however, extended to cover many commodities the conclusion remains unchanged. This is obvious from the generalised theorem which is based on Haberler's[2] treatment discussed below.

Let us assume that two countries X and Y trade in five commodities A, B, C, D and E. The per unit labour cost of production of each commodity in country X is m_1, m_2, m_3, m_4, and m_5 respectively while in country Y it is e_1, e_2, e_3, e_4, and e_5 respectively. The per unit supply price (money cost) of these commodities in country X is P_1, P_2, P_3, P_4, and P_5 while in country Y it is P_1', P_2', P_3', P_4', and P_5' respectively. Now let us suppose that the money wage in country X is W_x while in country Y it is W_y. The per unit supply price (money cost) of each commodity will be equal to the labour cost per unit of output (m and e) multiplied by the money wage W_x in country X and W_y in country Y, i.e.,

$$P_1 = W_x.m_1 \qquad\qquad P_1' = W_y.e_1$$
$$P_2 = W_x.m_2 \qquad\qquad P_2' = W_y.e_2$$
$$P_3 = W_x.m_3 \qquad\qquad P_3' = W_y.e_3$$
$$P_4 = W_x.m_4 \qquad\qquad P_4' = W_y.e_4$$
$$P_5 = W_x.m_5 \qquad\qquad P_5' = W_y.e_5$$

Thus, the relative prices of commodities in each country are determined by their unit labour cost of production such that

$$P_1 : P_2 : \qquad\qquad : P_5 = m_1 : m_2 : \ldots\ldots : m_5$$
$$\text{and} \quad P_1' : P_2' : \ldots\ldots : P_5' = e_1 : e_2 : : e_5$$

2. Gottfried Von Haberler, *The Theory of International Trade*, Third Impression, 1950, Chapter X.

If R is the rate of foreign exchange (currency units of country Y which exchange for one currency unit of country X), the price of one unit of each one of the five commodities expressed in terms of the currency unit of country Y in the two countries will be as shown below.

Table 2

Unit of Commodity	Price in Country X	Price in Country Y
A	$m_1 \times W_x \times R$	$e_1 \times W_y$
B	$m_2 \times W_x \times R$	$e_2 \times W_y$
C	$m_3 \times W_x \times R$	$e_3 \times W_y$
D	$m_4 \times W_x \times R$	$e_4 \times W_y$
E	$m_5 \times W_x \times R$	$e_5 \times W_y$

Country Y will export those commodities whose prices (which equal the unit money cost) are lower in country Y and will import those commodities whose prices are higher in country Y. For example, if commodity A is to be exported from country Y the following inequality must hold.

$$e_1 \times W_y < m_1 \times W_x \times R$$

Or $$\frac{e_1}{m_1} < \frac{W_x \times R}{W_y}$$

Similarly, if commodity B is to be imported by country Y the following inequality must hold.

$$e_2 \times W_y > m_2 \times W_x \times R$$

Or $$\frac{e_2}{m_2} > \frac{W_x \times R}{W_y}$$

Since $$\frac{e_1}{m_1} < \frac{W_x \times R}{W_y}, \text{ and } \frac{e_2}{m_2} > \frac{W_x \times R}{W_y}$$

Therefore, $$\frac{e_1}{m_1} < \frac{e_2}{m_2}$$

This shows that country Y enjoys a comparative cost advantage over country X in the production of commodity A. This is what the comparative cost advantage theory tells about the phenomenon of trade in the two commodities. What is true of two commodities is also true of many commodities. Arranging the above commodities in order of the comparative cost advantage possessed by country Y over country X we will have the following relationship.

$$\frac{e_1}{m_1} < \frac{e_2}{m_2} < \frac{e_3}{m_3} < \frac{e_4}{m_4} < \frac{e_5}{m_5}$$

If we draw a line dividing the commodities which country Y exports from those which she imports, all the former will be on one side of the line while all the latter will be on the other side of the line without changing the order in which they are arranged. For example, it will not be possible for country Y to export commodities A and C and to import commodity B.

The above conclusion is drawn on the basis of the cost data or supply conditions alone. The cost data alone is, however, insufficient to determine the position of the dividing line between a country's exports and imports. For determining the exact position of the dividing line, the knowledge of the relative strength of international demand is also essential. In this context, Haberler has stated that the quotient $\dfrac{W_x \times R}{W_y}$ in the relationship plays a significant role. In other words, any change in the relative wage level $\dfrac{W_x}{W_y}$ and/or the rate of foreign exchange R will alter the lines of production and the comparative cost situation. To elucidate this point, let us consider the following hypothetical cost of production of the commodities in the two countries. The units of the quantity of various goods have been so chosen as to make the per unit cost of every good in country X the same or equal.

Cost of Production	Commodities								
	A	B	C	D	E	F	G	H	I
Real cost per unit in country X expressed in labour hours (m_1, m_2 ....)	10	10	10	10	10	10	10	10	.10
Real cost per unit in country Y expressed in labour hours (e_1, e_2, ...)	30	25	20	16	10	8	7	6	4

The exact position of the dividing line between the exports and imports of country X is determined by $\dfrac{W_y}{W_x \times R}$. Any factor which causes change in either W_x, W_y, and/or R will cause the dividing line to shift from its original position. If $\dfrac{W_y}{W_x \times R} = 1$, i.e., if money wages are the same in both the countries, the unit money cost of production of goods from A to D (which are produced at a smaller absolute real cost in country X) will be lower in country X than in country Y. Consequently, country X will export commodities A, B, C and D and import commodities F, G, H and I. Commodity E, being on the dividing line showing equal real unit cost, will neither be exported nor imported and will be produced by both the countries. If the value of $\dfrac{W_y}{W_x \times R}$ is greater (less) than 1, the dividing line will shift to the right (left) of E. The exact position of the dividing line depends upon the condition of the international balance of payments of the countries concerned.

Starting from an initial equilibrium in the external balance of the payments of country X, suppose that the existing equilibrium is disturbed and the balance of payments of the country becomes adverse due to an increase in the demand for the goods of country Y. The price-specie-flow monetary mechanism will come into operation. Gold (assuming that both countries are on the gold standard) will be exported from country X to country Y. In consequence of this price-specie-flow mechanism, prices and wages will rise in country Y and fall in country X. Consequently, the value of W_x will fall while that of W_y will rise. In the process, the value of $\dfrac{W_y}{W_x \times R}$ will increase and the dividing line will shift to the right. Consequentiy, commodity E will be included in the list of commodities exported by country X and equilibrium in country's external balance of payments will be restored because (1) commodity E is now exported, (2) export commodities A, B, C and D of the country X having become cheaper their total export value increases, and (3) export commodities of country Y having become costlier the imports of country X from country Y decrease both in volume and value terms. However, if all this fails to restore equilibrium in the external balance of payments of country X, the outflow of gold from country X will continue and the dividing line will keep on shifting to the right expanding the list of the export goods and contracting the list of import goods of country X until equilibrium in the balance of payments of the country is eventually restored. Thus, the multi-commodity trade model enunciated by Gottfried Von Haberler is an improvement over the two-commodity Ricardian trade model explaining the same principle of comparative cost advantage. In this case, however, bulk of the burden of the balance of payments adjustment mechanism is borne by country X.

Multi-commodity and Multi-country Trade Model

By presenting a multi-commodity and multi-country trade model, Paul A. Samuelson has shown that trade along comparative cost lines is not only possible in such a case but is also very convenient. It has been diagrammatically illustrated in Figure 5.1 where four countries A, B, C and D and four commodities E, F, G and H are involved in trade. Country A exports commodity E to country B, country B exports commodity F to country C, country C exports commodity G to country and country D exports commodity H to country A.

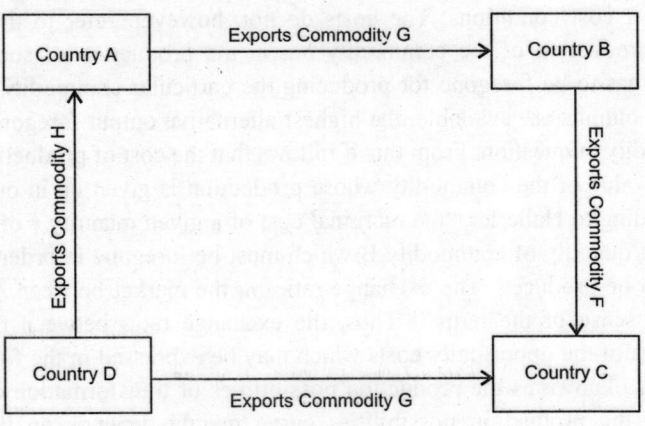

Figure 5.1

In Figure 5.1 the figure the arrows indicate the direction of exports of each country. The direction of arrows clearly shows that trade in this case is unilateral—one way affair—in as much as country A does not import anything from country B, country B does not import any good from country C, country C does not import any good from country D while country D likewise does not import any good from country A. It is obvious that country B pays for her imports from country A through her export earnings received from country C. Similarly, country C pays for her imports from country B through her export earnings received from country D. Country D in turn pays for her imports made from country C through her export earnings received from country A. Country A pays country D through the export earnings received from country B. But the condition that trade ultimately has to be balanced has been fulfilled so that for any country the total value of her exports is equal to the total value of her imports. The process of multilateral balancing of trade shown here is superior to the bilateral balancing of trade from the point of view of the community welfare.

Theory Expressed in Terms of the Opportunity Cost

The core of the classical comparative cost advantage theory of trade is the labour theory of value which is based on several assumptions. The most restrictive assumption is the assumption of homogeneous labour. In the real world, however, there is not one single class of labour but several non-competing groups of labour between which the tendency toward equalisation of wages is very weak and practically nonexistent. Moreover, goods are not produced by labour alone. Goods are produced through the combined productive activities of many factors such as land, capital equipment, natural resources, raw materials, technology, etc. Technically it is not possible to measure all the various factors used in production in terms of unskilled labour. Moreover, the factors of production are not combined with labour in any fixed proportion and different cooperating inputs are combined in varying proportions. Thus, it is erroneous to compare the relative prices of commodities in terms of their labour contents alone.

To overcome these problems, the 'opportunity cost' approach has been used by Haberler[3]. The basic contention of the opportunity cost theory is that the relative prices of commodities are determined by their cost conditions. The costs do not, however, refer to the amount of labour contained in the production of the commodity but in the production of some other alternative commodity which has to be foregone for producing the particular commodity in question. Where several alternative outputs are available, the highest alternative output foregone is the opportunity cost of the commodity in question. From this it follows that the cost of production of any particular commodity is the value of the commodity whose production is given up in order to produce that commodity. According to Haberler, "the marginal cost of a given quantity x of commodity A must be regarded as that quantity of commodity B which must be foregone in order that x, instead of x − 1, units of A can be produced. The exchange-ratio on the market between A and B must equal their costs in this sense of the term."[4] Thus, the exchange ratio between two commodities is determined in terms of the opportunity costs which may be expressed in the form of a substitution curve otherwise also known as the production possibilities or transformation curve.

The shape of the production possibilities curve mainly depends on the returns to scale operating in production. Under constant returns to scale or constant opportunity cost, the production possibilities curve will be linear as shown in Figure 5.2A. Under diminishing returns to scale or increasing opportunity cost, production possibilities curve will be concave viewed from the origin of the axes as shown in Figure 5.2B. Under increasing returns to scale or the decreasing opportunity cost of production, the production possibilities curve will be convex toward the point of origin of the axes as shown in Figure 5.2C. Under these conditions there are several possible outcomes. The two important cases have been shown in Figure 5.2D.

3. Gottfried Von Haberler, *Op. cit.,* Chapter XII.
4. Gottfried Von Haberler, *Op. cit.,* p. 177.

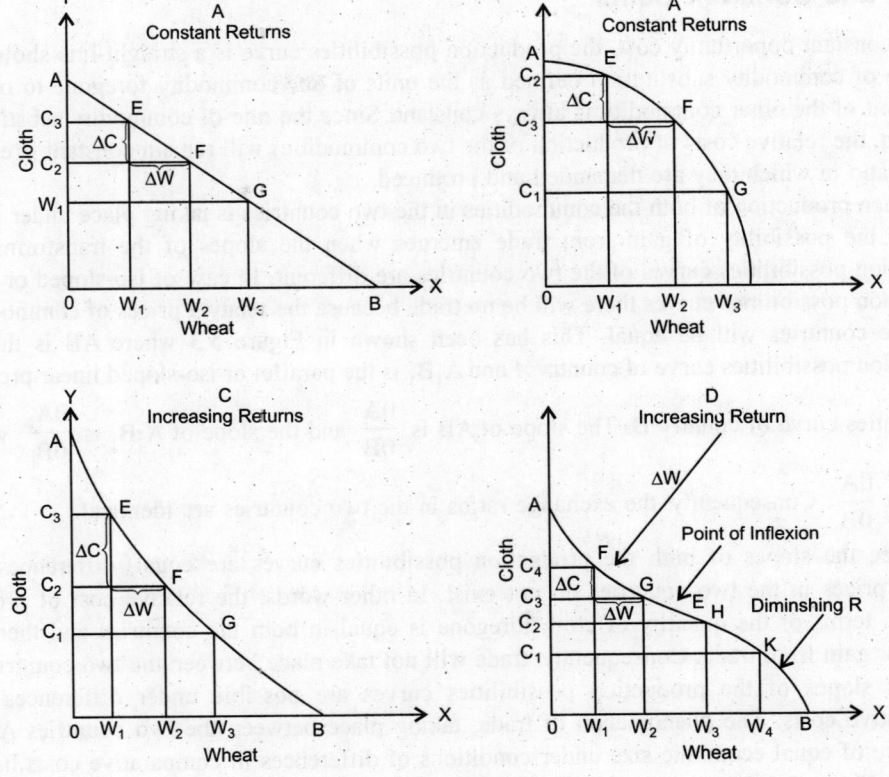

Figure 5.2

Under constant returns, the marginal rate of commodity substitution $\left(-\dfrac{\Delta C}{\Delta W}\right)$ is constant all along the production possibilities curve AB in Figure 5.2A. In the case of diminishing returns, the marginal rate of commodity substitution $\left(-\dfrac{\Delta C}{\Delta W}\right)$ is increasing showing that the opportunity cost of producing wheat in terms of the quantity of cloth foregone is increasing all along the production possibilities curve AB as has been shown in Figure 5.2B. Under increasing returns, the marginal rate of commodity substitution $\left(-\dfrac{\Delta C}{\Delta W}\right)$ is diminishing as shown in Figure 5.2C. In other words, as the output of wheat increases its unit cost of production in terms of the quantity of cloth sacrificed or foregone decreases. Figure 5.2D is a combination of Figure 5.2B and Figure 5.2C showing that over the AE range of the AB production possibilities curve, the rate of commodity of substitution is diminishing while beyond E this rate is increasing. E is the point of inflexion where momentarily the rate of change of the slope of the production possibilities curve is zero. A typical production possibilities curve will usually be of this shape displaying the operation of the law of varying proportions in production.

Trade and Constant Costs

Under constant opportunity cost, the production possibilities curve is a straight-line showing that the rate of commodity substitution defined as the units of one commodity foregone to obtain an extra unit of the other commodity is always constant. Since the rate of commodity substitution is constant, the relative costs of production of the two commodities will remain constant irrespective of the ratio in which they are demanded and produced.

When production of both the commodities in the two countries is taking place under constant returns, the possibility of gain from trade emerges when the slopes of the transformation or production possibilities curves of the two countries are different. In case of iso-sloped or parallel production possibilities curves there will be no trade because the relative prices of commodities in both the countries will be equal. This has been shown in Figure 5.3 where AB is the linear production possibilities curve of country A and A_1B_1 is the parallel or iso-sloped linear production possibilities curve of country B. The slope of AB is $\dfrac{0A}{0B}$ and the slope of A_1B_1 is $\dfrac{0A_1}{0B_1}$ which is equal to $\dfrac{0A}{0B}$. Consequently, the exchange ratios in the two countries are identical.

Since the slopes of both the production possibilities curves are equal, differences in the relative prices in the two countries do not exist. In other words, the relative cost of producing wheat in terms of the quantity of cloth foregone is equal in both the countries and there is no scope for gain from trade. Consequently, trade will not take place between the two countries. The different slopes of the production possibilities curves are possible under differences in the comparative costs. The phenomenon of trade, taking place between the two countries A and B which are of equal economic size under conditions of differences in comparative costs has been shown in Figure 5.4.

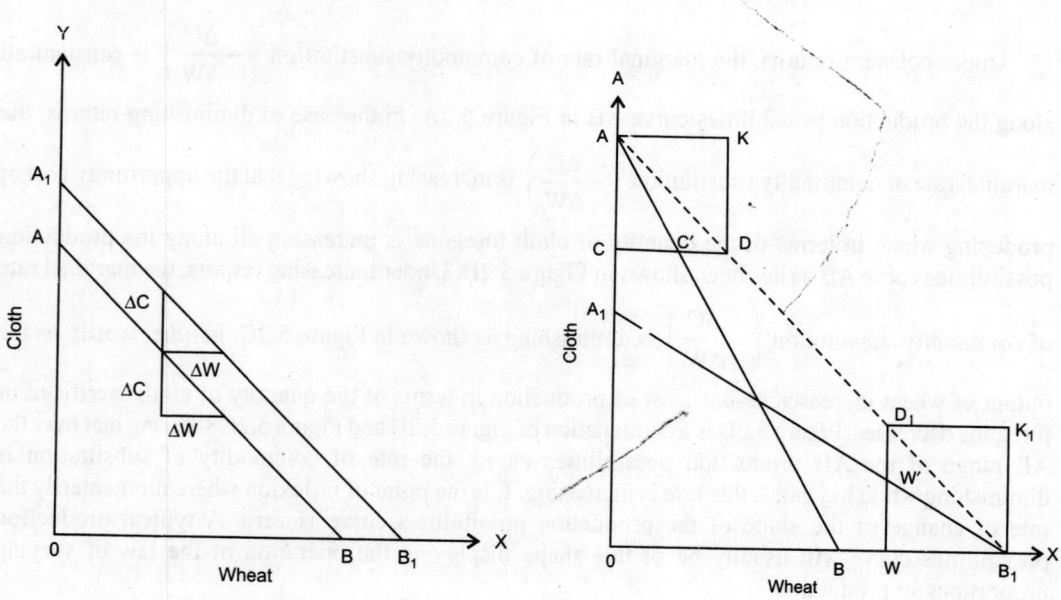

Figure 5.3   Figure 5.4

AB is the production possibilities curve of country A. It shows that with the given factors of production (specified factor endowments) country A can produce either the 0A output of cloth and no wheat or 0B output of wheat and no cloth or anyone of the many alternative combinations of the two commodities situated on country's production possibilities curve AB. Similarly, A_1B_1 is the production possibilities curve of country B. It shows that with given factors of production country B can produce either $0A_1$ amount of cloth and no wheat or $0B_1$ amount of wheat and no cloth or any one of the many other possible combinations of the two goods situated on the production possibilities curve A_1B_1.

The production possibilities curves show that country A has a comparative advantage in the production of cloth over country B while country B has a comparative advantage over country A in the production of wheat. If after trade country A specialises completely in the production of cloth and country B specialises completely in the production of wheat both the countries will exchange commodities in the ratio indicated by the dotted price line AB_1. Assuming that both the countries want to consume the two commodity bundles shown by point D for country A and by point D_1 for country B, country A will export CA (= DK) quantity of cloth and import CD (=AK) quantity of wheat. At this exchange ratio, country B exports WB_1 (= D_1K_1) quantity of wheat and imports WD_1 (= B_1K_1) quantity of cloth. Since the amount of cloth which A exports (DK) equals the amount of cloth which country B imports (WD_1) and the amount of wheat which country B exports (WB_1) equals the amount of wheat which country A imports (CD), AB_1 is the equilibrium terms of trade line. It is so because at the terms of trade or exchange ratio shown by the slope of AB_1, the balance of trade of both the countries is in equilibrium, i.e., the total value of imports equals the total value of exports of each one of the two countries. There are also other amounts of the two goods which could be traded, the requirement for equilibrium being that the total quantities of exports and imports should match exactly for each commodity.

Both the countries will gain from trade. The production possibilities curve AB of country A shows that in autarky AC quantity of cloth can be exchanged for CC' quantity of wheat in the domestic market but under free trade in the international market AC quantity of cloth is exchanged for CD quantity of wheat. Consequently, as a result of trade, country A obtains C'D (= CD – CC') additional quantity of wheat which is the net gain to the country from trade. Similarly, the gain of country B from trade is the $W'D_1$ (= WD_1 – WW') quantity of cloth.

Specialisation in both the countries will be complete if both the countries are of approximately equal size. If the two countries are of unequal size—one country is very small and the other country is very large—then only partial specialisation will take place in the large country while the small country will practise complete specialisation. If only partial specialisation takes place in the large country, the entire gain accruing from trade will be appropriated by the small country. It is so because the international exchange ratio will coincide with the pre-trade domestic exchange ratio of the large country which continues to produce both the commodities since even after the small country has specialised completely in one commodity there is still some excess demand for this commodity which is satisfied by production made under the domestic cost conditions in the large country. Consequently, the small country can buy or sell any quantities at the domestic exchange ratio of the large country. The small country is just a price taker. In this situation, the smallness of the country becomes a boon for the small country. Figure 5.5 shows such a situation.

We take two countries India and Nepal. India is very large in size while Nepal is very small. Nepal commands comparative cost advantage in the production of timber wood. Consequently, she completely specialises in the production of timber. Being, however, a very small country her total

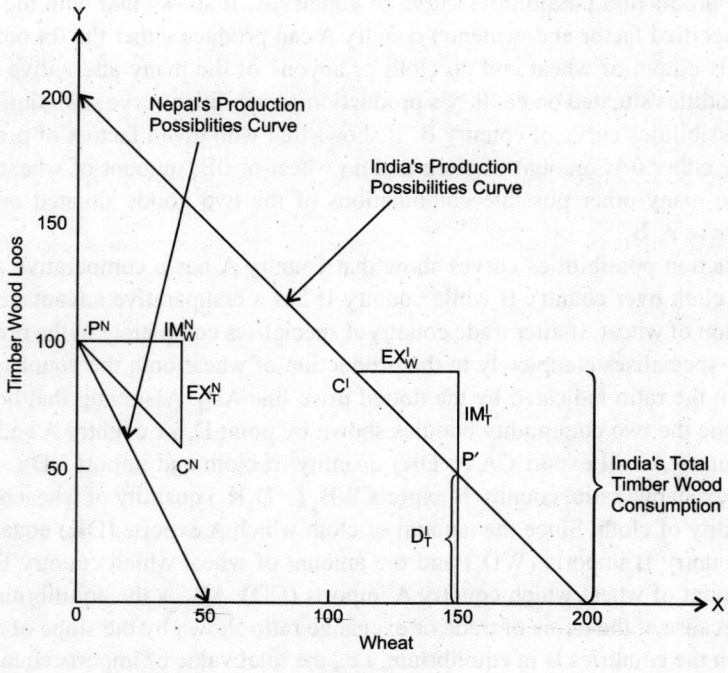

Figure 5.5

production of timber is insufficient to meet her own domestic demand and India's demand for timber wood. Consequently, India has to produce some timber in order to meet part of her domestic demand. She cannot, however, produce timber domestically unless it is exchanged against wheat at her pretrade domestic exchange ratio. In this situation, Nepal will sell (export) her timber to India and buy (import) wheat from India at India's pretrade domestic exchange ratio represented by the slope of India's production possibilities curve. In Figure 5.5, Nepal (small country) produces at point P^N specialising completely in the production of timber-wood logs. She trades part of her timber production against imports of wheat from India, permitting her to consume the commodity combination at point C^N. To achieve this commodity consumption pattern, Nepal imports IM_W^N amount of wheat from India in exchange for EX_T^N amount of timber-wood logs. India consumes at point C^I where her total consumption of timber-wood logs exceeds the total import of timber available to her from Nepal. Consequently, she supplements the total imports amounting IM_T^I from Nepal with D_T^I amount of timber produced in the country. Thus, India's production point is P^I where she produces D_T^I amount of timber-wood logs and 150 units of wheat. In short, while the small country Nepal completely specialises in the production of timber-wood logs and appropriates the entire gains from trade, being a large country India produces both the commodities and imports timber from Nepal at her domestic exchange ratio.

Trade and Increasing Costs

The above analysis was confined to the special case of constant costs. Constant cost is possible only under the assumptions of fixed factor proportions and equal efficiency in producing the relative outputs of the two goods. In the real world, however, the factors of production are combined not in any fixed proportion but in varying proportions. Resources released from the production of anyone commodity, such as cloth, may not be equally suited for the production of other commodity, say wheat. Many factors of production are specialised in particular uses, being suited to the production of certain commodities only. Such factors are referred to as product-specific factors. The existence of a product-specific factor of production is responsible for the phenomenon of increasing costs in production. Where increasing costs operate in production, the slope of the transformation curve varies throughout its length showing changes in the relative unit costs of production of the two commodities. Consequently, the relative prices of commodities will also vary and the price at which one commodity will be exchanged for another cannot be determined by the production possibilities curve alone. Consequently, in order to know the actual exchange ratio at which the two commodities will be exchanged, the demand conditions will have also to be taken into account. This situation has been illustrated in Figure 5.6 given below.

In Figure 5.6, the production possibilities curve of country X has been shown by AB while the domestic price or exchange-ratio has been shown by the slope of price line KL which is tangent to the production possibilities curve at point P. The condition of simultaneous tangency between the price-ratio line, the production possibilities curve and the community indifference curve shows that the rate of commodity substitution in consumption (RCS) equals the rate of commodity substitution in production or the rate of product transformation (RPT), i.e., at point P which is located on the production possibilities curve AB, the RCS = RPT. This ensures equilibrium in the sense of absence of either excess supply or excess demand for both the

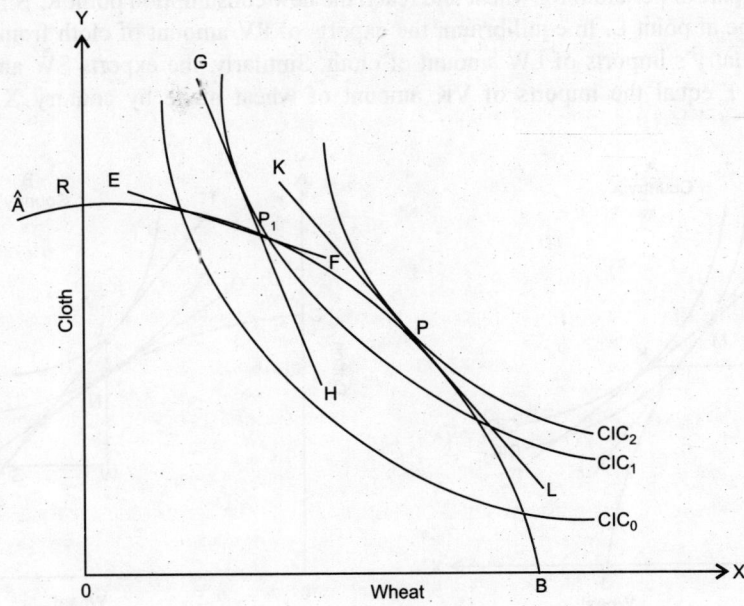

Figure 5.6

commodities in the country. If the production of goods takes place at point P_1 on the production possibilities curve of the country there will be excess supply of cloth and short supply (excess demand) of wheat in the market. The unsold supply of cloth and the unsatisfied demand for wheat would necessitate a change in the relative prices of the two commodities so as to keep the production of the two goods at point P. If the demand conditions do not change, producers will have to adjust their production of the two goods according to the demand for the two commodities. Since the factors of production are getting higher remuneration in the wheat-growing industry, resources will be released from the cloth-making industry and will be employed in the production of wheat. Eventually, equilibrium will be achieved at point P where the rate of commodity substitution in production (slope of the production possibilities curve) is equal to the price-ratio shown by the slope of line KL. At point P, the slopes of the production possibilities curve AB and of the community indifference curve CIC_2 are identical.

Now it is easy to demonstrate the phenomenon of trade between two countries. In Figures. 5.7A and 5.7B, the production possibilities curve of countries X and Y have been shown by AB and CD. In autarky, X country's production and consumption equilibrium takes place at point M where her domestic exchange or price-ratio line EF is tangent to her production possibilities curve AB. Similarly, Y country's production and consumption equilibrium takes place at point N where the domestic price-ratio line GH is tangent to country's production possibilities curve CD. After the commencement of trade, the international exchange ratio represented by the slope of line TT and parallel line T'T' is settled between the domestic exchange ratios of the two countries. Consequently, country X will increase her production of cloth as cloth has become dearer in terms of wheat in the international market. In the new situation, equilibrium in country X will be achieved at point R where the international terms of trade line TT is tangent to country's production possibilities curve. Similarly, equilibrium in country Y will be attained at point S where the international terms of trade line T'T' is tangent to her production possibilities curve. Country X will exchange part of her cloth for wheat and reach the new consumption point K. Similarly, country Y will consume at point L. In equilibrium, the exports of RV amount of cloth from country X are equal to Y country's imports of LW amount of cloth. Similarly, the exports SW amount of wheat from country Y equal the imports of VK amount of wheat made by country X. To conclude,

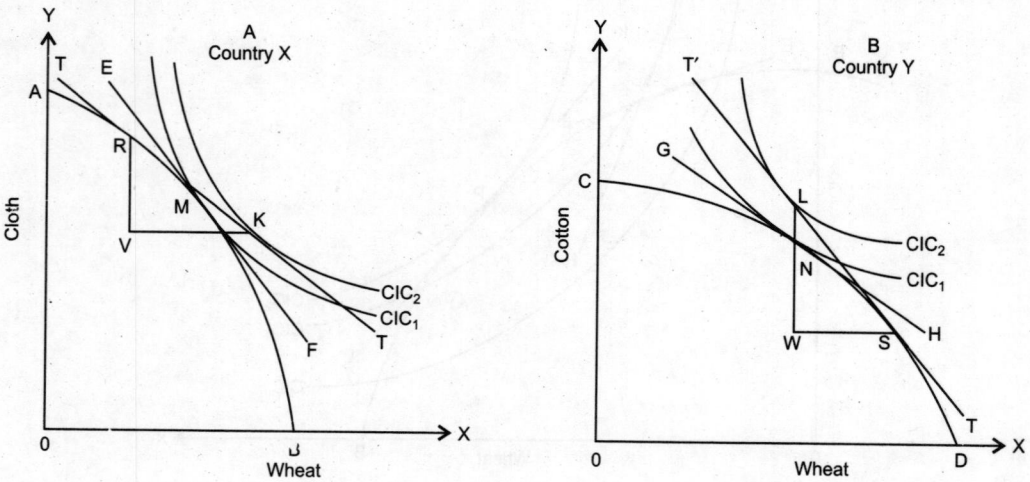

Figure 5.7

countries producing goods under the increasing cost of production will partially specialise in the production of that commodity in the production of which they have comparative cost advantage.

Trade and Diminishing Costs

Increasing returns or diminishing opportunity cost refers to a situation in which both the average and the marginal costs of production fall as output increases. The fall in the unit cost of production may be due to either the fuller exploitation of the internal economies[5] as the firm's output expands or the external economies of scale of production. Economists, however, object to diminishing costs due to the internal economies of scale as it would lead to breakdown of perfect competition and the emergence of monopolistic pricing pattern. According to Charles P. Kindleberger, however, "increasing returns based on internal economies have always played a role in discussion of international trade theory and policy, usually in connection with infant industry argument for a tariff."[6]

Under increasing returns, the production possibilities curve of the country will be convex to the origin of the axes as shown in Figure 5.8 where the production possibilities curve of country X, producing cloth and wheat, has been shown by AB. The pretrade equilibrium of the country in production takes place at point P where the domestic price-ratio line GH is tangent to the production possibilities curve. In equilibrium, the country produces 0C (= PW) quantity of cloth and 0W (= CP) quantity of wheat. If there are internal economies in the production of cloth, the country will use all her domestic resources in cloth and after complete specialisation production will take place at point A where the country will produce 0A quantity of cloth and nothing of wheat. Similarly, if internal economies are available in the production of wheat the entire domestic resources will be devoted to the production of wheat. Consequently, complete specialisation in production will take place at point B where the country will produce 0B quantity of wheat and no cloth. In case internal economies are equally enjoyable by both the industries, the pattern of production will be decided by the domestic exchange ratio (slope of the pretrade price-ratio line) for a closed economy and by the international exchange ratio (slope of international price-ratio line) for an open economy.

When production of both the commodities in both the countries is taking place under increasing returns, the production possibilities curves of both the countries will be convex to the origin. In Figure 5.9, the production possibilities curves of two countries X and Y, producing wheat and cloth, have been shown respectively by AB and A_1B_1. The production possibilities curves are convex toward the origin of the axes showing that the production of wheat and cloth in both the countries is taking place under increasing returns or decreasing opportunity cost. Let us assume that the domestic exchange ratio in country X is determined by the slope of the price-ratio line GH while that in country B is determined by the slope of her domestic price-ratio line CD. Consequently, in autarky equilibrium in production in country X takes place at point P and in

---

5. Internal economies are generally due to the presence of indivisibility of certain factors of production whose quantity does not need to be increased in order to increase the output till such time the factor of production is optimally used in production. Such economies may also arise from the more efficient resource utilisation due to greater specialisation and division of labour made possible consequent upon the expansion of firm's scale of production.

6. Charles P. Kindleberger, *International Economics*, Fifth Edition, 1973, p. 31.

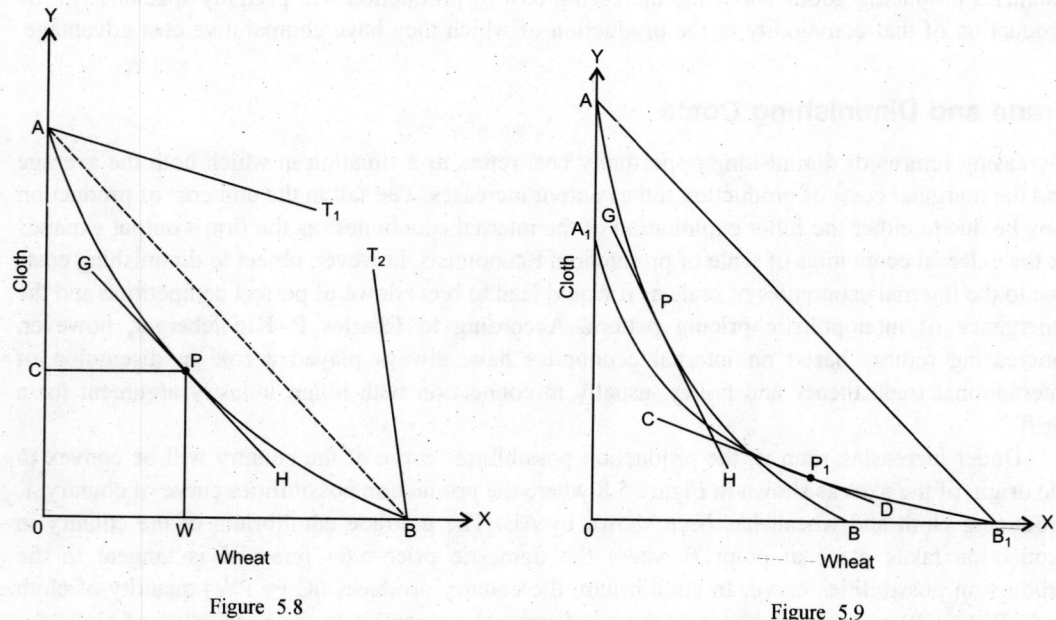

Figure 5.8   Figure 5.9

country B at point P_1 where the domestic exchange-ratio lines GH and CD are tangent to the production possibilities curves AB and A_1B_1.

The domestic exchange ratios shown by the slopes of price-ratio lines GH and CD show that country X has a comparative advantage over country *Y* in the production of cloth while country Y has a comparative advantage over country X in the production of wheat. If the international price-ratio line is AB_1, country X will completely specialise in the production of cloth as the flatter international exchange-ratio line AB_1 shows that cloth for her is dearer in the international market. For country Y, cloth is cheaper in the world market as the slope of the international price-ratio line AB_1 is steeper than the slope of the domestic price-ratio line CD. Consequently, country Y will specialise in the production of wheat. After trade takes place complete specialisation will take place in both the countries. Country X will export cloth to country Y and import wheat from her while country *Y* will export wheat to country X and will import cloth from country X. The equilibrium point in consumption of both the countries will lie somewhere on the international exchange-ratio line AB_1 and will represent the gain from trade for both the countries.

There is another possibility of increasing returns as shown in Figure 5.10 where AB is the production possibilities curve of country X. This production possibilities curve is neither uniformly concave nor uniformly convex toward the point of origin of its axes. While its upper segment beginning from point A upto the point of inflexion L is concave toward the origin, the lower segment beginning from point L upto point B is convex toward the origin. In other words, over the upper AB stretch of the production possibilities curve diminishing returns while over the LB lower stretch of the curve increasing returns operate in production. Such production possibilities curve may occur where one commodity is a primary product, such as wheat, whose production takes place under increasing opportunity cost or diminishing returns while the other commodity is a manufactured good, such as fountain-pens, whose production takes place under decreasing opportunity cost or increasing returns.

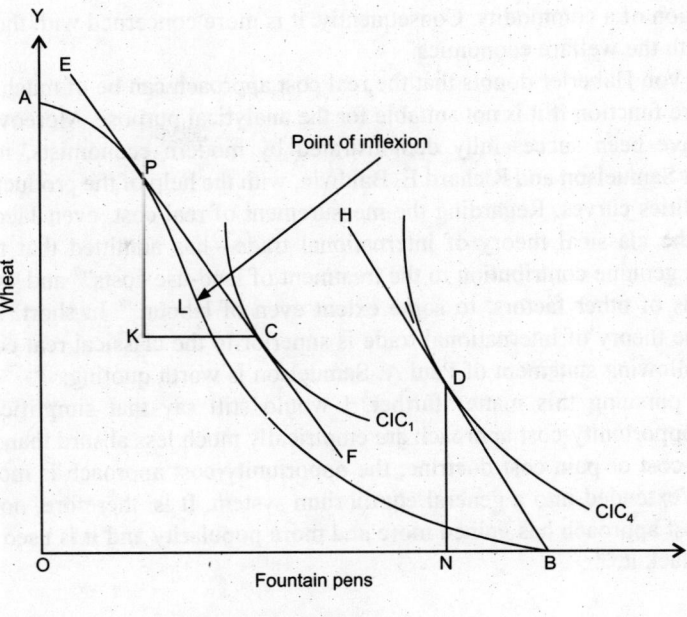

Figure 5.10

At the international exchange ratio shown by the slope of price-ratio line EF, the country produces at point P which lies on the concave part of her production possibilities curve AB while she consumes at point C. Consequently, the country exports PK quantity of wheat and imports KC quantity of fountain-pens. The consumption equilibrium takes place at point C. However, if the production of fountain-pens is expanded beyond the point of inflexion upto point B the country will trade along the international exchange-ratio line BH which is parallel to the international exchange-ratio line EF and will consume at point D which shows higher aggregate satisfaction to the consumers of country X because the community indifference curve CIC_4 which touches the international exchange-ratio line BH at point D is higher than the community indifference curve CIC_1 to which the international exchange-ratio line EF is tangent at point C. However, the expansion of output of fountain-pens from point L to point B is not easy to achieve under foreign competition. In order to cross the point of inflection L, the country may have to impose a temporary tariff on the imports of fountain-pens.

Evaluation

The 'opportunity cost' approach is superior to the classical 'real cost' approach. Apart from presenting a more correct, precise and scientific explanation of the pattern of international trade, the opportunity cost approach provides us with the indispensable analytical tools and the necessary insight into a system of general equilibrium. It also equips us with the necessary background for further theorising in the sphere of international trade.

However, it is generally argued by the supporters of the classical theory that while the opportunity cost approach is better for analytical purposes, the real cost approach is superior for the study of the welfare or policy formulation. In their support, they argue that the opportunity cost approach does not measure the real cost—pain, disutility or irksomeness of labour—involved in the production of a commodity. Consequently, it is more concerned with the positive economics rather than with the welfare economics.

Gottfried Von Haberler doubts that the real cost approach can be of much use for constructing a social welfare function if it is not suitable for the analytical purpose. Moreover, the welfare gains from trade have been successfully demonstrated by modern economists, including Murray C. Kemp, Paul A. Samuelson and Richard.E. Baldwin, with the help of the production possibilities and utility possibilities curves. Regarding the measurement of real cost, even Jacob Viner—a staunch supporter of the classical theory of international trade—has admitted that the opportunity cost approach is "a genuine contribution to the treatment of land-use costs"[7] and "surely, what holds of land also holds of other factors, to some extent even of labour."[8] In short, the opportunity cost approach to the theory of international trade is superior to the classical real cost approach. In this context, the following statement of Paul A. Samuelson is worth quoting:

"Without pursuing this matter further, I would still say that simplifications made in the (unqualified) opportunity-cost approach are empirically much less absurd than those resorted to by any other real-cost or pain cost doctrine; the opportunity-cost approach is more fertile because it can be readily extended into a general equilibrium system. It is, therefore, not surprising that the opportunity-cost approach has gained more and more popularity and it is used even by those who, in principle, attack it."[9]

Suggested Readings

W.R. Allen, *International Trade Theory, Hume to Ohlin*, 1965, Chapter 4.
Gottfried Von Haberler, *The Theory of International Trade*, 1950, Chapters 11 and 12.
H. Robert Heller, *International Trade*, Second Edition, 1973, Chapter 3.
Bo Sodersten, *International Economics*, Second Edition, Reprinted 1981, Chapters 1 and 2.

Questions

1. How will the Ricardian theory of comparative cost advantage be modified when the assumptions of
   (a) labour as the only factor of production, and (b) the law of constant cost is dropped?
2. Discuss Haberler's theory of the opportunity cost in the case of (a) constant cost, and (b) increasing cost?
3. Explain briefly the opportunity cost doctrine of international trade as propounded by Haberler?

7. Jacob Viner, *Studies in the Theory of International Trade*, 1937, p. 525.
8. Paul A. Samuelson, "Welfare Economics and International Trade", *The Economic Journal*, June 1948.
9. Paul A. Samuelson, *Ibid.*

CHAPTER 6

Heckscher-ohlin Theory of International Trade

Recent contributions in the sphere of pure theory of international trade have relied heavily on the 'factor proportions analysis' developed by the eminent Swedish economists Eli F. Heckscher[1] and Bertil Ohlin[2] who extended the application of the theory of general equilibrium or the mutual interdependence theory of pricing developed by Leon Walras, Vilfredo Pareto, Gustav Cassel and others to international trade.

The classical economists had accepted the validity of the general theory of price determination in the context of a single market on the assumption of perfect competition in both the product and factor markets. According to them, however, the theory of value which explained the exchange of goods in the domestic market was incapable of explaining the phenomenon of trade among the different nations on the ground that there was difference between interregional trade and international trade.[3] Consequently, they developed a separate theory of international trade.

According to Bertil Ohlin, the general equilibrium analysis applicable to interregional trade may also be used without substantial changes to deal with the problems of international trade. According to Ohlin, nations are only regions distinguished from one another by such obvious marks as national frontiers, tariff barriers, differences in language, customs, culture and monetary systems. In practice, however, political boundaries, tariff barriers, other differences and obstacles to free flow of goods and factors between two countries have not always been permanent. International frontiers change and tariff walls collapse. Consequently, regions may be identified with nations and the nature of the interregional trade may be the same as that of the international trade.

In Bertil Ohlin's view, the same fundamental principle of general theory of value propounded by the Austrian school of economists holds good for all trades irrespective of whether trade takes place between different regions within the country or between different countries. According to Ohlin, therefore, the most advantageous and natural approach to the theory of international trade is to start from the mutual interdependence theory of pricing, i.e., the general theory of value which has been briefly discussed below.

---

1. Eli F. Heckscher, "The Effects of Foreign Trade on the Distribution of Income", *Ekonomisk Tidskrift*, XXI, 1919, pp. 497–512 (in Swedish). Reprinted in translation by Svend and Nita Laursen in Howard S. Ellis and Lloyd A. Metzler (eds.), *Readings in the Theory of International Trade,* Philadelphia: the Blakiston Company, 1949, pp. 272–300.
2. Bertil Ohlin, *Interregional and International Trade,* Cambridge, Harvard University Press, 1933.
3. For detailed discussion of the difference between interregional and international trade, see the first part of Bertil Ohlin's book entitled *Interregional and International Trade.*

General Equilibrium Theory of Pricing

Bertil Ohlin's theory of international trade is based on and is an extension of the general equilibrium theory of value. According to this theory, the price of a commodity is determined by the interaction of the total demand for and total supply of that commodity. The demand for any good depends upon the wants and preferences of consumers, their incomes and the prices of all other goods: The supply of the commodity depends upon the supply of the factors of production and the physical and technical conditions of production.

Under perfect competition, in equilibrium the demand for a commodity will be equal to its supply and the price of the product will be equal to the average unit cost of production. The supply of goods ultimately depends up on the supply of factors of production and the physical conditions of production. The physical conditions—the natural and unchanging properties of the physical world which are everywhere similar—determine the pattern of combination of the factors of production, i.e., the technical process with due consideration of their prices. The cost of production of a commodity is composed of the prices of all those factors of production which enter into the production of that commodity. Factor prices which form the cost of production are determined by the demand for and the supply of factors of production. The demand for the factors of production will ultimately be determined by the demand for final goods since factors of production are required to produce consumer goods which directly satisfy the consumers' demand. The demand for consumer goods is determined by the wants of consumers and conditions of ownership of the factors of production which by affecting individuals' incomes affect the demand for final goods. High is the demand for the final goods, larger will be the demand for those factors of production which are needed to produce those products. The aggregate demand for a factor of production will be the sum total of the quantities of that factor demanded by all the industries in the economy. The supply of the factors of production, unless perfectly inelastic, will vary with changes in their prices. Figure 6.1 summarises the above discussion.

From a perusal of Figure 6.1, two important points emerge. Firstly, factor demand is the result of the demand for final goods. Secondly, consumers' incomes on which the demand for final goods eventually largely depends, depend on the (i) conditions of ownership the of factors of production, and the (ii) prices received by the factor owners for the productive services of factors of production. The prices of commodities depend on the prices of factors of production which, in turn,

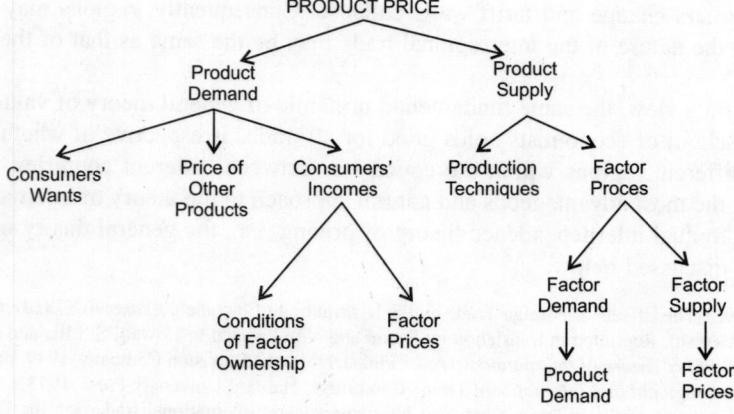

Figure 6.1

depend on the former, and so on. Ohlin neatly brings out the mutual interrelationship and interdependence between the prices of commodities, prices of factors of production, consumers' incomes, demand for the final goods, the demand for and the supply of the factors of production, etc. The whole thing constitutes a complex group of interacting and intimately interrelated factors, each exerting its influence simultaneously with all others—everything depending up on everything else. If we assume full employment and divisibility the of factors of production within each market and also further assume that the supply of factors of production is known and given, we can show that the price system of each market contains the following five sets of functional relationships.

1. The price of a commodity equals its unit cost of production.
2. The demand for each commodity is a function of its own price, prices of all other commodities, wants and incomes of consumers.
3. The income of an individual depends up on the conditions of factor ownership—on the quantity of factor services he owns and their prices.
4. The demand for a factor of production is equal to its supply which is assumed to be given and constant.
5. The quantity of any given factor of production demanded for the production of a commodity depends partly on the physical conditions of production and partly on the prices of other factors of production.

From the above set of five relationships, it can be inferred that in each region at any given time prices of all the goods and factors of production are ultimately determined by the four basic conditions, two of which relate to the demand side and the other two relate to the supply side. These four conditions are the (i) wants of consumers, (ii) conditions of ownership of the factors of production which by affecting the incomes of individuals affect the demand for goods and services, (iii) supply of the factors of production, and (iv) physical conditions of production. Given these four conditions, both the factor and commodity prices can be deduced from the above five sets of relationships. This is the fundamental fact analysed by the general theory of value.

The general equilibrium analysis is applicable to a single market in a region or a country. In fact, the general theory of price determination is a single-market theory. Ohlin observes that it considers only the 'time element' and ignores the 'space element' which plays a vital role in price determination. If the space element is also introduced into the general theory of value, the theory can be extended to determine the values in many markets involved in the trade between different countries or regions. Thus, the theory of international trade is simply a 'multi-market theory of value'.

The credit for first realising this truth goes to the eminent Swedish economist Eli F. Heckscher,[4] who pointed out that the mutual interdependence theory of pricing applies in the case of trade between two countries. This idea was elaborated by Heckscher's well-known student-colleague Bertil Ohlin[5] who developed the theory of international trade known in the literature as the Heckscher-Ohlin or the factor-proportions theory of international trade.

4. Eli F. Heckscher was Bertil Ohlin's distinguished teacher at the University of Stockholm. As Ohlin had himself admitted, his general theoretical background was entirely coloured by the ideas of the Stockholm group of economists among whom Ohlin had particularly mentioned Professor Heckscher's name.

5. Born in 1899 in Klippan, Sweden, Bertil Ohlin received his Doctorate from the University of Stockholm. He was Professor of Economics at the University of Stockholm from 1924 to 1929 and subsequently at the Stockholm School of Economics from 1929 to 1955. He was the leader of the Swedish Liberal Party during 1944–67 and Swedish Minister of Commerce during 1944–45. In the words of James W. Angell, "Ohlin's book published in 1933 was a signal contribution to the theory of international trade and one which will interest every student of economic problems." He shared the coveted Nobel Prize for 1977 for Economics with James E. Meade. Bertil Ohlin died on 3rd August, 1979 at the ripe age of 80 years.

Heckscher-Ohlin Theory

The theoretical explanation of international trade first developed by Eli F. Heckscher and subsequently elaborated by Bertil Ohlin has now become an integral part of the basic structure of the modern theory of international trade almost replacing the classical and the neoclassical approaches to the theory of international trade. Bertil Ohlin presents his most general exposition of the theory of international trade by means of a system of algebraic simultaneous equations. It does not, however, mean that Ohlin's theory invalidates the classical theory of comparative cost advantage; it rather supplements it. There is no real conflict between the Heckscher-Ohlin approach and the classical theory of comparative cost advantage. Differences in comparative costs, if properly stated, emerge from the Heckscher Ohlin trade model and this model goes behind the comparative costs doctrine. Both the theories state that the immediate basis for trade is the existence of relative differences in the prices of those commodities which are traded. If commodity prices differ between the countries, the self-interest of the traders will be served if they export the relatively cheap products of their own countries and import those that are relatively costly at home and cheap in the other countries.

This raises the important question: why do relative prices of the commodities differ from country to country? The classical economists failed to answer this question properly. Since this lacuna of the classical theory was filled by Ohlin, the Heckscher-Ohlin theory of trade is naturally the starting point of the classical theory.

The question of differences in the relative prices is: under what circumstances will relative prices of commodities be different in different countries? Differences in the relative prices will depend on the differences in the demand for and the supply of the commodity in the two regions. The single market theory of value states that the demand for a commodity depends upon the (a) wants and preferences of consumers, and (b) conditions of ownership the of factors of production which affect individuals' incomes and their demand for the final goods. The supply of the commodity depends upon the (a) supply of factors of production, and (b) physical conditions of production which, according to Ohlin, are everywhere the same and may consequently be left out of account. Consequently, differences in the commodity prices depend up on the demand conditions, i.e., on (a) and (b) and up on the supply of factors of production.

The relative prices of all goods and services will be identical in the two regions if the conditions determining the demand for goods and services, i.e., (a) the wants and preferences of consumers and their incomes are identical in both the regions, (b) the factors of production are available in identical proportions in both the regions, and (c) the differences in the supply factors of production are balanced by the compensating or offsetting differences in the demand conditions.

These three conditions may not, however, always exist and there may be differences in the relative factor supplies and in the demand conditions in the two isolated regions. Consequently, there may be differences in the relative factor prices and, therefore, in the relative commodity prices in the two countries. Basically, therefore, it is the differences in the relative 'scarcities of productive factors'—the supply of factors of production in relation to domestic demand—which is a necessary condition for the opening of trade between the different regions.

It should be noted that of the above stated three conditions, it is practically inconceivable to imagine that the first and the third conditions will coexist at any time, i.e., that the demand for goods in the two regions will be identical and that the differences in the demand conditions in the country will exactly offset the differences in the supply of factors of production. For example, it is inconceivable that an abundant factor supply will be offset by an extraordinarily large demand. Ordinarily, abundant factor supply and small demand will reinforce each other's influence creating a more favourable condition for trade. For example, Australia—a land-abundant country is also sparsely populated. Consequently, she has small domestic demand for wool, mutton, wheat etc.— commodities which she is best suited to produce as these require a large proportion of land which is relatively abundant and, therefore, cheap in the country. Bertil Ohlin's conclusion can be stated as follows:

1. The immediate cause of international trade rests in the differences in the relative prices of different goods services in the two regions. In other words, the key to the establishment of trade lies in the inequalities of relative commodity prices in the two isolated regions.[6]

2. Differences in the goods and prices services arise due to differences in the relative supply of different factors of production in the two regions, i.e., due to differences in the factor endowments causing differences in the factor proportions in the two regions. In other words, some countries have abundant land supply while others are capital-rich and still some others may have labour in abundance. Since the price of a factor varies inversely with its supply, it turns out that abundant factors will be relatively cheap while scarce ones will be relatively costly.

According to Heckscher and Ohlin, among the factors which cause differences in the commodity prices between different countries factor endowments of different countries exhibit great variations. While some countries, like Australia and Argentina, are endowed with abundant land, others, like Germany and the United Kingdom, possess relatively large accumulation of capital. Yet some other countries, like India and China, command abundant labour force. Variations in factor supplies will cause similar variations in the factor prices. Countries endowed with abundant land will have low rent, countries which are well-supplied with capital will have low interest rates while in the densely populated countries wages will be relatively low. However, relative differences in factor prices are not enough to guarantee relative differences in the commodity prices. In addition to differences in the factor prices, it is essential that different factor combinations should be required for the production of different commodities. It means that the production functions must be different for different commodities. Thus, the factor-proportions theory rests on the following two conditions.

1. There must be differences in the relative factor endowments of different countries.

2. Factor intensities or factor combinations required for the production of different goods and services must be different.

These two conditions together with the general theory of pricing form the core of the Heckscher-Ohlin theory of trade. Ohlin concludes thus: "Roughly speaking, abundant industrial agents are relatively cheap, scanty agents relatively dear in each region. Commodities requiring for their production much of the former and little of the latter are exported in exchange for goods that call for factors in opposite proportion. Thus, indirectly, factors in abundant supply are exported and factors in scanty supply are imported"[7] On this basis, Australia and Argentina where land abounds in plenty and is, consequently, cheap should produce and export land-intensive goods while

6. According to Bertil Ohlin, this simple statement which is the starting point for real analysis is evidently quite different from the classical doctrine of comparative costs because it runs in terms of price and expenses of production. The distinction between expenses and costs is fundamental to the classical theory. On the other hand, a statement in terms of price is very much the same thing as a reasoning in terms of the "opportunity costs".

7. Bertil Ohlin. *Op. cit.*, p. 92.

Germany and the United Kingdom where capital is cheap should produce and export the capital-intensive goods. Similarly, India and China, where labour is abundant and is, therefore, cheap should produce and export the labour-intensive goods.

Assumptions

In the real world, the phenomenon of international trade is exceedingly complex as it includes under its cover a multi-commodity, a multi-country and a multi-factor world. Abstraction from such a complex world of reality is needed in order to avoid total confusion. It is obvious that no theory of trade worth its name can be a perfect approximation to the real trading world. Consequently, every theory abstracts from complexities of real world trade by making several assumptions. The Heckscher-Ohlin theory of international trade is no exception to this rule of self-imposed restrictions of assumptions. The theory is based on the following assumptions. The theory assumes that

1. Trade takes place between only two countries and in only two commodities which are produced by only two factors of production. In other words, the Heckscher-Ohlin theory may be described as a two-country, two-commodity and two-factor trade model, i.e., as a $2 \times 2 \times 2$ trade model.
2. Production of both the goods involves the use of both factors and is subject to constant returns to scale. In other words, the production functions of both the commodities are linearly homogeneous.
3. The linearly homogeneous production functions are different for the two goods but are identical for each good in the two countries.
4. There is perfect competition in the goods and factor markets and resources are fully employed.
5. Transport costs, tariff and other barriers to trade are absent.
6. The relative factor endowments are different in the two countries.
7. Consumer tastes are fixed and identical in the two countries.
8. Production functions are such that the relative factor intensities are the same at all factor prices which are same in both the countries, i.e., the labour-intensive good remains labour-intensive in both the countries at all the different prices of labour. In other words, the theory is based on the assumption of strong factor intensity.
9. Factor supply (endowments) in each country is fixed, unchanging over time, homogeneous and qualitatively identical. In short, the theory abstracts from consideration of the effect on trade of differences in the qualities of factors of production in different regions.
10. International transactions are confined to only commodity trade. In other words, the theory ignores transactions arising from capital movements, remittances of interest or dividends, and other invisible items in the balance of payments.
11. Factors are mobile within each country but are immobile between the countries. At any rate, it is assumed that the factor mobility within the country is considerably greater than the international factor mobility.
12. Technology is fixed and information is costless and ubiquitous.

After stating the various above assumptions we can now explain the various tools of analysis employed in explaining the Heckseher-Ohlen theory of trade on the side of supply and on the side of demand.

Analytical Concepts (Supply Side)

1. Isoquants

The basic tool used to analyse the two-factor case is an isoquant or isoproduct curve. An isoquant is the locus of all those different possible combinations of the two factors which produce the same output of commodity. In Figure 6.2, the I_0, I_1, I_2 and I_3 negatively sloping isoquants have been shown. Labour units have been represented on X-axis while capital units have been shown on Y-axis. The isoquant map has been given in the input space. In general terms, the equation of an isoquant is $f(L, K)$ = constant. All the points located on an isoquant show the same output while the factor combination varies. For example, points M, P, on isoquant I_1 show those different alternative combinations of labour and capital which produce a given output of the commodity.

The slope of an isoquant is called the marginal rate of technical substitution (MRTS). The most efficient or optimal factor combination for the production of a given output is at that point where the isocost is tangent to an isoquant. In Figure 6.2, P is the point of optimal factor combination where the slope of the isocost or price-ratio line AB $\left(\dfrac{P_L}{P_K}\right)$ is equal to the slope of I_1 isoquant $\left(-\dfrac{\Delta K}{\Delta L}\right)$. The minus sign shows that the slope of the isoquant is negative.

Isoquants have the following salient characteristics.
1. Isoquants slope downwards from left to right showing that in order to produce any given amount of output a decrease in the quantity of one input (capital) entails an increase in the quantity of the other input (labour).
2. Isoquants do not intersect one another.
3. Isoquants are convex toward the point of origin of their axes showing the diminishing marginal rate of technical substitution (MRTS) which gives the quantity by which one factor must be increased when the other factor is decreased by a very small unit amount while the total output remains constant.

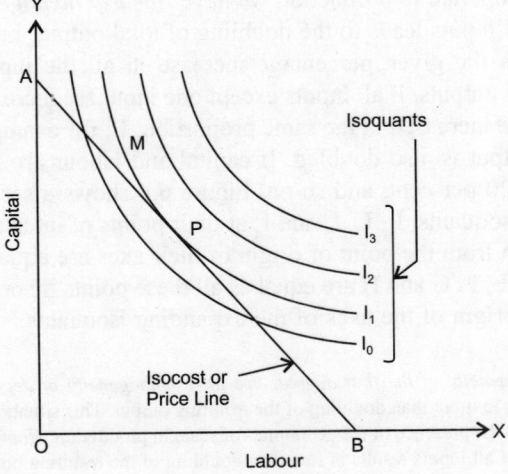

Figure 6.2

4. Isoquants situated farther away from the point of origin of their axes show higher output than those situated nearer the point of origin.

2. Homothetic and Homogeneous Isoquants

Homothetic and homogeneous isoquants have attracted much attention of trade theorists. Homothetic isoquants have equal slope or MRTS along any ray drawn from the origin as shown in Figure 6.3 where the slopes of isoquants I_1, I_2, I_3, I_4 and I_5 at points A, B, C, D and E where they are intersected by the 0K ray are equal. It shows that at the same relative prices of capital and labour, the two inputs are combined or used in constant proportion as the firm's scale of output expands.

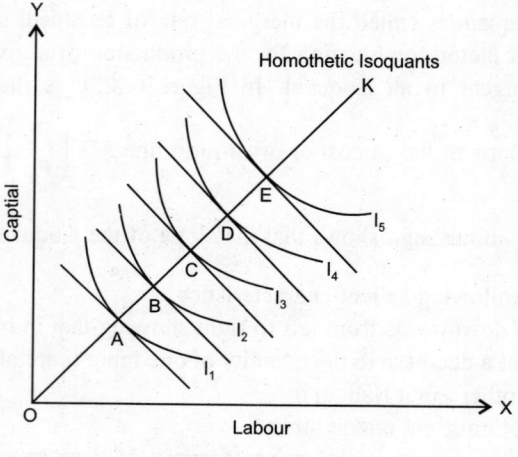

Figure 6.3

Homogeneous isoquants are a sub-set of all homothetic isoquants. They have identical slopes at their points of intersection with any straight-line drawn from the point of origin of their axes. If *constant returns to scale* operate in production we have *linearly homogeneous* isoquants.[8] In such a case, the doubling of all inputs leads to the doubling of total output, i.e., the percentage increase in the total output equals the given percentage increase in all the inputs. It means that taking combination of inputs and outputs, if all inputs except one input are increased in a given proportion that one input must also be increased in the same proportion. If, for example, capital and labour are doubled, the resulting output is also doubled. If capital and labour are increased by 20 per cent, output also increases by 20 per cent, and so on. Figure 6.4 shows a set of linearly homogeneous isoquants. The slopes of isoquants, I_1, I_2, I_3 and I_4 at their points of intersection A, B, C and D with the 0K straight-line drawn from the point of origin of their axes are equal. Similarly, the slopes of all the isoquants at points E, F, G and H are equal as all these points lie on the same 0L straight-line drawn from the point of origin of the axes of the expanding isoquants.

8. It is also known as *homogeneity of the first degree*. We have *homogeneity of degree greater than one* when the doubling of all inputs leads to more than doubling of the resulting output. This situation is characterised as *increasing returns to scale*. It is due to the presence of the economies of scale in production. *Homogeneity of degree less than one* exists when the doubling of all inputs results in less than doubling of the resulting output. This situation is known as *diminishing returns to scale*. It is due to the presence of the diseconomies of scale in production,

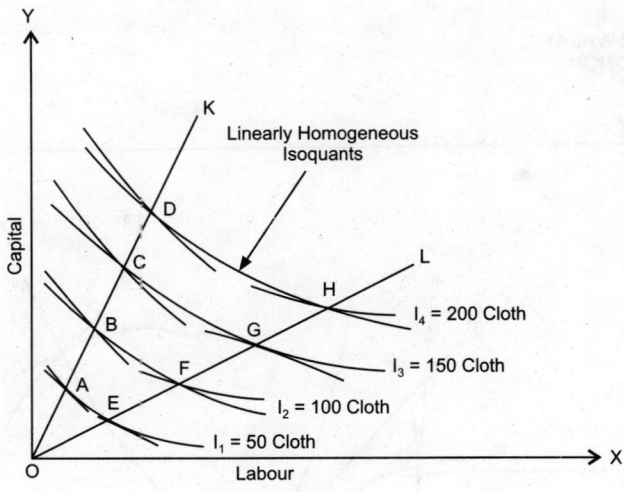

Figure 6.4

When the isoquants are *linearly homogeneous,* production takes place under constant returns to scale and the constant factor proportion line also becomes the constant proportion output expansion (scale) line. In other words, in terms of Figure 6.4 the distance of point B situated on isoquant I_2 from the point of origin of the axes of isoquant should be double that of point A situated on isoquant I_1 if isoquant I_2 represents twice the output represented by isoquant I_1.

3. Edgeworth Box Diagram[9]

In order to show the most efficient (optimum) combination of inputs used in the production of two commodities we employ the device of Edgeworth box diagram. The dimensions of the box show the total factor endowments (factor supplies) available in the country. In Figure 6.5, the vertical axis shows the total quantity of capital while the horizontal axis shows the total quantity of labour available in the country. The distance 0_wL (= 0_cL_1) shows the total supply of labour and 0_wK (= 0_cK_1) measures the total supply of capital.

Using 0_w as the point of origin of the axes, we draw a set of W_1, W_2, W_3, and W_4 isoquants for wheat. Similarly, starting from point 0_c on the upper right hand corner of the box we draw another set of C_1, C_2, C_3 and C_4 isoquants for cloth.

Due to the opposite curvature of these two sets of isoquants each isoquant of one set will be tangent at single point to an isoquant of the other set. All points of tangency of the two sets of isoquants will represent technologically the most efficient combinations of inputs. The line joining such points is called the *contract curve* or *efficiency locus.* In Figure 6.5, the line joining points E, F, G and H is the *efficiency locus* or *contract curve.* Any point located off the contract curve will show the full employment of resources; it will not, however, be the position of equilibrium in the sense of optimal or most efficient factor combination. For example, the marginal product of labour is greater at point D than at point G because the slope of tangent TT_1 drawn on the wheat isoquant W_2 is steeper than the slope of tangent PP_1 drawn at point G on the wheat isoquant W_3.

9. Christianed after the name of the famous Cambridge economist Francis Ysidro Edgeworth who first demonstrated the gains from trade through the use of indifference curve analysis.

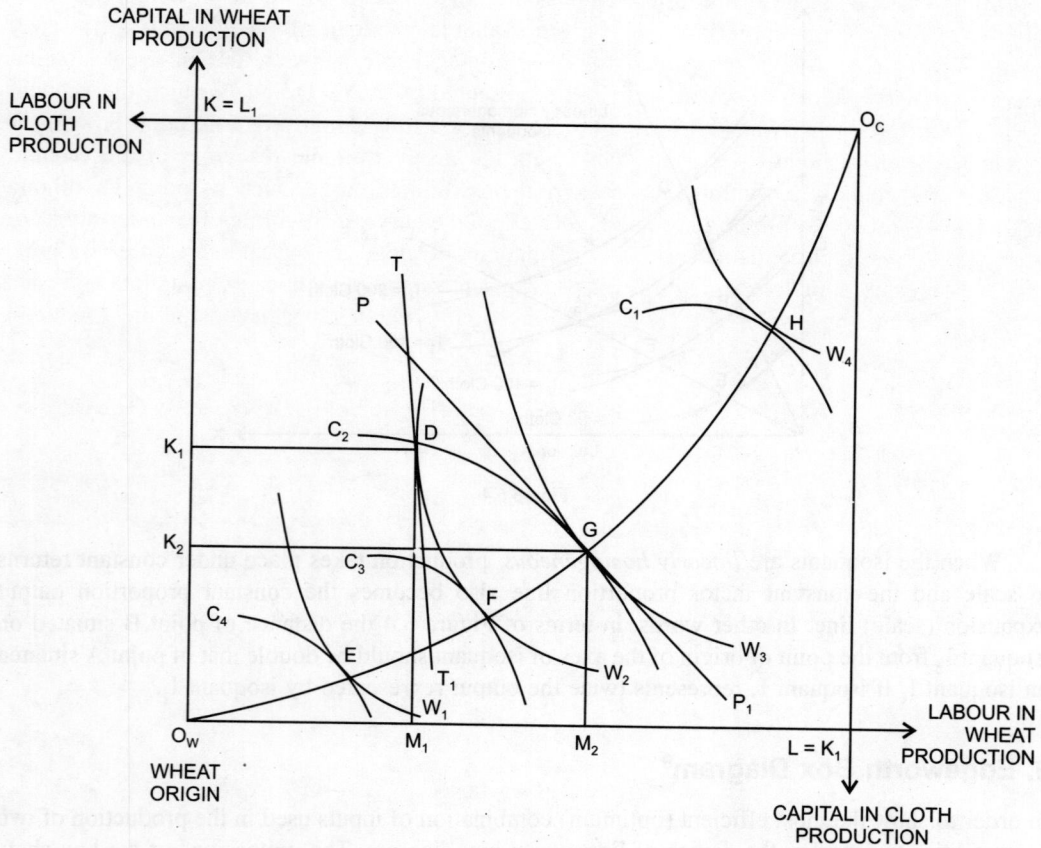

Figure 6.5

Consequently, more labour will be employed in the production of wheat while capital will be withdrawn from this industry. If M_1M_2 amount of additional labour is employed in the production of wheat and if $K_1 K_2$ amount of capital is withdrawn from this industry, production will take place at point G. At point G, the factor price-ratio line PP_1 is simultaneously tangent to the wheat and cloth isoquants showing that the factor price-ratio is equal to the marginal rate of technical substitution for both the commodities, i.e., at point G the ratio of factor prices equals the ratio of factor marginal products. Consequently, it is the equilibrium position in the sense of the most efficient factor combination.

4. Production Possibilities Curve

Every point situated on the efficiency locus shows the maximum attainable output-mix of wheat and cloth with full employment of given capital and labour in the country. If we plot the different output combinations of two goods, cloth and wheat, taken from the contract curve on a diagram whose co-ordinates measure the units of cloth and wheat we obtain a curve which is variously known as the production possibilities curve, production frontier, transformation curve or substitution curve. It shows the different possible combinations of wheat and cloth which are attainable with the optimum utilisation of given available resources of the economy.

In Figure 6.6, AB is the production possibilities curve showing the different optimum alternative combinations of wheat and cloth represented by A, C, D and B which the economy is capable of producing when all available resources are fully employed. Any point lying above this curve, such as point E, is unattainable with the given available resources of the economy whereas any point lying within the production possibilities curve, such as point F, although attainable would not be preferred to points located on the curve as it will involve under-utilisation of the country's total available inputs. Combinations of cloth and wheat represented by points lying below the production possibilities curve represent smaller quantities of both cloth and wheat compared with those which are represented by points which are situated on the production possibilities curve.

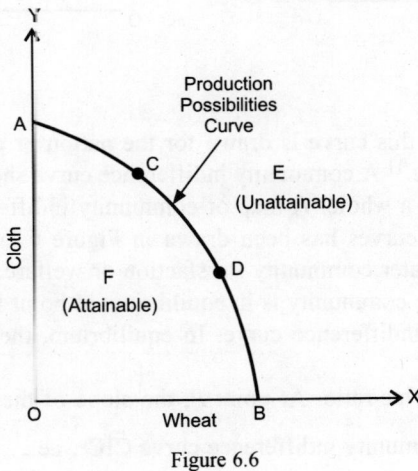

Figure 6.6

Analytical Concepts (Demand Side)

Recent developments in trade theory have made it difficult to handle the various theoretical cases by traditional means of numerical examples. The indifference curve analysis applied in the sphere of trade comes in handy for understanding some of the intricate problems of trade.

Community Indifference Curve[10]

An indifference curve is the locus of different combinations of the two commodities from which the consumer derives the same satisfaction. In other words, an indifference curve is an iso-utility or

10. The credit for introducing the concept of *the community indifference curve* in the theory of international trade goes to Francis Ysidro Edgeworth. More recently, this analytical concept has been employed in their analytical works, among others, by Nicholas Kaldor, Tibor Scitovsky, Abba P. Lerner and W.W. Leontief. Through its use, the theory of international trade has been linked to the utility and consumption theory, Paul A. Samuelson in his article published in *The Quarterly Journal of Economics* in February 1956 has presented a definitive clarification of "community indifference curves". According to him, it is impossible (except in a singular case) to derive from an individual's indifference map a community indifference map from which it is possible to derive the community offer or demand curve just as it is possible to derive an individual's demand or offer curve from the former.

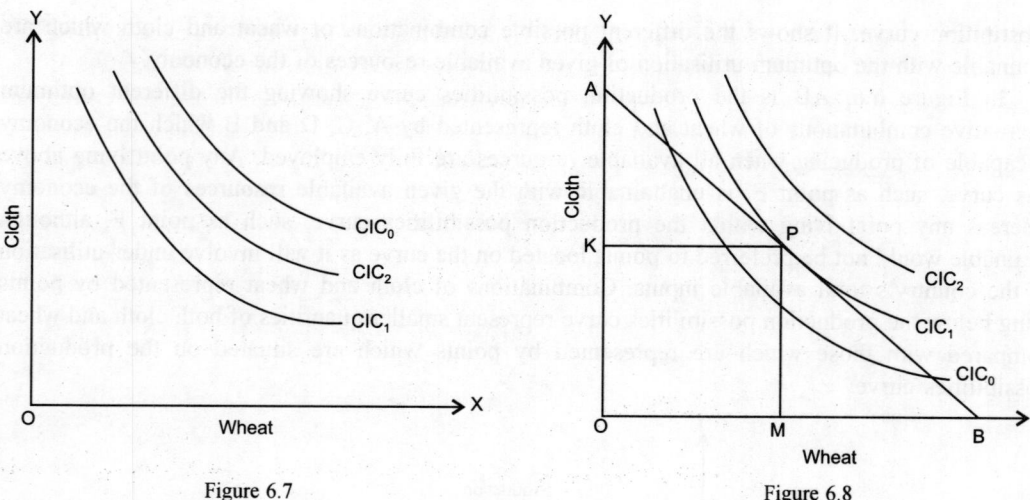

Figure 6.7   Figure 6.8

iso-satisfaction curve. When this curve is drawn for the nation or community it is known as the community indifference curve.[11] A community indifference curve shows the order of preference of the consumers of a nation as a whole. A map of community indifference curves shown by CIC_0, CIC_1 and CIC_2 indifference curves has been drawn in Figure 6.7 where the higher community indifference curve shows greater community satisfaction or welfare.

Figure 6.8 shows that the community is in equilibrium at point P where AB price-ratio line is tangent to CIC_1 community indifference curve. In equilibrium, the marginal rate of commodity substitution is equal to the price-ratio. At point P, the slope of the AB price-ratio line $\left(\dfrac{P_W}{P_C}\right)$ is equal to the slope of the community indifference curve CIC_1, i.e.,

$$\frac{P_W}{P_C} = -\frac{\Delta C}{\Delta W}$$

A community indifference curve has the following four important characteristics.
1. It slopes downward from left to right.
2. It is convex toward the point of origin of its axes.
3. Community indifference curves do not intersect each other.
4. Higher community indifference curve shows higher community satisfaction.

It must, however, be noted that notwithstanding the serious efforts made by the economists to apply the indifference curve analysis as a tool for explaining and evaluating the performance of individualistically organised economies as if these economies represented merely the complicated replica of a single household or firm, the analysis breaks down when we transcend from individual to society because it involves the knotty problem of social or collective utility which has so far defied any satisfactory solution.

The Heckscher-Ohlin theory assumes different factor intensities between different goods and different given factor endowments between different countries. Consequently, differences in the

11. For derivation and detailed discussion of the community indifference curve see Chapter 8 entitled "Equilibrium in International Trade". An individual indifference curve will be identical to the community indifference curve on the extreme assumption that there is only one consumer in society.

production possibilities curves of the two countries are the net result of the influence of these two factors. The different factor intensities between different goods determine the curvature of the production possibilities curve while the factor endowments together with demand determine the pattern of international trade. Consequently, the study of factor intensities and factor endowments is of paramount importance.

Factor Intensity and Factor Intensity Reversal

Factor intensity refers to the relative quantities of the two (or more) inputs used in the production of two (or more) commodities. In terms of an isoquant, the factor intensity is represented by the proportion in which the two inputs, say, capital and labour, are combined at each point on the curve. In order to understand the factor intensity, we take the two sets of homothetic isoquants for the two commodities cloth and wheat produced by the two inputs capital and labour shown in Figure 6.9. At the factor price-ratio shown by the slope of AB price-ratio line, the production of cloth is capital-intensive while the production of wheat is labour-intensive. It is so because the physical capital-labour ratio required to produce cloth is higher than that required to produce wheat, i.e.,

$$\left(\frac{K}{L}\right)_C > \left(\frac{K}{L}\right)_W$$

It is not the absolute factor intensities comprising absolute amounts of the factors used in the production of wheat and cloth but the relative factor intensities comprising the factor quantity ratios or proportions used in the production of the two goods which matter for it may be that the production of wheat may be absolute capital-intensive in the sense that more absolute amount of capital may be required to produce a given quantity of wheat. It may, however, be used in combination with a very large amount of labour so that wheat production may turn out to be labour-intensive as compared to the production of cloth. For our analysis, only the factor proportions are important. A large country may display identical factor proportions to a small country despite the fact that the absolute amount of her factors is much larger.

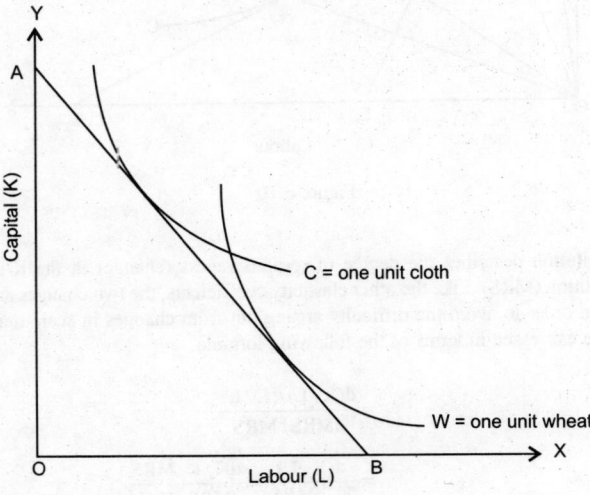

Figure 6.9

The cloth and wheat isoquants in Figure 6.9 show that there is no scope for factor intensity reversal i.e., the production functions reveal strong factor intensity which is a necessary assumption for the factor-proportions theory to explain the basis for trade. If the factor intensities of production functions are reversible, the factor-proportions theory will be invalidated. For complete absence of factor intensity reversal, it is necessary that the substitution elasticities[12] of the two production functions should be identical throughout so that these can be unambiguously identified as being either capital-intensive or labour-intensive at all the marginal rates of factor substitutions.

If the elasticities of substitution of the two production functions are not identical, the same production function may be labour-intensive at one marginal rate of factor substitution (MRS) and capital-intensive at the other marginal rate of factor substitution (MRS). It happens because in response to a given change in the marginal rate of factor substitution (MRS) the consequential change in the capital-labour ratio (K/L) will differ between industries. Consequently, it is possible that the initial K/L ratio involved in the capital-intensive cloth production changes so much more

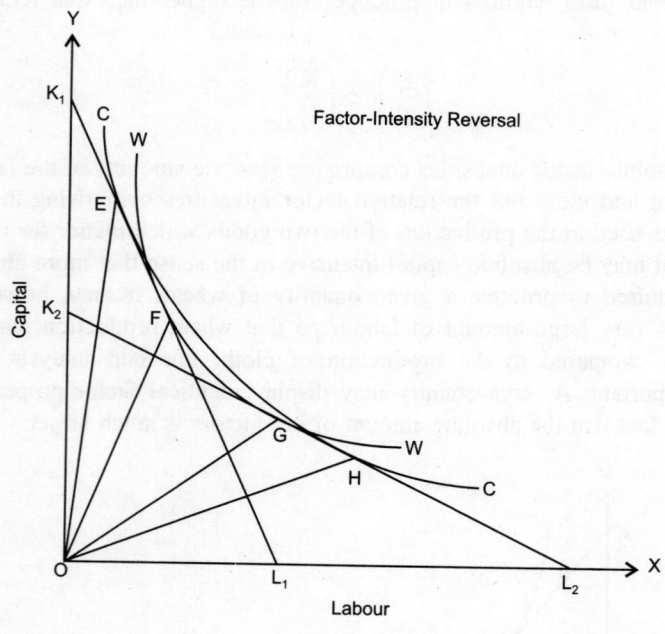

Figure 6.10

12. The elasticity of substitution describes the degree of responsiveness (change) in the K/L ratio to changes in the marginal rate of substitution (MRS). Like the other elasticity coefficients, the two changes are expressed in percentage or proportional terms in order to avoid the difficulty arising out from changes in scale units. Thus, the elasticity of substitution (e_s) may be expressed in terms of the following formula.

$$e_s = \frac{d(K/L)/(K/L)}{dMRS/MRS}$$

$$= \frac{d(K/L)}{\frac{K}{L} \cdot \frac{dMRS}{MRS}} = \frac{dK/L}{dMRS} \cdot \frac{MRS}{K/L}$$

than the K/L ratio involved in the labour-intensive production of wheat such that cloth industry no longer remains a capital-intensive industry; instead it becomes labour-intensive while the production of wheat becomes capital-intensive. Consequently, the factor intensities are reversed. Figure 6.10 shows the case of factor intensity reversal.

In Figure 6.10, at the MRS shown by the slope of line K_1L_1 the production of cloth is capital-intensive while that of wheat is labour-intensive. However, at the MRS shown by the slope of line K_2L_2 the order of relative factor intensities is reversed and the production of cloth becomes labour-intensive while the production of wheat becomes capital-intensive. In other words, the phenomenon of *factor intensity reversal* has occurred.

Factor Endowment

The factor endowment of any country consists of the available supplies of different factors of production used in the production of goods and services. Nature has distributed resources—inputs—between the different countries in a non-egalitarian manner bestowing plenty of one resource on some countries and the other resources on some others so that none of the world countries is 'exclusively' self-sufficient. This most unsocialistic pattern of resource distribution as between different countries on nature's part is reflected in the relative abundance of some factors of production and relative scarcity of some others in the country.

In the Heckscher-Ohlin theory, the concept of 'relative factor abundance' is used in the physical criterion and the price criterion which have been discussed below.

1. *Physical Criterion*—According to the physical criterion, a country is said to be relatively capital-abundant *if and only if* it is endowed with a higher proportion of capital to labour than is the other country. Consequently, country 1 is relatively capital-rich if the following condition holds, i.e., if

$$\left(\frac{\bar{K}}{\bar{L}_1}\right)_1 > \left(\frac{\bar{K}}{\bar{L}_2}\right)_2$$

where the terms $\bar{K}$ and $\bar{L}$ denote the fixed physical amounts of capital and labour in each country and subscripts 1 and 2 outside the parenthesis denote country 1 and country 2 respectively. The physical criterion ignores the effect of the demand conditions in determining the relative abundance or scarcity or factor endowment in a country.

2. *Price Criterion*—According to the price criterion, a country having capital relatively cheap and labour relatively dear is a capital-abundant country, irrespective of the ratio of the total physical quantity of capital to the total physical quantity of labour compared with the other country. Consequently, compared with country 2 country 1 will be capital-rich if

$$\left(\frac{P_K}{P_L}\right)_1 > \left(\frac{P_K}{P_L}\right)_2$$

where the terms P_K and P_L denote respectively the price of capital and the price of labour and subscripts 1 and 2 denote country 1 and country 2 respectively. In the above condition, country 1 is relatively capital-rich while country 2 is relatively labour-abundant. Consequently, country 1 will produce and export the capital-intensive goods—those goods in whose production the abundant-cheap–input capital is intensively used—and import the labour-intensive goods—those goods in whose production scarce–dear–input labour is intensively used. Country 2 will export labour-intensive goods and import capital-intensive goods. If, however, the relative factor prices are identical in both the countries, i.e., if

$$\left(\frac{P_K}{P_L}\right)_1 = \left(\frac{P_K}{P_L}\right)_2$$

there will be no comparative cost differences. Consequently, there will be no basis for trade between the two countries because no country will gain from engaging in trade.

Ohlin uses the price criterion of relative factor abundance in his trade model. He has, however, stated that differences in the relative factor prices are due to differences in the relative physical factor endowments in the two countries. Thus, for him the two criteria are identical. Ohlin assumes that the factor supply aspect plays a dominant role in determining the relative factor prices in a country.

Verification of the Theory (Physical Criterion)

Let us assume that on the basis of the physical criterion, country 1 is capital-rich while country 2 is labour-abundant. Consequently, for the fixed factor endowment in the two countries the physical capital-labour ratio in country 1 will be higher than in country 2, i.e., the following condition of physical criterion of the concept of relative abundance will hold.

$$\left(\frac{K}{L}\right)_1 > \left(\frac{K}{L}\right)_2$$

In Figure 6.11, the production possibilities curve of country 1 is A_1A_3 and of country 2 is A_2A_4. The production of cloth is capital-intensive while that of wheat is labour-intensive. If both the countries produce wheat and cloth in the same proportion and production takes place along the linear expansion path 0P, production in country 1 will take place at point L and in country 2 at point K. In this situation, the equilibrium price-ratio line of country 1 will be T_1T_2 while that of country 2 will be T_3T_4. Consequently, in country 1, the $0T_2$ quantity of wheat will be exchanged

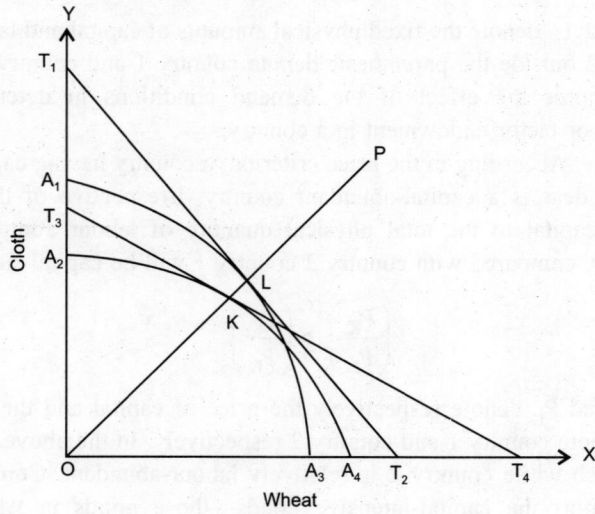

Figure 6.11

against $0T_1$ quantity of cloth while in country 2, the $0T_4$ quantity of wheat will be exchanged for $0T_3$ quantity of cloth. Since $0T_1 > 0T_3$ and $0T_4 > 0T_2$ cloth is relatively cheap in country 1 while wheat is relatively cheap in country 2. Since country 1 is capital-rich and the production of cloth is capital-intensive, she will extend her production of cloth. Similarly, country 2 will increase the production of wheat because wheat is labour-intensive and country 2 is labour-abundant.

This fact alone is not, however, sufficient to enable us to conclude that country 1 will export cloth and country 2 will export wheat since it is possible that the demand conditions in the two countries may be such that the above tendency may be reversed or nullified. In other words, on the basis of supply (cost) conditions alone we cannot reach the infallible conclusion that the two countries would necessarily engage in trading and that the actual direction of trade would follow the direction expected on the basis of cost considerations alone. Demand reversals may change the direction of trade and country 1 may import (rather than export) cloth and country 2 may import wheat if the community indifference curves in the two countries are such that lead to high price of cloth in country 1 and to high price of wheat in country 2. Possibility also exists that the pre-trade domestic exchange ratios in the two countries are equal eliminating any scope for trade. Consequently, both countries will remain self-sufficient. Figure 6.12 shows the case in which although the production possibilities curves of the two countries are different showing scope for trade between the two countries on cost (supply) considerations alone but due to the demand conditions in the countries reflected in their community indifference curves the pre-trade domestic exchange ratios in the two countries are equal. Consequently, no trade will take place between the two countries.

In Figure 6.12, the capital-rich country I has the cost advantage in the production of capital-intensive commodity cloth and one would expect her to export cloth on cost considerations alone. The labour-abundant country 2 commands the cost advantage in the production of labour-intensive commodity wheat. Accordingly, she should export wheat if cost considerations alone mattered. However, whether or not trade will take place and if it takes place in what direction it would occur—which good will be exported and which good will be imported—would depend on the domestic pre-trade exchange ratios in the two countries. The pre-trade domestic exchange ratios depend on, along with cost (supply) conditions, the demand conditions. In the figure, the

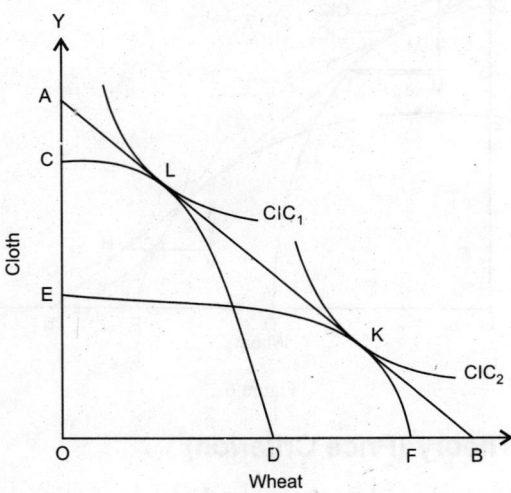

Figure 6.12

pre-trade domestic exchange ratios represented by the slope of common price-ratio line AB are, however, equal. Consequently, no trade results despite the cost (supply) conditions providing scope for trade.

In the case illustrated in Figure 6.12, no trade takes place between the two countries. A stronger situation invalidating the Heckscher-Ohlin theory will emerge if due to demand reversal trade takes place between the two countries such that the capital-rich country 1 exports the labour-intensive commodity wheat and imports the capital-intensive commodity cloth while the labour-abundant country 2 likewise exports the capital-intensive cloth and imports the labour-intensive commodity wheat. In short, the direction of trade shows a complete reversal of the direction of trade expected on considerations of cost or supply alone. Figure 6.13 shows the reversal of trade direction due to the phenomenon of demand reversal which the Heckscher-Ohlin theory ignores. The consideration of demand reversal phenomenon lies outside the static framework of the Heckscher-Ohlin theory.

Figure 6.13 shows that due to the demand reversal, country 1 which is capital-rich and on cost advantage consideration alone is expected to export the capital-intensive commodity cloth, actually imports it while she exports wheat. Similarly, the labour-abundant country 2 imports the labour-intensive commodity wheat which she is expected to export. Instead, she exports cloth in which her relatively scarce input capital is intensively used. In short, the direction of trade expected on the basis of cost advantage alone is *reversed* due to the demand conditions which show that people in country 1 have such a high demand for cloth that even though cloth is produced cheaply its price is very high and additional cloth is imported to satisfy the high demand of cloth. The same is true of country 2 with respect to the high demand for wheat. The Heckscher-Ohlin theory of trade is valid on the basis of the physical criterion if the consumption pattern in both the countries is identical, i.e., if consumer tastes in the two countries are identical and if the income elasticity of demand for each commodity is unity.

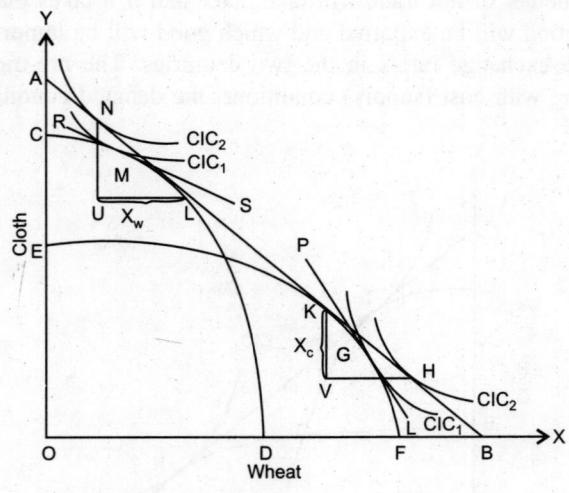

Figure 6.13

Verification of the Theory (Price Criterion)

Figure 6.14 verifies the Heckscher-Ohlin theory on the basis of price criterion. Let us assume that there are two countries—country 1 and country 2. The factor price-ratio line of country 1 is AB

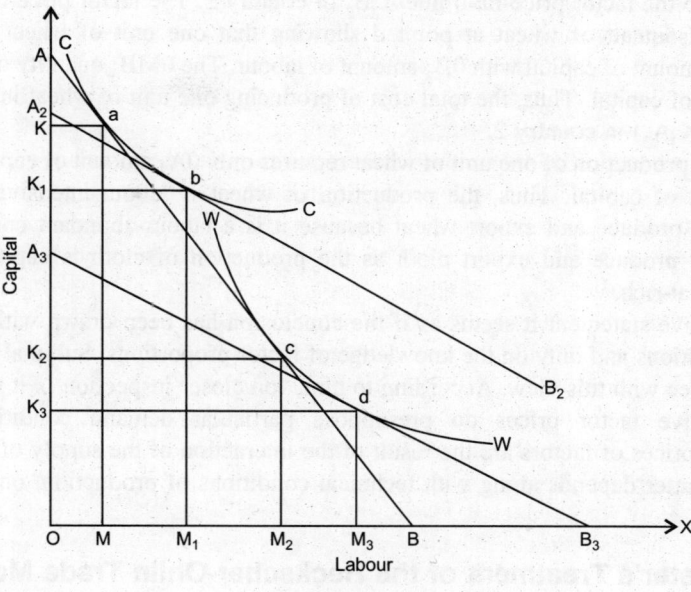

Figure 6.14

while that of country 2 is A_3B_3 $(= A_2B_2)$. The slope of the factor price-ratio line AB of country 1 is higher than the slope of the factor price-ratio line A_3B_3 $(= A_2B_2)$ of country 2 showing that capital is relatively cheap in country 1 while labour is relatively cheap in country 2. In other words, the interest-wage ratio $\left(\dfrac{P_K}{P_L}\right)$ in country 1 is lower than it is in country 2, i.e.

$$\left(\frac{P_K}{P_L}\right)_1 < \left(\frac{P_K}{P_L}\right)_2$$

The two countries produce the two commodities cloth and wheat. The production of cloth is capital-intensive while that of wheat is labour-intensive. We also assume that the production function for each commodity is identical in both the countries but it is different for the two commodities. Factor price-ratio line AB is tangent to CC isoquant of cloth at point a showing that country produces a given quantity of cloth (one unit of cloth) by combining 0K amount of capital with 0M quantity of labour. In terms of cost, however, this factor combination is identical to 0A amount of capital and zero amount of labour. In other words, in terms of cost the 0M amount of labour is equal to AK amount of capital. Thus, the total cost of producing one unit of cloth in terms of the amount of capital in country 1 is 0A. Factor price-ratio line A_2B_2 of country 2 is tangent to CC isoquant at point b showing that one unit of cloth in country 2 is produced with the combination of $0K_1$ quantity of capital and $0M_1$ quantity of labour. If labour is converted into capital then $0M_1$ amount of labour is equal to A_2K_1 amount of capital. Thus, the total cost of producing one unit of cloth in terms of capital in country 2 is $0A_2$.

Adopting the above technique, we can calculate that the total cost of producing one unit of wheat in country 1 is 0A amount of capital while in country 2 it is $0A_3$ amount of capital. Line

A_3B_3 is parallel to the factor price-ratio line A_2B_2 of country 2. The factor price-ratio line A_3B_3 is tangent to WW isoquant of wheat at point d showing that one unit of wheat is produced by combining $0K_3$ amount of capital with $0B_3$ amount of labour. The $0MB_3$ quantity of labour is equal to A_3K_3 quantity of capital. Thus, the total cost of producing one unit of wheat in terms of capital is $0A_3$ $(= 0K_3 + K_3A_3)$ in country 2.

In country 2, production of one unit of wheat requires only $0A_3$ amount of capital which is less than $0A_2$ amount of capital. Thus, the production of wheat is labour-intensive. Consequently, country 2 should produce and export wheat because it is a labour-abundant country. Similarly, country 1 should produce and export cloth as the production of cloth is capital-intensive and country 1 is capital-rich.

From the above statement it seems as if the conclusion has been drawn without considering the demand conditions and only on the knowledge of factor proportions. Subimal Mookerjee does not, however, agree with this view. According to him, "on closer inspection ... it will be clear that data about relative factor prices do presuppose particular demand conditions and factor proportions. For prices of factors are the result of the interaction of the supply of and demand for factors, and the latter depends along with technical conditions of production, on the demand for commodities."[13]

Kelvin Lancaster's Treatment of the Heckscher-Ohlin Trade Model

Using the physical criterion of factor endowment, Lancaster[14] has presented a geometrical treatment of the Heckscher-Ohlin trade model which has been discussed below. Lancaster has used the device of the Edgeworth box diagram depicted in Figure 6.15 for illustrating the allocation of the given amounts (fixed endowment) of two inputs labour and capital in the production of two commodities cloth and wheat.

The rectangle of the Edgeworth box diagram ADBC represents an economy with fixed amount of capital and labour. AD (= BC) shows the total fixed amount of capital and AC (= DB) shows the total fixed amount of labour available in the economy. Any point inside the box shows the allocation of the given amounts of the two factors (capital and labour) in the production of cloth (C isoquants) and wheat (W isoquants). At any particular point, the horizontal distance from AC and the vertical distance from AD measures the quantities of capital and labour used in the production of cloth while the horizontal distance from BD and the vertical distance from BC measures the quantities of capital and labour employed in the production of wheat. For example, point M in the box shows the AM_2 $(= M_4M)$ amount of capital and M_2M $(= AM_4)$ amount of labour used in the production of cloth. The balance amount MM_1 $(= M_2D)$ of capital and MM_3 $(= BM_1)$ amount of labour is employed in the production of wheat. The quantities of cloth and wheat can be read off from the respective isoquants of wheat and cloth such as isoquants W_1 and C_1 shown inside the box. If the production functions are homogeneous, all the other isoquants can be obtained from the unit isoquant by means of uniform radial expansion or contraction from the point of origin of the axes of the isoquants.

Under perfect competition, the most efficient allocation of the two factors which would optimise the total output of the two goods will occur when the ratio of the marginal physical products of labour and capital in the production of cloth is equal to the ratio of their marginal

13. Subimal Mookerjee. *Factor Endowment and International Trade,* p. 29.
14. Kelvin Lancaster, "The Heckscher-Ohlin Trade Model: A Geometric Treatment", *Economica,* Volume 24, February 1957, pp. 19–39.

physical products in the production of wheat. This ratio is measured by the slopes of isoquants. The slopes of the two isoquants with opposite curvatures will be identical at the points where they are tangential. The line joining the points of tangency of the isoquants is called the production efficiency locus or contract curve. In Figure 6.15, the line joining points A, M and B is the efficiency locus or contract curve. If the isoquants are smooth, continuous and convex to the origin, the efficiency locus will be smooth and continuous, will pass through points A, M and B and will lie wholly either on one or on the other side of diagonal AB depending on the relative factor intensities of the two production functions. If the production of cloth is always capital-intensive then at the same relative prices for the factors in both the industries the capital-labour ratio will be higher in the production of cloth than in the production of wheat. Consequently, the contract curve will lie to the south-east of diagonal AB as shown in the diagram.

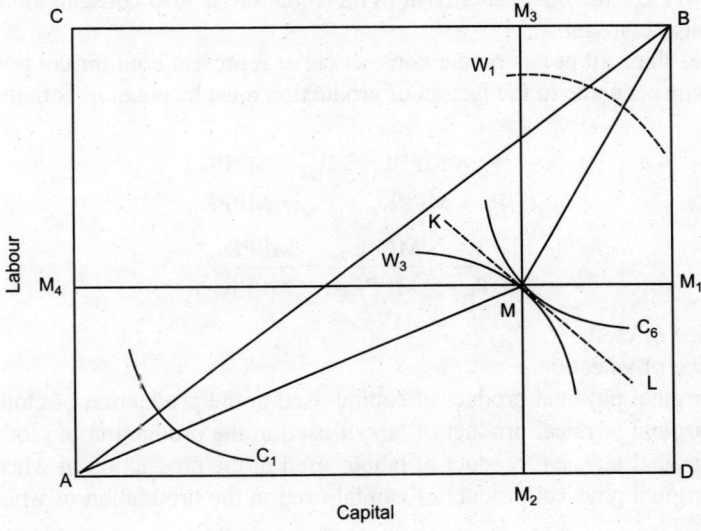

Figure 6.15

Let us now consider the ray AM drawn from the point of origin A of the axes of cloth isoquants and intersecting C and W isoquants. If the production function is linearly homogeneous, each isoquant will have an identical slope at its point of intersection with the straight line AM drawn from the point of origin A of the axes. Consequently, the ratio of the marginal physical products of labour and capital in the production of cloth will be constant all along the AM ray, i.e., $\dfrac{MPP_L}{MPP_K}$ is constant.

Under constant returns to scale, the additional output produced due to a given percentage increase in all the inputs will be equi-proportional regardless of the level of output. Consequently, the distance of any point located on the AM ray from the point of origin of the axes A will be equi-proportional to the factor quantities at that point. Moreover, the absolute values of the marginal physical products of the factors will remain constant all along AM. Since the production function is linearly homogeneous, Euler's Theorem holds so that

$$Q_C = (L_C) \times (MPPL_C) + (K_C) \times (MPPK_C)$$

where Q_C is the total output of cloth, L_C and K_C are respectively the labour and capital inputs employed in the production of cloth, $MPPK_C$ and $MPPL_C$ are the marginal physical products of capital and labour in the production of cloth.

By dividing the above equation by L_C and after rearranging, it can be written as

$$\frac{Q_C}{L_C} = MPPL_C \left(1 + \frac{K_C \times MPPK_C}{L_C \times MPPL_C} \right)$$

Now $\dfrac{Q_C}{L_C}$ is constant because the ratio of output to the quantity of either factor is constant along AM and the expression inside the bracket on the right-hand side of the equation is also constant. Consequently, $MPPL_C$, the only other term in the equation, is also constant along AM. It follows that $MPPK_C$ is also constant.

Furthermore, since all points on the contract curve represent equilibrium points under perfect competition, payments made to the factors of production must be equal in both the cloth and wheat industries, i.e.,

$$P_C \times MPPL_C = P_W \times MPPL_W$$
$$P_C \times MPPK_C = P_W \times MPPK_W$$

or

$$\frac{P_C}{P_W} = \frac{MPPL_W}{MPPL_C} = \frac{MPPK_W}{MPPK_C}$$

where, P_C = Price of cloth,
    P_W = Price of wheat,
    $MPPK_C$ = Marginal physical product of capital used in the production of cloth,
    $MPPL_C$ = Marginal physical product of labour used in the production of cloth,
    $MPPL_W$ = Marginal physical product of labour used in the production of wheat, and
    $MPPK_W$ = Marginal physical product of capital used in the production of wheat.

The above property is of great importance for the analysis of the Heckscher-Ohlin trade model which has been discussed below.

Simple Heckscher-Ohlin Trade Model

Let us start by assuming that there are two countries X and Y. The Edgeworth box diagrams of both these countries have been shown by ACBD and $A_1C_1B_1D_1$ respectively in Figure 6.16A and Figure 6.16B.

AD and A_1D_1 are the fixed amounts of capital in countries X and Y respectively while AC and A_1C_1 are the fixed amounts of labour in the two countries. Since $AD > A_1D_1$ and $A_1C_1 > AC$, country X is capital-rich while country Y is labour-abundant. The two countries produce the two goods cloth and wheat. A and A_1 are the points of origin of the axes of C isoquants of cloth while B and B_1 are the points of origin of the axes of W isoquants of wheat.

Let M be some point on the contract curve of country X. In box $A_1C_1B_1D_1$ in Figure 6.16B, line $A_1 F_1$ has been drawn such that it is parallel to line AM in box ACBD in Figure 6.16A. From the position of the line A_1F_1 it is obvious that angle DAM = angle $D_1A_1F_1$. Similarly, line B_1G_1 has been drawn parallel to line BM and it intersects line A_1F_1 at point M_1.

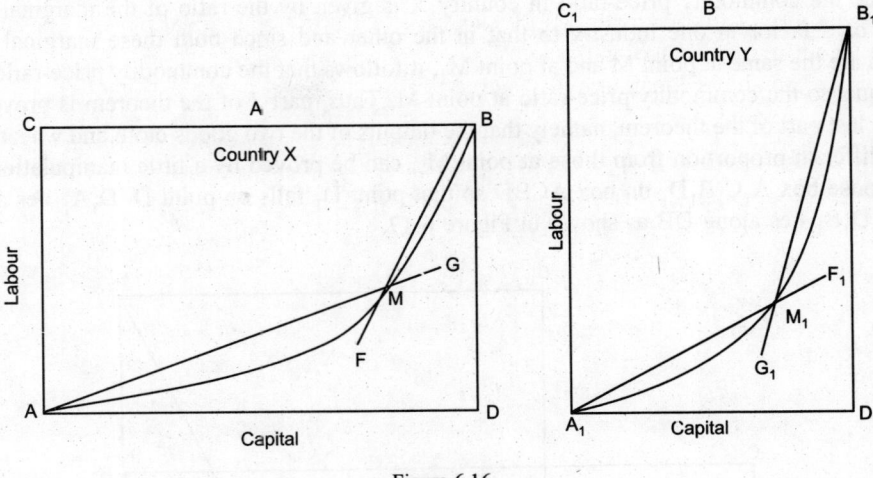

Figure 6.16

Applying the theorem of corresponding points Lancaster shows that points M and M_1 form a pair of points such that:

1. If point M is on the contract curve of country X, then point M_1 is on the contract curve of country Y.
2. The marginal physical products of capital and labour in the production of cloth and wheat are equal at points M and M_1.
3. The commodity price-ratio at point M is equal to the commodity price-ratio at point M_1.
4. The quantities of cloth and wheat produced at points M and M_1 are different, their proportion being such that the capital-rich country X produces more of the capital-intensive good cloth while the labour-abundant country Y produces more of the labour-intensive good wheat. Thus, each country produces relatively more quantity of that good which uses relatively large amount of the country's abundant (cheap) factor of production.

The theorem of corresponding points can be easily proved. Since the production function is linearly homogeneous in both the countries the set of C-isoquants of cloth spreading out from point A_1 in the $A_1C_1B_1D_1$ box is identical with the set of C-isoquants of cloth spreading out from point A in the ACBD box and since angle DAM = angle $D_1A_1M_1$ those properties which are relevant to the C-isoquants of cloth and which hold along AM are also applicable along A_1M_1. Similarly, properties relevant to W-isoquants of wheat holding along BM line also hold true along B_1M_1 line. Consequently, the marginal physical products of labour and capital in the production of cloth along lines AM and A_1M_1 are equal. Similarly, the marginal physical products of labour and capital in the production of wheat along BM and along B_1M_1 are equal. Again since M is the common point of AM and BM and M_1 is the common point of A_1M_1 and B_1M_1, the marginal physical products of labour and capital in the production of cloth and wheat are identical at points M and M_1. As point M lies on the contract curve of country X, the ratios of the marginal physical products of labour and capital in the production of cloth and wheat are equal. Since this property also holds true for country Y, point M_1 must lie on the contract curve of country Y. Thus, part 1 of the theorem is proved. Part 2 of the theorem simply follows from the property of a ray passing through the point of origin of the axes.

Since the commodity price-ratio in country X is given by the ratio of the marginal physical product of a factor in one industry to that in the other and since both these marginal physical products are the same at point M and at point M_1, it follows that the commodity price-ratio at point M_1 is equal to the commodity price-ratio at point M. Thus, part 3 of the theorem is proved.

The last part of the theorem, namely that the outputs of the two goods cloth and wheat, at point M will differ in proportion from those at point M_1, can be proved by a little manipulation. Let us superimpose box $A_1C_1B_1D_1$ on box ACBD so that point D_1 falls on point D. D_1A_1 lies along the DA and D_1B_1 lies along DB as shown in Figure 6.17.

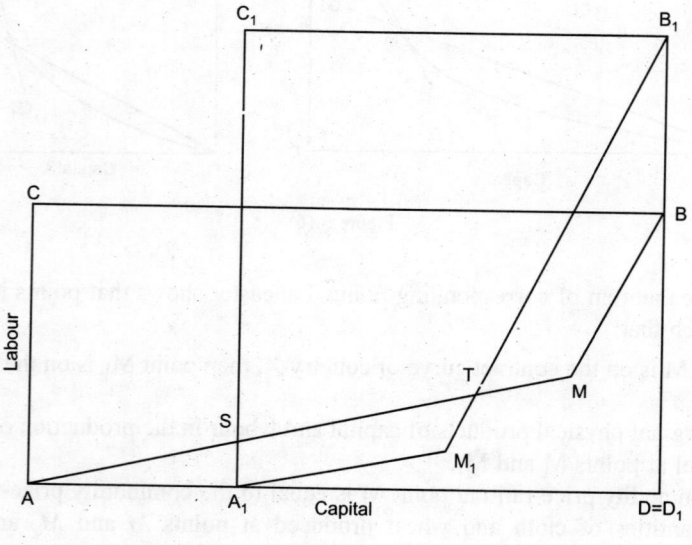

Figure 6.17

It can be seen that point A lies to the left of point A_1 because A_1D_1 is less than AD. Consequently, line AM lies above line A_1M_1. Since line BD is less than line B_1D_1 point B will lie below point B_1. Consequently, line BM will lie below line B_1M_1.

Considering points S and T we find that AM = AS + ST + TM. Since ST is parallel to A_1M_1, A_1STM_1 forms a trapezium with A_1S being the vertical side and M_1T being the non-vertical side. From this and with M_1T being inclined to the right of the vertical line A_1S it follows that ST > A_1M_1. Since AM > AS + TM + A_1M_1, it follows that AM > A_1M_1.

As stated, the production of cloth at point M in country X is proportional to AM while in country Y the production of cloth at point M_1 is proportional to A_1M_1. Since AM is bigger than A_1M_1, production of cloth is greater in country Y than in country X. Similarly, since BM is smaller than B_1M_1, production of wheat in country X is less than in country Y. Consequently, the production of cloth relative to the production of wheat is greater in country X than in country Y. As the production of cloth is capital-intensive, it can be concluded that each country produces relatively more of that good which uses intensively its abundant factor of production. Thus, part 4 of the theorem is proved.

The theorem of corresponding points neatly illustrates the important properties of the Heckscher-Ohlin trade model. Free trade equilibrium occurs in both the countries at the corresponding points M and M_1 since such a pair of points satisfies the internal conditions of equilibrium in each country. Trade occurs between two countries since they have identical tastes but produce two goods (cloth and wheat) in different proportions at any given price level (this follows from parts 3 and 4 of the theorem). Although trade equilibrium may take place at any point but at these two points the factor prices in both the countries will be equalised (part 2 of the theorem). Finally, it follows from part 4 of the theorem that each country will export that good which uses intensively its abundant factor of production.

Criticisms

Although the factor-proportions theory developed by Heckscher and Ohlin provides a more thorough-going and plausible explanation of international trade as compared to the classical comparative cost advantage theory, it is not free from criticisms. The Heckscher-Ohlin theory has been criticised on the following grounds.

1. The Heckscher-Ohlin theory is an oversimplified explanation of trade as it explains trade between only two countries in two commodities produced by only two inputs. In other words, it is a $2 \times 2 \times 2$ trade model. In practice, however, trade takes place between many countries and in many commodities which are produced by several inputs. Consequently, the theory cannot explain the actual complex trade pattern.

2. According to the factor-proportions theory, a country will produce and export that commodity in whose production relatively large amount of its abundant factor is used. Thus, the theory states that trade occurs due to differences in factor proportions between nations. According to the theory, trade will not occur between regions or countries endowed with similar relative factor endowments. For example, if the two countries are identical in capital and labour abundance they would not trade. The fact, however, is that a substantial part of world trade takes place between countries with similar factor endowments. For example, in 1960 the exports of manufactured goods to one another of the ten industrially developed countries—all these 10 nations are capital-rich and poorly endowed with labour—accounted for about 25 per cent of their total exports and for about 50 per cent of their exports of the manufactured goods. Although the relative factor endowments, especially with regard to labour, among these ten nations differ, yet these differences are rarely large enough (except the USA vis-a-vis the others) to explain specialisation by themselves. The factor-proportions theory is not competent to explain specialisation and trade in manufactures between these countries.

The fact is that like any trade, international trade also takes place due to differences in the prices of commodities between different regions which may be due to cost differences. As has been highlighted by the regional economics or location theory, differences in costs of different products between different regions arise from, besides the differences in factor proportions which cause factor price differences many factors including transport costs (whose influence on trade has frequently been either ignored or underrated by international trade theorists), the economies of scale and external economies. It was Ohlin's mistake to make the simplifying assumption that regional differences in factor proportions uniquely determined specialisation and trade. In short, the factor-proportions theory is faulty as an explanation of specialisation and trade because it ignores several other factors like transport costs, economies of scale, external economies, etc. which together with differences in factor proportions account for the differences in costs and consequently in the prices of products between different regions giving rise to trade between these regions. Ellsworth has correctly stated that "with several causes operating simultaneously upon costs, it becomes a matter of adding up the influence of all cost-reducing and increasing forces to arrive at a net result. More concretely, comparative advantage becomes a net cost advantage, in the reckoning of which one must take into account differences in labour costs, in capital costs, in natural resource costs, and also the costs of transporting raw materials (sometimes several from different locations), fuel, and the finished product, as well as the cost-reducing influence of economies of scale and of external economies."[15]. As the factor-proportions theory does not take into account all the various cost-influencing factors it is unable to develop a general equilibrium system. Consequently, it is a partial equilibrium analysis.

3. The theory ignores the role of product differentiation in international trade. If there is product differentiation, trade may take place even if the two countries are similarly endowed with the productive agents. For example, English cutlery finds a market in Germany and German cutlery is sold in England. Well-known Swedish economist Staffan Burenstam Linder has developed the theory of trade in manufactures which states that countries with similar tastes and with roughly the same level of income will trade substantially with each other exchanging one differentiated product against another. The bulk of the trade that takes place between the developed nations in automobiles, electronic and other manufactured goods is explained by this theory. Thus, Italian Fiats are exported to England while British Fords are sold to Italy. German Volkswagens find a good market in the USA while Peugeots are sold to Germany. The gains from trade in these exports and imports are, however, relatively limited and are measured by how much the prices of exports would fall or the prices of import-substitutes would rise if no trade took place in these differentiated goods.

4. Wijanholds[16] has criticised the Heckscher-Ohlin theory on the ground that the prices of commodities are not determined by the factor costs of production. He maintains that the relation is quite the reverse. The prices of commodities are determined by their utility to the consumers and the prices of raw materials and labour ultimately depend on the prices of final goods.

5. The factor-proportions theory does not mention anything about the by-products. Sometimes the by-products are more important than the main final products in influencing the structure and direction of international trade.

6. The theory rests on the static assumptions of fixed quantities of factors of production, given consumer incomes and tastes, given production functions, etc. The conclusions drawn from the static equilibrium analysis cannot be applied to a dynamic economy characterised by changes in tastes, technical knowledge and relative factor endowments over time. In other words, the theory neglects the problems of long-run historical developments and their impact on the nature and pattern of international trade. The assumptions of fixed factor endowments and unchanging technology (it follows from the assumption of given production functions) are difficult to defend. None of a country's factors is fixed for all time to come. A country's total labour force grows over time as result of growth of population and improvement in labour skills. In recent times, world's underdeveloped countries have experienced rapid population growth due to improved public health services. Similarly, a nation's capital stock is augmented due to increased saving and capital accumulation made possible through income redistribution in favour of the people with high marginal propensity to save and through increase in productivity. Even a country's natural resources are not fixed for ever as their supply may be either augmented due to new discovery and exploration or depleted through exhaustion. For example, India's known oil reserves have been substantially augmented as result of the discovery and exploitation of new oil wells in Bombay High and the eastern and western parts of the country creating a basis for self-sufficiency in oil in

15. P.T. Ellsworth, *The International Economy*, Fourth Edition, 1969, p. 147.
16. Wijanholds, H.W.J., "The Pure Theory of International Trade: A New Approach", *South African Journal of Economics*, September 1953. p. 236.

the near future. Similarly, Venezuela became among the world's principal oil exporters due to the discovery of oil in the early years of this century. On the other hand, exhaustion of iron ore deposits in the United States of America transformed that country's economy from one of self-sufficiency to one of increasing dependence on imports in the matter of iron ore. Even the land can expand in economic sense through changes in technology which may make abandoned once-exhausted mines useful and worth exploiting or the once-eroded land useful through afforestation through capital investment.

However, for international trade and specialisation what is important is not the indisputable fact that factor supplies can increase or decrease because it would be of no significance if the relative factor supplies of different countries remained unaltered. What is important is the high possibility that changes in the factor supplies of different nations may change the relative factor endowments of different countries over time. The effect of such changes in a country's factor endowments can be easily seen in terms of the shifts in her production possibilities curve which will influence the structure of international trade.

7. By assuming given production functions, the theory ignores the important influence of technological progress on trade. The assumption of identical production functions in both the countries is plausible only in a static world in which the technical conditions of production remain constant. Barring exceptional cases of very few well-guarded industrial secrets and patented production processes which their owners will not license, any production function is available for exploitation by any potential user. In the long run, there are no industrial secrets and after a certain time lapse even the patents and copyrights expire.

8. Jacob Viner and others have criticised Bertil Ohlin's assumption that different units of the factors of any particular category, such as 'labour' or 'factor A' are homogeneous in quality everywhere. It is from this assumption that a corollary stating that the production functions are everywhere identical if expressed in terms of common classifications of factors which are everywhere homogeneous in quality follows. It is, therefore, evident that the Hecksher-Ohlin theory leaves no scope in the analysis for any influence on trade of the qualitative differences between the factors in the different countries.

9. Some economists have questioned the validity of the premise of the absence of international factor mobility on which the theory rests. John H. Williams has argued that international factor mobility was far greater than was supposed to be. According to him, factor mobility was, in fact, greater *between* countries than *within* a single country. Williams has been supported in his view by Jonathan V. Levin who has argued that the recent development of "export economies—countries which have developed industries producing and exporting primary products—proves, if anything, the fact of international factor mobility. According to Levin, the entry of export economies into international trade was thrust upon them by a sudden invasion of foreign entrepreneurs who brought with them the necessary capital, managerial skill, technical and unskilled labour needed for the development of export industries. Levin has stated that before the War Burma's rice industry depended heavily on the seasonal Indian labour, Chettyar moneylenders and British rice millers. The extraction of petroleum in the Middle East, Sumatra and Venezuela also owes its development to foreign capital, enterprise, and skilled labour. Mineral industries in Mexico, and copper mining in Chile can also be cited as examples of export industries whose development has been due to international factor mobility and not immobility.

10. Despite identical factor proportions in different countries, trade may take place if the consumer preferences are not identical, i.e., if consumer tastes in the two countries are different for various reasons. The possibility of trade taking place between the two countries with identical factor proportions as reflected in a single production possibilities curve for the two countries but

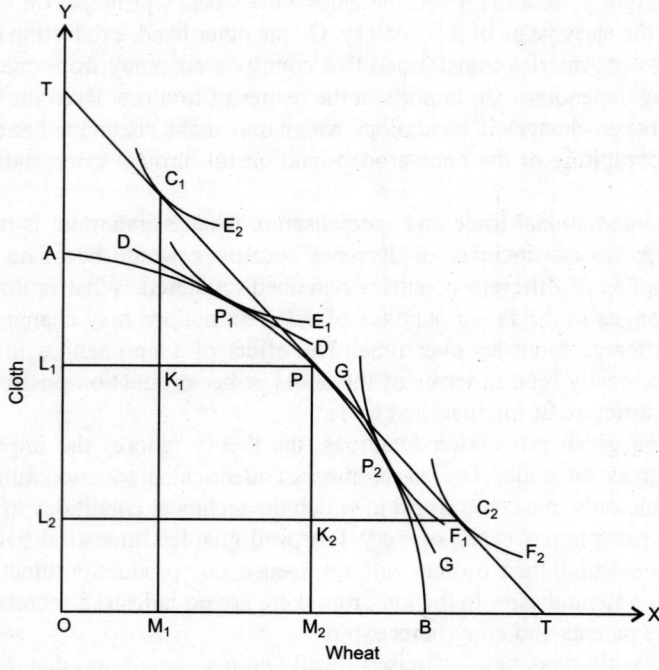

Figure 6.18

with different consumer preferences reflected in the different forms of the community indifference curves of the two countries have been illustrated in Figure 6.18.

There are two countries A and B producing two commodities cloth and wheat. The production possibilities curves of both the countries being identical are represented by a single production possibilities curve AB. In autarky, country A's production and consumption equilibrium takes place at point P_1 showing strong consumer preference and consequent high price for cloth while country B's consumption and production equilibrium occurs at point P_2 which shows strong consumer preference for and consequent high price of wheat in the country.

The opening of trade between the two countries establishes the international terms of trade or exchange-ratio line TT which provides superior alternatives to both the countries. In the new situation, the residents of country A attain higher consumption equilibrium at point C_1 consuming M_1C_1 quantity of cloth and $0M_1$ ($= L_1 K_1$) quantity of wheat while the residents of country B move on to the higher consumption equilibrium point C_2 consuming L_2C_2 quantity of wheat and $0L_2$ ($= M_2K_2$) quantity of cloth. Both the countries increase their specialisation in consumption and cease their specialisation in production moving to production equilibrium at P. It shows that for international trade to take place, differences in the factor proportions between different countries are not necessary.

11. If the two countries have different production possibilities curves and identical tastes, trade may be unorthodox if one cf the two goods has a negative income elasticity of demand, i.e., if one of the two goods is an inferior good. In such a situation, trade will follow an unorthodox pattern in the sense that the capital-rich country exports labour-intensive goods—a conclusion which refutes the factor-proportions theory. It is, therefore, evident that different relative factor endowments with

identical taste patterns do not ensure orthodox trade pattern envisaged by the Heckscher-Ohlin trade theory.

12. Critics have argued that since the factor-proportions theory is based on the assumptions of perfect competition, full employment of resources, identical production functions, absence of transport costs, absence of product differentiation, etc. it is unrealistic and has no relevance in the real world. Kelvin Lancaster does not, however, agree with this view. He argues that "the model occupies the very centre of international trade theory for reasons unconnected with its realism and indeed strengthened by the very properties which have been subject to so much criticism".

The Heckscher-Ohlin trade model is, however, too simple to explain the 'reason of trade' and the 'future of trade'. Talking about its simplicity Lancaster has correctly stated that "it is, in fact, the simple model of international trade with just as the two-commodity indifference curve is the simple model of consumer's behaviour."[17] According to him, the Heckscher-Ohlin trade theory deserves a place at the centre of international trade theory.

Empirical Validity of Heckscher-Ohlin Theory—Leontief Paradox

The utility of any theory is to be judged by its ability to explain the real phenomena and its capacity to forecast the future, i.e., by its empirical contents. In other words, in order to be valid a theory must have empirical contents and the more the better. Interest in verifying any trade theory through facts arose after the Second World War. The first ever such study was made by MacDougal in 1951 who tested the empirical validity of the classical comparative costs doctrine. Basing his findings on the data relating to the productivity of British and American industries along with the data on wages in the two countries MacDougal's study lent a strong support to the comparative cost advantage theory as the basis for predicting the flows of international trade.[18]

The study of the factor-proportions theory initially developed by Eli Heckscher and his worthy student-colleague Bertil Ohlin has been taken up by several economists for empirical evaluation. The first comprehensive study to test the empirical validity of the theory was made by the well-known American economist Wassily W. Leontief who was followed by others, including R. Bhardwaj, B.S. Minhas, A. Walters, Tatemoto and Ichimura. The findings of these studies have been discussed below.

The assumptions of identical production functions[19] and absence of reversibility of factor intensities for different commodities are an essential part of the Heckscher-Ohlin theory. Minhas and others computed the CES (constant elasticity of substitution) production functions for 24 industries using data for 19 countries and reached the conclusion that the production functions between countries were actually different.[20] But they differ only by a constant scale factor. This means that production for a commodity, say the telephone industry in the United States of America,

17. K. Lancaster, "The Heckscher-Ohlin Trade Model: A Geometrical Approach", *Economica*, Volume 24, February 1957, pp. 21–39.

18. G.D.A. MacDougal, British and American Exports: A Study Suggested by the Theory of Comparative Costs", *The Economic Journal*, Volume LXI, December 1951, pp. 679–724.

19. The assumption of identical production functions implies, apart from identical technical knowledge, skills and so forth, also identical climates, physical and social milieu, and so on. Bertil Ohlin thought it self-evident that the production function is everywhere the same. It followed, in his opinion, from the fact that the same causes produce everywhere (and at every time) the same effect. "The constancy of the laws of nature" must ensure identical production functions between different countries. However, to be meaningful, the concept of the production function should be conceived in terms of the well-defined variable inputs leaving social milieu and climate factors *extra commercialism* outside the function.

20. Kenneth Arrow, Hollis B. Chenery, B.S. Minhas and R.M. Solow, "Capital Labour Substitution and Economic Efficiency," *The Review of Economics and Statistics*, August 1961.

differs from the one in Japan by a fixed percentage. The scale factor will be different from industry to industry. Consequently, the production possibilities curves for any pair of countries and commodities will be different. This will not, however, influence the pattern of international trade.

For testing the factor intensity reversibility, Minhas[21] compared the ranking of 20 industries by capital intensity in the USA and Japan. The ranking, however, showed no correspondence at all. The correlation between the United States ranking and the Japanese ranking was very poor indicating the likelihood of factor intensity reversal which robs the factor-proportions theory of any predictive significance regarding the direction of trade.

Using the time series data, Walters[22] made 14 studies of the aggregate production functions for different countries and found the linearity of returns to scale which conforms to the constant returns to scale hypothesis of the Heckscher-Ohlin theory.

The most comprehensive empirical work done to examine the influence of factor endowment on international trade in the general framework of the Heckscher-Ohlin trade model was done by Professor Leontief.[23] Leontief constructed an input-output table for the United States based on 1947 data for 192 industries. He computed the requirements for one million dollar worth of exports whose percentage composition was the same as the total US exports for 1947, and the US production of one million dollar worth of import replacements. He found to his own and to every body else's great surprise that in 1947 the United States of America apparently exported labour-intensive goods and imported capital-intensive ones. According to his findings, the US export industries used relatively more labour than did the import-competing industries. The conclusion is epitomised by the following figures.

	Per Million Dollars	
	Exports	Import Replacements
Capital (dollars in 1947 prices)	2,550,780	3,091,337
Labour (man-years)	182,313	170,004

It is readily observed that the American import-replacement industries require more capital to labour than do the American export industries. Even after a more extensive and detailed enquiry conducted in part on 1951 data, as well as data for 1947, Leontief saw no reason whatever for changing this conclusion, We can also have the structure of the American foreign trade as depicted in Figure 6.19.

In the figure, the US exports are found more towards the right of the figure toward the X-axis with the imports concentrated nearer the Y-axis and more towards the upper part of the figure.

Criticisms of Leontief Paradox

Numerous criticisms have been levelled against Leontief's study. Several writers have questioned the accuracy and appropriateness of the data. Several others have argued that Leontief's procedure

21. B.S. Minhas "The Homohyphalogic Production Function, Factor Intensity Reversal and the Heckscher-Ohlin Theorem", *The Journal of Political Economy,* Volume 70, 1962, pp. 138-156.
22. A.A. Walters, "Production and Cost Functions: An Econometric Survey", *Econometrica,* January-April, 1963.
23. Wassily W. Leontief, "Domestic Production and Foreign Trade: The American Capital Position Re-Examined", *Proceedings of the American Philosophical Society,* Volume 97, September 1953, pp. 332 ff, reprinted in American Economic Association's *Readings in International Economics,* 1968.

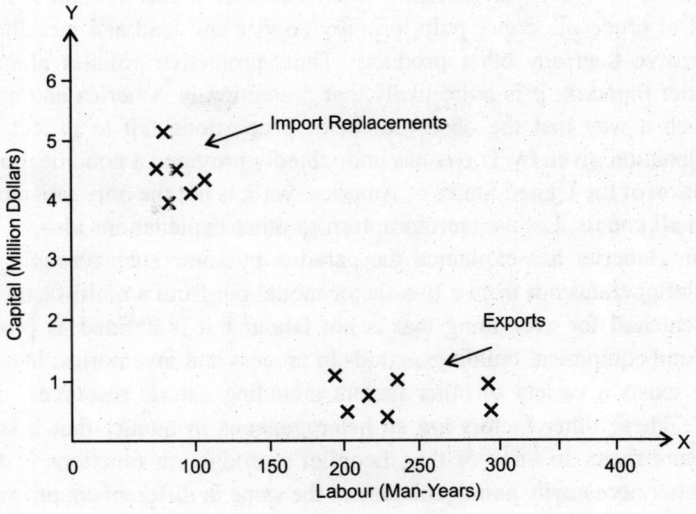

Figure 6.19

is not a logically valid method of testing the empirical validity of the Heckscher-Ohlin theory. N.S. Buchanan, for example, criticised Leontief's measurement of capital. In his opinion, Leontief's capital coefficients were 'investment requirement coefficients' which did not take into account the durability of capital.[24] G.A. Loeb argued that differences in intensity between the export sector and the import-competing sectors were not statistically significant.[25] Another serious charge came from B.C. Swerling who argued that 1947, the year selected for his study by Leontief, was an atypical year as far as trade flows were concerned.[26]

Related to the Leontief Paradox comes another interesting criticism which takes into account the resource endowments of America. An attempt to explain the paradox by taking into account the natural resources effect has been made by Eric Hoffmeyer.[27] He argues that the Leontief Paradox is due to Leontief's failure to deal adequately with the natural resources. If commodities that require a relatively large amount of natural resources are excluded from the capital-labour ratios, then Leontief's conclusion will be reversed and the paradox will disappear. Since Leontief was concerned only with the import replacements, it would be better to exclude from the calculations those commodities which were intensive in natural resources.

Another explanation to the question raised by Hoffmeyer's contention has been given by William P. Travis. Most critics who have examined the Leontief Paradox have so far implicitly assumed that the law of comparative advantage determines the existing pattern of trade. Travis does not subscribe to this view. He explains Leontief Paradox in the light of the American trade policy. According to Travis, it was the then-protective trade policy which gave rise to the

24. N.S. Buchanan, "Lines on the Leontieff Paradox", *Economia Internazionale*, Volume 8, November 1955. pp. 791 ff.
25. G.A. Loeb, "A Estrutura do comercio Exterior da America do Norte", *Revista Brasileira de Exonomia*, Volume 8, December 1954, pp. 81 ff.
26. B.C. Swerling, "Capital Shortage and Labour Surplus in the United States", *The Review* of *Economics and Studies*, Volume 36, August 1954, pp. 286 ff.
27. Eric Hoffmeyer, "The Leontief Paradox Critically Examined", *Manchester School of Economic and Social Studies*, Volume 26, May 1958, pp. 160 ff.

paradox.[28] According to Travis, at the time when Leontief made his study, most competitive imports consisted of crude oil, paper pulp, primary copper and lead and metallic ores which are more capital intensive than any other products. Thus, protective policies alone are enough to explain the Leontief Paradox. It is quite likely that protection in America and abroad might bias trade flows in such a way that the observed factor proportions fail to reflect the actual ones. Although the explanation given by Travis has undoubtedly provided a good deal of insight into the foreign trade position of the United States of America, yet it is not the only satisfactory explanation of the paradox on all counts. Let us, therefore, turn to other explanations also.

Gottfried Von Haberler has explained the paradox by some such reasoning as follows. He argues that Leontief operates not from a two-factor model but from a multi-factor model. "Capital for him is not a catchall for everything that is not labour but is defined as produced means of production, plant and equipment, buildings, goods in process and inventories. In addition to labour and capital, there exists a variety of other factors including natural resources, management and entrepreneurship". These other factors are so heterogeneous in quality that it is too difficult to identify and measure them. In view of this, Leontief's production functions in terms of labour-capital use are neither necessarily homogeneous nor the same in different countries. In view of this, no *a priori* generalisation regarding the composition of trade is possible.

Another interesting explanation of the Leontief Paradox runs in terms of the new fashionable concept of 'human capital'. The basic idea is simple. Human capital is created by schooling, i.e., by investment in human beings. It is well known that labour's productivity can be enhanced by either utilising more physical capital per unit of labour or by combining more human capital such as investment in education, public health etc. with each unit of basic labour. Leontief himself alludes to the fact that labour's productivity in the American export sector is considerably higher than in the import-competing sectors. This may be the result of a high rate of investment in human capital in the export sector. Thus, if capital is defined as physical plus human capital the paradox could be resolved.

A widely discussed criticism of the Leontief Paradox is that of P.T. Ellsworth who suggests that the comparison of factor-intensities should be confined to the case of American exports and foreign exports and not, as Leontief has done, to the case of American exports and American import replacements. For the imports from each country of origin he should have computed the capital labour ratio on the basis of input-output studies of these economies. As Leontief has not done this, only half of Leontief's calculations, those relating to capital-labour ratio of the American exports were relevant, while those half relating to the import replacements were irrelevant. However, most critics are not satisfied with Ellsworth's contention. They treat Ellsworth's argument as against the basic postulates of the Heckscher-Ohlin theory which assume that production functions are everywhere identical.

The discussion of the Leontief Paradox is meaningful from two angles. In the first place, it shows that it is not easy to reformulate simple theoretical models such that these will be suited to empirical testing since a model can be easily mis-specified. In the second place, empirical research is in itself fruitful as it will certainly give rise to reformulations and extensions of the model used as a starting point for the empirical testings. The Leontief Paradox provides an excellent illustration to these two propositions.

---

28.   William P. Travis, *The Theory of Trade and Protection,* 1964. Chapters 3–4.

Explanation in Terms of the Demand Factors

Most of the above explanations are mostly concerned with the supply side and little attention has been paid to the demand side. A country's demand pattern plays a crucial role in shaping her exports and imports pattern. If consumption in a capital-abundant country is heavily biased towards capital-intensive goods, it is conceivable that the export pattern of the country may not be as visualised by the Heckscher-Ohlin trade theory. It is quite possible that due to rapid social and economic development experienced in America, the demand has concentrated on services which require an extremely huge capital outlay and highly finished manufactured goods in whose production a large amount of expensive and intricate capital equipment is required. Several scholars including R. Robinson and R. Jones have held the view that the Leontief Paradox might be the result of American demand pattern being biased towards capital-intensive goods. However, this explanation which runs in terms of biased demand conditions is a theoretical possibility.

Another interesting consideration is the role of foreign-based American firms in America's trade structure. A considerable amount of the US imports comes from the American-owned firms located abroad. It is believed that these firms use a higher capital-labour ratio than do the native firms there. If imports' production under these conditions were excluded from the American import bills then production in the American import-replacement industries competing with the other imports will become labour-intensive. This is one way of explaining the Leontief Paradox.

Leontief's Explanation

Leontief himself endeavoured to resolve the paradox by giving different explanations. While giving different explanations to his findings, he always kept two things in mind. Firstly, he wanted to rescue the Heckscher-Ohlin theory from apparent refutation. Secondly, he supported the common belief that America has more capital per worker than any other country. His different explanations mainly ran along two different lines. In the one he gave priority to labour efficiency (productivity). Leontief argued that the American worker was three times as efficient as was a worker elsewhere. Consequently, the effective supply of labour in America must be regarded as a certain multiple of the apparent labour supply there. After adjusting for varying degrees of labour efficiency, America would be found to be relatively rich in labour and poor in capital. This is how he safeguarded the validity of the Heckscher-Ohlin theory.

The other explanation given by Leontief is connected with the two-factor framework and broad use of the term 'capital'. The only two factors considered in Leontief's study are labour and capital. However, the inadequacy of his procedure has been noted by Leontief himself. He observes: "Invisible in all these tables but ever present as a third factor or rather as a whole additional set of factors determining this country's productive capacity, in particular, its comparative advantage vis-a-vis the rest of the world, are natural resources, agricultural lands, forests, rivers and other rich mineral resources."[29]

If one takes into account this third factor, an explanation of the paradox can be obtained. However, empirical research in this field is not conclusive. Much still remains yet to be done. Nevertheless, it is true that till now no other trade models have proved to be as effective in predicting the trade flows as either the classical or the Heckscher-Ohlin trade model.

---

29. W.W. Leontief, "Factor Proportion and the Structure of American Trade: Further Theoretical, and Empirical Analysis," *The Review of Economics and Statistics*, Volume 38, November 1956, p. 346.

Stimulated by Leontief's paradoxical results, several studies have been made for other countries to test the empirical validity of the Heckscher-Ohlin theory. Employing the Leontief type approach for Japan, M. Tatemoto and S. Ichimura[30] found that the Japanese exports consisted of the capital-intensive goods. Since Japan is a labour-abundant country, the result contradicts the Heckscher-Ohlin hypothesis. Recognising, however, that Japan's capital-labour endowment is intermediate between that of the developed and underdeveloped countries, they suggested that Japan should possess comparative advantage in the labour-intensive goods in her trade with the developed countries and in the capital-intensive goods in her trade with the underdeveloped countries. Consequently, they classified Japan's exports according to destination and obtained results that revealed at least partial consistency with the factor-proportions theory in so far as the Japanese exports to the United States of America were relatively labour-intensive while those to the underdeveloped countries were relatively capital-intensive. Yet the capital-labour ratio of the US-destined exports was still higher than that of the import-competing goods.

Among the other studies made to test the empirical validity of the Heckscher-Ohlin theory, mention may be made of a study of the erstwhile East German trade[31] which showed that East Germany exported relatively capital-intensive goods while her imports consisted of the relatively labour-intensive goods. More than three-fourths of East Germany's trade was with the Communist block nations. Since East Germany was a capital-abundant country compared with the other countries of the Communist block with which she traded, these results were consistent with the Heckscher-Ohlin theory.

In a similar study, R. Bhardwaj[32] found that India exports the labour-intensive goods and imports the capital-intensive commodities. However, in her trading with the USA—the most capital-abundant country—India was found to export the capital-intensive commodities while she imported the labour-intensive commodities. D.F. Wahl found that Canada exported the capital-intensive goods and imported the labour-intensive goods.[33] The Canadian trade is mostly with the United States of America which is acknowledged to be the most capital-abundant country in the world. Thus, these empirical studies refute the Heckscher-Ohlin theory. It does not, however, mean that the theory cannot explain the direction of international trade. In practical life, we find that Australia—a land-abundant country—exports land-intensive commodities like wheat, meat, dairy products, wool etc. Similarly, Brazil and Columbia which are situated in the tropics and are richly endowed with land at medium altitudes, are among the world's largest exporters of coffee.

In short, we have yet no definitive answer about the empirical validity or otherwise of the Heckscher-Ohlin theory. While some studies have refuted the theory, some other studies have supported the theory. We will have to review whether the tools applied to test the validity of the theory are sophisticated and appropriate to fit in the framework of the assumptions of perfect competition and full employment. Richard E. Caves is correct in saying that the Heckscher-Ohlin trade model has not yet come close to having a full scale testing. Perhaps some day a supercharged economist with supersonic calculating equipment backed by supersaturated foundation will perform this onerous task.

30. M. Tatemoto and S. Ichimura, "Factor Proportions and Foreign Trade: the Case of Japan", *The Review of Economics and Statistics*, XLI November 1959.

31. W. Stolper and K. Roskamp, "An Input-Output Table for East Germany with Applications to Foreign Trade", *Bulletin of the Oxford University Institute of Statistics*, XXIII, November 1961.

32. R. Bhardwaj, *Structural Basis for India's Foreign Trade*, Bombay, 1962; "Factor Proportions and the Structure of Indo-US Trade", *The Indian Economic Journal*, Volume 10, October 1962.

33. D.F. Wahl, "Capital and Labour Requirements for Canada's Foreign Trade", *The Canadian Journal of Economics and Political Science*, XXVII, August 1961.

Suggested Readings

M.O. Clement, R.L. Pfister and K.J. Rothwell, *Theoretical Issues in International Economics,* 1967, Chapters 1–3.

P.T. Ellsworth and J. Clark Leith, *The International Economy,* Fifth Edition, 1975, Chapters 6 and 8.

J.L. Ford, *The Heckscher-Ohlin Theory of the Basis and Effects of Commodity Trade,* 1965.

H. Robert Heller, *International Trade,* Second Edition, 1973, Chapters 4–7.

H. Makower, "The Pure Theory of International Trade: The Teaching of It", *The Indian Economic Journal,* Volume XXI, No.1, July-September 1973, pp. 1–18.

Bertil-Ohlin, *Interregional and International Trade,* 1933, Chapters 1–12.

Bo Sodersten, *International Economics,* Second Edition, Reprinted 1981, Chapters 3, 4 and 6.

Questions

1. Explain the meaning of factor intensity reversal. How does this affect the conclusions of the Heckscher-Ohlin theory of international trade? Discuss.
2. What do you understand by the Leontief Paradox? Explain its importance and limitations.
3. Examine critically the Heckscher-Ohlin theory of international trade. Is the theory supported by empirical evidence?
4. On what assumptions is the Heckscher-Ohlin theory of international trade based? Explain the relative factor abundance and relative factor intensity.
5. Are demand patterns or factor endowments more important in determining the structure of international specialisation?
   Explain in this connection the views of Heckscher and Ohlin.
6. Explain the Leontief Paradox. How does it limit the empirical validity of the Heckscher-Ohlin theory of international trade?

CHAPTER **7**

Other Reasons for Trade

The basis of all trade, whether domestic or foreign, is the difference in the prices of goods and services as between different regions or countries. Consequently, any theory of foreign trade must be able to demonstrate the factors that are at the back of differences in the relative prices of goods and services between different countries. Both the classical comparative costs theory and the Heckscher-Ohlin theory of international trade ascribed the differences in the relative prices of goods and services between different countries to the differences in their relative costs of production. In other words, according to both these theories trade between two countries takes place because certain goods are cheaper in one country while certain others are cheaper in the other country. And certain goods are relatively cheap in one country compared with the other because their relative costs of production in one country are low compared with those in the other. Thus far, the two theories are alike or similar. However, the two theories differ on the explanation of lower comparative or relative costs of production of different goods in the two countries.

While the classical doctrine looked at the comparative differences in factor (labour) productivity (cost) in the production of two goods in the two countries, the Heckscher-Ohlin theory emphasised differences in the factor proportions between two countries as the basis of difference in the unit costs of production (and consequently in the prices) of goods between the two countries. The Heckscher-Ohlin theory stated that since different regions or countries are differently endowed in the supply of factors of production and since the production functions (input-output relationships) for different goods are different, commodities making greater use of the abundant factor while using the scarce factor in small amounts will be relatively cheap in the country while some other goods using little amounts of the abundant factor and large amounts of the scarce factor will be costly or dear in the country. If similar situation prevails in the other country such that goods costly in the first country are cheap in the second country and *vice versa* trade will take place to the mutual advantage of both the countries.

Both these theories, however, ignored the role of demand as a causative factor or as an alternative basis for trade. The Heckscher-Ohlin theory took it for granted that the forces of demand reinforced the forces of supply with the result that if people in country 1 have a relatively greater preference for good *B* while the country commands a technological advantage in the production of good *A* in autarky, the relative price differences between country 1 and country 2 would be further sharpened providing greater scope for trade.

The differences in the comparative costs and factor proportions are only two of the several logically possible reasons for trade. These several alternative reasons for trade have in recent years attracted increasing attention of the trade theorists. Some of these alternative bases for trade have been discussed here.

Vent (Outlet) for Surplus

Most of the colonial countries of yester years were drawn into trade because these countries had surplus production to trade. The vent or outlet for surplus approach assumes that a country has 'free' or surplus commodity or some unused resources which it can use to generate the export earnings. This approach is specially applicable to developing countries. It states that a less developed country often has some good for which the domestic demand is already satisfied and which it can easily export. Central to this approach, is the idea that in autarky the country is producing well inside her production possibilities curve. It is not simple and easy to explain why it is so. We explain it in terms of Figure 7.1 which shows that the country is producing the two goods cloth (handicrafts) and food at point A which lies inside its production possibilities curve EF. It shows low productivity both in absolute and relative terms in the production of cloth (handicrafts). Since food is relatively cheap, producers do not produce the maximum attainable amount of food. Since land is plentiful and fertile, people produce enough for their requirements by producing at point A leaving part of the resources unused. They bask in the sun in spare time although the economy is clearly capable of producing more of both the goods if and when the economic conditions changed to the advantage of the economy.

The opening up of trade with a mature economy offers the economy opportunity for such a change. Since food is cheap it is exported and its price rises above the domestic exchange-ratio. Instead of basking in the sun, people now produce more and move on to point B situated on the production possibilities curve. By trading at the improved terms of trade shown by the slope of new exchange-ratio line BC, the economy moves on to point C which shows considerable economic gain from trading compared with the pre-trade point A. Here trade gives rise to an outlet or vent for the surplus land and labour in the colonial economy. Through the improvement in the terms of trade brought about by trade, the economy's unused resources become gainfully employed while enjoying substantial increase in consumption which is represented by the movement from point A situated on a lower community indifference curve to point C situated on the highest feasible community indifference curve.

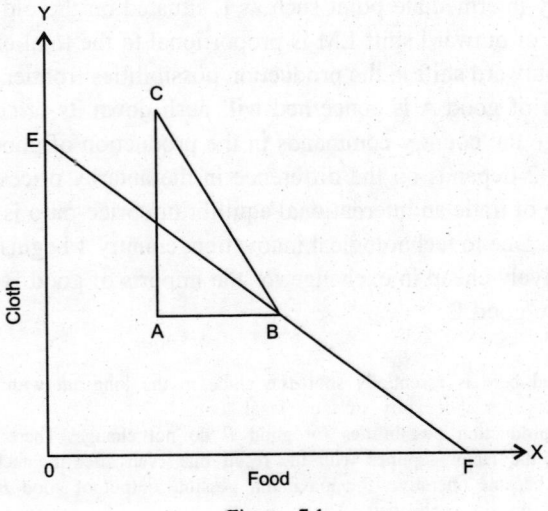

Figure 7.1

The vent-for-surplus approach to trade offers an explanation for trade which is not easily explicable in terms of the Heckscher-Ohlin approach which emphasises the differences in factor proportions as the cause for international trade. The vent for surplus approach also points to some risks connected with trade. If a country's traditional handicrafts sector is wiped out due to international competition and if the increase in population consumes the exportable surplus of food, trade will eventually contract and development will come to a grinding halt.

Differences in Technology

Ceteris paribus, differences in the production possibilities or transformation curves between countries will result in the differences in the prices of goods in autarky. Such price differences in autarky may arise either from the differences in factor proportions between different countries or from the differences in technology between different countries.

Innovation which introduces new products and new production methods is at the heart of modern production and exerts profound influence on the pattern of international trade. In order to show how differences in technological efficiency may result in trade between the two countries by causing price differences in autarky let us assume that in a two-country model, in country 1 due to technological innovation practised in the production of good A an input combination yields a given additional proportion p of output of good A. This fact will entail the renumbering of isoquants of good A in country 1's box diagram. While technological innovation improves good A's output in country 1, it is assumed that producers of good A in country 2 do not have access to technological improvement due to patent rights[1] held by the producers of good A in country 1.

With everything else remaining the same, as result of the technological innovation the production possibilities curve of country 1 will shift upward in a manner that while the intercept on the Y-axis showing the production of good B remains unchanged,[2] the intercept on the X-axis showing the output of good A shifts outward by proportions as shown in Figure 7.2 where the total output of good A expands from $0A_1$ to $0A_2$ such that $0A_2 = 0A_1 (1 + p)$. It is evident that while the country's production possibilities curve shifts outward on the A-good axis, no shift takes place on the B-good axis. At any intermediate point such as L situated on the old production possibilities curve A_1B_1 the amount of outward shift LM is proportional to the total output KL.

The effect of this outward shift in the production possibilities frontier of country 1 in so far as the possible production of good A is concerned will push down its price in the country thereby increasing the advantage the country commands in the production of good A. The basis for trade between country 1 and 2 depends on the difference in the autarky prices of goods. If now trade starts, as a consequence of trade an international equilibrium price-ratio is established between the two autarky price ratios. Due to technological innovation, country 1 begins to export good A which she now produces relatively cheap in exchange for the imports of good B while country 2 obtains good A by exporting the good B.

1. The problem considered here is essentially short-run since in the long-run with the lapse of time even the patent rights also lapse after the expiry of their legal term.
2. It is so because the production possibilities for good *B* do not change. There is no change either in the production function or the factor supplies with the result that even after the technological innovation in the production of good *A* became operative, the maximum possible output of good *B* shown by the intercept on the *B*-good axis would remain unchanged.

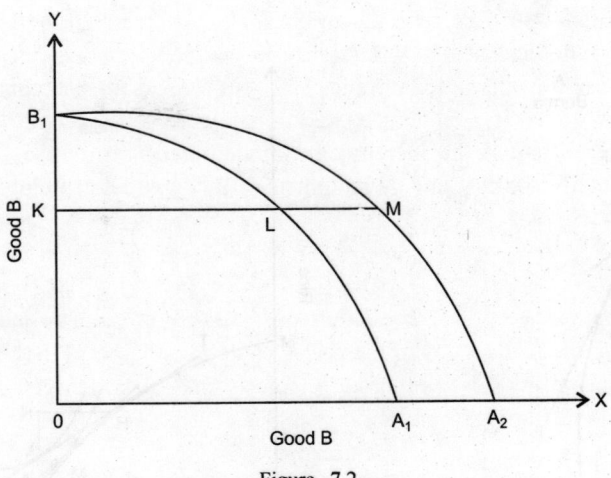

Figure 7.2

Differences in Demand

The influence of differences in the demand conditions has either been overlooked or assumed to reinforce the influence of the forces of supply on trade. In other words, the possibility of demand conditions reversing the direction of trade expected on that basis of cost considerations alone remained outside the orbit of both the classical and the Heckscher-Ohlin theories of trade. In fact, demand reversals may cause a change in the direction of trade such that a country may import a good in spite of the fact that she commands a comparative advantage in its production based on cost considerations alone. This will happen because the consumers may have a strong liking (preference) for the particular commodity. For example, countries of South-East Asia (Japan and China) may import rice to meet the high demand for rice by their residents in spite of the fact that these countries produce rice at low cost. It happens because due to the high population density although the unit cost of production of rice is low but its price in the country is high due to high demand. Figure 7.3 shows the influence of demand conditions in as much as the demand conditions change (reverse) the direction of trade from the one expected on the basis of cost advantage or supply conditions alone.

In Figure 7.3, there are two countries Burma and India producing two commodities rice and wheat. In spite of the fact that Burma is a low cost producer of rice (as shown by her production possibilities curve) and India is a cheap producer of wheat (represented by her production possibilities curve), due to the strong demand conditions at home the autarky price of rice is relatively high in Burma and that of wheat is high in India. On the basis of comparative cost advantage the pattern of trade between Burma and India should have been such that Burma exported rice to India while India exported wheat to Burma. Due to the domestic demand conditions in the two countries the pattern of trade is, however, reversed. Trade takes place between Burma and India with Burma importing rice from India and India importing wheat from Burma—a phenomenon which would not be expected on cost advantage considerations alone. In the theoretical literature, this situation in which a good which a country is expected to export on the basis of her comparative cost advantage or supply consideration is actually important because its demand and following it its price at home is so high that the country has to supplement her domestic production by imports is known as the phenomenon of *demand reversal*.

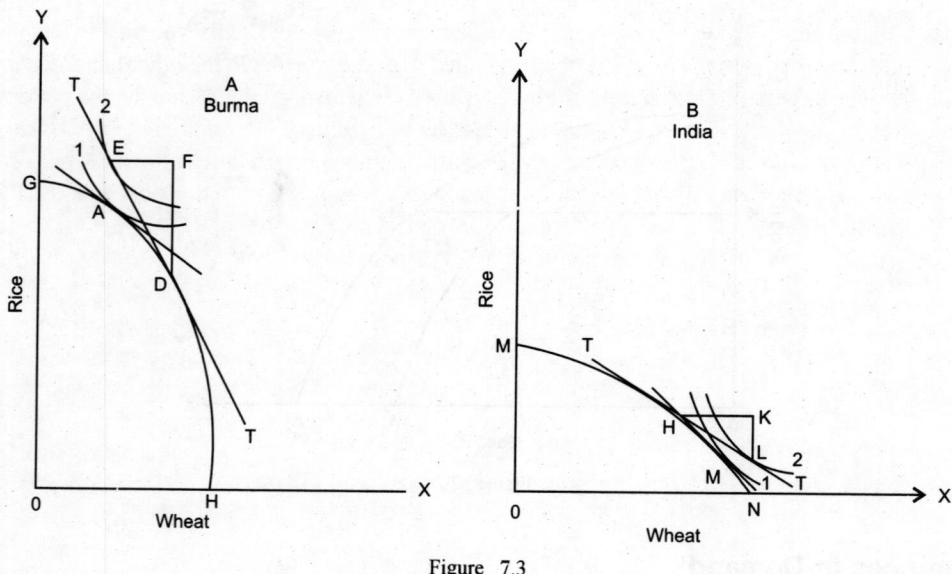

Figure   7.3

Even when the cost conditions are similar in the two countries (represented by identical or similar production possibilities curves of both the countries) trade will emerge if due to the differences in the demand conditions the autarky prices in the two countries differ. Figure 6.18 in Chapter 6 shows such a situation in which although the production possibilities curves of both the countries are identical, due to differences in the autarky prices trade takes place between the two countries.

Product Differentiation

The comparative cost advantage and the Heckscher-Ohlin theories explain the phenomenon of trade on the assumption that the goods traded were homogeneous products. In other words, these theories abstract from the fact of product differentiation. A substantial part of international trade is, however, accounted for by exchange—imports and exports—of the differentiated products. For example, Italian Fiats are exported to England while British Fords find a market in Italy. Similarly, German made Volkswagens exchange for the American Peugeots. Notwithstanding that the gains from trade in these exports and imports are relatively limited being determined by the extent to which the prices of export goods would have fallen or the prices of import substitutes would have risen if the trade was cut off, bulk of the trade that takes place between the developed countries relates to the imports and exports of the differentiated products.

Natural Resources

The assumption of two sets of homogeneous factors capital (K) and labour (L) which is one of the many assumptions of the Heckscher-Ohlin theory ignored the role of natural resources and entrepreneurship in explaining the basis of trade. In international exchange of goods, one can spot out many commodities in the making of which natural resources enter as an important input. The inclusion of natural resources as an input presents substantial difficulties since they are numerous in varieties.

Some natural resources like petroleum, zinc, tin, mica, nickel, copper and even forest resources (teakwood, fir and spruce) are most unevenly distributed over the world. On the other hand, fisheries cover a different category of natural resources which are widely dispersed. Obviously, the pattern of distribution of the natural resources is bound to influence the pattern of international trade. To cite an example, Japan is world's important producer and exporter of tinned fish. A large bulk of the pulpwood used for paper manufacturing is obtained from Canada, Norway and Sweden. Consequently, these countries are among the world's major exporters of newsprint paper.

Entrepreneurship

Neither the classical theory nor the Heckscher-Ohlin theory recognised entrepreneurship or enterprise as an explicit input. In a competitive market economy, entrepreneurial ability being almost evenly distributed, does not matter much as an independent factor of production which influences the nature and scope of trade. However, in monopolistic competition administrative skill and organisation-building ability play a crucial role in the success of large firms and are consequently vitally important in the emergence and growth of modern giant multi-nationals which dominate the international trade. Highly developed and specialised organising ability is, however, a rare phenomenon although its availability offers a special productivity advantage to the country compared with the other country which does not possess this rare ability.

Availability

According to the availability approach to trade, a country exports some goods simply because these are available in abundance at cheap costs. For examples, Saudi Arabia exports oil, or Chile exports copper because these goods are available in abundance (cheap) in these countries, as Saudi Arabia is replete with rich oil fields while Chile has copper deposits in abundance. The well-known American economist I.B. Kravis has stated the availability approach as a rival doctrine to the theory of comparative cost advantage.

Other Factors

Apart from the above reasons for trade, the existence of many commodities, transport costs, varied character of labour and capital also influence the nature and size of international trade. Increasing returns to scale can also significantly influence the location of production both in the static and dynamic ways. The presence of these incresing returns to scale can significantly influence production through trade and will lead a country to specialise in one or more products. The existence of innovation gap may also profoundly influence production and trade. A country having innovative advantage may export new and technically advanced products. Changing technology can also prove an important determinant of trade. It points to the fact that human knowledge and technical ingenuity are important determinants of international trade beside land, labour and capital.

Suggested Readings

P.T. Ellsworth and J. Clark Leith, *The International Economy,* Fifth Edition, 1975, Chapter 7.
Charles P. Kindleberger, *International Economics,* Fifth Edition, 1973, Chapter 4.

Staffan Burenstam Linder, *An Essay on Trade and Transformation,* 1961.
Bo Sodersten, *International Economics,* Second Edition, Reprinted 1981, Chapter 7.
R. Findlay, *Trade and Specialisation,* 1970, Chapter 4.

Questions

1. What factors other than the differences in comparative costs and factor proportion between countries are responsible for international trade? Discuss.
2. To what extent product differentiation and technological differences can explain the phenomenon of international trade? Discuss.
3. What is meant by demand reversal? How does it change the direction of trade and conclusions of the Heckscher-Ohlin theory of international trade? Discuss.

CHAPTER 8

International Trade and Factor Prices

The factor-proportions theory of international trade developed by Eli F. Heckscher[1] and Bertil Ohlin not only replaced the classical and the neo-classical theories, it also opened a new chapter in the pure theory of international trade. The Heckscher-Ohlin theory provided great scope to trade theorists to deal with the several problems in the realm of international trade. One such problem is concerned with the effects of foreign trade on the functional distribution of national income, i.e., on the relative incomes of the factors of production and more important on their absolute real income.

Since the classical theory assumed a single factor of production (labour) and an unchanging composite dose of factors, it did not seriously deal with the problem of trade and factor prices. Although some neoclassical economists, including Charles F. Bastable and John Elliot Cairnes, had made attempts to study the relationship between trade and the distribution of national income, they offered only a partial explanation of the problem as they could not divorce their analysis from the assumption of a single factor of production (labour). In this connection, Charles F. Bastable[2] used the concept of 'specific' and 'non-specific' labour. He explained that as a consequence of trade, 'specific' factor suffers a loss of real income if it is employed in the import-competing industry but if employed in the export-expanding industry its real income will increase. Similarly, John Elliot Cairnes[3] based his study of the effect of trade on the factor prices on 'competing' and 'non-competing' groups among labour. Thus, the neoclassical economists could not satisfactorily analyse the effects of foreign trade on the functional distribution of national income.

Trade, Factor Movements and Prices

While presenting a different explanation of international trade, both Eli Heckscher and Bertil Ohlin showed that trade tended to equalise the factor prices between the trading countries. In the absence of transport costs or any other obstacle to free movement of goods between different nations (regions to use Ohlin's terminology), the immediate effect of trade is to make the commodity prices identical in

---

1. The outlines of this theory were first sketched by Eli F. Heckscher, who was one of Ohlin's distinguished teachers at the Stockholm University, in his pioneering article "The Effects of Foreign Trade on the Distribution of Income", published in Swedish in 1919 in *Ekonomisk-Tidskrift*, 21. The article has now been translated into English and has been reprinted in the American Economic Association's publication *Readings in the Theory of International Trade*, 1949.
2. Charles F. Bastable, *The Theory of International Trade: With Some of its Applications to Economic Policy*, 1887. Chapter VI.
3. John Elliot Cairnes, *Some Leading Principles of Political Economy Newly Expounded*, 1874, Chapter III

all the regions. Trade, however, also influences the prices and the combination and use of the factors of production between different regions.

In a two-region and two-factor simple case, each region finds it advantageous to import those goods which require for their production more amount of the scarce factor as these can be produced cheaply abroad and export those goods which require for their production more amount of the abundant factor as these can be produced cheaply at home. Consequently, instead of producing goods requiring much of the scarce factor, factors of production in the country will be directed in the production of those goods that require greater proportion of the abundant and consquantly factor. Consequently, each region tends to expand the production of goods requiring more of the abundant factor and curtail the production of goods requiring more of the scarce factor. This tends to increase the demand for the abundant factor and to decrease the demand for the scarce factor in both the regions. Given the fixed factor supply, the increase in the demand for the abundant (cheap) factor raises its price while the decrease in the demand for the scarce (costly) factor reduces its price. As a result, "the relative scarcity of the productive factors is made less different in the two regions."[4]

Bertil Ohlin emphasises that apart from equalising the commodity prices, international trade also equalises the factor prices between the different regions or countries because "from each region goods containing a large proportion of relatively abundant and cheap factors are exported, whereas goods containing a large proportion of scantily supplied and scarce factors are imported, the latter becoming less scarce. The same result could have been obtained by a transfer of the factors. As it is, interregional trade serves as a substitute for such interregional factor movements."[5] Hidden behind the commodity movements lie the factor movements. Behind the veil of internatioı.al trade, factors in abundant supply are exported while factors in scant supply are imported. Indirectly, exports of commodities involve the exports of abundantly supplied cheap factors of production while commodity imports virtually amount to importing in the country the factors of production which, being scantily supplied, are dear iı. the country.

The movement of commodities takes the place of movement of the factors of production between regions. But although trade tends to equalise factor prices by reducing the prices of scarce factors and raising those of the abundant ones, nevertheless prices even of those factors which are made relatively less scarce may increase in terms of the commodities because the total volume of commodities increases as result of the more efficient use of the productive factors made possible through the international trade.

The tendency towards complete factor-price equalisation may, however, be thwarted by the various obstacles to free trade. One such obstacle is the cost of transport which prevents complete equalisation of the commodity prices and consequently of the factor prices. Tariffs, differences in taxation and other social conditions of production between different countries may also prevent complete factor-price equalisation.

Stolper-Samuelson Theorem

In 1941, Wolfgang F. Stolper and Paul A. Samuelson derived from the Heckscher-Ohlin theory a simple clear-cut conclusion regarding the impact of trade on the functional distribution of national income. They stated that "international trade necessarily lowers the real wage of the scarce factor expressed in terms of any good."[6] In other words, if the commencement of trade increases the

4. Bertil Ohlin, *Interregional and International Trade*, 1933, p. 35.
5. Bertil Ohlin, *Op. cit.,* p. 49.
6. W.F. Stolper and Paul A. Samuelson, "Protection and Real Wages", *The Review of Economic Studies*, Volume 9, 1941, pp.58–73.

output of one commodity in the country, the relative and absolute income shares of the factor which is employed with relatively greater intensity in the production of this commodity will increase, i.e., free trade will benefit the relatively abundant and hurt the relatively scarce factor of production. This is the famous Stolper-Samuelson theorem which is based on the following assumptions.

1. The country is producing only two goods, wheat and cloth, with only two factors of production, labour and capital.
2. The production functions are homogeneous of the first degree, i.e., production of both the commodities is subject to constant returns to scale.
3. The supply of both the factors of production in the country is fixed or given.
4. Labour is an abundant and consequently cheap factor while capital is scarce and consequently the costly factor in the country.
5. The production of wheat is labour-intensive while that of cloth is capital-intensive.
6. Perfect competition prevails both in the factor and in the commodity markets.
7. The terms of trade are fixed.
8. There are no transport costs.

Proof of the Theorem

The fixed amounts of labour and capital in the country have been shown in the Edgeworth-Bowley box diagram in Figure 8.1B where the dimensions of the box represent the given amounts of labour and capital in the country. The $0M$ ($= 0_1N$) side of rectangle $0M0_1N$ represents the total amount of labour and the $0N$ ($= 0_1M$) side represents the total amount of capital available in the country. These factor supplies are homogeneous in nature and are fixed in amount. The country produces the two commodities wheat and cloth. The production of wheat is labour-intensive while that of cloth is capital-intensive.

In the figure, the origin of the wheat isoquants W is 0 and the origin of the cloth isoquants C is 0_1. In autarky, country's production equilibrium takes place at point P in Figure 8.1A where the community indifference curve CIC_0 is tangent to the production possibilities curve AB. The

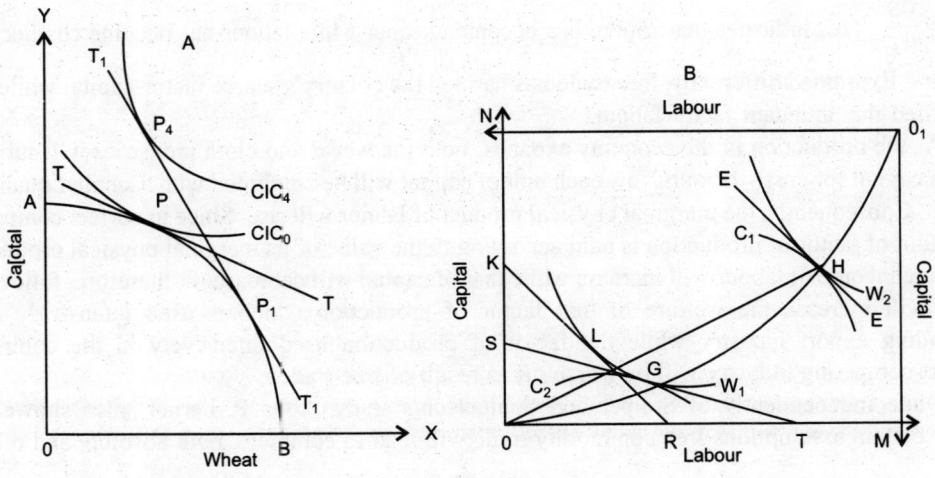

Figure 8.1

corresponding point to point P in the Edgeworth-Bowley box is point G in Figure 8.1B which lies on the efficiency locus 00_1. At point G, the wheat isoquant W_1 and the cloth isoquant C_2 are tangential. The slope of the common tangent line ST at point G shows the ratio at which labour can be substituted for capital.

When trade begins, the country exports wheat whose production is abundant factor (labour) intensive and imports cloth whose production is scarce factor (capital) intensive in the country. Trade will increase the relative price of wheat which is exported by the country and decrease the relative price of cloth which is now imported by the country. Consequently, the production of wheat in the country will increase while that of cloth will decrease. This will mean a movement down along the production possibilities curve AB from point P to P_1 showing that due to trade the production of wheat has expanded while that of cloth has shrunk in the country. The new price-ratio line $T_1 T_1$ in Figure 8.1A touching the production possibilities curve AB at point P_1 is steeper than the pre-trade price-ratio line TT showing that due to trade wheat has become costlier (in terms of cloth) while cloth has become cheaper (in terms of wheat) in the country.

As the production of wheat in the country expands and the production of cloth shrinks, resources are transferred away from the shrinking cloth making industry to the expanding wheat growing industry. However, since the production of cloth is capital-intensive, the cloth industry will release relatively more capital than will be absorbed at the existing remuneration rate in the expanding wheat industry. Consequently, capital's remuneration—interest rate—will have to fall sufficiently in order to absorb the substantial amount of capital released from the shrinking cloth making industry. On the other hand, the expanding wheat industry will want to employ more workers than can be hired in the labour market at the going wage rate. Consequently, labour's wage relative to capital's interest will rise. This is evident from the slope of tangent EE to the wheat and cloth isoquants W_2 and C_1 at point H. The tangent drawn at point H is steeper than the tangent drawn at point G. The EE tangent drawn at point H is parallel to KR line which has been drawn tangent to W_1 isoquant of wheat at point L. The slope of KR tangent drawn at point L is $\dfrac{OK}{OR}$ while the slope of ST tangent drawn at point G is $\dfrac{OS}{OT}$. It is obvious that $\dfrac{OK}{OR}$ is greater than $\dfrac{OS}{OT}$, i.e., $\dfrac{OK}{OR} > \dfrac{OS}{OT}$. This indicates that capital has become cheaper while labour has become costlier in the country. Expressed differently, free trade has harmed the country's scarce factor-capital while it has benefited the abundant factor-labour.

As the production in the economy expands, both the wheat and cloth industries will substitute cheap capital for costly labour. Now each unit of capital will be combined with a smaller quantity of labour. Consequently, the marginal physical product of labour will rise. Since in perfect competition each unit of factor of production is paid according to the value of its marginal physical product, the absolute income of labour will increase while that of capital will decrease. It, therefore, follows that trade will increase the welfare of that factor of production which is used intensively in the expanding export industry while the factor of production used intensively in the contracting import-competing industry will be worse off as result of free trade.

Quite independently of Stolper and Samuelson's study, Abba P. Lerner[7] also showed that under certain assumptions free commodity trade will lead to complete, both absolute and relative,

7. Abba P. Lerner, "Factor Prices and International Trade", *Economica*, February 1952.

factor-price equalisation between the trading nations. Jan Tinbergen, James E. Meade and S. Laursen further developed this proposition.[8] All these studies reached the conclusion that free trade was a complete and not merely a partial substitute for international factor mobility.

This conclusion is at variance with the classical theory of international trade. The Ricardian doctrine of comparative costs implies that free trade equilibrium is perfectly compatible with large and lasting differences in real wages or per capita real income between countries, i.e., it stresses that the equalisation of factor prices is not brought about by free trade, except in very special cases.

The Stolper-Samuelson theorem was subsequently amended and elaborated by Kelvin Lancaster, Llyod A. Metzler and Jagdish Bhagwati. Dropping the assumption of fixed terms of trade, Metzler has shown that the terms of trade effect will cause the internal price of country's exports to rise. Consequently, the return to country's scarce factor will fall as a consequence of trade. If the foreign trade offer curve is inelastic, the terms of trade as a consequence of imposing an import tariff may improve so much that the domestic price of imports falls in the tariff-imposing country causing fall in the domestic production of import-substitutes and a redistribution of earned income in favour of the factor used relatively intensively in the production of exports.[9] Lancaster does not accept the Stolper-Samuelson theorem as a universal proposition. He argues that "protection raises the real wage of the factor which is used relatively more intensively in the imported good."[10] Jagdish Bhagwati also does not accept the universal validity of the theorem. According to him, "protection (prohibitive or otherwise) will raise, reduce or leave unchanged the real wage of the factor intensively employed in the production of good according as protection raises, lowers or leaves unchanged the relative price of that good."[11]

Factor-Price Equalisation Theory

The most important and fascinating corollary of the 'factor-proportions theory' is the famous 'factor-price equalisation theory' which has "held the same fascination for trade theorists as scrabble or rock and roll for the populace at large."[12] The corollary states that international trade in commodities acts as a substitute for the international mobility of factors of production. The free exchange of commodities between any two countries eliminates the pre-trade differences in comparative costs. Equilibrium is achieved at that point where comparative cost differences disappear and the relative commodity prices in both the countries are equalised. The relative commodity prices would, however, become equal only when the relative factor prices are equalised. Thus, the factors of production which were assumed immobile between countries are supposed to move through the commodities.

According to the factor-price equalisation theory, when a country tends to specialise and export those commodities whose production requires relatively large amount of the abundant

8. Jan Tinbergen, "Equalisation of Factor Prices between Free Trade Areas". *Metroeconomica*, April 1949; James E. Meade, "The Equalisation of Factor Prices: The Two-Country, Two-Factor and Three-Product Case", *Metroeconomica*, December 1950; S. Laursen, "Production Functions and Theory of International Trade", *The American Economic Review*, September 1952.
9. Llyod A. Metzler "Tariffs, the Terms of Trade and the Distribution of National Income", *The Journal of Political Economy*, February 1949, pp. 1–29.
10. Kelvin Lancaster, "Production and Real Wages: A Restatement", *The Economic Journal*, June 1957.
11. Jagdish Bhagwati, "Production, Real Wages and Real income", published in *Readings in International Economics*, Penguin Series, pp. 281.
12. Richard E. Caves, *Trade and Economic Structure*, 1960, p. 79.

factor, she indirectly exports its abundant (cheap) factor abroad. Due to the additional pressure of world demand the abundant factor becomes scarce. Consequently, its price rises. Similarly, when a country imports those commodities in which her scarce factor is used intensively, the domestic pressure of demand on such factor is reduced as a result of which the costly factor becomes cheap. This process will continue until the factor prices are equalised in both the countries engaged in the international exchange of goods and services. Thus, free trade in commodities is a substitute for international factor mobility bringing about equalisation in factor prices between the trading countries. The factor-price equalisation theory is based on the following assumptions.

1. There are two countries producing only two goods with only two factors of production (labour and capital).
2. There is free competition in both the countries in the factor and goods markets.
3. Production functions in both the countries are identical and homogeneous of the first degree, i.e., a given percentage change in the quantity of all the inputs causes an equal percentage change in the resulting output.
4. Production functions for both the goods are different such that the production of one commodity is always labour-intensive and that of the other is always capital-intensive regardless of the relative supply of factors and the ratio of factor prices. In other words, there is complete absence of factor intensity reversal. Further, the production function for each good is identical in both the countries.
5. There is absence of complete specialisation, i.e., both the commodities continue to be produced in both the countries even after trade has begun.[13]
6. The factors of production in both the countries are qualitatively identical but are available in different quantities.
7. The number of factors is not greater than the number of commodities, i.e., in a two commodity case there must not be more than two factors of production.
8. The marginal physical product of factors is diminishing, i.e., the production possibilities curve of each country is concave towards the point of origin of the axes.
9. The factor supply curves are perfectly inelastic for each country.
10. Tariff and transport costs are ignored so that the prices of goods are equal between different countries.

Diagrammatical Illustration

Since several writers have worked on the factor-price equalisation theory and different presentations of the theory are available in economic literature, the easiest and simplest method has been discussed below.

For a diagrammatical demonstration of the factor-price equalisation theory let us refer to Figure 8.1A and Figure 8.1B relating to the Stolper-Samuelson theory. In Figure 8.1A, the movement along the production possibilities curve AB is uniquely related with the movement along the contract curve 00_1 in Figure 8.1B. The movement along the production possibilities curve changes the composition of output which in turn changes the commodity price-ratio. Similarly, the movement along the contract curve changes the resource allocation which in turn changes the factor price-ratio. Thus, there is a

---

13. In a two-commodity case this condition does not seem overly restrictive. However, in a multi-commodity model it would require the production of every commodity in every country making the situation very unrealistic indeed.

unique relationship between the commodity price-ratios, factor price-ratios and physical resource endowment ratios. The information collected from Figure 8.1A and Figure 8.1B has been condensed in Figure 8.2A and Figure 8.2B. In Figure 8.2A, the Y-axis measures the commodity price-ratio $\left(\dfrac{P_W}{P_C}\right)$ while the X-axis measures the factor price-ratio $\left(\dfrac{P_L}{P_K}\right)$. These two ratios always change in the same direction. Points P and P_1 show the product price-ratios corresponding to points P and P_1 in Figure 8.1A and the factor price-ratios corresponding to points G and H, in Figure 8.1B. By plotting all such points we generate line RR which shows the commodity-price and the factor-price relationship in the economy.

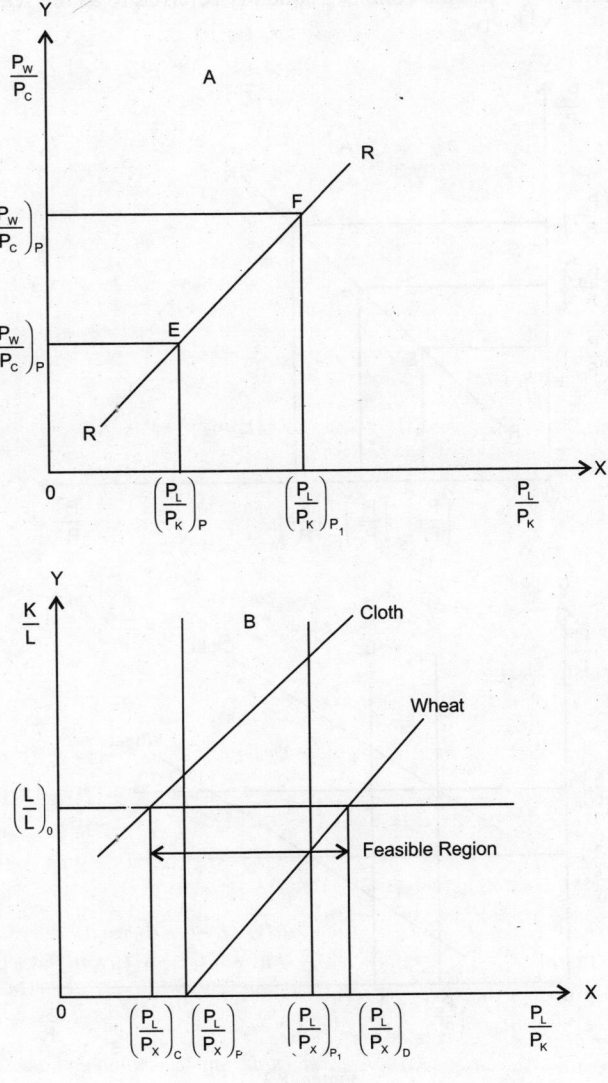

Figure 8.2

Figure 8.2B shows the efficient capital labour ratios (K/L) for different labour/capital price-ratios (P_L/P_K) for the two commodities cloth and wheat. The line for cloth lies above the line for wheat indicating that the production of cloth is more capital-intensive than the production of wheat. The overall capital/labour ratio for the economy as a whole is $(K/L)_0$.

If the factor price-ratio was equal to or lower than $\left(\dfrac{P_L}{P_K}\right)_C$, it would be advantageous for the country to produce only cloth, while at the $\left(\dfrac{P_L}{P_K}\right)_D$ factor price-ratio the country would specialise only in the production of wheat. Factor prices are free to move between the limits set by the factor price-ratios $\left(\dfrac{P_L}{P_K}\right)_C$ and $\left(\dfrac{P_L}{P_K}\right)_D$ This zone is frequently referred to as the feasible region for factor

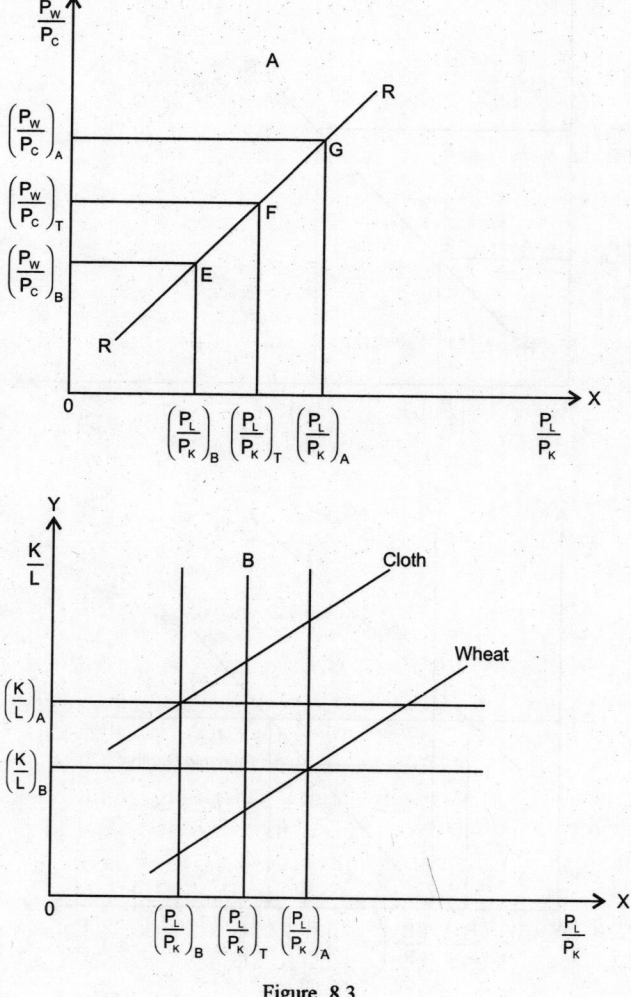

Figure 8.3

price changes, i.e., factor prices can change within this range. The factor-price equalisation theorem can now be proved by drawing Figure 8.3A and Figure 8.3B following the method established by Figs. 8.2A and 8.2B.

There are two countries A and B which are respectively capital-rich and labour-abundant. Both the countries produce the two goods cloth and wheat. The production of cloth is capital-intensive while that of wheat is labour-intensive. The pre-trade wheat-cloth price-ratios in country A and country B as shown in Figure 8.3A are $\left(\dfrac{P_W}{P_C}\right)_A$ and $\left(\dfrac{P_W}{P_C}\right)_B$ respectively. In Figure 8.3B, the capital/labour endowment ratios for country A and country B have been shown by $\left(\dfrac{K}{L}\right)_A$ and $\left(\dfrac{K}{L}\right)_B$ respectively. These two conditions imply a set of factor price-ratios shown along the horizontal axis of both the figures by $\left(\dfrac{P_L}{P_K}\right)_A$ and $\left(\dfrac{P_L}{P_K}\right)_B$. Under these conditions, cloth is cheap in country A while wheat is cheap in country B. Consequently, country A will export cloth while country B will export wheat. International trade will equalise product prices. Since the product prices are directly related to the factor prices, these too will be equalised as a consequence of free trade. The final world product price-ratio with trade taking place between the two countries is $\left(\dfrac{P_W}{P_C}\right)_T$ and the corresponding factor price-ratio is $\left(\dfrac{P_L}{P_K}\right)_T$. As the physical capital-labour ratio remains unchanged on our assumption of given factor supplies in both the countries, the only responsive variables which adjust themselves to product price changes are the factor prices. It is clear that international trade under certain assumptions will lead to complete factor-price equalisation.

Obstacles to Factor-Price Equalisation

The factor-price equalisation theory is based on the following assumptions which are very seldom met with in the real world.

1. The theory assumes complete absence of tariffs and transport costs. In real life, these two factors, however, play a significant role in determining the direction and composition of international trade. Their mere existence will prevent the complete specialisation and equalisation of product prices that free international trade will otherwise bring about. Even Bertil Ohlin who, following Eli Heckscher, enunciated the so-called *Heckscher-Ohlin law of factor-price equalisation,* admitted that actually only a *partial* factor-price equalisation will occur. According to him, except in very special cases, complete equalisation of factor prices would take place only if the factors of production were themselves internationally perfectly mobile.

2. One of the conditions required for the complete equalisation of factor prices is that the countries engaged in international trade should produce all the commodities. In other words, partial specialisation in production is inevitable for the theory to hold good. Thus, incomplete specialisation will rule out the possibility of complete factor-price equalisation.

3. The factor-price equalisation theory is based on the crucial mythical assumption of perfect competition. In the real world, there is either monopoly, oligopoly or monopolistic competition in the product and factor markets. Moreover, if the production of anyone good obeys to the law

increasing returns to scale, the assumption of perfect competition will breakdown invalidating the factor-price equalisation theory.

4. In the case of multi-country and multi-commodity trade, if the number of factors of production exceeds the number of commodities, the system would remain undetermined. Consequently, for complete factor-price equalisation to occur, the number of factors of production should not exceed the number of commodities.

5. The factor-price equalisation theory will breakdown if the relationship between the product prices and factor prices is not identical in all the countries. It will happen if the production functions are not identical in the different countries.

6. In the case of factor intensity reversal the factor-price equalisation theory will be invalidated because a country with abundant capital will export capital-intensive commodity while the labour-abundant country may also export the same commodity produced by using the labour-intensive technique of production. Consequently, factor prices will differ between the two countries. The case of factor intensity reversal has been discussed below.

Assume the two countries A and B. In Figure 8.4, the factor price-ratio line of country A is AB while that of country B is A_1B_1. The slope $\dfrac{0A}{0B}$ of the factor price-ratio line AB is greater than the

slope $\dfrac{0A_1}{0B_1}$ of the factor price-ratio line A_1B_1 showing that capital is cheap in country A while labour is cheap in country B. Country A may produce cloth with the factor proportion expressed by $0T$ and wheat with the factor proportion represented by $0S$. Thus, the production of cloth is capital-intensive while that of wheat is labour-intensive. But below and to the right of the radiant of tangency $0P$, the capital-intensive commodity cloth is labour-intensive and the labour-intensive commodity wheat is capital-intensive at the factor price-ratio line A_1B_1. In this case, it is not clear

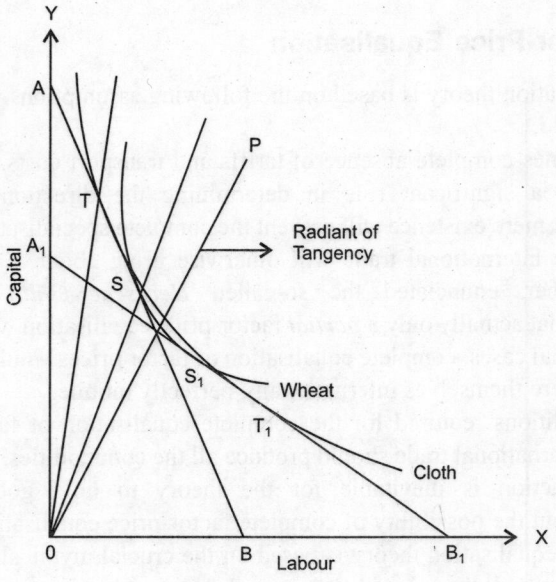

Figure 8.4

which country will produce and export which commodity. This is the case of factor intensity reversal. The factor-price equalisation theory assumes strong factor intensity resulting in the complete absence of factor intensity reversal. Consequently, it cannot explain such situations.

This happens when there is a wide range of factor substitution in at least one of the commodities. In Figure 8.4, the range of substitution for cloth is TT_1 which is wider than the range of substitution SS_1 for wheat. The substitution elasticities of the two production functions are not identical. In the case of the non-identical substitution elasticities factor intensity reversal will occur.

7. The factor-price equalisation theorem is a static type of theory. It only states what the effects of trade will be with given factor supplies and technology in the two countries. In the real world, however, all sorts of changes occur and it is very seldom, if at all, in a state of given equilibrium for ever. Factor supplies and technology change over time. For example, if capital is country's abundant factor, trade will raise its return. Trade will also raise the real income of the entire community. As a result of higher real income of the community and higher reward for capital brought about by trade, savings—capital supply—in the country will increase. The increase in capital's supply by bidding down the interest rates will encourage the growth of capital-intensive industries in the economy. Such a development is conducive to technological progress which will cause further rise in the community's real income.

It suggests that the capital-rich countries will find themselves becoming still richer in capital with capital becoming available at relatively cheaper cost. Conversely, countries with abundant labour and scarce capital may find their poverty increased due to labour's wages being reduced as a result of high rate of population growth. The increase in wages as a result of trade in the labour-abundant underdeveloped countries combined with workers' ignorance and illiteracy would result in rapid population growth. Consequently, even with the increased specialisation in the production of labour-intensive goods, wages would rise very slowly. It suggests that under the conditions of changing factor supplies and technology, factor prices may become more unequal rather than become more equal as a consequence of trade. This conclusion seems to be supported by the historical fact of growing inequality in the per capita income between the rich and poor countries although in both the sets of countries the per capita income and wages have been rising. It is so because rise in the per capita income has been faster in the rich countries than in the poor ones.

The real world is so dynamic that most variables frequently change and many other factors at work may tend to increase, rather than to narrow, the existing factor price differences. In a fast moving and changing economy, international trade will at best merely tend to equalise the factor prices. This is very different from asserting that as a result of international exchange of goods and services complete equalisation of factor prices between different countries will occur. It could also be that while trade may have the effects on factor prices in conformity with the prediction of the factor price equalisation theorem, some other factors not considered by this theory may work in the opposite direction thwarting the conclusion of the factor-price equalisation theory.

Moreover, the factor-price equalisation theory does not say that trade will equalise incomes. Wages do not constitute the only source of income. People must also own capital and capital also commands price. If a country has more capital for a given amount of labour, higher will be the average income. Consequently, as long as capital stocks in different countries differed, incomes would also differ even if the factor rewards were completely equalised. Robert Heller has correctly stated that "the effect of international trade may be considered as a "leaning against the wind", in the sense that factor price differentials would be even larger in the absence of trade."[14]

14. H. Robert Heller, *International Trade Theory and Empirical Evidence*, Second Edition, 1973, p. 139.

Suggested Readings

Bertil Ohlin, *Interregional and International Trade,* Cambridge, Mass., Harvard University Press, 1933, especially Chapters 5–8.

R.W. Jones, "Factor Proportions and the Heckscher-Ohlin Theorem", *The Review of Economic Studies,* January 1956.

Richard E. Caves, *Trade and Economic Structure,* 1960, Chapter 3.

P.T. Ellsworth and J. Clark Leith, *The International Economy,* Fifth Edition, 1975, Chapter 6, pp. 100–103.

H. Robert Heller, *International Trade,* Second Edition, 1973, Chapter 7.

Harry G. Johnson, "Factor Endowments, International Trade, and Factor Prices", *The Manchester School of Economics and Social Studies,* Volume XXV, September 1957.

Charles P. Kindleberger, *International Economics,* Fifth Edition, 1973, Appendix B.

Abba P. Lerner, "Factor Prices and International Trade", *Economica,* February 1952.

Paul A. Samuelson, "International Trade and the Equalisation of Factor Prices", *The Economic Journal,* Volume LVII, June 1948.

Paul A. Samuelson, "International Factor Price Equalisation Once Again", *The Economic Journal,* Volume LIX, June 1949, reprinted in American Economic Association's *Readings in International Economics,* 1968.

Paul A. Samuelson, "A Comment on Factor Price Equalisation", *The Review of Economic Studies,* 1951–52.

Paul A. Samuelson, "Prices of Factors and Goods in General Equilibrium", *The Review of Economic Studies,* Volume 21, No. 1, 1953.

Bo Sodersten, *International Economics,* Second Edition, Reprinted 1981, Chapter 4.

Questions

1. Under what conditions will free international trade in commodities tend to equalise the factor prices in the two trading countries? Explain fully.
2. Discuss critically the factor-price equalisation theorem.

CHAPTER 9

Supply and Demand Equilibrium in International Trade

In this chapter we shall deal with the problem of integration of different variables relating particularly to supply and demand forces which determine the equilibrium in international trade under the assumption that trade takes place between two countries and in two commodities which are produced by the two factors of production, capital and land. The discussion of equilibrium in the previous chapters was restricted since a few variables were assumed to be constant and given in order to analyse the effects of other variables. In this chapter, we shall study the interaction of several variables within the framework of static general equilibrium analysis.

To analyse the conditions of demand, we have already developed the concept of the community indifference curve. We shall now present a detailed discussion of the community indifference curve. On the supply side, we have already developed the concept of the efficiency locus and the production possibilities curve. However, the concept of the trade offer curve, the most important tool of analysis, has been simply touched upon in Chapter 6. We shall now exhaustively dwell upon the trade offer curves in order to elucidate the equilibrium between imports and exports. We shall also discuss the concept of the trade indifference curve developed by James Meade.[1] Lastly, we shall assemble together all the tools to analyse the general supply and demand equilibrium in international trade.

Community Indifference Curves

Community indifference curves have been extensively used in the theory of international trade to demonstrate how the countries reach equilibrium before trade and after trade and how they are benefited by trade. When there is only one consumer in society there is no difference between an individual's indifference curve and a community indifference curve. However, when there are more than one consumer in the economy there is a fundamental difference between these two curves. The derivation of a community indifference curve, assuming that there are two consumers X and Y and two commodities wheat (W) and cloth (C), has been explained below and has been illustrated in Fig 9.1.

In Figure 9.1A, the total quantities of wheat and cloth available in the country have been shown by rectangle XC′YW (or YCXW′). The XW (or YW′) side of the rectangle shows the total quantity of wheat while the XC′ (or YC) side represents the total amount of cloth in the country.

---

1. James E. Meade, *A Geometry of International Trade,* Chapter 2; and Charles P. Kindleberger, *International Economics,* Fifth Edition, 1973, Appendix C.

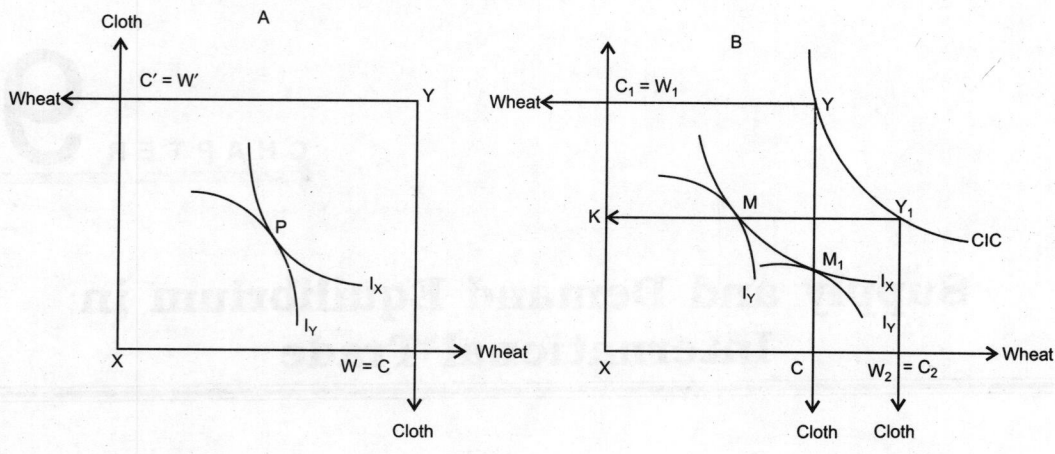

Figure 9.1

The indifference curves I_x of individual X and I_y of individual Y are tangential at point P which shows the equilibrium conditions of the two consumers.

Let us slide the indifference curves I_x and I_y such that the two coordinate systems stay parallel to each other as shown in Figure 9.1B. Keeping the coordinate of X's indifference curve I_x constant, let us slide Y's indifference curve I_y whose initial position is Y. In the initial position, both the consumers are in equilibrium at point M which is located on indifference curves I_x and I_y. Let us now move the coordinate system and the indifference map downward to the right in such a way that Y's indifference curve I_y is tangent to X's indifference curve I_x at point M_1. At point M_1 although the total satisfaction derived by the two consumers is the same but the commodity combination has changed. The moving points Y and Y_1 generate a path known as the community indifference curve (CIC). This curve shows the various alternative possible combinations of cloth and wheat between which the two consumers are indifferent. Thus, a community indifference curve shows all those different commodity combinations which yield constant utility to the members of the community collectively.[2]

It is also possible to draw some inference about the slope of the community indifference curve. In Figure 9.lB, the two individual indifference curves have identical slopes at points M and M_1. The marginal rates of substitution of cloth for wheat are identical for both the consumers individually and, therefore, they are also the same as the marginal rate of substitution for the two consumers collectively. Thus, the slope of the community indifference curve CIC of individuals X and Y, which together constitute the community, at points Y and Y_1 are identical to the slopes of the individual indifference curves at points M and M_1.

2. While the indifference map of an individual is conceptually satisfactory, the same cannot be said about the community indifference map. Difficulties arise in the community indifference curve from the problem of interpersonal comparison of loss and gain of satisfaction experienced by the different individuals in the community. For example, while one individual may say meaningfully that he is better off than he was before by possessing 5 more cigarettes and 3 biscuits less, the same cannot legitimately be said about the community because while some members of the community lose biscuits others gain cigarettes. There is no method free from objection by which it can be said that the increase in the satisfaction of one group of consumers is greater than the decrease in the satisfaction of the other group of consumers. It is not possible to make any meaningful interpersonal comparison of welfare.

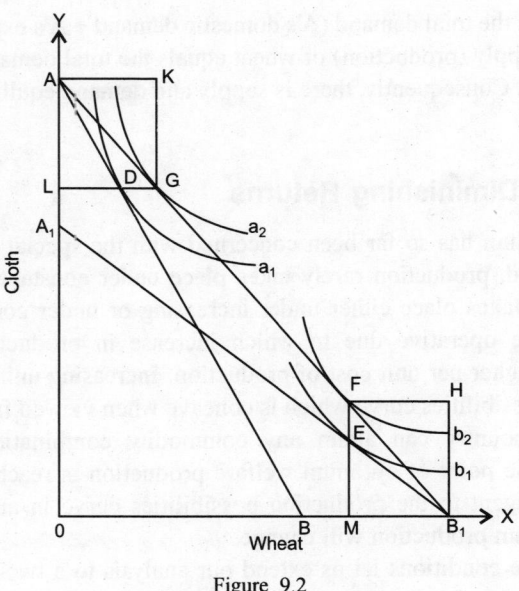

Figure 9.2

Equilibrium in Terms of the Community Indifference Curves and the Terms of Trade

Given the production possibilities curves and the terms of trade, the community indifference curves determine the autarky or pre-trade and the post-trade equilibrium. The condition of equilibrium under constant cost has been discussed in the following pages.

There are two countries A and B which produce two commodities cloth and wheat under conditions of constant costs. In Figure 9.2, AB is the production possibilities curve of country A while that of country B is A_1B_1. In autarky, the equilibrium point of country A is D where her production possibilities curve AB is tangent to her community indifference curve a_1. Similarly, the pre-trade equilibrium point of country B is E where country B's production possibilities curve A_1B_1 is tangent to her community indifference curve b_1.

After the opening of trade, country A specialises completely in the production of cloth because she commands comparative cost advantage in the production of cloth. For similar reason, country B specialises in the production of wheat. After complete specialisation takes place in both the countries in production, the production points of countries A and B are respectively A and B_1. By joining these two points we obtain the line AB_1 as the international exchange-ratio (terms of trade) line. After trade takes place, equilibrium is achieved when country A consumes at point G and country B consumes at point F. At these two points the international terms of trade AB_1 is tangent to the higher community indifference curve a_2 of country A and the commuunity indifference curve b_2 of country B. The exports of each country are just sufficient to pay for her imports. In other words, at these terms of trade the exports of one country are equal to the imports of the other country. For example, the exports of country A are AL (= KG) quantity of cloth which is equal to MF (= HB_1) imports of cloth of country B. Similarly, country A imports AK (= LG) quantity of wheat which is equal to MB_1 (= FH) exports of wheat of country B. In other words, at the international terms of trade determined by the slope of line AB_1 the total supply (production) of cloth in country *A* equals the total demand (A's domestic demand + A's exports) for cloth. Similarly, in country B her total supply (production) of wheat equals the total demand (B's domestic demand + B's exports) for wheat. Consequently, there is supply and demand equilibrium in the hypothetical two-country world.

Equilibrium under Diminishing Returns

The analysis of equilibrium has so far been concerned with the special case of constant costs or returns. In the real world, production rarely takes place under constant costs and even if in the early stages production takes place either under increasing or under constant returns, eventually increasing costs become operative due to which increase in production is only possible by incurring successively higher per unit cost of production. Increasing unit cost of production gives rise to the production possibilities curve which is concave when viewed from the point of origin of the axes. Although a country can attain any commodity combination along its production possibilities curve but the point of optimum welfare production is reached when the community indifference curve is tangent to the production possibilities curve in autarky. When trade takes place the point of optimum production will change.

To analyse the above conditions let us extend our analysis to a two-country, two-commodity and two-factor model. Let us assume that there are two countries F and G with different factor endowments but with identical consumer tastes. Let us further assume that they produce two commodities wheat and cloth and that due to her position of factor endowments country F has comparative cost advantage in the production of wheat while country G has comparative cost advantage in the production of cloth. The production possibilities curves of both the countries have been shown by AB in Figure 9.3A and in Figure 9.3B by A_1B_1.

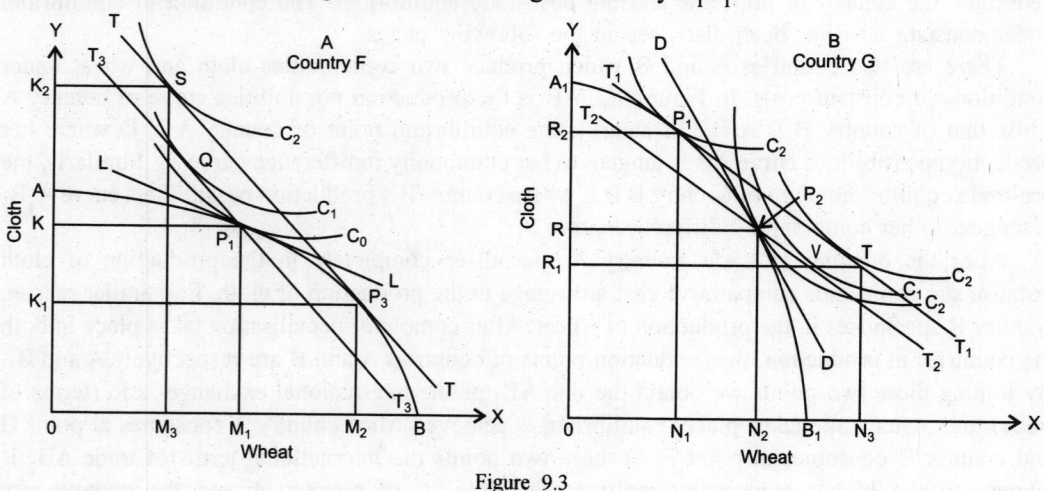

Figure 9.3

In autarky, as shown in Figure 9.3A, F country's equilibrium takes place at point P_1 where the community indifference curve C_0 is tangential to the production possibilities curve AB. Thus, there is simultaneous equilibrium both in the factor and the commodity markets. In the factor market, the ratio of the marginal products of labour and capital in each industry is identical and the price of each factor equals the value of its marginal product. In the commodity market, the price ratio equals the marginal cost ratio as shown by the equality of the slope of the domestic price-ratio line LL with the slope of the production possibilities curve AB at point P_1). At point P_1, the LL price-ratio line is also tangent to the community indifference curve C_0 which shows that the marginal rate of commodity substitution is equal to the commodity price ratio. Thus, at point P_1 the rate at which the consumers want to 'trade off' cloth against wheat is identical to the commodity exchange ratio. Similar condition as shown in Figure 9.3B holds for country G at point P_2. However, the slope of the autarky price-ratio line DD of country G is different from the slope of the autarky price-ratio line LL of country F showing that cloth is cheaper in country G while wheat is cheaper in country F. The existence of relative cheapness of the two commodities in each country provides scope for international trade.

International trade establishes a new exchange ratio line TT which is parallel to T_1T_1 in Figure 9.3B. At the international terms of trade, consumers in both the countries will gain by mere exchange of commodities (ignoring the reaction of producers). Country F will move along the exchange-ratio line T_3T_3 which is parallel to the international terms of trade line TT and which passes through the initial production equilibrium point P_1. Similarly, country G will move along the exchange-ratio line T_2T_2 which is parallel to T_1T_1 and which passes through the initial production equilibrium point P_2. Due to the international exchange of commodities, both the countries will move from their initial pre-trade equilibrium points P_1 and P_2 to the new equilibrium points Q and V which are located on the higher community indifference curve C_1. This movement from the lower community indifference curve C_0 to the higher community indifference curve C_1 shows the gain from trade accruing to both the countries with no change in production—the total output mix in both the countries remains pegged to the autarky position.

The producers in both the countries will, however, react to the new situation by re-allocating resources to earn the maximum profit. Since after trade the price of wheat is higher in country F, producers will expand the production of wheat by shifting resources away from the production of cloth to the production of wheat till the new equilibrium point P_2 where the international exchange-ratio is equal to the marginal rate of substitution, is reached. Similarly, producers in country G will expand the production of cloth and will move to point P_1. After exchange both the countries will move to the consumption equilibrium points S and T which are located on the still higher community indifference curve C_2.

Thus, the movement from Q to S and from V to T is the gain from trade due to the increased efficiency in production resulting from specialisation. At point S, the total exports of country F— M_2M_3 quantity of wheat—are equal to the total imports of country G—N_1N_3 quantity of wheat— and the imports of country F—K_1K_2 quantity of cloth—are equal to the exports of country G— R_1R_2 quantity of cloth. At points S and T the marginal rate of transformation is equal to the marginal rate of commodity substitution, and both these rates are equal to the international exchange-ratio. Consequently, both the countries are in equilibrium. As a consequence of trade, country F consumes M_3S ($=OK_2$) quantity of cloth and $0M_3$ ($= K_2S$) quantity of wheat while country G consumes R_1T ($= 0N_1 + N_1N_3$) quantity of wheat and TN_3 ($= 0R_1$) quantity of cloth.

Equilibrium in Terms of the Trade Offer Curves

So far we have assumed that the international equilibrium terms of trade were given. In fact, they are not given but are determined by the interaction of the trade offer curves of the countries concerned. A trade offer curve, on the one hand, denotes the willingness of a country to offer a certain amount of a particular commodity for a given amount of another commodity and on the other, it shows the willingness of a country to demand the product of the other country at different possible terms of trade or exchange ratios determined by the slope of the different price-ratio lines. Thus, a trade offer curve combines the elements of both the supply and the demand. This is why it is known on the one hand as an offer curve and on the other as a reciprocal demand curve. Figure 9.4 shows country A's trade offer curve 0A which shows the different amounts of wheat which country A offers at different prices (barter or real exchange ratios) for the given amounts of cloth. The derivation of trade offer curve, otherwise known as offer curve, under constant cost conditions shown in Figure 9.4 has been discussed further.

In Figure 9.4, the X-axis measures country A's exports (wheat) while the Y-axis measures country A's imports (cloth). The slopes of the price-ratio lines 0E, 0D, 0C and 0B respectively show the different relative prices of wheat in terms of cloth. In autarky, country A consumes at point M_1 which shows that for a very small amount of imports country A is indifferent to trade. However, if country B offers her better terms of trade as shown by the exchange-ratio line 0D, the consumers in country A can increase their aggregate satisfaction by exchanging M_2X_2 quantity of cloth against $0X_2$ quantity of wheat. At still more favourable terms of trade represented by the slope of 0C exchange-ratio line the consumers of country A would be willing to offer $0X_3$ quantity of wheat for M_3X_3 quantity of cloth. Ultimately, a point such as M_4, will be reached when country A will be unwilling to offer (export) any more wheat for any additional imports of cloth. The 0A curve generated by joining the several different equilibrium points like M_1, M_2, M_3 and M_4 is known as the trade offer curve of country A. The trade offer curve of a country shows the different quantities of her exports which a country is willing to exchange for the different quantities of imports at the different barter terms of trade.

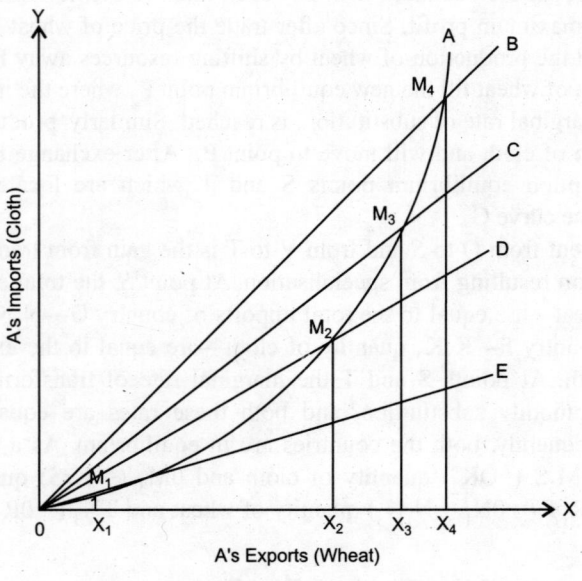

Figure 9.4

Similarly, we can derive the trade offer curve of country B as shown in Figure 9.5 where the different combinations of exports (cloth) and imports (wheat) at different relative prices shown by the slopes of price-ratio lines 0E', 0D', 0C' and 0A' have been shown by points M'_1, M'_2, M'_3 and M'_4. The locus of all these points—0B' curve—is known as the trade offer curve of country B. The trade offer curves of country A and country B represented by 0A and 0B′ in Figure 9.4 and Figure 9.5 respectively intersect each other at point P as shown in Figure 9.6. The line obtained by joining the points 0 and P is the equilibrium international terms of trade line. At the equilibrium terms of trade shown by the slope of the exchange-ratio line 0P, i.e., at the $X_1P/0X_1$ commodity exchange-ratio, country A is willing to export the same amount of wheat which country B is willing to import at this exchange-ratio. Similarly, at this exchange-ratio the total amount of cloth which

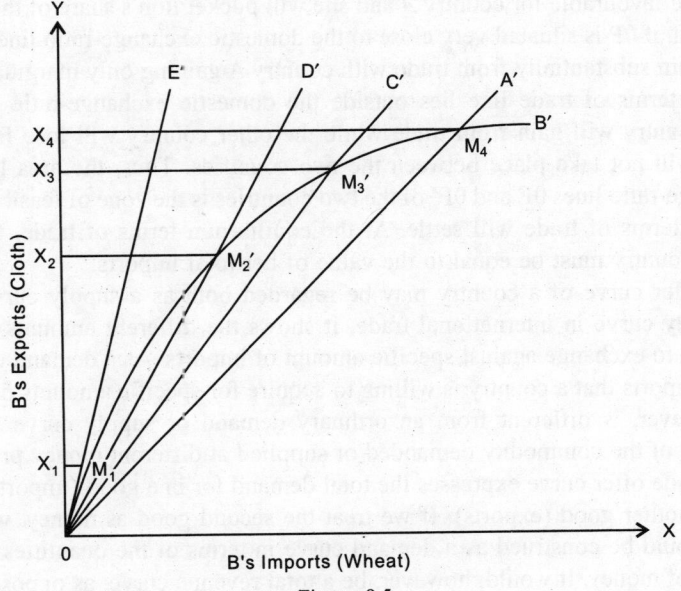

Figure 9.5

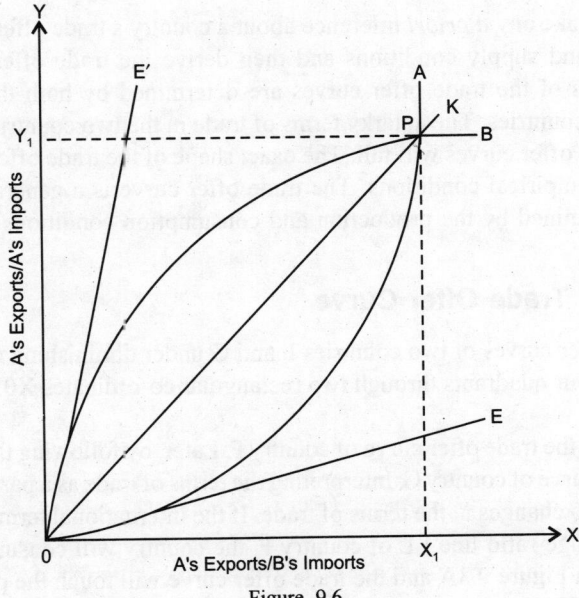

Figure 9.6

country B would import equals the total amount of cloth which country B would export, i.e., $0X_1$ (= Y_1P) units of wheat are equal to $0Y_1$ (= X_1P) units of cloth. When the international terms of trade line lies between the domestic exchange-ratios of both the countries, as represented by the slopes of price-ratio lines 0E and 0E', both the countries will gain from trade although one country may get a larger share of the total gain resulting from trade than the other country. For example, if the international terms of trade line 0P lies closer to B's domestic exchange-ratio line 0E', the terms of trade will be more favourable for country A and she will pocket lion's share of the gain from trade. On the other hand, if 0P is situated very close to the domestic exchange-ratio line 0E of country A, country B will gain substantially from trade with country A gaining only marginally from trade. If the international terms of trade line lies outside the domestic exchange-ratio lines of both the countries, one country will gain from trade while the other country will lose from trade. In this situation, trade will not take place between the two countries. Thus, the area lying between the domestic exchange-ratio lines 0E and 0E' of the two countries is the zone of feasibility within which the international terms of trade will settle. At the equilibrium terms of trade, the value of total exports of each country must be equal to the value of her total imports.

The trade offer curve of a country may be regarded both as a supply curve and a demand curve. As a supply curve in international trade, it shows the different amounts of exports that a country is willing to exchange against specific amount of imports. As a demand curve, it indicates the amounts of imports that a country is willing to acquire for specific amounts of exports. A trade offer curve, however, is different from an ordinary demand or supply curve which shows the various quantities of the commodity demanded or supplied at different money prices, *other things being equal.* A trade offer curve expresses the total demand for one good (imports) in terms of the total supply of another good (exports). If we treat the second good as money, which is possible, the offer curve could be construed as a demand curve in terms of the quantities of goods against the total amount of money. It would, however, be a total revenue curve, as opposed to an ordinary demand curve which relates the different quantities of a good to different prices or average revenue per unit.

It is difficult to make any *a priori* inference about a country's trade offer curve. We must study a country's demand and supply conditions and then derive the trade offer curve in the manner described. The shapes of the trade offer curves are determined by both the supply and demand conditions in the two countries. The autarky terms of trade in the two countries determine the limits within which the trade offer curves will fall. The exact shape of the trade offer curve will depend on the prevailing given empirical conditions. The trade offer curve is a general equilibrium concept which is jointly determined by the production and consumption conditions.

Derivation of the Trade Offer Curve

To derive the trade offer curves of two countries F and G under diminishing returns let us construct a plane divided into four quadrants through two rectangular co-ordinates X0X' and Y0Y' as shown in Figure 9.7.

First, let us derive the trade offer curve of country F. Later, by following the same pattern we can derive the trade offer curve of country G. Interpreting the terms of trade as a parameter we consider the reaction of country F to changes in the terms of trade. If the international terms of trade are identical to the domestic exchange-ratio line LL of country F, the country will consume and produce at the equilibrium point P_1 in Figure 9.3A and the trade offer curve will touch the point of origin 0 of the quadrant in Figure 9.7. If, however, the international terms of trade differ from the domestic exchange-

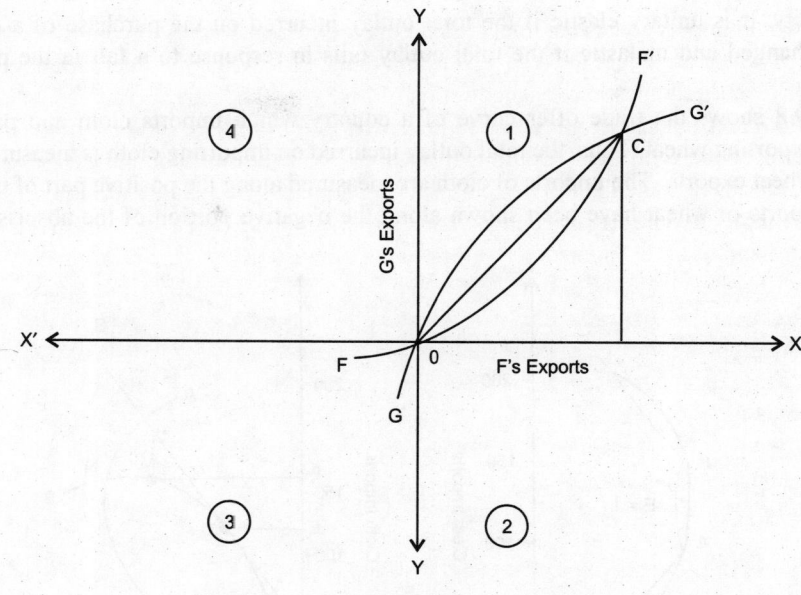

Figure 9.7

ratio shown by the domestic price-ratio line *LL*, the pattern of equilibrium consumption and production of the country will be altered. If the international terms of trade line is steeper than the domestic exchange-ratio line, country F will expand the production of wheat and export it and will contract the production of cloth and import it. If, however, the international terms of trade line is less steep than the domestic exchange-ratio line LL the position of exports and imports in country F will be reversed. If we plot the different quantities of exports and imports which become available due to steeper slope of the international price-ratio line in the first quadrant of Figure 9.7 and the different quantities of exports and imports that become available due to less steep slope of the international price-ratio line in the third quadrant, we will obtain FF′ curve which is known as the trade offer curve of country F. Similarly, we can derive from Figure 9.3B, the GG′ trade offer curve of country G shown in Figure 9.7. The two trade offer curves intersect in the first quadrant at point *C* where the total exports of country F are equal to the total imports of country G and the total exports of country G equal the total imports of country F. Since the trade offer curves of country F and country G will never intersect in the third quadrant, the portions of the trade offer curves abutting in the third quadrant of the figure are redundant. The 0C line joining the point of intersection C of the two trade offer curves with the point of origin 0 is the equilibrium terms of trade line.

Elasticity of the Trade Offer Curve

A trade offer curve is different from the conventional price demand curve which shows the relationship between changes in the price and amount demanded of the commodity. It does not by itself reveal the total outlay incurred on the commodity. A trade offer curve, however, shows the total outlay (exports) made in order to obtain the different amounts of imports at different relative prices or terms of trade. Thus, in a sense a trade offer curve is a total outlay curve.

It is easy to make statement about the elasticity of a trade offer curve. A demand curve for a commodity is elastic when a given price reduction results in the increased total outlay incurred on the commodity. It is unitary elastic if the total outlay incurred on the purchase of a commodity remains unchanged and inelastic if the total outlay falls in response to a fall in the price of the commodity.

Figure 9.8 shows the trade offer curve of a country which imports cloth and pays for her imports by exporting wheat. Thus, the total outlay incurred on importing cloth is measured in terms of the total wheat exports. The imports of cloth are measured along the positive part of the ordinate while the exports of wheat have been shown along the negative portion of the abscissa.

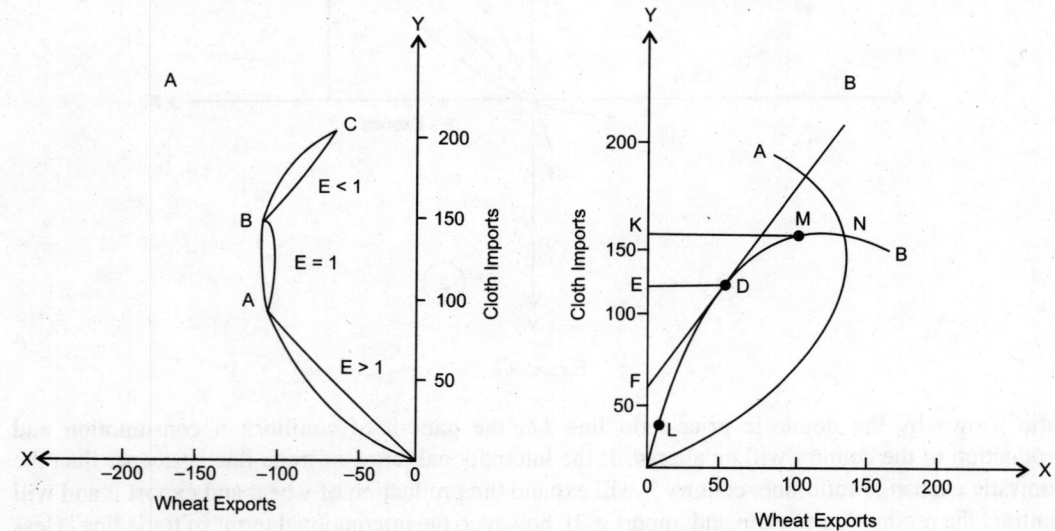

Figure 9.8

In Figure 9.8A, the 0C trade offer curve is elastic over the 0A range because over this range (portion) the total exports of wheat (outlay incurred on imports of cloth) increase as the quantity of cloth imported increases in response to a fall in the relative price of cloth. Over the AB portion, the trade offer curve is unit-elastic because the total exports of wheat (total outlay) remain constant. It shows that the country's elasticity of demand for cloth imports is unitary. Beyond B the trade offer curve is inelastic because increasing amount of imports of cloth can be obtained for an increasingly smaller total outlay measured in terms of the total exports of wheat. By treating the trade offer curve as the total outlay curve, the definition of elasticity becomes very clear. Moreover, the total expenditure incurred on cloth imports is measured in terms of wheat—the commodity exported— and not in money terms. The elasticity of 0C trade offer curve in Figure 9.8A over the three different parts may be expressed in the following form.

Elasticity of the Trade Offer Curve

Range	If on fall in cloth price the total outlay in terms of wheat	The trade offer curve will be
0A	Increases	Elastic
AB	Constant	Unit Elastic
BC	Decreases	Inelastic

Figure 9.8B depicts the method of measurement of the elasticity of a trade offer curve. In the figure, 0A is the trade offer curve of country A while 0B is the trade offer curve of country B. We measure the elasticity of trade offer curve 0B at different points. At point D, DE is the perpendicular drawn from point D on the Y-axis and DF is the tangent on the trade offer curve at point D. The elasticity at point D on the 0B trade offer curve is given by the value of 0E/0F, i.e., the elasticity of trade offer curve 0B at point D is equal to the

$$\frac{\text{Distance from poi}}{\text{Distance from the point of origin of axes t}}$$

The elasticity of trade offer curve 0B at point M where the perpendicular and the tangent overlap each other is unity. At point L the elasticity of the trade offer curve is infinity because the perpendicular is divided by zero. The concept of elasticity here is based on the following familiar formula:

$$\text{Elasticity (E)} = \frac{\text{AR}}{\text{AR} - \text{MR}}$$

At point D, AR (average revenue) is equal to 0E/ED and MR (marginal revenue) is equal to EF/ED. Thus, the elasticity of 0B trade offer curve at point D is equal to

$$E_D = \frac{\dfrac{0E}{ED}}{\dfrac{0E}{ED} - \dfrac{EF}{ED}}$$

$$= \frac{0E}{0E - EF}$$

$$= \frac{0E}{0F} > 1.$$

The elasticity of the trade offer curve at point M is unity while it is infinity at point L. Thus, the elasticity of a normal trade offer curve falls as we move up along the trade offer curve.

Equilibrium in Terms of the Trade Indifference Curves

While discussing the equilibrium in terms of the trade offer curves, we have shown the relationship between the consumption indifference curves, production possibilities curves and trade offer curves of the two countries and have dealt with the problem of equilibrium in terms of the trade offer curves. In this part, the technique of the trade indifference curves developed by James E. Meade has been used to deal with the problem of general equilibrium in international trade.

A trade indifference curve shows the different amounts of foreign trade, i.e., the different export-import combinations which yield the same utility or satisfaction to the residents of a country. In other words, it is the schedule of all those foreign trade (export-import) combinations between which the residents of a country are indifferent since all these foreign trade combinations are iso-utility combinations. To illustrate the above, let us construct a plane divided into four quadrants through two rectangular coordinal axes X0X' and Y0Y' as shown in Figure 9.9. On the horizontal axis, we measure the quantity of wheat with points situated to the left of the point of

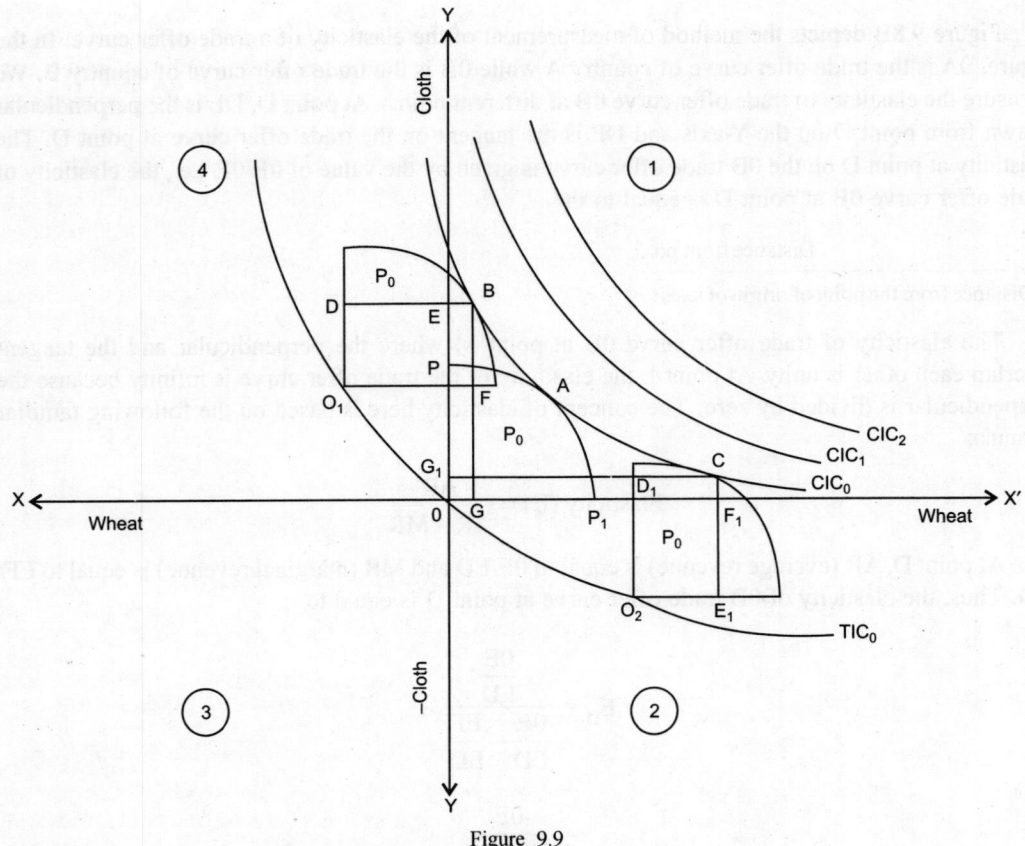

Figure 9.9

origin of the axes 0 showing the negative quantities of wheat. Similarly, the vertical axis measures the quantity of cloth with points below the point of origin of the axes 0 indicating the negative quantities of cloth. The negative quantities denote the exports of the country.

A set of community indifference curves CIC_0, CIC_1 and CIC_2 has been drawn in the first quadrant of the figure. A production block shown by P_0 which is tangent to the highest possible indifference curve CIC_0 at point A which shows the equilibrium consumption and production pattern of the country in autarky has also been drawn.

Now slide the production block P_0 such that it remains tangent to the community indifference curve CIC_0 while the co-ordinates of the moving production block remain parallel to the original coordinate system. In the figure, the sliding production block P_0 is tangent to the community indifference curve CIC_0 at points A, B and C. Since all these points lie on the same community indifference curve CIC_0, consumers in the country remain indifferent between these three positions. Corresponding to these three points, there are the three points of origin 0, 0_1 and 0_2 of the ordinate system which show the different combinations of exports and imports between which consumers of the country are indifferent in the sense that the different combinations of exports and imports represented by these points enable the residents of the country to attain the same aggregate utility or satisfaction. The locus of these three and all other such points is called the trade indifference curve. At all the corresponding points located on the production possibilities curve, on the community indifference curve and on the trade indifference curve the marginal rate of transformation in production equals the marginal rate of substitution in consumption which equals the marginal rate of export-import substitution. From this it follows that the slope of the community indifference curve CIC_0 at points A, B and C equals the slope of the production possibilities curve at these points. The slope of the community indifference curve and the slope of the production possibilities curve at points A, B and C equals the slope of the trade indifference curve TIC_0 at the corresponding points 0, 0_1 and 0_2, Similarly, we can draw the trade indifference curve corresponding to the community indifference curves CIC_1 and CIC_2. Since corresponding to each community indifference curve there is a particular trade indifference curve, we can generate the trade indifference curves corresponding to the community indifference curves CIC_1 and CIC_2.

At point B, the residents of the country consume GB quantity of cloth. Of this, the GF quantity is imported and the FB quantity is domestically produced. The residents of the country consume EB quantity of wheat which is domestically produced. At point B, the country exports DE quantity of wheat since it produces the BD quantity of wheat which is far in excess of her total domestic consumption (EB) of wheat. At point C, the residents of the country consume G_1C quantity of wheat. Of this, the G_1D_1 quantity is imported and the balance quantity D_1C is domestically produced. The residents of the country consume CF_1 quantity of cloth which is produced in the country. The country exports F_1E_1 quantity of cloth at point C since her total domestic production of cloth (CE_1) exceeds her total domestic consumption of cloth (CF_1).

Derivation of the Trade Offer Curve from the Trade Indifference Curves

We have seen that there is a trade indifference curve corresponding to each community indifference curve. Consequently, corresponding to a community indifference map there is a trade indifference map. The country is better off as a result of trade if after trade she reaches on to a higher trade indifference curve. Given the term of trade and the trade indifference curves, we can construct a trade offer curve of the country as shown in Figure 9.10.

In Figure 9.10, the trade indifference curves of country A have been shown by T_0, T_1, T_2 and T_3. The terms of trade have been shown by the slopes of lines D_0, D_1, D_2 and D_3 passing through the point of origin 0. As the terms of trade or price-ratio lines become steeper, country A obtains a larger quantity of imports for a given quantity of her exports. It means that the exports of the country become costlier in terms of her imports or that her imports become cheaper in terms of her exports. The optimal combinations of exports and imports of the country consistent with her consumption and production pattern as well as with her trading opportunities are represented by points M, M_1, M_2 and M_3 where the terms of trade lines D_0, D_1, D_2 and D_3 are tangent to the trade indifference curves T_0, T_1, T_2 and T_3 respectively. By joining the points of optimum combinations of exports and imports M, M_1, M_2 and M_3 with the point of origin 0, we obtain the trade offer curve $0A$ of country A. Similarly, we can derive the trade offer curve 0B of country B. Equilibrium in trade will be achieved at that point where the trade offer curves of the two countries intersect. As shown in Figure 9.11, the trade indifference curves of country A are D_1, D_2 and D_3 and that of country B are E_1, E_2 and E_3 with opposite curvatures. Line KK_1 joining the points of tangency P_1, P_2, and P_3 of these two sets of trade indifference curves is known as the contract curve. The equilibrium terms of trade lies on the line which joins the point of origin 0 to the point of intersection P_2 of the trade offer curves of the two countries. At the equilibrium terms of trade shown by the slope of line OP_2, the total value of exports of country A is equal to the total value of her total imports.

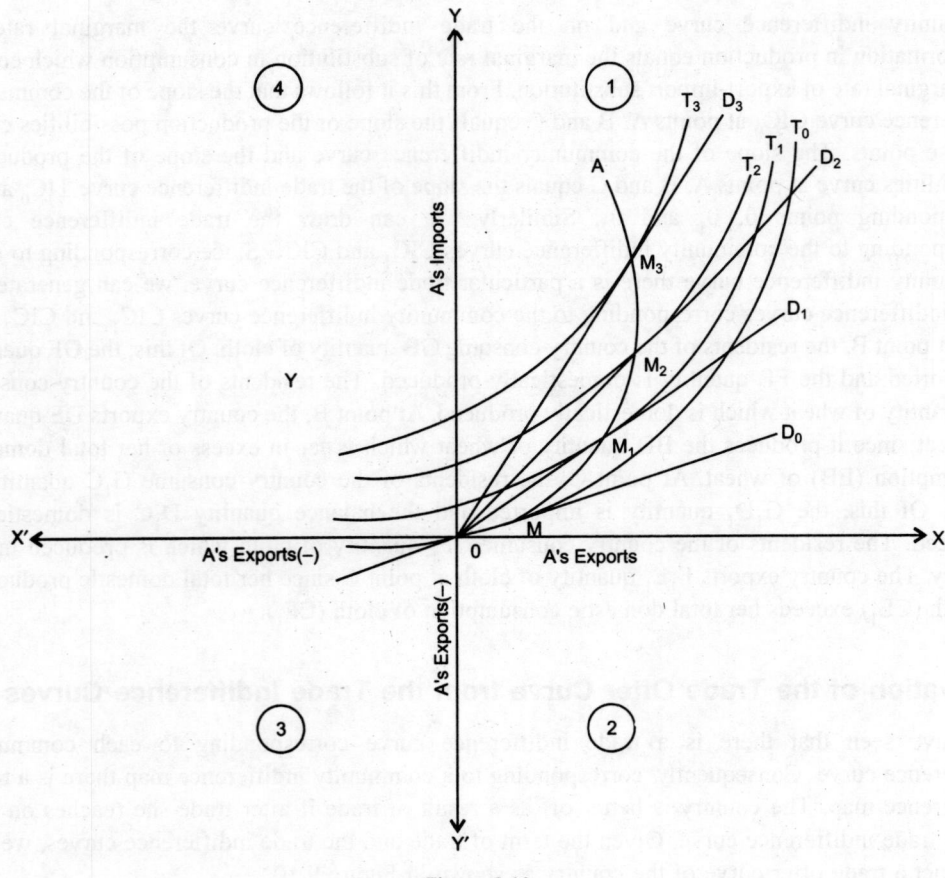

Figure 9.10

Equilibrium in Trade

In order to study the general equilibrium in trade, the analytical tools of the production possibilities curve, community indifference curve, trade indifference curve and trade offer curve are assembled together to analyse the general equilibrium in trade between two countries A and B whose individual balance of trade is in equilibrium, i.e., there is neither deficit nor surplus in any country's balance of trade.

In Figure 9.12, Y0Y′ axis measures the quantity of imports (cloth) and X0X′ axis measures the quantity of exports (wheat). The consumption of cloth in country A, which is the sum of her total domestic production and her imports from country B, is measured on Y-axis and the consumption of wheat, which is wholly met from domestic production is measured on the X-axis. Corresponding to these consumptions there are several community indifference curves, one of which is U_0A. After setting the position of the community indifference curve U_0A, let us place the production block P_A of country A such that it is tangent to U_0A at point M. Showing country A's exports (wheat) horizontally and her imports (cloth) vertically we move it such that corner Q of the production block P_A reaches the highest possible trade indifference curve T_0A of country A. Now we should draw from the point of origin 0 the various price-ratio lines which are tangent to different trade

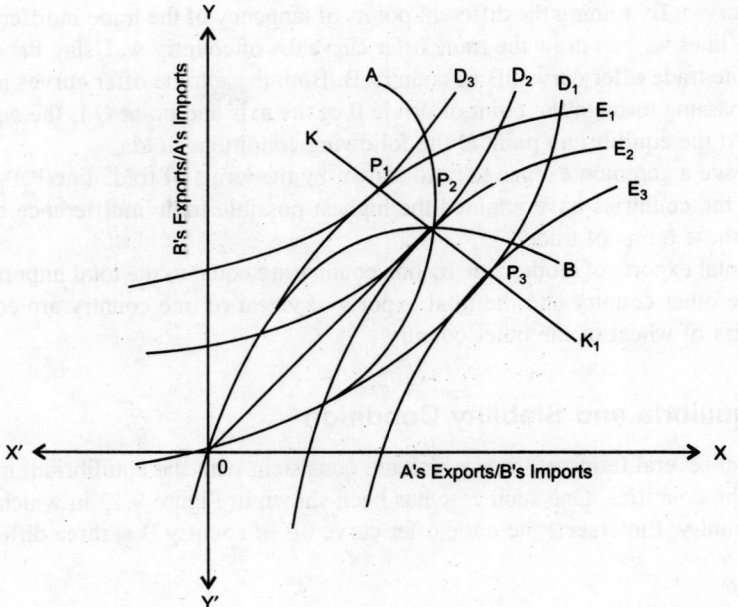

Figure 9.11

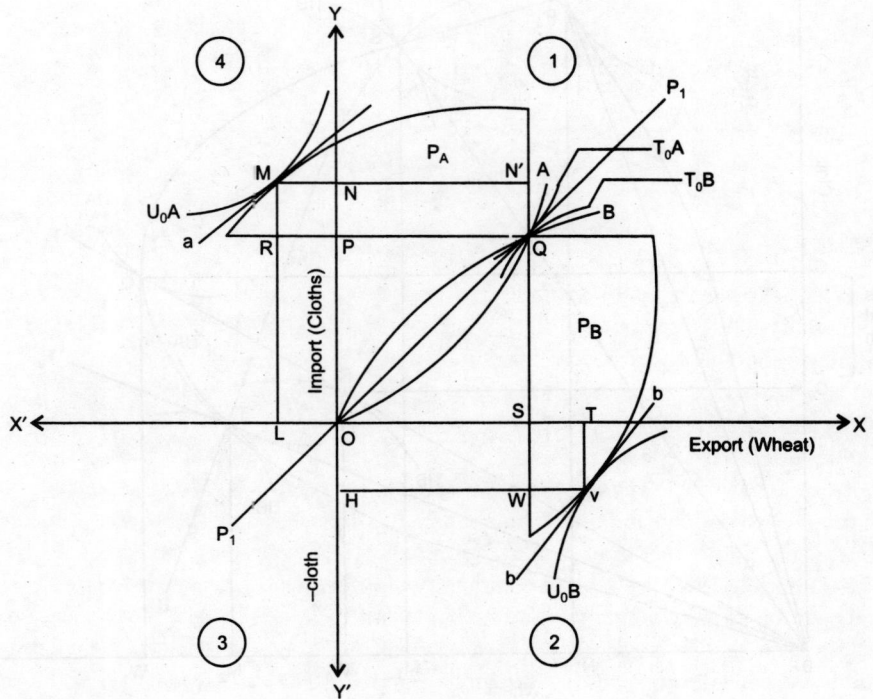

Figure 9.12

indifference curves. By joining the different points of tangency of the trade indifference curves and the price-ratio lines we can draw the trade offer curve 0A of country A. Using the same technique, we can draw the trade offer curve 0B of country B. Both these trade offer curves intersect at point Q. Line P_1P_1 passing through the point of origin 0 of the axis and point Q is the equilibrium terms of trade line. At the equilibrium point Q the following conditions hold.

1. We have a common exchange ratio shown by the terms of trade line P_1P_1.
2. Both the countries have attained the highest possible trade indifference curve consistent with these terms of trade.
3. The total exports of cloth made by one country are equal to the total imports of cloth made by the other country and the total exports of wheat of one country are equal to the total imports of wheat of the other country.

Multiple Equilibria and Stability Condition

There may exist several terms of trade which are consistent with the equilibrium in the balance of trade of both the countries. One such case has been shown in Figure 9.13 in which the trade offer curve 0A of country A intersects the trade offer curve 0B of country B at three different points P, S

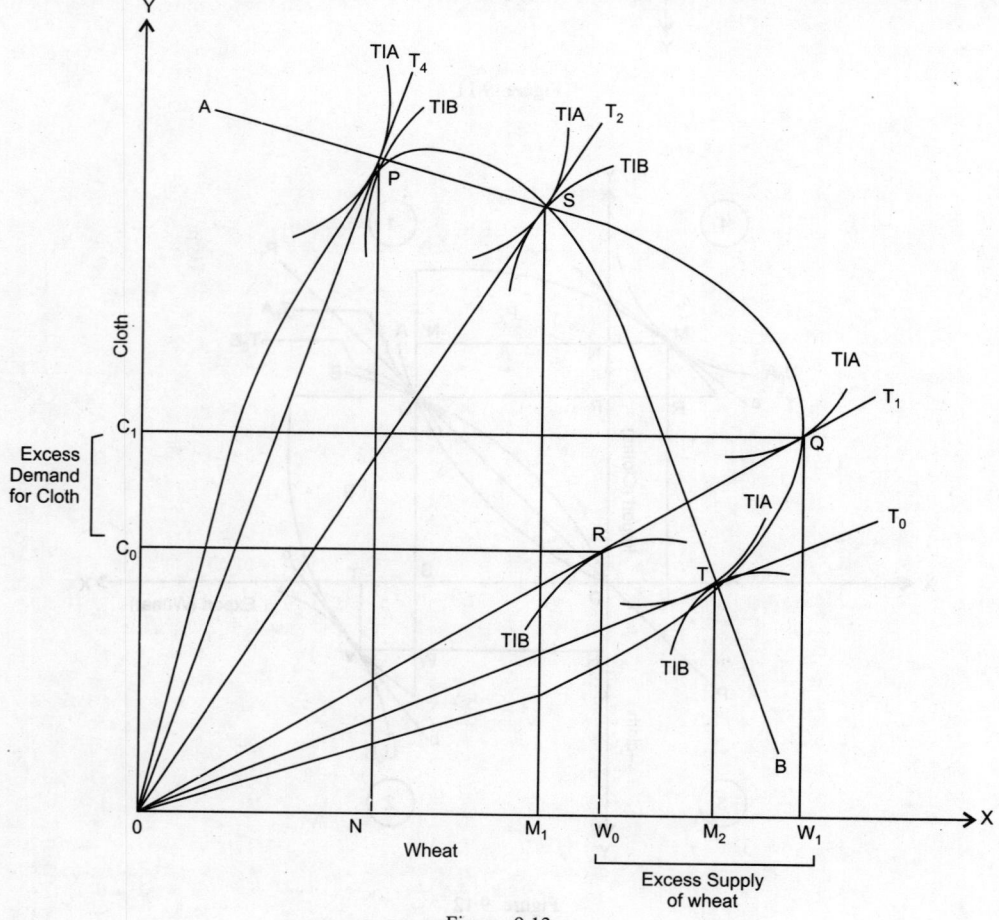

Figure 9.13

and T. At all these three points, the trade indifference curves of the two countries are tangent to each other. Consequently, the equilibrium terms of trade are shown by the slopes of lines $0T_0$, $0T_2$ and $0T_4$ and points P, S and T are all the equilibrium points.

It is, however, significant to note that all the three points P, S and T are not the points of stable equilibrium. There is substantial difference between the stability conditions established at these three points. To find out the stable equilibrium point, let us draw the terms of trade line $0T_1$. The trade indifference curve TIA of country A is tangent to this line at point Q while the trade indifference curve TIB of country B is tangent at point R. The points of tangency Q and R show that at the terms of trade represented by the slope of $0T_1$ terms of trade line, country A wants to export $0W_1$ quantity of wheat against QW_1 imports of cloth while country B is willing to import only $0W_0$ quantity of wheat against RW_0 exports of cloth. Consequently, there is an excess supply of wheat amounting to W_0W_1 ($= 0W_1 - 0W_0$) quantity of wheat in country A. Expressed differently at the $0T_1$ terms of trade country B wants to export only $0C_0$ ($= RW_0$) quantity of cloth while country A wants to import $0C_1$ quantity of cloth. Consequently, there is an excess demand for cloth amounting to C_0C_1 ($= 0C_1 - 0C_0$) quantity of cloth in country A. Due to the excess supply of wheat and the excess demand for cloth in country A, the price of wheat in terms of cloth will fall and stability will be attained at point T. Similar reasoning can be applied for the stable equilibrium point P. It is, therefore, obvious that while P and T are the points of stable equilibrium, S is a point of unstable equilibrium.

Suggested Readings

H. Robert Heller, *International Trade,* Second Edition, 1973, Chapter 6.

James E. Meade, *A Geometry of International Trade,* 1952, Chapters 1 and 4.

Bo Sodersten, *International Economics,* Second Edition, Reprinted 1981, Chapters 2 and 3.

Jaroslav Vanek, *International Trade,* 1962, Chapter 14.

Questions

1. Show and explain the trade equilibrium of the two trading countries with the help of Marshall-Edgeworth trade offer curves.
2. Explain the following graphical tools of international trade theory.
   (i) trade offer curve;
   (ii) trade indifference curve; and
   (iii) community indifference curve.
3. Graphically illustrate the trade offer curve of a country and explain the elastic and inelastic parts of this curve.
4. Explain the law of reciprocal demand with the help of Marshall-Edgeworth offer curves. On what grounds has it been criticised?

CHAPTER 10

Imperfect Comptetion and Intraindustry Trade

The trade models from David Ricard to Bertil Ohlin and Paul A Samuelson which we have explained and discussed in previous chapters are no doubt remarkable achievements. They help us to understand and solve many complexities of international trade. However, despite their immense values, these models rely heavily on weak assumptions which are inappropriate for dealing with many important questions. The main assumptions of these models are that the factors of production do not move from country to country, that there are no economies of scale (internal and external), which are costless inputs and that the factor endowments in the trading countries are different. If one country is capital rich, the other trading partner is either labour abundant or land rich; there is also perfect competition both in the factor and the commodity markets. The pattern of trade is interindustry (food for manufacture or manufacture for food) or there is the two way exchange of the two different commodities. The trade between the countries takes place due to the relative difference in the commodity prices which is the result of the differences in the factor endowments. In other words, the theory of comparative cost advantage holds good. However, in the the real world perfect competition is a myth and the most prevalent form of market is imperfect competition which includes monopoly, duopoly, oligopoly and monopolistic competition. In most industrialized countries a large part of the manufacturing sector is dominated by very small number of firms. For instance, the number of firms producing automobiles in Japan, America and Europe may be counted on fingers. Although firms compete among themselves but competition is far from being perfect because there are many barriers to entry. First, products and technologies are protected by patents; second, these firms have cost and reputational advantages, third, it can be very expensive to build and start a new plant big enough to reap the benefits of the economies of scale which the established frims already enjoy. These firms are nearly monopolists in their respective countries. These firms produce identical and homogeneous products differing in physical shape, quality and style and trade heavily among themselves. When manufacture is exported or imported for manufacture or when there is two way exchanges of the same product this is known as intraindustry trade which is not based on comparative cost advantage as in the case of interindustry trade. Nearly one fourth of world trade is intraindustry trade. For examples Germany imports cutlery sets made in England and Britain imports cutlery sets made in Germany. Fiat cars made in Italy are imported in Germany and Volkswagen cars made in Germany are imported in Italy. There are several other examples like this. This is perhaps on account of the preferences of the counsumers.

Intraindustry Trade And Its Importance

When there is a two way exchange of the same product between the two countries, it is known as intraindustry trade. Intraindustry trade does not reflect comparative advantage. When the home country and the foreign country are similar in their capital-labour ratios, then there will be little interindustry trade and intraindustry trade will dominate. The firms would continue to produce the differentiated products and the demand of consumers for these products from abroad would countinue to generate intraindustry trade. Thus intraindustry trade is a two way exchange of goods within the standard industrial classification. In this respect Table 10.1 is a good example.

Table 10.1 is concerned with the exports and imports of USA in the nine industrial categories.

Table 10.1 Intraindustry Trade in the US in 1984

Cateogery	Exports Million US Dollers	Imports Million US Dollers	Value of T
Motor Vehicle Engines	1000.4	2962.7	50.5
Harvesting Machinery	635.9	323.9	67.5
Metal Cutting Machine Tools	235.8	1117.8	39.8
Welding Equipments	206.6	198.6	98.0
Air Conditioning Machinery	536.9	216.0	58.0
Radio and T.V. Transmitters	591.7	306.0	68.0
Electronics Micro Receivers	1930.0	6887.4	48.0
Printed Books	662.0	468.4	83.0
Sound Recordings Disk & Tapes	856.2	918.2	96.0

Source : U.N. Year Book of International Trade Statistics, 1984.

The values of T lie between 96 and 39.8 Befor we calculate T-Index or level of intraindustry trade, we should have a brief idea of the method of calculation of T. The best known method was developed in 1975 by Grubel and Lloyd. Which is as follows:

$$Tj = \frac{(Xj + Mj) - (Xj - Mj)}{(Xj + Mj)} \times 100$$

The above equation is generally presented in the aggregative from as below and is widely used at present[1].

$$Tj = \left[1 - \frac{(Xj - Mj)}{(Xj + Mj)} \right] \times 100$$

Where Tj = index of level of intraindustry trade for the Jth industry
Xj = exports of the Jth industry, Mj = imports of the Jth industry, Here,

$$O < Tj \leq 100$$

(Exports-Imports) means the absolute value of trade balance. If a country's value of exports is equal to the value of her imports the value of T will be equal to 100. If a country either exports or imports not both then the value of T will be equal to zero.

1. N.G. Grubel and P.T. Lloyd, Intraindustry Trade, Macmillan, 1975.

Now returning to Table 10.1 we can compute T for US trade in sound recording, disks and tapes.

$$T = 1 - \frac{[856.5 - 918.2]}{[856.2 + 918.2]} \times 100 = 96$$

which reveals that trade in this product group comes close to being pure intraindustry trade. Similarly, T for motor vehicle engines equals to

$$T = 1 - \frac{[1000.4 - 2962.7]}{[1000.4 + 2962.7]} \times 100 = 5(}$$

50 percent of trade is intra industry. The values of 'T' for the 9 industrial categories are given in Table 10.1 The values of 'T' lie between 96 and 39.8.

Table 10.2   Index of Intraindustry Trade in U.S. in 1993

Cateogry	Index T
Irorganic Chemicals	99
Power Generating Machinery	97
Electrical Machinery	96
Organic Chemicals	91
Medical and Pharmaceuticals	86
Office Machinery	81
Telecommunications Equipment	69
Road Vehicles	65
Iron and Steel	43
Clothing and Apparel	27
Foot wear	0.00

Krugman & Obstfeld, *International Economics*, P. 132.

In Table 10.2 have been given values of T for the 11 manufacturing industries for 1993 for the US economy. The value of T ranges from 99 in the inorganic chemicals, an industry is which US exports and imports are nearly equal; and 0.00 for the foot-wear in which US has large imports and virtually no exports. The industries in Table 10.2 are ranked by their relative importance in the intraindustry trade. Those with higher intraindustry trade are higher ranking order. Industries with high levels of intraindustry trade tend to be the sophisticated manufactured goods, such as the inorganic chemicals, power generating machinery, electrical machinery etc. At the other end of the scale with very little intraindustry trade are typically the labour intensive products such as the footwear and apparel. These are the goods that US imports from the developing countries where the comparative cost advantage is clear-cut.

Diagrammatic Presentation of the Intraindustry Trade

Intraindustry trade is a special feature of the imperfect competition. Imperfect competition is characteristics both of those industries in which there are only few major producers and of those industries in which the producer's product is seen by consumers strongly differentiated from those of his rivals. In imperfectly competitive firms are aware that they can influence the prices of their products and that they can sell more by reducing their products prices. The most important feature of imperfect competition is monopoly. Consquntly, first of all we shall deal with the intraindustry trade when both the national markets are monopolised. The pattern of analysis developed here will be helpful for further analysis.

Trade Between Monopolised National Markets

Let us assume the two national markets of America and Japan which have similar factor endowments and identical cost structure. There are two large firms, Alpha in America and Beta in Japan which are producing automobiles at large scale. To keep the analysis simple suppose that they have constant marginal cost curves but the fixed cost of making automobiles including the costs of design and development are high enough to prevent additional firms from entering the automobiles market. Consequently Alpha and Beta function as the monopolist until trade is opened between the two markets. Let us first take the case of Beta in Japan before trade is opened. What is true in the case of Beta in Japan is also true for Alpha in America due to the assumption of similar factor endowments and similar cost structure Figure 10.1 describes the case of Beta in Japan before the trade is opened.

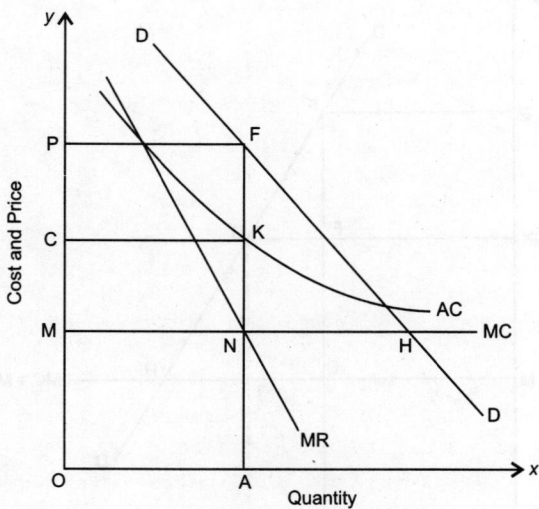

Figure 10.1 Beta's Japanese Market Before Trade.

In Figure 10.1 the quantity of oputput is measured on the Ox-axis and costs and prices are measured on the Y-axis. The negatively sloping demand curve DD shows that more sale will take place at the lower prices. The marginal revenue curve (MR) lies below the DD demand curve. The DD demand curve is a straight line. The slope of MR curve is half of the slope of the DD curve.The average cost curve AC is falling downward on account of the economies of scale. When the output is zero average fixed cost is infinity, when the production is very large, the average fixed cost will approach the MC, the AC curve will also approach the MC but it will never touch the MC, but approach it. In Figure 10.1 the MR intersects the MC at point N.

According to Cournot solution this is the point of profit maximization. Beta produces OA number of automobiles and sells them in the national market at the OP price. Beta earns profit equal to CKFP which is the difference between the price and average cost per unit multiplied by the total sale which is equal to CK. It is evident that MC intersects MR at N, half way of MH (in the further diagrams MR is omitted). The situation which Beta is enjoying in the Japanese market before trade, Alpha enjoys the same situation in the American market. Alpha also produces the OA number of automobiles and sells them at the OP price.

Opening of Trade

When trade takes place between the two countries (Japan and America) Beta and Alpha remain no longer monopolists. With monopoly the individual firm does not have to worry about the behaviour of the other, because there is no other competitor while planning its own move. Cosequently, the situation will be like oligopoly where large firms dominate the market. We have dealt with monopolistic competition as a special case of oligopoly in the next chapter in detail. At present we assume though umcalistically that each firm expects its foreign rival to keep selling the same quantity of automobiles even when its own behaviour leads to a change in price. The effect of this assumption has been explained in Figure 10.2.

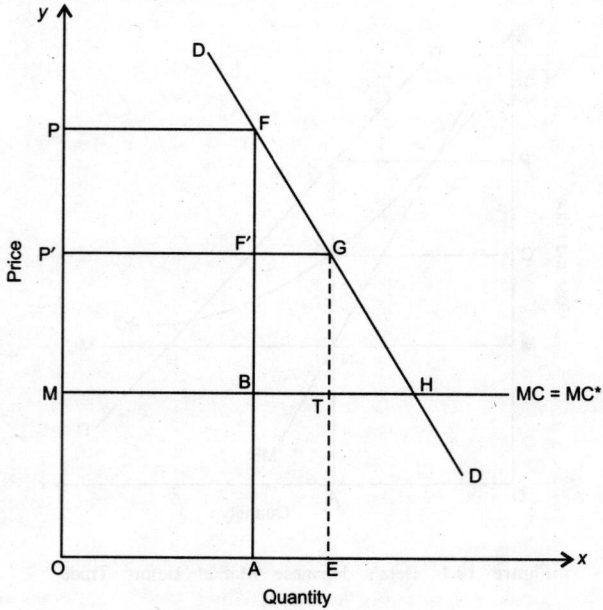

Figure 10.2 Alpha Enters Beto's Market in Japan.

Figure 10.2 reproduces the main features of Figure 10.1 but assumes that Beta and Alpha have the same marginal cost, curves, Beta has been selling OA automobiles in the Japanese market and Alpha expects it to keep on doing so. Alpha knows that it cannot sell automobiles in the Japanese market if the price remains at OP. At any lower price, howeve it can invade the Japanese market.

In effect, Alpha faces that port:on of the demand curve that lies to the right of the line AF and thus a marginal revenue curve (not shown in the figure) crosses the MC* at point T, half way along BH, therefore Alpha will ship AE automobiles to the Japanese market. This will drive down the price to OP. Consequently Alpha's profit will rise to BTGF' and Beta's profit will fall by PFP'F' until it changes the volume of its sale. What Alpha is doing in the Japanese market, Beta will do the same in the American market. Thus the pattern of trade will depend upon their reaction functions. The derivation of the reaction curves has been shown in Figure 10.3.

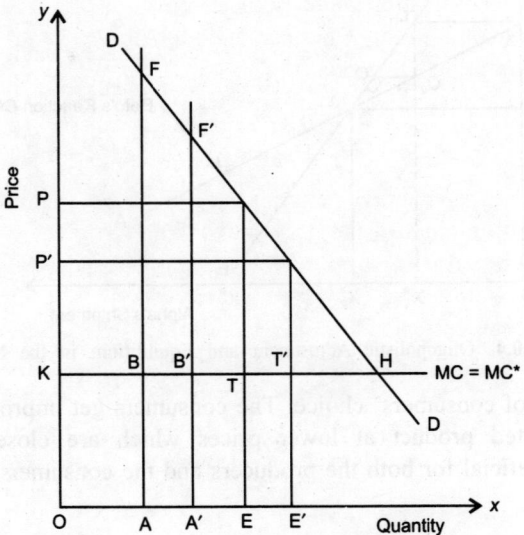

Figure 10.3 Derivation of Reaction Curves.

Suppose Beta decides to sell OA automobiles in the Japanese market; OA is not the same as in Figure 10.2. Alpha will face the demand curve right to AF and will go to T, half way along BH just as it did in Figure 10.2. It will sell AE automobiles in the Japanese market. Now suppose Beta decides to sell OA' output driving down the price to OP', Alpha will face that part of the demand curve that lies right to A'F' and will to T' half way along B'H. It will sell A'E' number of automobiles in the Japanese market. Now let us see what is the relationship between the sales of the two firms? When Betra sells OA, Alpha Sells (½) BH; when Beta sells OA', Alpha sells (½) B'H. Consequently, the increase in Beta's sales AA' leads Alpha to change its sales by (½) (B'H-BH). But B'H = BH-BB' where BB' = AA', so the change in Alpha's sale is (½) [(BH-AA')-BH] = (½) AA'. The sale of Alpha falls by half of any increase in Beta's sale, Reversing the same process in Alpha's market we will get the same result. On the basis of this information we can draw the reaction curves of both firms as shown in Figure 10.4.

In Figure 10.4 AA is the reaction curve of Beta and EE is the reaction curve of Alpha. Beta was selling OX_0 automobiles before the opening of trade. When trade is opened Alpha will go to Q_1 on its reaction curve EE and sell OX_1^x automobiles in the Japanese market (This is the quantity AA' in Figure 10.2) Assuming that Alpha will hold its sale at OX_1^x Beta will go to Q_2 on its reaction curve, reducing its sales by half of Alpha's initial sale and that will cause Alpha to adjust its sale, it will go to Q_3 and thus raise its sales by half of the reduction in Beta's sales. Ultimately both the firms will reach Q which is the equilibrium point. The equilibrium point at Q has two characterstics. First, the total car market is evenly divided between the two firms. Second, the price in the Japanese market is lower at Q than it was before trade. The same will also happen in the American market.

From the above discussion it is clear that intraindustry trade enlarges the size of the market

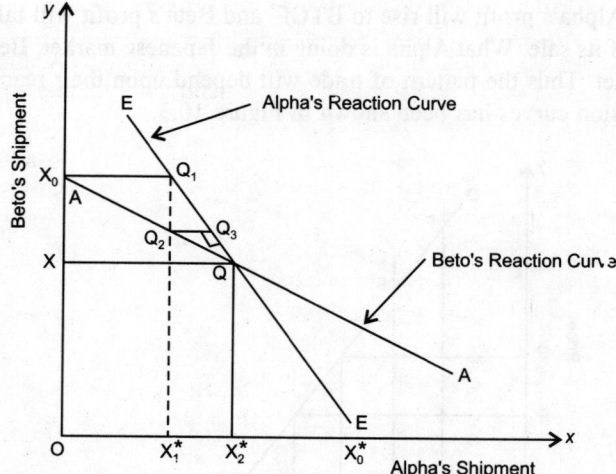

Figure 10.4 Oligopolistic Adjustment and Equilibrium in the Market.

and widens the horizons of consumers' choice. The consumers get improved quality and greater variety of the differntiated product at lower prices which are close substitues. Thus the intraindustry trade is beneficial for both the producers and the consumers.

Importance of the Intraindustry Trade

Intraindustry trade produces extra gain from international trade over and above those derived from the comparative cost advantage because intraindustry trade allows the trading countries to benefit from the larger markets. Nearly one-fourth of the world trade consists of the intraindustry trade. Itraindustry trade plays a large role in trade in the manufactured goods in the advanced industrial nations which account for most of world trade. The average value of T (Index for the level of intraindustry trade) for the manufactured goods lies between 56 to 75 for most industralised countries[1]. In 1957 European Economic Community (EEC) established free trade area in the manufactured goods. Trade with in EEC grew twice as fast as the world trade as a whole. The growth in trade, however, was almost entirely intraindustry rather than interindustry. In 1964 the United States of America and Canada ageed to establish the free trade area in automobiles. In 1962 Canada exported 16 million of automobiles products to USA while importing 519 million worth of automobiles. By 1968 the numbers were $ 2.4 billion and $ 2.9 billion respectively. In otherwords both exports and imports increased sharply. By the early 1970, the Candadian industries were comparable to the US industries in productivity.[2] The intraindustry trade in the United States of America is of paramount significance as shown in Tables 10.1 and 10.2.

Factors Affecting the Intraindustry Trade

One of the most important empirical studies made by Bela Balassa and LucBauwens covering 38 countries and 157 product groups is summerised in Table 10.3. Fifteen variables have been listed in the table which have positive and negative effects on intraindustry trade.

1. See, David Greenaway and Chris Milner, Intraindustry Trade, Oxford, Blackwell, 1986, T. 5.3
2. Krugman and Obstfeld, International Economics VII Ed.p. 134

Table 10.3   Factors affecting the Intraindustry Trade

1.	Average of countries incomes per capita	Positive
2.	Difference in income per capita	Negative
3.	Average of countries total incomes	Positive
4.	Difference in total incomes	Negative
5.	Average trade orientation of countries	Positive
6.	Distance between countries	Negative
7.	Common border between countries	Positive
8.	Common language between countries	Positive*
9.	Membership in common trade block	Positive
10.	Product differentiation within industry	Positive
11.	Economics of scale for firms in industry	Negative
12.	Industrial concentration in industry	Negative
13.	Multinational production by firms in industry	Negative
14.	Average tariff level for industry	None
15.	Traiff dispersion within industry	Negative

* Positive for English, German, French and Portuguese but not for Spanish and Scandin-avian language

Source : Bela balass a and Luc Bauwens, '*Intra Industry Specillsation in a Multi-country and MultiIndustry Framework*' World Bank Discussion Paper DRD 116, Dec. 1984 Table-1.

The first five variables listed in The Table pertain to the general characterstics of the trading countries. The next four variables relate to the characteristics of the bilateral relationship between each pair of countries. The last six characterstics relate to the industrics and product groups rather than the countries. If we look at the first five varables it may safely be concluded that ccounties with high incomes are expected to trade heavily in intraindustry in manufactured but countries with difference in per capite income are expected to discourage intraindustry trade. Similary countries large in size are expected to have positive effect on intraindustry trade. The countries that are outwardly oriented with low trade barriers are expected to engage heavily in intraindustry trade.

Looking at the next four variables, it is found that distance discorages intraindustry trade while common borders and common language promote intraindustry trade. Membership in a common trade block encourages intraindustry trade.

Looking at the last six variables it is expected that measures of product differentiation have the positive effect while the economies of scale and industrial concentration apear to have the perverse effect. Finanlly, tariffs and other trade barriers might be expected to limit the intraindustry trade.

Suggested Reading

P.B. Kenen, *Internation Economics*, Second Edition, Chapter 7. pp. 111-131

J. Brander, IntraIndustry Trade in Identical Commodities, *Journal of International Economics,* 11 February 1981.

David Greenaway and Chris Milner, *The Economics of Intraindustry Trade,* Oxford Blackwell, 1986.

P.R. Krugman, *Import Prevention as Export Promotion.*

P.R. Krugman and M, Obstfeld, *International Economics,* VIIth Edition.

Questions

1. What is Intraindustry Trade? How does it take place between the two monopolized national markets?
2. How the index of the level of intraindustry trade is calculated? Explain with suitable examples.
3. Distinguish between interindustry and intraindustry trade. What is the importance of intraindustry trade.
4. What is intraindustry trade? Discuss the factors affecting the intraindustry trade.

CHAPTER 11

Monopolistic Competition and International Trade

Monopolistic Competition contains elements of both monopoly and perfect competition. It is akin to perfect competition in the sense that the number of sellers is sufficiently large so that the actions of an individual seller have no perceptible influence upon his competitors. The firms are not price takers like those in perfect competition, but they are price setters. Products are differentiated, although they are close substitutes but not perfect substitutes as in the perfect competition. It is akin to monopoly and differentiated oligopoly as each seller possess a negatively sloping demand curve. Like the monopoly, the firms in the monopolistic competition reap the benefits of the economies of scale. Since monopolistic competition is akin to monopoly we should briefly review the monopoly because the tools developed to analayse the monopoly will be widely used to deal with the complex problem of monopolistic competition.

In a monopoly market the monopolist is a king without crown. If he fixes the quantity of the product, the price of the product is determined by the consumer. In this situation $P = f(q)$, where P is the price and q is the quantity. On the other hand, if the monopolist fixes the price the quantity is decided by the consumer or $q = f(P)$. In our example let us assume that the monopolist fixes the price, the demand curve is as following:

$$q = a - b'P \qquad (11.1)$$

where q = quantity of the firm, P = price per unit, a and b' are constants, b' is the slope parameter of the demand curve or $dq/dp = -b'$, the demand curve is sloping downwards.

Let us consider the relationship between the marginal revenue, MR and the price, of the product P. In Figure 11.1 X-axis measures amount of sale and Y-measures the cost and price per unit at which the sales take place. Both MC and AC are falling downwards, this is on account of the internal economies of scale. According to Cournot solution the equilibrium of the firm is achieved when MC = MR. In Figure 11.1. MC intersects the MR at point K. The amount of sale is OM and price per unit is $P' = OP' = MN'$. The marginal revenue MR = MK = MN' − KN' = P' − KN', where KN' is quantity of sale divided by the slope parameter of demand curve b' or KN' = q/b' hence.

$$MR = P - q/b' \qquad ...(11.2)$$

Implying $P - MR = q/b'$

Equation (11.2) reveals that the gap between price and the marginal revenue depends on the initial sale of the firm and slope parameter of the demand curve. If the sales quantity is higher the marginal revenue is lower and vice versa. The greater b' means that more sales fall for any given increase in price.

Corresponding to the price and marginal revenue, there is the related average and marginal cost. Suppose the cost function takes the form of $C = F + cq$ (11.3), where F is the fixed cost, c is the marginal cost and q is the quantity of sale, then the

Average cost is equal to

$$C/q = F/q + c \qquad (11.4)$$

It means that AC declines as quantity q increases. This is the result of the economies of scale. In Figure 11.1 the equilibrium is at point K. When P > AC, the firm earns profit. In Figure 11.1, the profit is equal to PP′ N′N or the quantity of sale multiplied by the difference of the price and average cost or q x (P – AC)

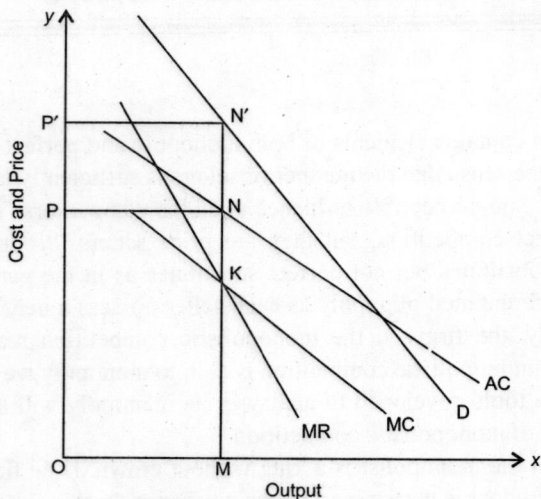

Figure 11.1 Monopolistic Pricing and Production Deasion.

Monopolistic Competition

The tools developed in the analysis of pure monopoly will be utilized in the analysis of monopolistic competition. Monopoly profits never go uncontested. A firm earning profit will attract competitors. Consequently the situation of pure monopoly is rarely found in practice. Instead, the usual market structure is one of oligopoly. The general analysis of oligopoly is complex and a controversial subject because in oligopoly the pricing policies of firms are interdependent. However, there is a special case of oligopoly known as monopolistic competition which is relatively easy to analalyse. Since 1980 monopolistic competition models have been widely used in international trade.

In monopolistic competition models two key assumptions are made to get around the problem of interdependence. First, each firm is assumed to differentiate its products from that of its rivals. Second, each firm is assumed to take the prices charged by its rivals as given. The important question is –are there any monopolistically competitive industry in real world? Some industries may be a reasonable approximation.

For instance, the automobiles industry in Europe where the major producers of automobiles are the Ford, General Motors, VolksWagen, Peugot, Fiat, Nissan etc. which produce the differentiated products, which are different in shape, quality, style, service after sales etc. but these are the close substitutes of each other and they all serve the same need. Similarly cement, mobiles, cigarette, tooth paste etc. are good examples of monopolistically competitive industries. The demand for a particular product depends on the demand of the other similar products available in the market and the prices of other firms in the industry.

Assumptions of the Monopolistically competitive Models

In general, there would a firm to sell more, the larger is the demand for its industry's product and higher the prices charged by its rivals. On the other hand, we expect to sell less, the greater the number of firms in the industry and higher its own price. The demand curve for a firm facing these products is[1]

$$q = S \times \left[1/n - b (P - \dot{P}) \right] \tag{11.5}$$

or
$$q = S/n - S \times b (P - \dot{P})$$

Where q is firms sale, S is the total sale of the industry b, responsiveness of firm's sales to its price or slope parameter of demand curve, P is price charged by the firm $\dot{p}$ the average price charged by its competitors and 1/n is the share of the firm in total sale, S. If $P = \dot{P}$ then q = S/n, if $P > \dot{p}$, q is < S/n and if $P < \dot{p}$, q is > S/n

Market Equilibrium

We know that in the monopolistic competition the long run equilibrium is achieved when the average cost AC is equal to average revenue AR or price. Let us assume that in the monopolistically competitive industry all firms are symmetric, i.e. demand and cost functions are identical for all the firms. Consequently we need not know the detailed features of each and every firm. We need only to know how many firms n, are there in the industry and what price the typical firm is charging. There are certain characteristics of the monopolistically competitive industry. The more the number of firm are, the lower will be output/sale for each firm and higher is the cost per unit of output. As a result, the average cost curve will be upward sloping as shown in Figure 11.2. The more are the number of firms the more intense will be the competition among them. As a result the average revenue curve will be downward sloping as shown in Figure 11.2 Finally, when AR or price per unit exceeds the average cost AC, additional firms will enter the industry. On the other hand when the cost per unit is higher, some firms will suffer loss and exist. Ultimately, equilibrium will be achieved when average cost will be equal to the average revenue or price. Consequently in the long run, equilibrium will be achieved when average cost curve intersects average revenue curve which will determine the number of firms, n, and the equilibrium price P.

In Figure 11.2, when the number of firms is n_1 the price per unit of time is P_1 which is more than average cost AC. Consequently the firms are earning abnormal profits, more firms will enter the industry. Suppose the number of firms after entry is n_3, price P_3 is lower than the average cost AC_3,

1. For development of this approach, see Stephen Salop, Monopolistic Competition with Out Side Goods, Bell Journal of Economies 10 (1979) pp. 40-156

the industry is incurring loss, so some firms will leave the industry. When the number of firms is reduced to n_2, equilibrium will be achieved because price equals average cost per unit. Consequently the long run equilibrium will be at point F where price p_2 is equal to the average cost AC_2.

Now let us determine two things; (1) what is the relationship between the number of firms n and the average cost and (2) what is the relationship between the number of firms and price. Since we know that all firms are symmetric, they will charge the same price or $P = \dot{P}$. From equation (11.5) it is evident that when $P = \dot{P}$, $q = S/n$ and the share of each firm is $1/n$ of total sale S. When $P = \dot{P}$, $q = S/n$ so the average cost as shown in equation 11.4 is

$$AC = F/q + c, \text{ when } q = S/n$$
$$AC = F/q + c = n \times F/S \qquad \qquad ...(11.6)$$

Where S is the total sale of the industry.

Equation (11.6) reveals that more the number of firms, n, in the industry, the higher is the average cost, the upward sloping relationship between the number of firm n and average cost has been shown in Figure 11.2.

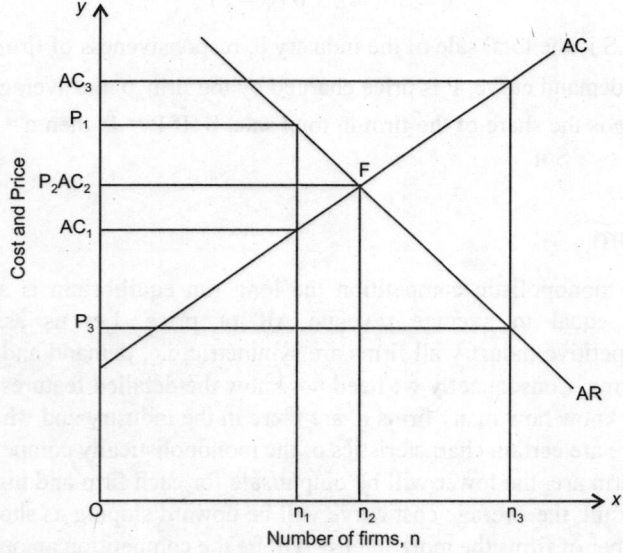

Figure 11.2 Equilibrium in a Monolistically Competitive Market.

So far as the relationship between number of firms and price is concerned, we know that in monopolistic competition model the firms are assumed to take each others prices as given; in other words, each firm ignores the possibility that if it changes its price the other firms will also change them. If each firm treats P as given, we can rewrite the demand equation (11.5) in the following form:

$$q = (S/n + S \times bx \, \dot{P}) - S \times b \times P \qquad \qquad ...(11.7)$$

This equation is in the same from as (11.1) with $(S/n + S \times bx P)$ in place of constant term 'a' and $(S \times b)$ in place of slope coefficient b'. If we place these values in the formula 11.2, the marginal revenue for the typical firm is:

$$MR = P - q/(S \times b) \qquad \qquad ...(11.8)$$

Profit maximization point will be achieved when MC = MR

$$MR = P - q/(S \times b) = c$$

or

$$P = c + 1/(b \times n) \qquad \qquad ...(11.9)$$

(Since in equilibrium q = S/n)

Equation (11.9) show that more the firms are in the industry the lower the price each firm will charge. On the other hand, more the firms are there in the industry the share of output of each firm will be less and the average cost curve will be sloping upward. These facts are evident from Figure 11.2.

Monopolistic Competition and International Trade

The monopolistic competition approach to international trade is attractive because it increases the market size and makes available the differentiated goods and products at lower prices to consumers at a large scale. Integrating the markets through international trade, therefore, has the same effect as the growth of market within a single country. The number of firms in a monopolistically competitive industry and the price they charge are affected by the size of the market. Consumers in a large market will be offered both lower prices and greater variety of the products than consumers in a small market.

In Figure 11.2 the demand and curve AR is falling downwards from left to right. The definition of the curve as given in equation (11.9) is as following.

$$P = c + 1/(b \times n)$$

The size of the market S does not enter into the equation, it does not shift the demand curve AR. If the number of firms n is larger, the value of price p will be lower. Similarly, if the slope parameter b, which is constant will be higher the price P will be lower.

The average cost curve in Figure 11.2 is higher when the number of firms is greater. The definition of AC as given in equation (11.6) is as following.

$$AC = n \times F/S + c$$

From the above equation it is evident that if number of firms n remains constant but size of the market S increases, it will reduce the average cost of production because the firms will increase the output and the average cost of production of each firm will decline. If the number of firms n increases, the AC will be increasing with S remaining constant.

Using the above information we show the effect of an increase in size S of the market on the long run equilibrium through Figure 11.3. Initially, the equilibriums is at point a with price P_1 and the number of firms is n_1. An increase in the size of the market by industry's sale S, shifts the average cost curve from, AC_1 to AC_2. The new equilibrium point is at b. Since the existing firms are earning profits the number of firms will increase from n_1 to n_2 and price will fall from P_1 to P_2. Clearly consumers will prefer to be part of a larger market than smaller market. At the equilibrium point b greater varieties of product are available at lower prices than at point a.

Gains from an Integrated Market : A Hypothetical Numerical Example

Let us assume that there are two markets — home and foreign-where monopolistically competitive industries are producing automobiles (cars) and facing the identical cost and revenue curves. Industry in the home market is producing 900 automobiles and in the foreign market the industry is

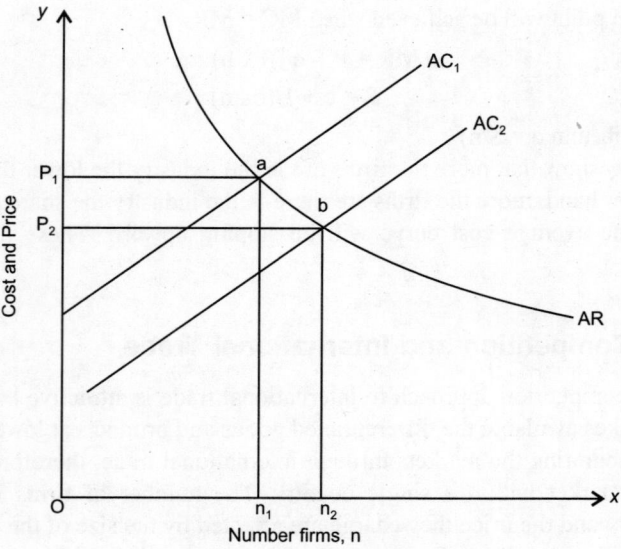

Figure 11.3 Effecti of Large Market.

producing 1600 automobiles. Figure 11.4 a reveals that in the home market 6 (six) firms will produce 900 automobiles and sell them at Rs.. 10,000 each before trade because at this point cost is equal to price. Thus each firm will produce 150 automobiles. Similarly in the foreign market, as shown in Figure 11.4b, Eight (8) firms will produce 1600 automobiles and sell them at price of Rs. 8,750 each. Each firm will sell 200 automobiles because at this point cost is equal to price or AC = AR and the long term equilibrium is achieved.

Now suppose that it is possible to trade automobiles with one another with zero transport cost. The integrated market, as shown is Figure 11.4c, supports 10 firms (AC = AR here) the total production is 2,500 automobiles and sale price per unit is Rs. 8,000. Thus the sale price is much lower as compared to the home and foreign markets individually. The gains from hypothetical integrated market is presented in Table 11.1

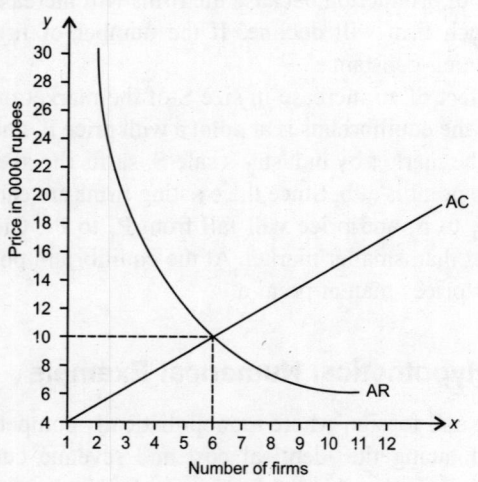

Figure 11.4a Home Market.

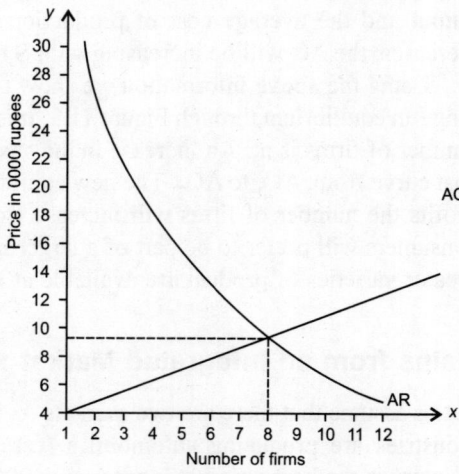

Figure 11.4b Foreign Market.

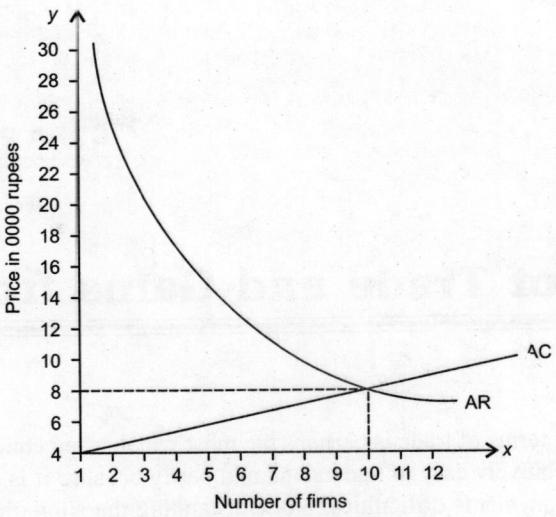

Figure 11.4c   Integrated Market.

Table 11.1 Grains from hypothetical Integrated Market

Item	Home Market before Trade	Foreign Market before Trade	Integrated Market after Trade
Total sales of autos	900	1600	2500
Number of firms	06	08	10
Sale per firm	150	200	250
Average cost per unit	Rs. 10000	Rs. 8750	Rs. 8000
Price per unit	Rs. 10000	Rs. 8750	Rs. 8000

Table 11.1 compares each market with the integrated market. The integrated market supports more firms, each producing at a larger scale and selling at the lower prices. Clearly every one is better off after the international trade. The firms are reaping the benefits of the economies of scale and consumers satisfy their 'love of variety'.

Suggested Readings

P.R. Krugman and M. Obstfeld, *International Economics* Eighth Edition, Chapter 6, pp. 117-129.

P.R. Krugman, "Monopolistic Competition and International Trade", *The Journal of Political Economy,* Volume 39, pp. 959-73.

H. Kierzkowski (ed.) *Monopolistic Competition in International Trade,* 1983. Oxford University Press.

E. Helpman and P.R. Krugman, *Market Structure and Foreign Trade,* MIT Press, 1985.

Questions

1. Define monopolistic competition and discuss international trade under monopolistic competition.
2. Explain how does international trade under monopolistic competition benefit both the consumers and producers?

CHAPTER 12

Terms of Trade and Gains from Trade

In international trade, terms of trade is perhaps the most widely used concept by the policy-makers partly because it is relatively easy to understand and partly because it is convenient to explain the country's balance of payments difficulties. Notwithstanding the similarity of the technicalities of the terms of trade for both the developed and the developing economies, due to the greater sensitivity of prices of raw materials over the business cycle of the developed economies, the greater dependence of the developing economies on the fewer export goods and the pivotal role of international trade in promoting the economic growth, in recent years the policy discussions and academic writings about the terms of trade have been associated more with the problems of the developing countries than those of the developed countries.

When trade takes place between two open economies, certain goods are offered for sale by both the countries. The physical exchange-ratio at which the goods are exchanged for one another between these countries is termed as the 'terms of trade'.[1] Since many commodities are exported and imported simultaneously from both the sides, we can no longer use the physical quantities such as kilograms of wheat or metres of cloth, etc. But we can use as a measure of the terms of trade the ratio between the index of export prices and the index of import prices. The 'ratio' of export prices to import prices is nothing but the 'cost-ratio' which is simply the ratio of the factor costs of production involved—quantities of the real efforts or labour embodied–in the production of goods and services in question. Thus, the 'terms of trade'—the physical exchange-ratio—is an indicator of the interaction of many forces (prices, wages, productivity, etc.) governing the economic relationship and welfare of the countries engaged in trade.

Terms of Trade as Index of the Gains from Trade

From the beginning of the classical period, the trend of the commodity terms of trade has been accepted as an index of the direction of change in the amount of total gain from trade. Consequently, economists regard a rise in the prices of country's exports relative to the prices of her imports as a "favourable" movement of terms of trade indicating an increase in the total gain from trade. It has, however, been recognised by the economists that the proposition that improvement in the commodity terms of trade is an index of increase in the total gain from trade is only valid subject to certain

---

1. The term 'terms of trade' was first used by Alfred Marshall in his book entitled *Money, Credit and Commerce* published in 1923. Marshall had suggested the term 'rate of exchange' as a possible alternative. The well-known American economist Frank William Taussig speaks of the 'barter terms of trade'. See his *International Trade,* Second Edition, 1927, p. 8. Similarly, Arthur Cecil Pigou speaks of the 'real rate of exchange'. See his *Essays in Applied Economics,* 1930, Second Edition, p. 149.

important qualifications. While David Ricardo recognised that, of itself, an increase in the amount of imports which a country could obtain for a given amount of exports was a favourable development, he pointed out that wheather or not an improvement in the terms of trade could be treated as a genuine improvement in the country's position depended on how it was caused or brought about. Ricardo doubted if an improvement in the terms of trade could be brought about through government interference although he reluctantly admitted that an import tariff would improve a country's terms of trade but that it would be accompanied by the offsetting disadvantages in the form of diminished production of the domestic goods, a high price of labour and low rate of profit.

John Stuart Mill was more emphatic than David Ricardo about the relationship between the terms of trade and the gain from trade. He, however, pointed out that a favourable movement of the commodity terms of trade did not necessarily indicate an improvement in the amount of gain from trade. For example, while a protective import tariff caused an improvement in the terms of trade, this advantage was more than offset by the loss of benefit which had earlier accrued from importing those goods which are now produced at home under tariff protection. Both Alfred Marshall and Francis Ysidro Edgeworth considered changes in the "consumer's surplus" rather than changes in the terms of trade as a better index of changes in the amount of gain from trade. Frank William Taussig mentioned the specific circumstances in which the commodity terms of trade would be a misleading index of gain from trade. The general position of the leading economists was that an improvement in the commodity terms of trade, in the absence of offsetting factors, indicated an increase in the amount of the gain from trade. Alfred Marshall mentioned increase in the cost of export commodities while Taussig pointed out a decrease in the desire for import goods as examples of offsetting factors.

Different Concepts of the Terms of Trade

Several concepts of terms of trade have been given by the economists. These concepts have been placed by Gerald M. Meier[2] under the following three broad groups.

1. Those terms of trade which relate to the real ratio of international exchange between the commodities. In this group are discussed the
   A. Net barter terms of trade,
   B. Gross barter terms of trade, and
   C. Income terms of trade.
2. Those terms of trade which relate to the interchange between the productive resources. In this group we have the
   A. Single factoral terms of trade, and the
   B. Double factoral terms of trade.
3. Those terms of trade which interpret the gain from trade in terms of the utility analysis. In this group we have the
   A. Real cost terms of trade, and the
   B. Utility terms of trade.

We may now briefly explain each one of the above-mentioned concepts of terms of trade.

Net Barter or Commodity Terms of Trade

Frank William Taussig[3] introduced the concept of the net barter or the commodity terms of trade.

2. Gerald M. Meier, *International Trade and Development*, 1963, p. 41.
3. Frank William Taussig, *International Trade*, 1927, pp. 113–114.

His net barter terms of trade, popularly called 'the commodity terms of trade', is the ratio between the import-prices and export-prices and can be written as

$$T_C = \frac{P_X}{P_M}$$

where T_C stands for the net barter terms of trade, P_X stands for the price of export commodity and P_M stands for the price of import commodity.

When this concept of net barter terms of trade is applied to more than one export and import commodity we use the export price and the import price indices instead of using the prices of particular export and import goods. The change or movement in the net barter terms of trade in any given year over the base year would be written as

$$T_C = \frac{P_{X1}/P_{X0}}{P_{M1}/P_{M0}}$$

$$= \frac{P_{X1}}{P_{M1}} \cdot \frac{P_{M0}}{P_{X0}}$$

where T_C stands for the net barter terms of trade, P_{X1} and P_{M1} are the price indices of exports and imports for the given year and P_{X0} and P_{M0} are the price indices of exports and imports in the base or initial year respectively.

As the price index of imports and exports for the base year will always be equal to 100, the term $\frac{P_{M0}}{P_{X0}}$ in the formula will be equal to $\frac{100}{100} = 1$. The net barter terms of trade will, therefore, move in accordance with the movement of the given year's price index of imports and exports. Let us suppose that the price index of imports and exports in India in the base year 2010 is 100 and in 2011 the price indices of imports and exports are 120 and 160 respectively. Consequently, the net barter terms of trade will be $\frac{160}{120} \times \frac{100}{100} = 1.33$. It means that in 2010 the net barter terms of trade show an improvement of 33 per cent over the base year. Consequently, a given amount of exports will fetch 33 per cent more in terms of imports in 2011 compared with that in 2010. If the price index of imports rises relatively to that of exports, the net barter terms of trade will become unfavourable for the country.

In order to measure the short term changes in the trading position of a country, the use of the commodity terms of trade has been accepted as the most useful device. The application of the commodity terms of trade is so popular and general that whenever we talk of the terms of trade we mean the commodity terms of trade.

In spite of its general applicability and usefulness, the concept of the net barter terms of trade suffers from several drawbacks. According to Taussig, this ratio is appropriate only when the balance of payments of the country in question includes nothing other than the payments received and made for the exports and imports of goods and services. If the balance of payments of a country also includes capital transactions and unilateral payments and receipts represented by the presence of unrequited exports or imports so that there is an excess of either exports or imports, we should then consider the gross and not the net barter terms of trade in forming a judgement about the amount of gain which the country derives from its international trade.[4]

At times, an unfavourable movement in the commodity terms of trade might mislead us into believing that due to trade the country's position has worsened while, in fact, it may have improved. For example, the import price index remaining unchanged, a fall in the export price index would indicate an unfavourable change in country's commodity terms of trade. This means that the country is receiving smaller amount of imports than before for a given bundle of exports. However, if the unit cost of production of producing export commodities in the country has fallen more than the fall in the prices of her export goods, then despite the worsening of the commodity terms of trade the country derives a gain from her foreign trade. For example, if the index of export prices has recorded a fall of 10 per cent whereas the unit cost of production of export goods has fallen by 15 per cent, still the country gains to the extent of 3.5 per cent in the sense that she receives 3.5 per cent more in terms of import goods compared to the initial year in exchange for a given physical bundle of exports. In such a situation, the commodity terms of trade may fail to provide a satisfactory guide even of the direction of the trend of gain from trade much less to serve as a measure of gain from trade. Where the productivity in the export goods producing sector increases more than the deterioration in the commodity terms of trade, the country does not suffer as result of the trade; she rather gains. It only shows that the country is passing only a part of her productivity gains to her trading partner although she would have been in a happier position if she could retain the whole of the gains to herself.

Moreover, the criterion of the net barter terms of trade may be inappropriate to decide the issue of the distribution of gains from trade between the developed and developing countries. In resolving the issue, the terms of trade on merchandise alone should be distinguished from the terms of trade on goods and services in combination. This distinction is important because a part of the "price" in the export price level in the less developed countries is attributed to the value added by foreign factors of production. Consequently, if following the rise in the prices of the export goods of the developing countries, the profits on foreign investments also rise sufficiently to absorb the entire increase in export prices, the domestic economy of the country will not gain at all although improvement in the commodity merchandise terms of trade gives the misleading impression that the country is gaining from her foreign trade.

Expressed differently, if the prices of exports fall and the profits on foreign investments also fall, then the country's position is not worse off although the commodity merchandise terms of trade would make us believe that it was so. For a deeper analysis, it is necessary to focus attention on the net barter terms of trade pertaining to the current account of the balance of payments as a whole and not merely on the net barter terms of trade.

Gross Barter Terms of Trade

The gross barter terms of trade is the ratio of the total physical quantity of imports to the total physical quantity of exports of a country. The higher is this ratio the more favourable are the gross barter terms of trade. Since the total exports and total imports are available in terms of the total money value of exports and the total money value of imports, both the totals are obtained by correcting the crude figures by the relevant price index in order to eliminate changes due to merely change in the exports' and imports' price levels. This way of reckoning enables us to compare the total (corrected) value of the exports and the total (corrected) value of the imports.

4. Gottfried Von Haberler, *The Theory of International Trade*, English Translation, Third Impression, 1950, pp. 162–163.

Fransk william Taussig introduced this concept to correct the commodity or net barter terms of trade for the unilateral transactions or unrequited exports or imports such as tributes, gifts, immigrants' remittances, etc. The gross barter terms of trade can be expressed as

$$T_g = \frac{Q_M}{Q_X}$$

where T_g stands for the gross barter terms of trade, Q_M for the total quantity of imports and Q_X for the total quantity of exports.

For comparing changes in the gross barter terms of trade between two time periods we use the index numbers of the quantities of exports and imports in the two time periods instead of the quantities alone. The ratio is then expressed as

$$T_g = \frac{Q_{M_1}}{Q_{X_1}} \cdot \frac{Q_{X0}}{Q_{M0}}$$

where the terns Q_X and Q_M stand for the index of quantity of exports and imports of the country and the subscripts 0 and 1 stand for the base year and the given (current) year respectively.

Let us suppose that the index of total imports and total exports in the base year 2010 is 100 and that in 2011 the export index rises to 137 while the import index rises to 181. In this situation, the gross barter terms of trade will be equal to $\frac{181}{137} \cdot \frac{100}{100} = 1.32$. It means that there is an improvement of 32 per cent in the gross barter terms of trade of the country in 2011 compared with the base year 2010. In practical terms, it means that a given quantity of exports in 2010 will bring 32 per cent more imports in 2011. Conversely, if the import index rises to 131 and the export index rises to 136 in 2011, the gross barter terms of trade become unfavourable for the country because for a given bundle of exports the country secures 4 per cent $\left\{ \frac{131}{136} \div \frac{100}{100} = 0.96 \right\}$ less imports in 2011 than she did in 2010. Consequently, the country suffers a relative loss from international trade.

This argument is not, however, very sound because the consumers may derive more satisfaction from a smaller bundle of imports of different goods due to change in their tastes and habits. In this situation, the concept of gross barter terms of trade as an index of the measurement of gain from trade will be misleading, and even if the gross barter terms of trade are unfavourable there will be gain from international trade because had this not been so the country would not have imported the goods. Similarly, there may be gain from trade in case of unfavourable gross barter terms of trade if the factor productivity in the export sector has increased and as a consequence the cost of production per unit of export output has decreased. Moreover, the gross barter terms of trade do not reflect the capital movements and their effects on the economy of the country. On account of these drawbacks, economists have preferred the use of the net barter terms of trade to gross barter terms of trade for measuring the amount of gain which accrues to a country from international trade.

If we consider only the balance of trade and if the balance of trade is balanced, i.e., if $Q_X P_X = Q_M.P_M$, then obviously the gross barter terms of trade and the net barter terms of trade will be identical, i.e $\frac{Q_M}{Q_X} = \frac{P_X}{P_M}$. When, however, the trade is not balanced and $\frac{P_X}{P_M} \leq \frac{Q_M}{Q_X}$, the net barter terms of trade will differ from the gross barter terms of trade. Since the gross barter terms of trade consider only the merchandise or at most the current account of the balance of payments they do not portray the true price changes which may become necessary to correct the balance of payments deficit even when the balance of trade, being in equilibrium, does not necessitate price changes.

Income Terms of Trade

D.S. Dorrance and H. Staehle[5] have refined the concept of the commodity terms of trade and developed a new concept known as the 'income terms of trade'. The income terms of trade is the ratio of the total value of exports divided by the price index of imports and can be written as

$$T_i = \frac{P_X Q_X}{P_M}$$

where the terms T_i, P_X, P_M and Q_X denote the income terms of trade, the price index of exports, the price index of imports and the quantity of exports respectively. It is obvious from the formula that given the import prices, if the export prices rise and the volume of country's exports falls equally, the net barter terms of trade will improve while the 'income terms of trade' will show no change. This concept is sometimes called the 'capacity to import'.[6] Imports of a country are her receipts while exports are her payments made for imports. The statement that in the long run payments should be equal to receipts means that $P_X \cdot Q_X$ (payments) $= P_M \cdot Q_M$ (receipts). Dividing the term

$P_X \cdot Q_X = P_M \cdot Q_M$ by P_M we get $\dfrac{P_X Q_X}{P_M} = Q_M$ which is the total volume of imports which a country

can import. The concept of the 'income terms of trade' is of great relevance in the context of the developing countries whose capacity to import is very low.

The total imports which a country can make will in the long run be determined by the total exports which she can make multiplied by the net barter terms of trade. In other words, a country's total capacity to import depends upon her total exports weighted by the net barter terms of trade. It is argued on behalf of the developing countries that since they cannot change either the prices of their exports (P_x) or the prices of their imports (P_M) or the amount of their exports (Q_x) their total capacity to import (Q_M), is determined for them by others. In the context of the developing countries, the concept of income terms of trade is superior to the concept of the commodity terms of trade.

However, the income terms of trade may mislead in cases where the commodity terms of trade furnish the right answer. For example, assuming the balance of trade and the balance of payments to be in equilibrium, if the export prices rise by 20 per cent and exports also fall by 20 per cent, then the income terms of trade will show no improvement while the country is better off as she buys the same imports with smaller exports. The direction of change in country's welfare is correctly shown by the improvement in her commodity terms of trade. If we assume that the export prices have fallen by 20 per cent and exports have increased by 20 per cent, then although the income terms of trade show no change the country's position has deteriorated since for the same total imports she has to export larger quantities of goods. The worsening of the commodity terms of trade correctly shows the direction of change in the country's welfare. It is, therefore, obvious that there may arise

5. G.S. Dorrance, "Income Terms of Trade", *Review of Economic Studies*, XVI, 1950, p. 52; H. Staehle, "Some Notes on the Terms of Trade", *International Social Science Bulletin*, Spring 1951. Imlah has called it the "Export Gain from Trade". See A.H. Imlah, "The Terms of Trade of the United Kingdom", *Journal of Economic History*, November 1948.
6. Jacob Viner, *Studies in the Theory of International Trade*, 1937, pp. 558–564.

situations in which the commodity terms of trade serve as a better indicator of the welfare gain to the community from trade.

Single Factoral Terms of Trade

With the changing factor productivity, the concept of the net barter terms of trade is distinctly misleading as a measure of the gain from trade for a country without keeping in view the improvement in the factor productivity which results from trading. The concepts which used commodity technical coefficients index were developed by Jacob Viner.[7] The concepts developed by him are known as the 'single factoral' and the 'double factoral' terms of trade.

The single factoral terms of trade is the ratio of the export price index and the import price index adjusted for changes in the productivity of country's factors of production engaged in the production of export commodities. If the commodity terms of trade index is multiplied by the reciprocal of the export commodity technical coefficients index, the resultant index is known as the 'single factoral terms of trade' which can be expressed as

$$T_s = \frac{P_X}{P_M} . Z_X$$

where T_s denotes the single factoral terms of trade, $\frac{P_X}{P_M}$ is the net barter terms of trade earlier denoted by T_C and Z_X denotes the index of factor productivity in exports. By substituting T_C for $\frac{P_X}{P_M}$ in the above formula, the single factoral terms of trade can also be written as

$$T_S = T_C . Z_X$$

The single factoral terms of trade form the basis of Alfred Marshall's theory since his representative bales or bundles of goods are so chosen that each bale contains a constant quantity of productive resources. By taking account of changes in the factor productivity in exports, the single factoral terms of trade removes the shortcoming of the net barter terms of trade as a measure of gain from trade. Let us assume that export price index falls by 10 per cent but export cost has fallen by 20 per cent. In this situation, the single factoral terms of trade will be $\frac{90}{100} \times 120 = 108$. Thus, the single factoral terms of trade for the country have improved by 8 per cent and the country is better off although the net barter terms of trade have deteriorated by 10 per cent. It is so because the fall in the export price index is more than compensated by an increase in the factor productivity in the export sector. In this condition, the use of the single factoral terms of trade is considered to be more logical and rational than the commodity terms of trade. When economic development takes place, the new techniques of production are introduced which reduce the cost of production per unit of output. Consequently, the use of the commodity terms of trade as a measure of the gain for a country from international trade is apt to be misleading.

However, the concept of the single factoral terms of trade suffers from the shortcoming that it does not take into consideration the potential domestic cost of production of imports. To overcome this weakness of the concept of single factoral terms of trade Jacob Viner has constructed another index known as the 'double factoral terms of trade'.

---

7. Jacob Viner, *Ibid.*

Double Factoral Terms of Trade

The concept of the double factoral terms of trade takes account of the productivity of factors of production entering into the production of country's exports as well as of the productivity of foreign factors of production producing country's imports. The concept of the double factoral terms of trade can be expressed as

$$T_d = \frac{P_X}{P_M} \cdot \frac{Z_X}{Z_M}$$

where the term T_d represents double factoral terms of trade, Z_M is the index of factor productivity in imports, Z_X is the index of factor productivity in exports, P_X and P_M stand for the price indices of exports and imports.

If the factor productivity of exports increases by 20 per cent while the factor productivity of imports increases by 30 per cent, the single factoral terms of trade which were favourable will become unfavourable. Consequently, the concept of the double factoral terms of trade is a more scientific and logical index than the concept of the single factoral terms of trade. According to Kindleberger, however, the concept of the single factoral terms of trade is more useful than the concept of the double factoral terms of trade because we are interested in knowing what our factors can earn in terms of goods from abroad and not in knowing what our factor services can command in terms of the services of foreign factors.[8]

In practice, however, both these concepts of the single factoral and double factoral terms of trade have little utility due to the difficulties involved in the measurement of changes in factor productivity. It is difficult to calculate these terms in practice because it is almost impossible to define operationally and to measure statistically the concept of a 'unit of productive factor' and, therefore, that of a productive index. Moreover, these concepts do not measure the real gain from trade as they neglect the relative utility obtained from imports as compared with the utility lost as a result of the denial to natives of the consumption of goods which could have been produced with the resources employed in the production of exports that are needed to pay for imports. This shortcoming is removed in the 'real cost terms of trade'.

Despite the difficulty in calculating the factoral terms of trade, many economists still prefer the concept of the factoral terms of trade because the interpretation of the long-run historical changes depends largely on the particular causes and circumstances which cause them. For example, it is frequently alleged by the developing countries that, barring some temporary interruptions, since the 1870s the secular terms of trade have moved against them. In other words, their complaint is that while the world prices of manufactured goods which they import have risen substantially, the prices of raw materials and agricultural products which constitute bulk of their exports have either remained unchanged or risen only marginally. However, even assuming that their assertion is correct, it does not at all follow from this that the developing countries are any worse off today or that they derive today less advantage from foreign trade than previously, or that the circumstances which have led to the alleged deterioration in their terms of trade have affected them adversely. It would depend on the nature of the causes of the deterioration in their terms of trade.

Let us assume that the commodity terms of trade have moved against country X because for some reason country Y's demand for X's goods has diminished. It might be either due to a fall in Y's national income or due to greater foreign competition faced by country X's exports in Y country's market, or due to the import-competing industries in country X having become mature. In

---

8. Charles P. Kindleberger, *International Economics*, Fifth Edition, 1973, p. 76.

these cases, deterioration in the terms of trade is undoubtedly unfavourable for country X because beside the commodity terms of trade, the single and perhaps even the double factoral terms of trade have moved against country X. The commodity terms of trade may, however, also deteriorate due to increase in the productivity of X's export industries as a consequence of which prices of her export goods fall. Provided the fall in the prices of export goods of a country is less than the increase in her factor productivity in the export sector, the position of the country is better compared with what it was before the change occurred. This is because the single factoral terms of trade have not deteriorated; these have rather improved. The country would, of course, have enjoyed the entire gain resulting from the increase in factor productivity in the export goods sector had country Y's demand for country X's exports been perfectly inelastic.

Real Cost Terms of Trade

The real cost terms of trade are obtained by multiplying the single factoral terms of trade by the index of the amount of disutility (pain, irksomeness. etc.) per unit of the productive resources employed in producing the exports. The real cost terms of trade can be expressed as

$$T_t = T_s.R_x = \frac{P_X}{P_M}.Z_X.R_X$$

where T_r denotes the real cost terms of trade and R_X stands for the index of the amount of disutility suffered per unit of productive resources employed in producing the exports. The terms P_X, P_M and Z_X have the same meaning as in the formula of the single factoral terms of trade.

The real cost terms of trade is a better measure of the real economic welfare achieved by the country from international trade. It, however, ignores the real cost involved in the production of imports. To overcome this defect the concept of the 'utility terms of trade' has been introduced by Jacob Viner.

Utility Terms of Trade

Demis HolmeThe utility terms of trade are obtained by multiplying the real cost terms of trade with the index of the relative desirability or utility of imports as compared with the goods that could have been produced for domestic consumption with those factors of production which are now used in the production of export goods (U_M). The utility terms of trade may be written as

$$T_u = T_t.U_M = \frac{P_X}{P_M}.Z_X.R_X.U_M$$

Robertson[9] calls it 'the true terms of trade'. The concepts of the 'real cost terms of trade' and the 'utility terms of trade' refer to the subjective costs (disutility) whose measurement is not possible. Consequently, these concepts are of no significance in practical life. In practical life, the concept of the commodity terms of trade is generally employed to measure the total gain that accrues to a country from international trade.

Factors affecting the Terms of Trade

The terms of trade of a country are determined by the relative intensities of her import demand and export demand compared with those of the other country. In other words, changes in the reciprocal

9. Dennis Holme Robertson, "The Terms of Trade", *International Social Science Bulletin,* Spring 1951, p. 29.

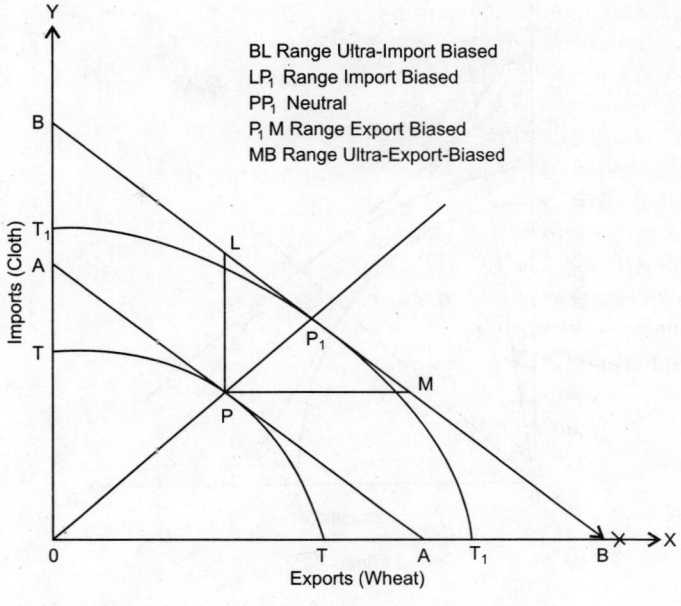

Figure 12.1

demand elasticities will cause changes in the terms of trade of a country. Changes in the elasticity of demand are the result of several factors such as population growth, nature of the goods imported and exported, government's trade policy, tastes and capacity to import, etc. Any change in anyone or all of these factors will cause a change in the relative intensity of demand of a country for other country's goods which in turn will affect her terms of trade. The major factors affecting the terms of trade of a country are:

1. Economic growth,
2. Shifts in the demand for country's exports and/or imports,
3. Tariff,
4. Devaluation, and
5. Availability of substitutes.

We may now discuss each one of the above five factors which affect the terms of trade of a country.

Economic Growth and the Terms of Trade

Economic growth will cause an outward shift in the production possibilities curve of a country allowing for larger aggregate output. The upward shift in the production possibilities curve may occur due to increase in the supply of the factors of production and technological improvement allowing the production of larger aggregate output with the given quantity of resources. The growth pattern and its impact on a country's terms of trade has been shown in Figure 12.1 where production takes place under the conditions of increasing opportunity costs, i.e., production of both the goods (wheat and cloth) in the country obeys the universal law of diminishing returns.

Given the initial production possibilities curve TT, the country will produce at point P and the

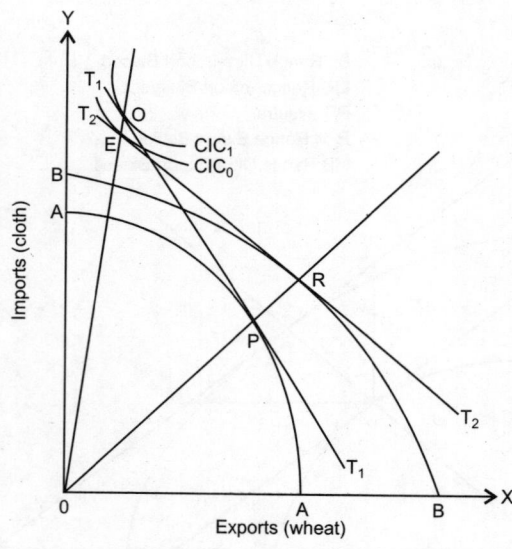

Figure 12.2

terms of trade will be shown by the slope of the exchange-ratio line AA. Due to economic growth, the production frontier of the country will shift outward as shown by the higher production possibilities curve T_1T_1. Consequently, the new terms of trade will be shown by the slope of exchange-ratio line BB. The right-angled triangle PLM marks the boundary within which growth can be defined to be import-biased, neutral or export-biased. If as a consequence of economic growth, production in the country takes place to the right of point M on the new terms of trade line BB, economic growth is ultra-export-biased. If production takes place to the left of point L on the new terms of trade line BB, economic growth will be ultra-import-biased. If economic growth is, however, neutral, then production will take place on point P_1 which is situated on the terms of trade line BB. In this situation, the relative shares of cloth and wheat will stay constant and the terms of trade will remain unaffected (in the special case the terms of trade will be unfavourable, it has been discussed in the following paragraph). If economic growth is ultra-export-biased, the country will produce more quantity of the export good and less quantity of the import good. This condition will necessarily worsen the terms of trade of the country. On the other hand, if growth is ultra-import-biased, the production of the import commodity will increase so much that the production of the other commodity will decrease in absolute terms and the terms of trade will be favourable for the country.

If the country is significantly large to influence the international terms of trade by her action, then even with the neutral-trade-biased economic growth there exists the possibility that the terms of trade may turn so strongly against the growing country that she will actually wind up on a lower community indifference curve than she could attain without economic growth. Jagdish Bhagwati[10] has described this situation as a case of "immiserizing growth". This has been shown in Figure 12.2.

Figure 12.2 shows that initially the production in the country takes place on the production possibilities curve AA at point P where the terms of trade line T_1T_1 is tangent to country's

10. Jagdish Bhagwati, "Immiserizing Growth: A Geometrical Note", *The Review   of Economic Studies*, Volume 25, June 1958, pp. 201–205; and "International Trade and Economic Expansion", *The American Economic Review*, December 1958, pp. 941–953.

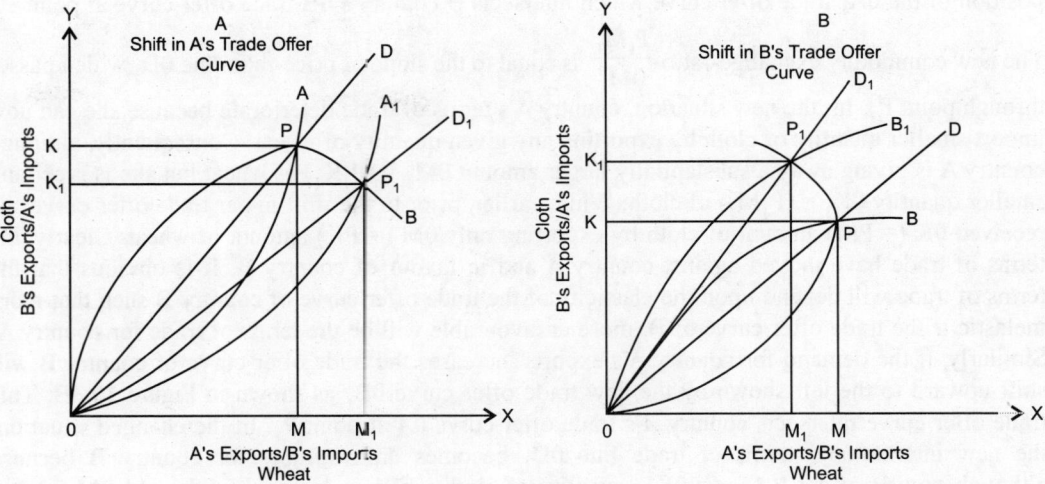

Figure 12.3

production possibilities curve AA. The country consumes the commodity combination shown by point D which is located on the community indifference curve CIC_1. After economic growth in the economy, the production equilibrium changes from point P to point R and the terms of trade line changes from T_1T_1 to T_2T_2. The equilibrium in consumption in the new situation is shown by point E which is located on the lower community indifference curve CIC_0 showing that the country has suffered a fall in the community welfare as she has been forced to step down from the higher community indifference curve CIC_1 to the lower community indifference curve CIC_0. Economic growth and the consequent change in the terms of trade forces the community in the country to descend to a lower community indifference curve. Consequently, the community is worse off due to economic growth.

Shifts in the Demand for Exports, Imports and the Terms of Trade

In equilibrium, the total value of exports of one country at any given particular terms of trade should be equal to the total value of her imports. Other things remaining the same, if the demand for imports of the country increases, the prices of imports relative to those of exports will rise. Consequently, the commodity terms of trade of the country will deteriorate in as much as for a given quantity of imports the country will have to give away a larger quantity of exports. Similarly, if the demand for the exports of the country increases, the prices of her exports relative to the prices of her imports will rise and the commodity terms of trade of the country will improve. This has been illustrated in Figure 12.3A and Figure l2.3B.

In Figure 12.3A, 0A is the initial trade offer curve of country A while 0B is the initial trade offer curve of country B. The two offer curves intersect at point P where the exports and imports of country A are equal to the imports and exports of country B. The cloth-wheat international

exchange ratio $\dfrac{PM}{OM}$ has been shown by the slope of line 0D which passes through point P. When country A's demand for imports increases, A's trade offer curve shifts downward to the right to the position of the $0A_1$ trade offer curve which intersects B country's 0B trade offer curve at point P_1. The new commodity exchange-ratio $\dfrac{P_1M_1}{0M_1}$ is equal to the slope of price-ratio line $0D_1$ which passes through point P_1. In the new situation, country A's terms of trade deteriorate because she can now import smaller quantity of cloth by exporting any given quantity of wheat. Consequently, although country A is giving away a substantially larger amount $0M_1 = (P_1K_1)$ of wheat but she is receiving smaller quantity $0K_1 = (P_1M_1)$ of cloth. While earlier, prior to the shift in her trade offer curve she received 0K (= PM) amount of cloth by exporting only 0M (= PK) amount of wheat. Clearly, the terms of trade have moved against country A and in favour of country B. It is obvious that the terms of trade will depend upon the elasticity of the trade offer curve of country B such that more inelastic is the trade offer curve of B, more unfavourable will be the terms of trade for country A. Similarly, if the demand for country A's exports increases the trade offer curve of country B will shift upward to the left shown by the new trade offer curve $0B_1$ as shown in Figure 10.3B. This trade offer curve intersects country A's trade offer curve $0A$ at point P_1. In the changed situation, the new international terms of trade line $0D_1$ becomes unfavourable for country B because although now B offers P_1M_1 (= $0K_1$) quantity of cloth which is larger than the old PM (= 0K) quantity of cloth but in exchange she receives only $0M_1(= P_1K_1)$ amount of wheat which is smaller than the old 0M amount of wheat. Consequently, the new international exchange-ratio $\dfrac{P_1M_1}{0M_1}$ is unfavourable for country B. Bigger the angle between the exchange-ratio line and the X-axis, more favourable will be the gross barter terms of trade for country A and *vice versa*.

Tariff and Terms of Trade

Let us assume that a country levies an import tariff in order to improve her terms of trade. When a country imposes tariff on her imports, its willingness to import is reduced and for a given quantity

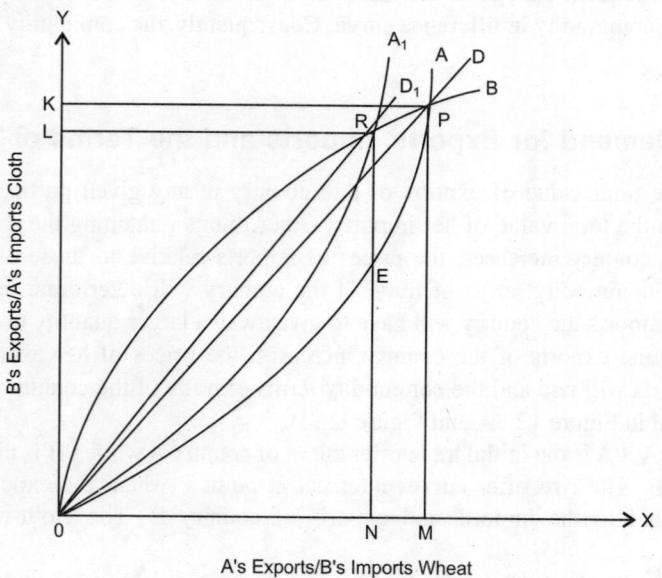

Figure 12.4

of exports the importers require a larger quantity of imports, a part of which is to be surrendered to the customs authorities. This has been shown in Figure 19.4.

Before the imposition of import tariff, the trade offer curves $0A$ and $0B$ of countries A and B intersect at point P and the equilibrium terms of trade are given by slope of line $0D$. Now tariff is imposed on her imports by country A. After imposing a 50 per cent *ad valorum* tariff on her imports, the trade offer curve $0A$ of country A shifts upward to the left to the position of $0A_1$ in such a way that the vertical distance between the original trade offer curve $0A$ and the new trade offer curve $0A_1$ everywhere is half of the distance between the X-axis and the original trade offer curve. In the figure, the new trade offer curve $0A_1$ of country A intersects country B's trade offer curve $0B$ at point R.

Consequently, the new equilibrium terms of trade line is $0D_1$. The slope $\dfrac{RN}{ON}$ of the new equilibrium terms of trade line $0D_1$ is greater than the slope $\dfrac{PM}{0M}$ of the old equilibrium terms of trade line $0D$. After the imposition of import tariff, the RE amount of import tariff is collected by the customs authorities. The terms of trade become favourable for the tariff-imposing country. However, the specific effects of tariff will depend on the elasticities of the two trade offer curves. Tariff shall improve the terms of trade for the tariff-imposing country if the elasticity of the other country's trade offer curve is greater than unity and less than infinity. The effect of an import tariff may, however, be either nullified or more than nullified if the country B retaliates the tariff moves of country A by simultaneously imposing an equal or more than equal amount of tariff on her imports.

Devaluation and the Terms of Trade

Devaluation lowers the value of the home currency unit expressed in terms of the currency of foreign country. After the foreign currency value of the home currency has been reduced as a result of devaluation, it is natural to think that the terms of trade of the currency-devaluing country will deteriorate because the prices of country's exports fall in terms of the foreign currency while the prices of her imports rise in terms of the home currency. But this is not a bare fact. As a matter of fact, the effects of devaluation will depend on the elasticities of demand for and supply of imports and exports of the country. Devaluation will tend to improve the terms of trade if the product of the demand elasticities for the country's imports and exports is greater than the product of the supply elasticities of her imports and exports.

According to the ultra-classicists, including Frank D. Graham and others, the terms of trade are likely to be unchanged because a country typically transacts at world prices on which it has no control.[11] As opposed to the ultra-classicists, the neoclassical economists are of the opinion that currency devaluation worsens the terms of trade while currency appreciation improves them. It is so because most countries specialise in the exports of a few commodities the foreign demand for which is relatively inelastic while these countries import several commodities from several different countries. Consequently, the supply of their imports is relatively elastic. According to Mrs. Joan Robinson, "... one country plays a more dominant role in the world supply of those goods which it exports than it plays in the world market for those goods which it imports."[12] According to Charles P. Kindleberger, "it is particularly true of larger developing countries."

11. Charles P. Kindleberger, *Op. cit.*, p. 264.
12. Joan Robinson, "Beggar My Neighbour Remedies for Employment", *Readings in the Theory of International Trade*, 1953, p. 40.

This is not, however, true in all cases. If a country enjoys the monopsony power in imports she will specialise in imports and generalise in exports. In this situation, devaluation of currency will improve the terms of trade. The outstanding example of this situation is offered by England whose terms of trade improved after the depreciation of the pound-sterling in 1931 because the world prices of British imports fell more than did the world prices of British exports.

In general, if the elasticities of supply of exports and imports are large in proportion to the elasticities of demand for exports and imports, devaluation of a country's currency will worsen its terms of trade while revaluation will improve its terms of trade. Expressed differently, if the product of the supply elasticities is greater than the product of the demand elasticities, devaluation will cause deterioration in the country's terms of trade and *vice versa*. If the elasticities of supply of exports and imports are equal to the elasticities of demand for exports and imports so that the product of the supply elasticities equals the product of the demand elasticities, the terms of trade will remain unchanged consequent upon currency devaluation. The relationship between the terms of trade and devaluation can be stated thus:

The terms of trade for the country will deteriorate, remain unchanged or improve as a result of currency devaluation if

$$\text{(i)} \quad S_X . S_M > D_X . D_M$$
$$\text{(ii)} \quad S_X . S_M = D_X . D_M \text{ and}$$
$$\text{(iii)} \quad S_X . S_M < D_X . D_M$$

Import Substitutes and the Terms of Trade

If close substitutes of import goods (import substitutes) are available in large quantity in the country, the terms of trade will be unfavourable for the exporting country. In the absence of the availability of close substitutes, the bargaining power of the exporting country will be strong. Consequently, the terms of trade will be favourable for the exporting country.

In addition to the above major factors, there are several other minor factors such as transfer problem, depression, political situation, etc. which also affect the terms of trade of a country.

Developing Countries and the Terms of Trade

For world's most developing countries which are caught in the vicious circle of low income, low employment and low technical know-how resulting in mass poverty, foreign trade is life and blood of the economy. Consequently, 'Trade or Perish' is an appropriate slogan in the context of the need for their rapid economic development. It is, however, argued by the spokesmen of underdeveloped countries that international trade which had served as an engine of economic growth for the open lands in the nineteenth century is no more an engine of economic growth for them today. According to the less developed countries, today they face unfavourable conditions which are very different from the favourable ones faced by the open lands in the nineteenth century. Raul Prebisch has argued that the terms of trade of the less developed countries have been secularly deteriorating.[13]

In his well-known work entitled *Towards a New Trade Policy for Development,* Raul Prebisch maintains that there is a long-run tendency for the prices of primary products to deteriorate relative to the prices of manufactured goods. Prebisch's argument is based on the fact that the

13. Raul Prebisch, "Commercial Policy in Underdeveloped Countries", *The American Economic Review, Papers and Proceedings,* May 1959, pp. 291–294.

underdeveloped countries are the net producers of primary products such as coffee, copper, tea, rice, sugar, fats etc. These products are roughly the same goods today that they were a century ago. On the other hand, the quality of manufactured goods produced by the developed countries such as automobiles, radios, petroleum refinery equipments, trucks, etc. has improved tremendously. Consequently, these goods fetch higher prices relative to those fetched by the primary products. This argument has, however, been rejected by Richard Lipsey and Jacob Viner on the ground that it does not corroborate with the empirical evidence.[14]

Singer has given a different explanation of the secularly deteriorating terms of trade of the primary products producing countries.[15] According to Singer, the fruits of technical progress can either be retained by the producers in terms of the high incomes or passed on to the consumers in the form of low prices. The fruits of technical progress in the developing countries have been passed on to the consumers in the developed countries in the form of lower prices of the primary products whereas these have been retained by the producers in the developed countries in the form of higher incomes of producers. Singer does not, however, give any explanation in support of his argument. McLeod has stated that this has happened due to the presence of greater degree of monopolistic control prevalent in the production of finished goods in the developed countries.[16] The less developed countries more or less produce the same (identical) goods and search the markets for these goods abroad. As these countries are less organised as sellers of their goods in the world markets, they compete among themselves for the export markets for their products and in the process give away a substantial part of their gain in productivity to the developed countries in the form of reduction in the prices of their exports.

According to Jagdish Bhagwati, the deterioration in the terms of trade of underdeveloped countries is due to the "immiserizing growth". The growth of underdeveloped countries is ultra-export-biased because that is all these countries know how to do. Lack of complementary resources, or other frictions prevent the resources from being readily shifted in the import-competing industries in the underdeveloped countries. Another reason given by Linder for the worsening terms of trade of underdeveloped countries is that import kills off import-competing industry. For example, cheap mill-made British textile exports made to India killed off completely the Indian handicrafts. Similarly, free trade between England and Ireland after 1801 resulted in the collapse of Irish industry forcing Irish workers to move across the Irish Channel in search for employment. In this difficult situation when it becomes impossible to make a living from import-competing industry, growth takes place in the export industry until it goes too far and in the process all the gains from foreign trade are lost to the country.

Another reason given for the deteriorating terms of trade of the developing countries runs in terms of Engel's law. As the national income of a country rises, the total expenditure as a proportion of national income incurred on the manufactured goods increases while that incurred on the agricultural products and minerals declines. To the extent that the income elasticities of demand determine the volume of exports and to the extent that the less developed countries grow food products in a world of growing per capita income, the demand for their exports grows more slowly than does the demand for the manufactured goods.

14. Harry G. Johnson, *Economic Policies Towards Less Developed Countries*, 1967, Appendix A, pp. 249–250.
15. H.W. Singer, "The Distribution of Gains between Investing and Borrowing Countries", *The American Economic Review, Paper and Proceedings*, May 1950, pp. 477–479.
16. A.N. McLeod, "Trade and Investment in Underdeveloped Areas: A Comment", *The American Economic Review*, June 1951, p. 141.

The less developed countries do not, however, produce only those goods whose income elasticity of demand is low but they also export products with high income elasticity of demand such as oil, rubber, non-ferrous metals, diamonds, artwares etc. If the underdeveloped countries can produce these goods more cheaply exports can grow very easily. For example, the exports of Japanese textiles grew not on the basis of their high income elasticities of demand but on the basis of their relative cheapness.

Technological change and factor growth in the developed countries seem to be biased against the less developed countries. In the developed countries capital has grown at a faster rate than population and technology makes it possible to substitute capital for land and labour which developing countries have in plenty. Production of synthetic rubber, artificial silk, cotton, rayon, fertiliser, etc. in the developed counties are part of this movement. This type of development in the developed countries causes deterioration in the terms of trade of the less developed countries as the dependence of the developed countries for raw materials on the less developed countries is reduced.

Linder[17] has emphasised the strategic role of scarcity of intermediate goods in the underdeveloped countries in the deterioration of the terms of trade of these countries in the long period. The lack of availability of the intermediate goods blocks the diversification and transformation of the economies of underdeveloped countries. If sufficient essential raw materials such as fuel, iron and steel, cement, etc. are not imported from the developed countries the domestic resources of the underdeveloped countries are not fully utilised. Since the possibilities of substituting domestic resources for imports are limited the need for larger imports due to higher population growth makes the terms of trade unfavourable for the underdeveloped countries.

The exports of the less developed countries have not been growing very fast and even today their share in the total world trade is insignificant. In 1996 the five countries which included the United States of America, Germany, Japan, France, United Kingdom, Italy and Canada accounted for one half of world merchandise trade. India, which has about 18 per cent of world's population, accounted for only 0.7 per cent in world merchandise trade. In short, the less developed countries have been marginalised in world trade. Contrary to expectations these countries have not been able to sell their exports at increasing prices. Consequently, trade has failed to transmit growth from the developed industrial countries to the less developed countries.

While the *per capita* income in the developed countries has recorded rapid rise, it has failed to generate a proportional increase in the demand for primary products. It seems plausible that the low income elasticity of consumer demand for many agricultural commodities which constitute bulk of exports of the less developed countries, growing agricultural protectionism in developed countries affecting the imports of primary products adversely, substantial economies achieved in industrial uses of natural materials and growing displacement of natural raw materials by synthetic and other man-made substitutes have been mainly responsible for this pathetic situation of the less developed countries.

Importance of the Terms of Trade in the Policy-making

Terms of trade are of great importance both in the national and international policy-making decisions. There are, however, several technical questions which should be resolved before the terms of trade can be used as a basis for policy-making. Firstly, the terms of trade should be

17. S.B. Linder, *Trade and Trade Policy for Development,* 1967.

measured in foreign currency prices. They can, however, also be measured in local currency prices if the conversion factors are the same for both exports and imports. Secondly, a marked shift over a period in the commodity-mix of exports and imports creates complications. Thirdly, there is the problem of choice of the base year. If any given particular year chosen as the base year is one in which export prices are relatively high, the terms of trade will appear to be unfavourable in the subsequent period. Conversely, if the export prices are low in the base year, the terms of trade will be unrealistically favourable. While there may be special reasons for favouring the choice of one year over another, the base year problem could partly be avoided by fitting trend lines to export prices and import prices and comparing the two trends. Even here, however, there is the problem of deciding what overall period one should consider and how far back should one go to make such a comparison.

Although the policy objective of not allowing a country's terms of trade to deteriorate is commendable, however, the means to achieve this objective are neither simple nor clear-cut. Judging from the plethora of speeches made at international conferences, policy-makers in the developing countries mainly rely on using the more developed trading countries to maintain the demand for raw material exports from their countries while taking appropriate steps to prevent the prices of industrial goods which underdeveloped countries import from rising. Although such exhortations may have some propaganda value, these have failed in preventing the developing countries' terms of trade either from fluctuating in the short period or from secularly deteriorating according to changes in the international market conditions.

More recently, the oil producing countries had succeeded in improving their terms of trade by restricting output and raising the export prices of oil substantially. Their success was partly the result of solidarity shown by the oil producing countries who formed an international cartel known as the OPEC (Organisation of Petroleum Exporting Countries) and was partly due to the inelastic world demand for oil, at least in the short run. While the other developing countries would also like to take similar measures to improve their terms of trade, international cartels for other commodities are not likely to meet the same success as the OPEC because close substitutes are available for other raw materials. Even the OPEC's bargaining power has considerably weakened.

The problem of the secular deterioration of the terms of trade being faced by the developing countries can be considerably solved by means of stabilisation of international commodity prices. Exports of primary products are by far the most important source of foreign exchange earnings for most developing countries accounting for 85 to 90 per cent of their export earnings. But the prices of primary products fluctuate widely leading to instability in foreign exchange earnings. The instability gives rise to many problems, including hurdles in planned economic development of these countries. If, somehow, the prices of primary products could be stabilised, an important obstacle to development could be eliminated. So far, however, the efforts of the UNCTAD and other trade bodies notwithstanding, precious little has been achieved. There must be a long-term stabilisation of the terms of trade between developing countries' primary exports and their principal industrial imports.

Difficulties in the Measurement of the Terms of Trade

It has been discussed earlier that Jacob Viner's concepts of the single factoral and double factoral terms of trade are not of much use in practical life because of the manifold difficulties involved in the calculation of variations in the factor productivity. Similarly, the concepts of the *real cost terms of trade* and the *utility terms of trade* cannot also be employed because of the formidable difficulties involved in the measurement of the utilities and disutilities. Consequently, we generally make use of only the net barter or commodity terms of trade as an indicator for measuring the gains from trade. This concept is, however, confronted with many statistical problems.

To measure the relative changes in prices, index numbers of export and import prices are utilised. For the construction of index numbers, selection of some suitable base year is of vital importance. The particular base year chosen should be a normal year judged from various considerations. A base year which is normal from all considerations is, however, very rare to come across in real life. Generally, countries choose a favourable year and not a representative year as a base year. For example, English economists always measure changes in England's terms of trade selecting 1938, when their goods were the most preferred ones in the world, as the base year. On the other hand, the developing countries today choose a base year of 1950 when the prices of primary products rose sky-high due to the Korean war. This tendency to select a biased base year gives deflated index numbers.

Secondly, index numbers of import and export prices make no allowance for changes in the quality of goods traded or for changes in the composition of goods entering into international trade. Changes in the quality and composition of exports and imports frequently take place. For example, nylon stockings of today cannot be compared with cotton socks of the past and mercury vapour lamp cannot be compared with kerosene lamp. The present almost ubiquitous use of cooking gas in the kitchen and of washing machines for cloth laundering was unknown three decades ago. Many commodities which were unknown a few years back cannot be given suitable weight today. Consequently, secular changes in the prices of goods and the volume of commodities cannot be measured with precision and accuracy. Moreover, the net barter or commodity terms of trade cannot serve as a reliable measure of the welfare gains from trade without taking changes in the factor productivity into account. With changing efficiency, the net barter terms of trade are distinctly misleading as a measure of welfare. We must also take into account improvements in productivity. Moreover, the time lag between exports and imports also poses a problem in the exact measurement of the terms of trade. If a country exports when the terms of trade are favourable for her due to lower prices of imports (export prices remaining constant) and imports when the terms of trade are favourable due to the higher prices of exports (prices of imports remaining constant) her actual terms of trade remain constant although the indices show an improvement in her terms of trade.

Gains from Trade

Different countries engage in trading with one another because they gain from trade. In short, the main *raison d'etre* of trade resides in the gains which countries reap from trading with one another. Looking at the world as a whole, free trade leads to a situation of higher level of community welfare compared to a situation of zero free trade. Consequently, it is advantageous for the world as a whole to engage in international trade. A very clear, vivid and lucid description of gains from trade is traceable in the writings of Adam Smith, the founder member of the classical school of economists. The following classic passage in Adam Smith's magnum opus *The Wealth of Nations* reveals the complex phenomenon of gains from trade.

"Between whatever places foreign trade is carried on, they all of them derive two distinct benefits from it. It carries out that surplus for which there is no demand among them, and brings back in return from it something else for which there is a demand. *It gives values to their superfluities,* by exchanging from something else, which may satisfy a part of their wants and increase their enjoyments. By means of it, the narrowness of home market does not hinder the division of labour in any particular branch of art or manufacture from being carried to the highest perfection. By opening a more extensive market for whatever part of the produce of their labour may exceed the home consumption, *it encourages them to improve its productive powers,* and to augment its annual produce to the utmost and thereby to increase the real revenue and wealth of society."[18]

From a perusal of Adam Smith's above passage following two important facts emerge:

1. International trade enlarges the horizon of market and creates an outlet for the surplus products over domestic consumption of the isolated economy. If the isolated economy, about to enter into trade, has surplus capacity (surplus capacity is not merely a matter of surplus land by itself, but of surplus land combined with surplus labour and the surplus labour is linked with the concept of unproductive labour) suitable for export markets, it has 'costless' means of acquiring imports and expanding the aggregate domestic economic activity.

When trade takes place between the old countries where total resources are given and are fully employed (as was assumed by comparative cost advantage theorists), the function of trade is to reallocate these given resources more efficiently between the production of domestic and export goods in the light of new set of prices now open to the countries. In this situation, each country gains when the total output increases as a result of the more extensive application of the division of labour and specialisation which become possible when goods are produced on a larger scale as a consequence of trade. David Ricardo had considered the gains from trade arising through the availability of cheaper commodities in each country engaged in trading. According to Ricardo, international trade contributed to the increase in the "sum of enjoyments" by contributing very powerfully "to increase the mass of commodities". In his *Notes on Malthus,* David Ricardo had stated that if two regions engage in trade with one another "the advantage to both places is not that they have any increase in value but with the same amount of value they are both able to consume and enjoy an increased quantity of commodities."[19] The gains from trade accruing to the countries are in the form of greater magnitude of more kinds and varieties of goods that become available for consumption in each country as a result of international trade. Thus, as was rightly observed by Thomas Robert Malthus, the gain from trade consists of "the increased value which results from exchanging what is wanted less for what is wanted more" and international trade "by giving us commodities much better suited to our wants and tastes than those which had been sent away, has decidedly increased exchangeable value of our possession, our means of enjoyment and our wealth."[20] This view is much nearer to Adam Smith's above-cited statement. John Stuart Mill calls it the 'direct gains' from trade. Modern economists refer to it as the 'static gains' from trade.

2. By increasing the size of the market, international trade improves the scope of division of labour in production and consequently raises the general level of productivity within the country. This is often referred to as the 'productivity' theory of international trade. "The productivity doctrine looks upon international trade as a dynamic force which, by widening the extent of the market and the scope of division of labour raises the skill and dexterity of the workmen, encourages technical innovations, overcomes technical indivisibilities and generally enables the trading countries to enjoy increasing returns and economic development."[21]

---

18. Adam Smith, *An Inquiry into the Nature and Causes of the Wealth of Nations,* Volume I, Cannan Edition, p. 413.
19. David Ricardo, *Notes on Malthus,* 1820, p. 215.
20. Thomas Robert Malthus, *Principles of Political Economy,* 1820, pp. 461–462.
21. Hla Myint, "The Classical Theory of International Trade and the Underdeveloped Countries", *The Economic Journal,* June 1958, reprinted in *The Readings in International Economics,* Volume XI, p. 320.

John Stuart Mill regarded this gain as "indirect effects" of trade which must be counted as benefits of the higher order. This type of gain is considered as 'dynamic gains' from trade by the modern writers.

From the above description it is clear that we can classify the gains accruing to a country from foreign trade into (1) static gains, and (2) dynamic gains.

Static Gains and Their Measurement

In this case each country gains from trade when the total output of goods increases as result of the extension of division of labour and specialisation. The producers are stimulated to make a more efficient reallocation of economy's given resources under given technology in the light of changing sets of relative prices as result of the commencement of international trade. Under the static gains from trade, although as result of trade the producers in the country move along the production frontier; the frontier, however, itself remains unchanged, i.e., the production possibilities curve of the country is assumed as *given*. As a result of trade, only the consumption frontier expands because the consumers are enabled to reach on a higher community indifference curve. Consequently, consumers enjoy a higher aggregate satisfaction which becomes possible partly due to the more favourable terms of trade on which goods are exchanged and partly due to the more efficient use of the given productive resources of the economy.

Static gains from trade are measurable. Jacob Viner[22] has developed the three different methods for measuring the national gains from international trade. These three methods are the: (1) economy in the cost of obtaining a given real income measured with the help of comparative cost advantage principle, (2) increase in the national income of the country, and (3) improvement in the terms of trade as an index of the international distribution and trend of gains from trade. The first two methods are identical when referred to *a point of time.* Any specific amount of gain can be expressed either as a reduction in the cost per unit of the real income or as an increase in real income per unit of cost. These two methods require a lot of necessary information which is not easy to collect. Even if the necessary massive information is somehow collected, its computation requires immense labour and dexterity. Consequently, instead of measuring the absolute gain from trade *at any given point of time,* economists content themselves with estimating the direction of the movement of that gain. For this purpose they adopt the third method of measuring the gains from trade, i.e., improvement in the terms of trade as an indicator of the gains from trade.

The incarnation of welfare economics in international trade theory has made it possible to measure the gains from trade in terms of the improvement in aggregate welfare. Firstly, we shall start with an investigation of the effects of trade on world's welfare without paying attention to the individual countries. Subsequently, we shall proceed to analyse the effects of trade on individual country's welfare.

World Welfare

International trade leads to more efficient allocation of world's total productive resources. An optimally efficient resource utilisation maximises the aggregate output which increases welfare of the world as a whole. To demonstrate this, let us suppose that there are only two countries A and B in the world producing only two commodities wheat and cloth. In other words, we assume,

22. Jacob Viner, *Op. cit.,* Chapter 1.

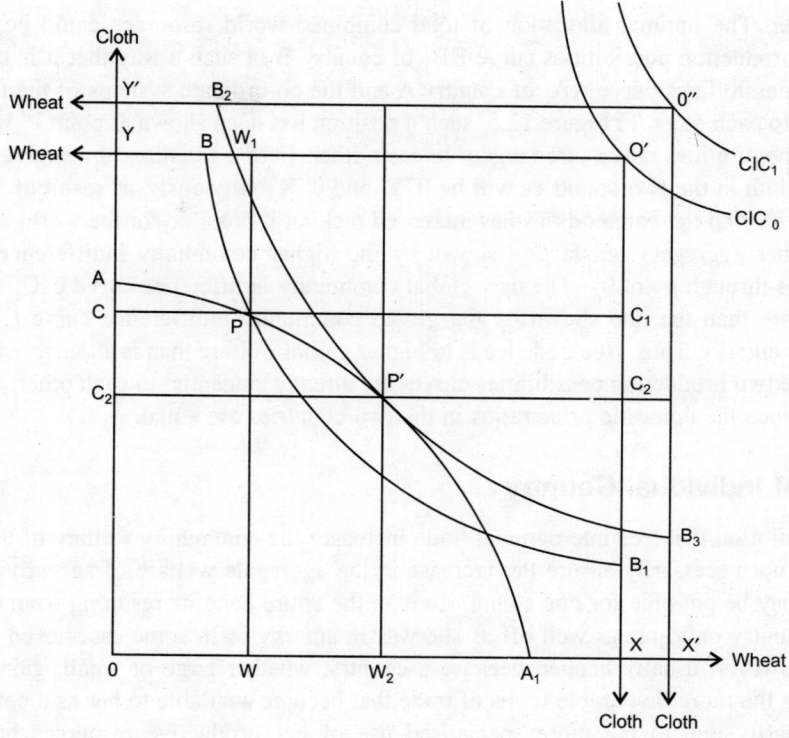

Figure 12.5

though unrealistically, a two-country and two-commodity world. In Figure 12.5 wheat has been measured the on 0X (or 0′Y) and cloth has been measured on the 0Y (or 0′X).

In the figure, the production possibilities curve of country A has been shown by AA₁. The consumption pattern of people in the country is such that the production possibilities curve and the community indifference curve (not shown in figure) touch one another at point P. Consequently, equilibrium occurs at point P in the sense that the marginal rate of substitution or transformation in production equals the marginal rate of substitution in consumption. Consequently, country A is producing and consuming 0W quantity of wheat and 0C quantity of cloth. A similar production frontier BB₁ of country B has been drawn and is superimposed on the original diagram upside down and with sides reversed. It has been placed in such a way that country B's consumption and production point coincides with country A's production and consumption point P. Consequently, country B is producing and consuming C₁0′ quantity of cloth and 0′W₁ quantity of wheat. The aggregate production of wheat and cloth in this hypothetical two-country world is respectively 0′Y (= CP + PC₁) and 0′X (= WP + PW₁). The two countries together can attain a combined welfare level indicated by the global community indifference curve CIC₀ which passes through point 0′.

At point P, however, the marginal rates of transformation in production in the two countries are not equal. Consequently, the resource allocation of world's given resources is not optimal. For an optimal resource allocation, the marginal rates of transformation in both the countries should be identical and for this to happen both countries' production possibilities curves should be tangent to each other. The optimal allocation of total combined world resources could be achieved by sliding the production possibilities curve BB_1 of country B in such a way that it is tangent to the production possibilities curve AA_1 of country A and the co-ordinate systems of the two countries are parallel to each other. In Figure 12.5, such a position has been shown at point P′ where the two production possibilities curves are tangent to each other. In this situation, the aggregate output of wheat and cloth in the two countries will be $0''Y'$ and $0''X'$. Obviously, as result of free trade the total output of both the commodities has increased making it possible for the world community to enjoy a higher aggregate satisfaction shown by the higher community indifference curve CIC_1 which passes through point $0''$. The new global community indifference curve CIC_1 shows higher global welfare than the one shown by the global community indifference curve CIC_0 which is attainable in autarky. Thus, free trade leads to higher global welfare than is attainable in autarky. If, however, the two production possibilities curves are already tangential to each other, trade will not take place since the domestic price ratios in the two countries are equal.

Welfare of Individual Country

The demonstration that free international trade increases the community welfare of the world as a whole does not necessarily ensure the increase in the aggregate welfare of an individual country because it may be possible for one country to reap the entire benefits resulting from trade leaving the other country only just as well off as she was in autarky or in some cases even worse off. It does not, however, usually happen because a country, whether large or small, gains from trade partly due to the more favourable terms of trade that become available to her as a consequence of trade and partly due to the more specialised use of her productive resources brought about through the increase in the area of trade resulting in the economies of scale through large scale operations. The untouched part of its resources which a country reaches by enjoying a given level of welfare with a smaller resource use than would be necessary under autarky has been illustrated in Figure 12.6.

Figure 12.6 shows that in autarky the country produces and consumes at point H where her community indifference curve CIC_0 is tangent to her production possibilities curve AB. After the opening of trade, the international terms of trade have been shown by the slope of the price-ratio line TOT. The price-ratio line TOT_1 has been drawn such that it is parallel to the international terms of trade line TOT and tangent to the community indifference curve CIC_0 at point R showing that the country can attain the same aggregate welfare (satisfaction) which was achieved in autarky if the production takes place anywhere along the TOT_1 line. All points located on the TOT_1 price-ratio line and lying below point D are attainable by the producers. But all points situated on this part of the price-ratio line TOT_1 lie inside the production possibilities curve AB showing that these can be reached without making full use of country's productive resources. The resources thus saved can be used for other productive purposes. The saved domestic resources represent a net gain which accrues to the country from trade.

Paul A. Samuelson on the Gains from Trade

Paul A. Samuelson[23] in his two well-known articles highlighted the gains resulting from trade. The first article entitled 'Gains from International Trade' was published in May 1939 in *The Canadian*

---

23. P.A. Samuelson, "Gains from International Trade", *The Canadian Journal of Economics and Political Science*, May 1939; and "The Gains from International Trade Once Again", *The Economic Journal*, Volume 72, 1962, pp. 820–829.

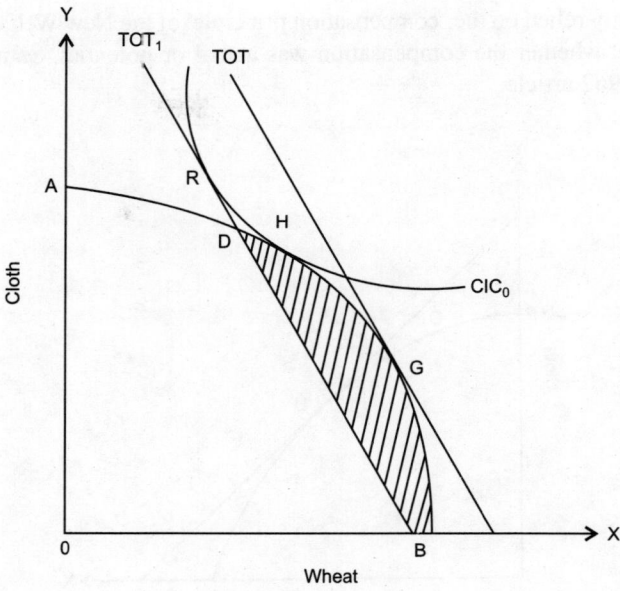

Figure 12.6

Journal of Economics and Political Science while the other article entitled 'The Gains from International Trade Once Again' was published in 1962 in *The Economic Journal*. Although the 1962 article is a considerable improvement over the 1939 article, the two articles are complementary.

Samuelson's 1939 article, based on the special assumption that the country under consideration is too small to influence the international terms of trade and that the price ratios abroad differ from those at home, shows that some international trade makes the country better off than if the country restricted herself to autarky. Samuelson recognises that free trade may harm some people but the harm which free trade inflicts on one group of people is necessarily less than the gain it confers on the other group or groups of people. Consequently, it is always possible to bribe or compensate those who suffer as a consequence of trade either by means of subsidy or through other redistributive devices so as to leave all better off as a result of trade. But what will happen after trade if such ideal redistributions are not perfected in actual life? Can we then say that free trade increases the national welfare? This question was answered by Samuelson in his article published in *The Economic Journal* in 1962.

Before discussing Paul A. Samuelson's 1962 article let us explain diagrammatically his 1939 classic article which shows that "free trade is superior to no trade". This has been illustrated in Figure 12.7 where PG is the production possibilities curve of the country. In autarky, the country produces and consumes at point D which is located on her production possibilities curve PG. Consequently, the consumption possibilities frontier PDG and the production possibilities frontier PG are identical. After some trade, with international prices differing from the domestic prices, the terms of trade line EF is tangent to the production possibilities curve PG at point U. The line EVUF represents the free trade consumption frontier. Since the free trade consumption frontier lies everywhere (except at the tangency point U) above the autarky consumption frontier, the society can have more of all goods (and less of all irksome inputs) with some trade. It is in this sense that trade makes a country potentially better off. This is the simple demonstration of Samuelson's 1939 article which implicitly relied on the 'compensation principle' of the New Welfare Economics. There was, however, doubt whether the compensation was actual or potential. Samuelson discussed all these issues in his 1962 article.

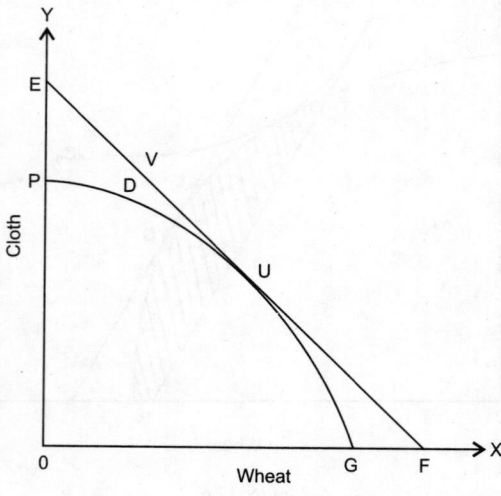

Figure 12.7

In his 1962 article, Samuelson dropped the assumption of a small country and treated the case of a large country. The very fact of 'largeness' of the country means that she is in a position to influence the international terms of trade. As soon as the large country enters the international market her trading activities will tend to raise the prices of her imports and lower the prices of her exports. In this situation, the terms of trade will no longer be represented by a straight line but by a curve known as the 'society-cum-trade consumption possibility frontier' which is formed by the envelope of the foreign trade offer curves which have their origin at various points on the home country's production possibilities curve. This has been shown in Figure 12.8.

In Figure 12.8A, AA_1 offer curve of country A has been shown at the various terms of trade, i.e., T_1, T_2, T_3 etc. In Figure 10.8B, PG represents the production possibilities curve of home country B. To find out the trade possibility curve of country B, let us choose some arbitrary terms of trade such as T_1 and place the origin of the co-ordinate system of country A in such a way that its origin is tangential to the production possibilities curve PG. For the T_1 terms of trade it is true at point R and for the T_2 terms of trade it is true at point R_1 and so on. The trade offer curve of country A shows her willingness to trade at the difference terms of trade. At the T_1 terms of trade she wants to move to point N and at the T_2 terms of trade she wants to move to point M. If the process is repeated for every conceivable terms of trade like T_1, T_2, T_3 etc. it is possible to generate a whole set of points showing the willingness of country A to trade with country B. The line connecting all such points E, N, M, Q and F represents the trade possibility curve of country B showing the different commodity combinations that are attainable by country B while trading with country A. The new trade consumption frontier ENMQF lies uniformly (except at point Q) outside the autarky consumption frontier PG. Consequently, Samuelson concludes that the "society is potentially better off in the sense that there is a way of reallocating the enlarged totals of goods so as to make every person better off."

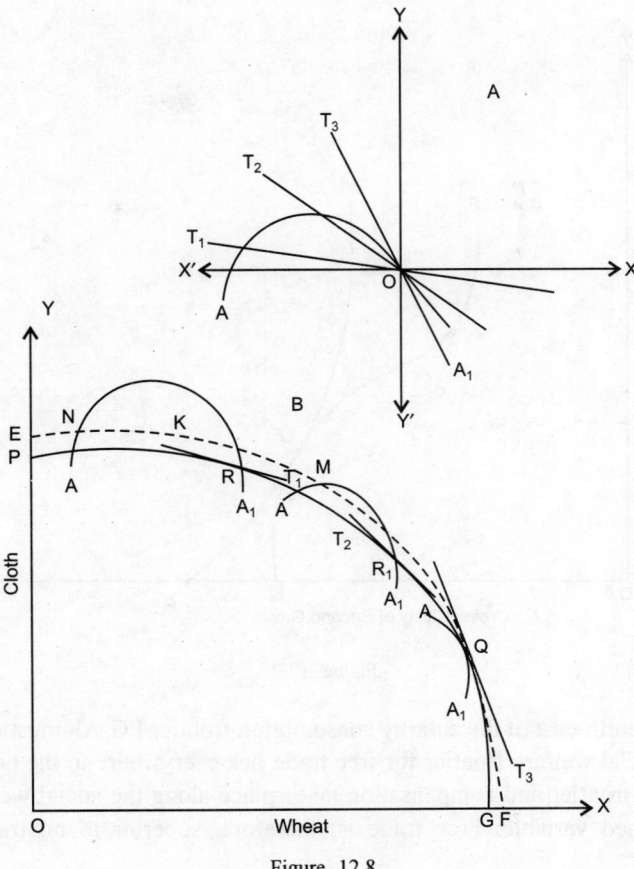

Figure 12.8

Now we have to show that the trade-lovers are theoretically able to compensate the trade-haters for the harm done to them as result of trade thereby making everyone better off. For this purpose we should construct the utility frontier from the trade possibility frontier. This has been done in Figure 12.9.

Let us suppose that there are only two groups of identical citizens in the country. The horizontal axis measures the ordinal utility of the first group of residents and the vertical axis measures the ordinal utility of the second group of residents in the country. Point D in Figure 12.9 corresponds to point D of Figure 12.7. The broken locus BDB_1 shows the varying utility combinations if the fixed goods totals of D are allocated in favour of one group or the other by 'ideal sum transfers' so that there is no dead-weight loss or inefficiency involved in the transfer. The envelope curve to all the utility possibility curves arising from the points located on the production possibilities curve PG in Figure 12.7 has been shown by PQ. This curve is the social welfare frontier for the no-trade situation. Similarly, VW is the utility possibility curve for the commodity bundle at some point on the trade possibility curve such as point K in Figure 12.8B and AF is the utility possibility frontier traced out for all the points on the free trade consumption frontier ENMQF. This is tangential at point T to the dotted locus VW representing the ideal reallocation of the goods at the post-trade point K. Since the free trade consumption frontier

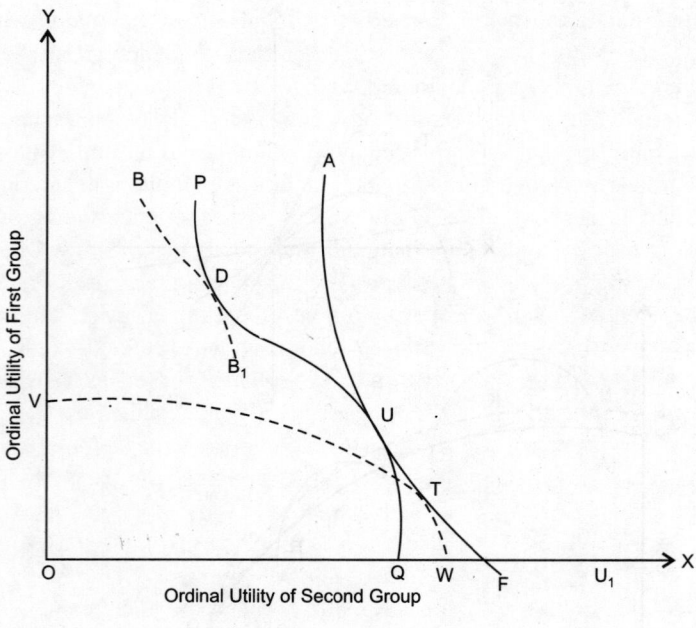

Figure 12.9

ENMQF lies to the north-east of the autarky consumption frontier PG, AF must lie to the north-east of PQ. Thus, the social welfare frontier for free trade lies everywhere to the north-east of the no-trade social welfare frontier and compensation takes place along the social welfare frontier when production is assumed variable. Free trade is, therefore, superior to no trade for all income distributions.

Dynamic Gains

We have so far discussed the static gains from trade. Consumers were given the opportunity to buy a good more cheaply which was costlier in the country before the commencement of trade. They were not, however, introduced to any new good as result of trade. Similarly, producers were stimulated to reallocate their productive resources in accordance with the new set of commodity prices that were now open to them as result of trade. No new technology which could bring surge in their production frontiers was, however, introduced in the analysis. In diagrammatical terms, the community's production frontier remained unchanged and only the community's consumption frontier expanded due to trade.

But this is not all what international trade brings about. International trade also brings fundamental measurable changes in the domain of consumption and production in the concerned countries. International trade introduces new consumer wants which sometimes work as a sort of industrial revolution in the early stages of economic development. The inauguration of trade with backward regions introduces the so-called "incentive" goods in the backward areas which prove sufficiently attractive for the local people. In order to buy these goods, they spare a part of their current gross natural product (GNP) for exchange against these goods. When the meagre quantity of new goods fails to satisfy their new wants they are induced to increase their output of agricultural and allied goods substantially in order to acquire these new goods. It spurs them to introduce new farm technology—mechanisation of agriculture—and ultimately they transform their subsistence agriculture into modern commercial farming.

Commercial farming needs new marketing centres and developed marketing facilities. Agriculture becomes more specialised employing new techniques of farming. Consequently, a vast network of railways, roads and allied infra-structural facilities are built in the interior of the country widening the range and influence of trade. In this way, as past experience shows, by generating the demand for modern transport facilities international trade facilitates the flow of capital from the developed to the developing regions of the world. The transporting facilities of railways and roads in the interior of the economy attract capitalists to invest their capital there. The establishment of new and the expansion of old centres of trade in the economy eventually create a market large enough to warrant the establishment of several market-oriented industries such as the breweries, commercial laundries, bottling plants, creameries, etc. These industries generate additional employment and income in the economy. Additional employment acts as a forceful stimulant for the establishment and growth of many industries like house-building, food and raw materials etc. It may also bring with it substantial external economies in the form of availability of trained labour and management. In short, one growth process is multiplied by the other and economic growth is generated creating its vital impact on the entire economy.

The development of the means of transport in the backward areas completely changes the outlook of the people living in these areas. They shed away their backwardness, modernising in the process their modes of production and consumption. The entire village economy is metamorphosed. The old bullock-cart is replaced by horse-drawn carts fitted with pneumatic rubber wheels. The link-road to the village removes the isolation of the village from other parts of the economy which are directly influenced by the impact of trade. In this way, the blessings of trade in the form of availability of new goods, flow of new ideas and new technology become accessible to the remote areas of the country. People's attitude toward life undergoes a sea-change. They no longer resign their lot to destiny and begin to believe in hard work and enjoyment of the material fruits resulting from it. Tradition and old values are replaced by new ones based on rational thinking and scientific reasoning. The change from the subsistence agriculture to commercial farming carried through practising the modern techniques of farming galvanises the entire rural sector of the economy. The farmers shift from old mud-dwellings to modern farm houses equipped with modern facilities. The case of progressive farmers in the states of Punjab, Haryana and Maharashtra proves the point.

After 1950 the growth of world trade has been strikingly rapid. It is evident from the fact that total value of world's merchandise export which in 1950 was merely US $ 376 billion expanded to US $ 7300 billion in 2003, and to still higher amounts in subsequent years. However, this rapid rate of growth has been lopsided and the expansion in world trade has been primarily concentrated on the exports of the industrial developed countries, the oil producing countries and few less developed manufacturing countries. More than one-half (over 58%) of the total merchandise exports was concentrated in only ten developed countries. The industrial countries have benefited from the rapid increase in world trade which has taken place during the past five and half decades.

Most undeveloped countries however have not experienced a similar stimulus from trade and the terms of trade have generally moved against these countries. However, as an exception, only a few of the fortunate less developed countries have succeeded in recording a rapid increase in the exports of manufactured goods. Among these countries are South Korea, Hong Kong, Singapore and Mexico. The majority of the less developed countries, however, still continue to export the traditional products, such as raw materials and agricultural products, the demand for which has grown at only a relatively low rate with the consequent continous fall in their share of the total world trade.

Experience shows that very often the existence of trade has promised high returns from mineral deposits or oil reserves or fertile farm-lands attracting foreign enterprise, capital and sometimes even labour to exploit these resources. To cite as an example, the development and commercial exploitation of the vast oil reserves in the Middle-East belt–Iran, Kuwait, Saudi Arabia and Iraq– which attracted foreign enterprise and capital has also provided additional employment and income to people in these countries which in the absence of oil exploitation would have remained poor. It has also resulted in the establishment of several need supplying industries such as house-building, road construction, food and raw materials. It has also brought with it external economies in the form of training of the native labour and management.

Dynamic gains from trade accrue to an underdeveloped economy in the form of an unending perennial income flows which continue even after the economy has become integrated into the world economy. "Trade is a dynamic force that stimulates innovation. New ways of producing and organising production are spread to the local economy through trade, and the competitive force of trade stimulates adoption of cost-saving techniques. Trade also makes possible economical local production of many goods that would otherwise be prohibitive to produce locally."[24]

Suggested Readings

P.T. Ellsworth and J. Clark Leith, *The International Economy,* Fifth Edition, 1975, pp. 89–92 and pp. 187–191, Chapters 9 and 11.

H. Robert Heller, *International Trade,* Second Edition, 1973, Chapters 8 and 11.

Charles P. Kindleberger, *International Economics,* Fifth Edition, 1973, Chapter 5, pp. 74–75. Gerald M. Meier, *International Trade and Development,* 1963, Chapter 3.

Bo Sodersten, *International Economics,* Second Edition, Reprinted 1981, Chapters 5 and 12.

Jaroslav Vanek, *Studies in the Theory of International Trade,* 1962, Chapter 16.

Jacob Viner, *Studies in the Theory of International Trade,* 1937, Chapters VIII and IX.

Questions

1. What do you understand by 'terms of trade'? How are they determined? What are the limitations of using a terms of trade index as an indicator of the gains from trade?
2. Explain the meaning of terms of trade and distinguish between the net and gross barter terms of trade. In what situation, will these two terms of trade be identical? Discuss those factors which determine the terms of trade.
3. Discuss fully the static and dynamic gains resulting to a country and world as a whole from free trade.
4. Explain the different indices of the terms of trade. What is the importance of the concept of terms of trade? Discuss fully.
5. What are the gains from international trade? Explain the argument which states that free trade maximises the gains from international trade.

24. P.T. Ellsworth and J. Clark Leith, *The International Economy,* Fifth Edition, 1975, pp. 190–191.

PART THREE

FOREIGN EXCHANGE RATE
AND
THE BALANCE OF PAYMENTS

CHAPTER 13

Foreign Exchange Rate

Meaning of Foreign Exchange Rate

When goods and services are exchanged, there has to be determined some rate or exchange ratio between them. In other words, there must be some 'price' at which the goods and services can be exchanged. In the case of domestic exchange of goods and services, the exchange ratios between different goods and services exchanged are determined by taking the ratios of their prices that are determined in the market and are expressed in terms of the domestic currency. The determination of exchange ratios is, however, not so simple when the different commodities and services exchanged are produced in the different countries having different currencies. Before the ratios of exchange between different goods and services traded between any two sovereign and independent countries can be known we must know the exchange rate between the currency units of the two countries.

A distinguishing feature of international trade resides in the involvement of the use of foreign currencies. If a seller in Bombay sells goods to a buyer in Calcutta he is paid in rupees. Despite considerable physical distance between the buyer and seller they use the same currency. However, when the buyer and seller belong to two different countries the problem of foreign exchange arises because the buyer wants to pay in his country's money while the seller wants to be paid in his home currency. Foreign exchange rates link the different national currencies and make the international costs and prices comparisons possible.

Foreign exchange rate between the currency units of two countries means the number of units of one national currency that are needed to buy one unit of the other national currency. Either country's money or currency unit may be used as the unit for expressing the price of the other country's money unit. For example, when 2 Indian rupees exchange for one Euro, the foreign exchange rate could either be stated as 1 Euro = 2 Indian rupees or as 1 Indian rupee = 0.5 Euro. Similarly, a change in the foreign exchange rate from 1 Euro = 2 Indian rupees to 1 Euro = 1 Indian rupee can be expressed either as a fall in the exchange rate on the Euro relative to the Indian rupee or as a rise in the exchange rate on the Indian rupee relative to the Euro.

Importance of the Foreign Exchange Rate

By linking together the currencies of different countries, foreign exchange rates render the comparison of international costs and prices possible. Consequently, they play a dominant role in determining the volume, composition and direction of international trade. For an Indian importer wanting to import refrigerators from Germany or Japan the price quoted in Euro or Japanese yens means nothing unless the Euro-rupee or the Yen-rupee exchange rate is known. Exchange rates are not only the prices at which different national currency units are exchanged; they are also the converters or translators of the prices of goods and services which are stated in terms of the domestic currencies into the currencies of other countries. Thus, *ceteris paribus,* foreign exchange rates play an important role in determining the volume, composition and direction of the flow of international trade. For example, the German price of a photo camera is 5 Euro. To the Indian buyers, basing their decisions on the prices stated in terms of the Indian rupees, this price which has been expressed in terms of the German currency is meaningless unless they know the rupee-price of Euro and it can be known if the rate of foreign exchange—rupee price of Euro—is determined. The conversion of the price of camera expressed in terms of the Euro into the price expressed in terms of the Indian rupees is as follows:

	Rupee-price of camera		Euro-price of camera	Rupee-price of Euro
Situation 1	Rs. 5	=	5	1 Euro = Re. 1
Situation 2	Rs.10	=	5	1 Euro = Rs. 2
Situation 3	Rs.15	=	5	1 Euro = Rs. 3

Suppose that the price of the camera in Germany remains unchanged at 5 Euro. At the foreign exchange rate of 1 Euro = 1 Indian rupee, which is a very low exchange rate on the Euro in terms of the Indian rupee or a very high exchange rate on the Indian rupee in terms of the Euro, the rupee-price of the imported German camera to the Indians will be five rupees. At foreign exchange rate of 1 Euro = 2 Indian rupees, which is a higher exchange rate on the Euro in terms of the rupee or a lower exchange rate on the Indian rupee in terms of the Euro, the price of German camera to Indians will be ten rupees. When the foreign rate on the Euro in terms of the rupee rises further to 1 Euro = 3 Indian rupees, the rupee-price of the camera, its price in Euro remaining unchanged, rises from ten rupees to fifteen rupees. Thus, given the level of domestic prices in Germany, higher the exchange rate on the Euro in terms of the Indian rupee, costlier will be the German exports in terms of the Indian rupees. To the extent the Indian demand for imports is price elastic, smaller will be the German exports to India. By depreciating the Euro in terms of the Indian rupee, German exporters can succeed in lowering the rupee-prices of the exports and can successfully sell a larger quantity of their goods in India.

Changes in the foreign exchange rate on the Euro shift the demand curve for German exports. Let us assume that during any given time period incomes, prices, and other conditions in India are such that the Indian demand for German goods stated as a function of their prices expressed in rupees is as shown by DD demand curve in Figure 13.1A. The DD demand curve shows that the amount of German goods demanded in India is a negative function of their prices expressed in Indian rupees such that any fall in the rupee-prices of German goods increases the demand for these goods in India and *vice versa.*

Figure 13.1B shows this demand curve translated into the demand curves when the prices of German exports are expressed in terms of Euro at different foreign exchange rates. When the foreign exchange rate is 1 Euro = 1 Indian rupee, the demand for German exports expressed as a function of their prices expressed in terms of Euro is exactly the same as is their demand expressed as a function of their prices expressed in terms of the Indian currency. For example, Indians buy $0Q_1$

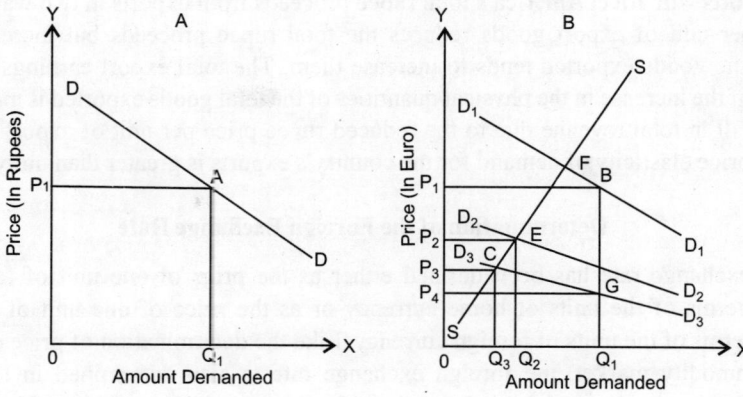

Figure 13.1

amount of German goods at the $0P_1$ rupee-price. They will, according to our assumption, buy the same $0Q_1$ amount of German goods at $0P_1$ Euro price as is shown by D_1D_1 demand curve in Figure 13.1B. But what will happen if the foreign exchange rate is 1 Euro = 3 Indian rupees showing that Euro has become three times costlier in terms of the Indian currency ? Unless India's demand curve for German exports shifts upward to the right, Indians will buy the old $0Q_1$ amount only if the Euro price of German goods is reduced to one-third of the old price. This reduced price has been shown by $0P_4$ in Figure 13.1B. This has been shown by the downward shift of demand curve D_1D_1 to the position of D_3D_3 demand curve in Figure 13.1B where for the different amounts of German goods demanded in India the Euro price is only one-third of the old price relevant on the D_1D_1 demand curve.

Thus, higher the foreign exchange rate on the Euro, *ceteris paribus*, lower is the Euro price at which Germany can sell any given amount of her exports and smaller is the amount of her exports which she can sell at any given Euro price. On the other hand, lower the foreign exchange rate on the Euro, higher will be the Euro price at which Germany can sell any given amount of her exports and higher will be the amount of her exports which she can sell at any given Euro price.

A country can make her exports cheaper for the foreigners without reducing their domestic prices by reducing the foreign exchange rate on its money, i.e., by devaluing her currency. By devaluing her currency a country can simultaneously succeed in cheapening her exports for foreign buyers and in raising the prices in domestic currency received by the home exporters. Suppose that the supply function of German exports is as shown by SS supply curve in Figure 13.1B and that Germany lowers the exchange rate on the Euro from 1 Euro = 3 Indian rupees to 1 Euro = 2 Indian rupees. Consequently, India's demand curve for German exports will shift upward from D_3D_3 to D_2D_2. The devaluation of the Euro simultaneously decreases the rupee price of German exports increasing in consequence the amount demanded of German goods by Indians from $0Q_3$ to $0Q_2$ and raising the export price expressed in terms of the Euro received by the German exporters from $0P_3$ to $0P_2$.

It may, therefore, be concluded that, *ceteris paribus*, unless the foreign demand for a country's exports is perfectly price inelastic, a fall in the foreign exchange rate on her currency will (a) increase the total physical quantity of her exports, and (b) incr+ease the total value of her exports measured in terms of the domestic currency. But how will a given fall in the exchange rate affect the total value of a country's exports in terms of the foreign currency—its total export proceeds in foreign currency? It will depend upon the elasticity of foreign demand for her exports. For example, a fall in the rupee price of the US exports will affect America's total rupee proceeds from exports in two ways. The reduced rupee price per unit of export goods reduces the total rupee proceeds but increase in the total quantities of the goods exported tends to increase them. The total export earnings will, therefore, increase only if the increase in the physical quantities of the total goods exported is more than enough to offset the fall in total revenue due to the reduced rupee price per unit of export goods; in other words, if the price elasticity of demand for the country's exports is greater than unity.

Determination of the Foreign Exchange Rate

The foreign exchange rate has been defined either as the price of one unit of foreign currency expressed in terms of the units of home currency or as the price of one unit of home currency expressed in terms of the units of foreign currency. Like the determination of price of a commodity in a free commodity market, the foreign exchange rate is also determined in the free foreign exchange market by the demand for and supply of foreign money. To make the demand and supply functions of foreign exchange look like the familiar conventional market demand and supply functions, we define the rate of foreign exchange as the price of one unit of foreign currency expressed in terms of the units of the home currency.

Demand for Foreign Exchange

The functional relationship between the amount of foreign exchange demanded and the foreign exchange rate is expressed in the demand schedule for foreign exchange. A demand schedule for foreign exchange is a schedule showing the different amounts of foreign exchange demanded at different rates of foreign exchange, *ceteris paribus*. It follows from the properties of this demand schedule that the amount of foreign exchange demanded and the rate of foreign exchange are inversely related such that a fall in the rate of foreign exchange is followed by an increase in the amount of foreign exchange demanded and *vice versa*. The main reason for this inverse or negative relationship is that a higher rate of foreign exchange by rendering the imports more expensive reduces the demand for them. Consequently, it also reduces the total amount of foreign exchange demanded which is required to pay for the imports. On the other hand, a lower foreign exchange rate by making the imports cheaper increases the demand for them. Consequently, it increases the demand for foreign exchange needed to pay for higher imports. Thus, ordinarily the demand for imports and the amount of foreign exchange demanded to pay for them change in the direction opposite to that of changes in the rate of foreign exchange. The relationship between the amount of foreign exchange demanded and the foreign exchange rate has been shown in Figure 13.2.

In Figure 13.2, when the foreign exchange rate (price of Euro expressed in terms of Indian rupees) is $0R_1$, the total amount of foreign exchange (Euro) demanded is $0Q_1$. When the foreign exchange rate falls from $0R_1$ to $0R_2$, i.e., when the rupee price of the Euro falls, the total amount of the foreign exchange (Euro) demanded increases from $0Q_1$ to $0Q_2$. This happens because due to the Euro becoming cheaper in terms of the Indian rupee, the prices of German goods expressed in terms of the Euro remaining unchanged, their prices expressed in terms of the Indian rupee fall. Consequently, the demand for German goods in India increases unless the extreme assumption is made that such demand is perfectly price-inelastic. Conversely, the total amount of foreign exchange demanded will decrease when the foreign exchange rate rises, i.e., when foreign currency becomes costlier in terms of the domestic currency.

The demand curve for foreign exchange has been shown in Figure 13.2 where the foreign exchange rate and the amount of foreign exchange demanded have been shown on the Y-axis and

X-axis respectively. The negatively sloping DD demand curve shows that the foreign exchange rate elasticity of the demand for foreign exchange is less than infinity and greater than zero. The demand for foreign exchange arising from the imports of goods and services has the same foreign exchange rate elasticity as is the elasticity of demand for the imported goods and services with respect to their prices expressed in terms of the home currency.

So far the demand for foreign exchange has been regarded as wholly arising from the imports of goods and services. Although the imports of goods and services constitute the single largest component of the total demand for foreign exchange, it is not the only source of demand for foreign exchange. Foreign loans and investments and outward unilateral transfers are also sources of the autonomous demand for foreign exchange which, apart from speculative capital movements based on anticipated changes in the foreign exchange rate, are not normally influenced by the level of foreign exchange rate.

Autonomous capital outflows are induced by the prospects of obtaining higher returns in the form of higher interest and dividend incomes on capital investments obtainable from abroad compared to those that can be obtained at home. While the foreign exchange rates affect the foreign money value of a given amount of loan or investment in the home currency, the rates of interest and dividends are affected in a like manner and to the same extent leaving the *rate of return* unchanged. This means that the elasticity of demand for foreign exchange for financing the capital outflows is zero. The addition of this autonomous demand for foreign exchange to the aggregate demand for foreign exchange will shift the demand curve for foreign exchange upward without affecting its slope.

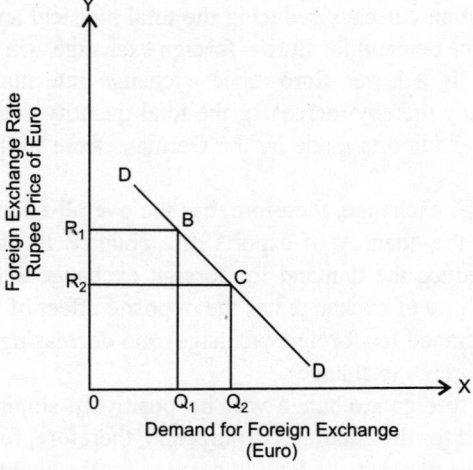

Figure 13.2

Unilateral transfer payments also affect the demand for foreign exchange in an identical manner shifting upward the entire demand curve for foreign exchange without changing its slope. As an exception, however, the demand for foreign exchange to finance the unilateral transfer payments would be unit-elastic if the transfer payments consist of a fixed sum expressed in the remitting country's currency but are remitted in foreign currency. Aggregating the various autonomous components of the demand for foreign exchange will yield the aggregate demand curve for foreign exchange which will resemble the shape of the demand curve for foreign exchange arising exclusively from the imports of goods and services as shown in Figure 13.2.

Supply of Foreign Exchange

The supply schedule or curve of foreign exchange shows the different amounts of foreign exchange that are available at different rates of foreign exchange in the foreign exchange market. Autonomous components of the aggregate supply of foreign exchange are the counterparts of the autonomous components of the aggregate demand for foreign exchange. The sources of supply of foreign exchange appearing on the credit side of the international balance of payments of a country are the exports of goods and services, capital inflows and inward unilateral transfer receipts. These sources of supply of foreign exchange depend largely on the decisions of foreigners.

The total amount of different goods and services which a country can export and, therefore, the total amount of foreign exchange which it can acquire through exports depends on the amount of different goods and services which the residents of foreign countries are willing to import from a particular country. The amounts of capital inflows and unilateral transfer receipts depend on the decisions of the foreign investors to invest in any given country. These decisions are based on the same considerations on which are based the home decisions relating to the movements of domestic capital in the opposite direction. Thus, *ceteris paribus,* a country's exports—which are imports of other countries—are a function of the foreign exchange rates because these together with their domestic prices determine the prices of exports expressed in terms of the foreign currencies.

Since the foreign exchange rates in the different countries are reciprocal of each other, the effects of any given change in the foreign exchange rates are opposite in the different countries. A higher rate of exchange in India on the Euro causes the German goods and services to be costlier in India in terms of the Indian currency reducing the total physical amount of German exports to India.[1] As a consequence, the demand for Euro—foreign exchange—in India also falls. But a higher rupee-Euro exchange rate is a lower Euro-rupee exchange rate making the Indian goods and services cheaper in Germany thereby increasing the total quantity demanded of the Indian export goods. To pay for the larger imports made by the Germans from India they demand larger rupee foreign exchange.

A higher rate of foreign exchange, therefore, has the overall effect of decreasing the quantity of imports and increasing the quantity of exports of a country. In other words, a higher rate of exchange tends both to reduce the demand for foreign exchange and to increase the supply of foreign exchange. A lower rate of exchange has the opposite effect of increasing the imports (and, therefore, increasing the demand for foreign exchange) and decreasing the exports (and, therefore, decreasing the supply of foreign exchange).

The supply curve of foreign exchange will be positively sloping from left to right if the elasticity of foreign demand for the country's exports and, therefore, for its currency is greater than unity; it will be vertical if the elasticity of foreign demand for the country's exports is unity while it will be backward bending if the elasticity of foreign demand for the country's exports is less than unity. We can illustrate it with the example of India and West Germany.

The total amount of foreign exchange supplied in India at any given rate of foreign exchange is the same as the total amount of Euro spent for rupees at that rate of foreign exchange. Suppose that at an exchange rate of 5 rupees per Euro in Mumbai, which is equivalent to a rate of 0.2 Euro per Indian rupee in Frankfurt, 1,000 Indian rupees are demanded in exchange for the Euro. Since each rupee commands 0.2 Euro in exchange, 200 Euro are offered in exchange for Rs. 1,000.

---

1. This must be so unless it is unrealistically assumed that India's demand for her imports from Germany is perfectly price-inelastic. Even though the demand for an imported good may be inelastic, nevertheless the demand for the imported goods from a single country is elastic.

Consequently, the total amount of Euro foreign exchange supplied in Mumbai at an exchange rate of 5 Indian rupees per Euro is 200 Euro. Thus, the supply of foreign exchange in a country is directly related to the demand of foreign countries for her currency which itself depends on the demand for her goods in the foreign countries.

What will be the total amount of Euro (foreign exchange) supplied in Mumbai at an exchange rate of 10 Indian rupees per Euro? In Frankfurt, the rate on the rupee will then be 0.1 Euro. Consequently, the Indian goods would become cheaper for the Germans and they would demand more of these goods unless the German demand for the Indian goods was perfectly price-inelastic.[2] The total amount of Indian rupees demanded in Germany will, therefore, be greater than before. But since for each rupee purchased by the Germans only 0.1 Euro is offered instead of 0.2 Euro offered earlier, whether or not the total amount of Euro supplied at foreign exchange rate of 10 Indian rupees per Euro will increase and depend on the elasticity of demand for the rupees over the range of foreign exchange rates from 0.2 Euro to 0.1 Euro per rupee. If the elasticity of demand is greater than unity, the percentage increase in the Indian rupees demanded will be greater than the percentage fall in the Euro-rupee foreign exchange rate. Consequently, the total amount of Euro supplied will increase.

Let us assume that the German demand for the Indian rupees over the range from 10 to 15 rupees per Euro is unit-elastic. In this situation, a 25 per cent fall in the foreign exchange rate is accompanied by an equal percentage increase in the amount of Indian rupees demanded against the Euro. Consequently, the total amount of Euro offered for rupees will be the same as was at an exchange rate of 10 rupees per Euro. Thus, when the German demand for Indian rupees is unit-elastic, the supply of Euro in India will be zero-elastic.

Let us now assume that between the foreign exchange rate of 15 and 20 rupees per Euro the demand for rupees is inelastic (less than unity). In such a situation, a 20 per cent fall in the foreign

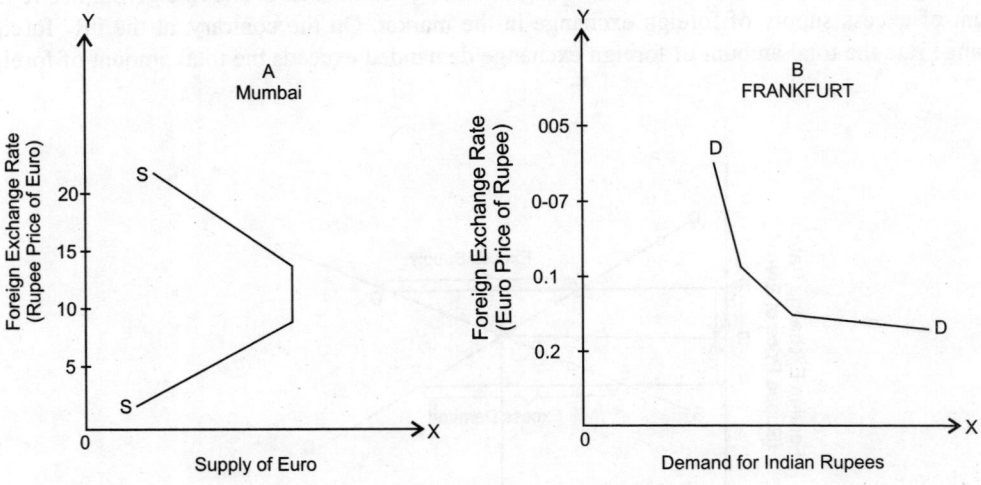

Figure 13.3

2. The assumption is that the rupee-price of the Indian export goods remains unchanged in India. In other words, the assumption is that when the foreign exchange rate falls making the Indian rupee cheaper in terms of the Euro, prices in India, particularly those of export goods, do not rise.

exchange rate will be associated with a less than 20 per cent increase in the amount of rupees demanded against Euro. Consequently, the total amount of Euro offered for rupees will be less at an exchange rate of 15 rupees per Euro. Thus, when the foreign demand for rupees is inelastic, the supply of Euros in India will have a negative elasticity, i.e., the supply curve of the Euro will be backward bending as shown in Figures. 13.3A and 13.3B.

In Figure 13.3A, the *SS* supply curve of foreign exchange has a positive elasticity—rising from left to right—between the foreign exchange rates ranging from 5 to 10 Indian rupees per Euro. It has a zero elasticity between the foreign exchange rates ranging between 10 and 15 rupees per mark. It has a negative elasticity at all the foreign exchange rates above 15 rupees per Euro.

Equilibrium Foreign Exchange Rate

The equilibrium foreign exchange rate in the free foreign exchange market is determined at the point of intersection of the supply and demand curves for foreign exchange as shown in Figure 13.4 where the foreign exchange rate has been shown on the Y-axis and the total amount of foreign exchange which will be demanded and supplied at different foreign exchange rates has been shown on the X-axis. The equilibrium rate of foreign exchange is $0R_1$ where the total quantity of foreign exchange demanded $(0Q_1)$ equals the total quantity of foreign exchange supplied $(0Q_1)$. Consequently, at this foreign exchange rate the foreign exchange market is in equilibrium in the sense that at the $0R_1$ rate of foreign exchange the total amount of foreign exchange demanded by the buyers (importers) equals the total amount of foreign exchange offered for sale by the sellers (exporters). At any rate of foreign exchange other than $0R_1$, the foreign exchange market will be in disequilibrium. Consequently, there will be either excess supply or excess demand for foreign exchange in the foreign exchange market. For example, when the foreign exchange rate is $0R_2$, the total amount of foreign exchange supplied at this rate in the market is greater than the total amount of foreign exchange demanded at this rate giving rise to BC amount of excess supply of foreign exchange in the market. On the contrary, at the $0R_3$ foreign exchange rate the total amount of foreign exchange demanded exceeds the total amount of foreign

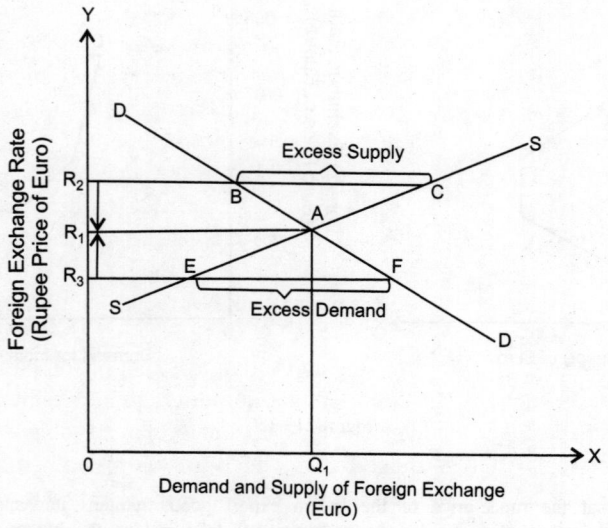

Figure 13.4

exchange supplied at that rate. Consequently, the foreign exchange market is in disequilibrium as has been shown by the presence of EF amount of excess demand for foreign exchange. Thus, the equilibrium rate of foreign exchange will be established at $0R_1$ where the total demand for and the total supply of foreign exchange are equal. In a free foreign exchange market, the rate of foreign exchange will not deviate from this equilibrium rate and even if it does deviate from this rate, soon the forces of correction will restore it at this equilibrium level.

The typical free market response to an excess demand situation will be reflected in a rise in the rate of foreign exchange while an excess supply situation will manifest itself through a fall in the rate of foreign exchange. Thus, a rise or a fall in the foreign exchange rate in the foreign exchange market shows the presence of disequilibrium in the market. The equilibrating movements in the foreign exchange rate may, however, be obstructed by government through foreign exchange control or exchange pegging.

Functions of the Foreign Exchange Market

The foreign exchange market performs the three important functions of transferring the purchasing power, providing credit for financing foreign trade, and furnishing facilities for the hedging of foreign exchange risks. Of these three functions, the most important function is transfer of the purchasing power from one country to another and from one national currency to another. The purchasing power is transferred through the use of credit instruments. The chief credit instrument used for transferring the purchasing power is the telegraphic transfer or the cable order by one bank in a country to its correspondent abroad to pay to the named individual funds out of its deposit account to a designated account or order. The telegraphic transfer is a sort of cheque which is wired or faxed rather than sent by post. Purchasing power may also be transferred through bank drafts. There is also the commercial bill of exchange or acceptance through which even today a considerable amount of foreign payments in international trade is made. A bill of exchange is an order written by the exporter of goods directing the importer to pay the stated amount to the exporter or the party's bank, discount house or other financial institution with whom the exporter has discounted the bill of exchange.

Apart from performing the most important function of transferring payments, the foreign exchange market also provides credit for foreign trade. Like all trades, international trade also requires credit. It takes time to move the goods from seller to purchaser and during this period the transaction must be financed. Even if the exporter does not need credit for the manufacture of export goods, credit is necessary for the transit of goods. In the nineteenth century, London was the hub of international finance and the world used to finance its trade in pound-sterlings since sterling commanded the position of international reserve and vehicle currency. In general, when the special credit facilities of the foreign exchange market are used, the foreign exchange department of a bank or the bill market of one country or the other extend the credit facilities to finance the foreign trade.

Finally, by providing the facilities of buying and selling the spot or forward foreign exchange, the foreign exchange market enables the exporters and importers to hedge their foreign exchange risks arising from unexpected and sudden changes in the foreign exchange rate. The forward foreign exchange market also enables those banks which are unlikely to run any considerable exchange position to cover their foreign exchange commitments. The forward foreign exchange market also enables the speculators to establish positions in the foreign exchange.

Determination of the Equilibrium Foreign Exchange Rate

The equilibrium foreign exchange rate is determined differently under different monetary systems. When the two countries are either on the gold or silver standard, the foreign exchange rate between their currency units is determined on the basis of their metallic contents, i.e., on the basis of parity between the mint ratios of currencies of the two countries. The theory explaining the determination of foreign exchange rate under the gold standard is known as the Mint Parity Theory of foreign exchange rate.

When the monetary systems of the two countries are not on the gold standard and their currencies are inconvertible paper currencies, the foreign exchange rate cannot be determined on the basis of mint parity because such a parity does not exist. Under inconvertible paper money standard, the basis of exchange rate determination is the parity of the purchasing powers of the two currencies. The theory explaining determination of the equilibrium rate of foreign exchange under inconvertible currency is known as the Purchasing Power Parity Theory.

When the currencies of the two countries are dissimilar but convertible, i.e., when one country is on the gold standard and the other is on the silver standard, the exchange rate between two such currencies will be determined in the same way as is determined under the gold standard as long as the market ratio between gold and silver remains stable, i.e., on the basis of mint parity. But if the market ratio between the two metals changes, the mint parity forming the basis of determining the exchange rate between the two countries' currencies will also change. It, therefore, follows that when there are dissimilar currencies—gold and silver—in the two countries, the equilibrium foreign exchange rate is determined on the basis of mint parity that changes according to changes in the market ratio between the two metals in the countries concerned. The rate of foreign exchange can neither remain for long above the moving parity because silver will be shipped from the gold standard country nor can it remain for long below the moving parity because gold will be shipped from the silver standard country. We may now discuss the different theories of the determination of equilibrium foreign exchange rate in detail.

1. Mint Parity Theory

The mint parity theory explains the determination of foreign exchange rate between the currency units of the gold standard countries. When the two countries are on the gold standard, their currency units are either made of gold of specified purity and weight or are freely convertible into gold of given purity at fixed rate. Under the gold standard, countries maintain the value of their currencies in a fixed relationship to the value of gold by committing themselves to buy and sell gold at fixed prices. A nation's currency is said to be fully on the gold standard if the government of that country (1) buys and sells gold in unlimited quantity at an officially fixed price, and (2) permits unrestricted gold flows into and out of the country. The practical meaning of (1) and (2) is that an individual who holds domestic currency knows in advance how much gold he can obtain in exchange for it and how much foreign currency this gold will buy when exported to another country. Under these conditions, the exchange rate between two gold standard countries' currencies will fluctuate within the narrow limits around the fixed mint parity. By mint parity is meant that the foreign exchange rate is determined on the weight-to-weight basis of the metallic contents of the two money units, allowance being given to the purity of the metallic contents. The mint parity theory of foreign exchange rate is applicable only when the countries are on the same metallic standard which in practice has taken the form of either the gold or the silver standard. Thus, there can be no fixed mint parity between a gold and a silver standard country.

Before World War I and during the later 1920s, England and the United States of America were simultaneously on the gold standard. The British sovereign contained 113.0016 grains and the US gold dollar consisted of 23.2200 grains of gold of standard purity. Since the mint parity is the reciprocal of the gold contents ratio between the two currencies, the exchange rate between the US dollar and the British sovereign based on the mint parity was 113.0016 ÷ 23.2200, i.e., 4.866 showing that 4.866 US dollars were equivalent to one British gold sovereign.

The foreign exchange rate determined on the basis of the gold contents of the two currency units, i.e., on the basis of mint parity is a norm which would prevail in the long run. The market rate of exchange may and does differ from the long-run mint parity equilibrium rate of exchange within well-defined limits. These limits are fixed by the cost of shipping gold, including insurance, packing and interest income charges for the period spent in the transit of gold from one country to another. When the two countries, say the United States of America and England, are on the gold standard an American importer desiring to convert the US dollars into pound-sterlings to pay to his English counterpart can do this in two ways. He can either buy the pound-sterling foreign exchange by tendering dollars in the foreign exchange market or he can export gold from America to England. If the second method did not cause any expense in exporting gold from America to England an effective foreign exchange rate of US $ 4.866 = £1 would be established in the free foreign exchange market.

The export of gold, however, involves certain expenses. Before gold can be exported it must be carefully packed. Besides the expenses involved in packing, freight and insurance costs have also to be incurred. Similarly, some interest income is lost while gold is in transit. There is also some small loss resulting from abrasion. Furthermore, the central bank in the gold-receiving country may charge a fee for minting. Thus, there are a host of expenses incurred in exporting gold from one country to another. The American importer will, therefore, export gold only if the market rate of exchange is higher than the mint parity exchange rate plus the expenses of exporting gold from America to England. If the cost of exporting gold from America to England and *vice versa* is 2 cents per pound-sterling, no gold would be exported from America for making foreign payments if the exchange rate in the foreign exchange market was either lower than or equal to US $ 4.886 = £1. In other words, so long as the American importer of British goods wishing to make payment to his British counterpart can buy the sterling exchange in the foreign exchange market either at or below an exchange rate of US $ 4.886 = £1 he would not bother to export gold to make payment for his imports of goods from England.

Thus, there are the upper and lower limits which are determined by the cost of exporting and importing gold between the gold standard countries within which the market rate of foreign exchange will fluctuate. These limits of fluctuations in the market foreign exchange rate are known as the upper and the lower gold points. The upper or the gold-export point is obtained by adding the cost of exporting gold to the mint parity rate of exchange while the lower or the gold-import point is arrived at by deducting the cost of importing gold from the mint parity rate of exchange. Thus, the market rate of exchange between the US dollar and the pound-sterling may be anywhere between $ 4.886 and $ 4.846 to one pound-sterling when the cost of shipping gold worth one pound-sterling between America and England is 0.2 cents. The day-to-day rate of foreign exchange within the upper and the lower gold points will be determined by the demand for and the supply of foreign exchange.

There would be no shipment of gold so long as the market rate of exchange stays within this range which is determined by the gold points. If the market rate of exchange exceeded the gold export point, i.e., if it exceeded $ 4.886 = £1, it would be economical for the American importer to export gold to England rather than to purchase the pound-sterling bills of exchange in the foreign exchange market because in exporting gold he does not have to incur more than 2 cents as transport charges which when added to the mint parity exchange rate of $ 4.866 = £1 comes to only $ 4.888 = £1. This critical rate above which gold will be exported is called the 'gold-export point'. Conversely, the market rate of foreign exchange will not fall below more than two cents from the mint parity rate of exchange, i.e., below $ 4.846 = £1 because below this rate of exchange gold would be imported. This rate is known as the 'gold-import point'. A market foreign exchange rate beyond these two gold points is impossible because so long as the more economical possibility of gold shipment existed, no American importer would pay more than $ 4.886 for a pound-sterling when buying the pound-sterling draft in the New York foreign exchange market and no British exporter would accept less than $ 4.846 for a pound-sterling when selling the dollar draft in the London foreign exchange market.

The gold points are important in the entire discussion of the determination of the equilibrium foreign exchange rate under the gold standard because they determine the maximum fluctuations to which the market rate of foreign exchange in the free foreign exchange market is subject from day-to-day. The gold points also tell us why the market rate of foreign exchange differs from the normal or equilibrium rate of foreign exchange which is determined on the basis of mint parity. The determination of the gold-export and the gold-import points and the market rate of foreign exchange on the basis of the demand for and the supply of foreign exchange between these gold points has been illustrated in Figure 13.5.

It is obvious that the day-to-day or the market rate of foreign exchange fluctuates in accordance with the changes in the supply and demand curves of foreign exchange within the range of 4 cents determined by the gold-export and gold-import points. The market rate of foreign exchange fluctuates within the limits determined by the cost of shipping gold represented by the

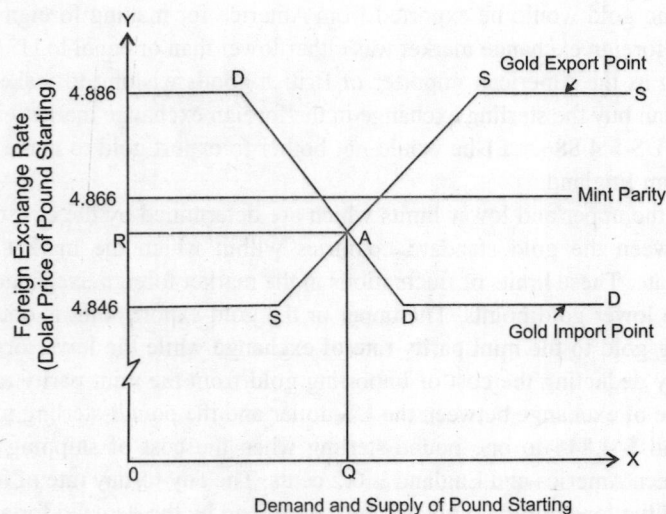

Figure 13.5

upper and the lower specie points. Between the gold-export and the gold-import points the foreign exchange rate is determined by the demand and supply forces as shown in Figure 13.5. The figure shows the situation under conditions of stabilising speculation because the demand curve is negatively sloped while the supply curve is positively sloped. It is obvious from the diagram that the foreign exchange rate shows no pull towards the mint parity which serves no more than the basis for calculating the gold points. The actual or market rate of foreign exchange as shown in Figure 13.6 is below the mint parity rate of foreign exchange although it could be anywhere within the range set by the gold-export and the gold-import points depending on the position of the demand and supply curves of foreign exchange.

At and above the gold-export point the normal supply curve while at and below the gold-import point the normal demand curve are non-operative. It is because at these two points the supply of and the demand for foreign exchange become perfectly elastic. When the rate of foreign exchange reaches the gold-export point of $ 4.886 = £ 1 it will not rise any further even when at this rate of exchange the demand for foreign exchange may be more than the supply of foreign exchange at that rate because any shortfall in the supply of foreign exchange in the foreign exchange market shall be met through gold exports. Conversely, the rate of foreign exchange cannot fall below the gold-import point of $ 4.846 = £ 1 because at this rate of foreign exchange gold shall be imported. Thus, the market rate of foreign exchange shall fluctuate within the range set by the gold-import and the gold-export points, i.e., within the range of 2 + 2 = 4 cents only.

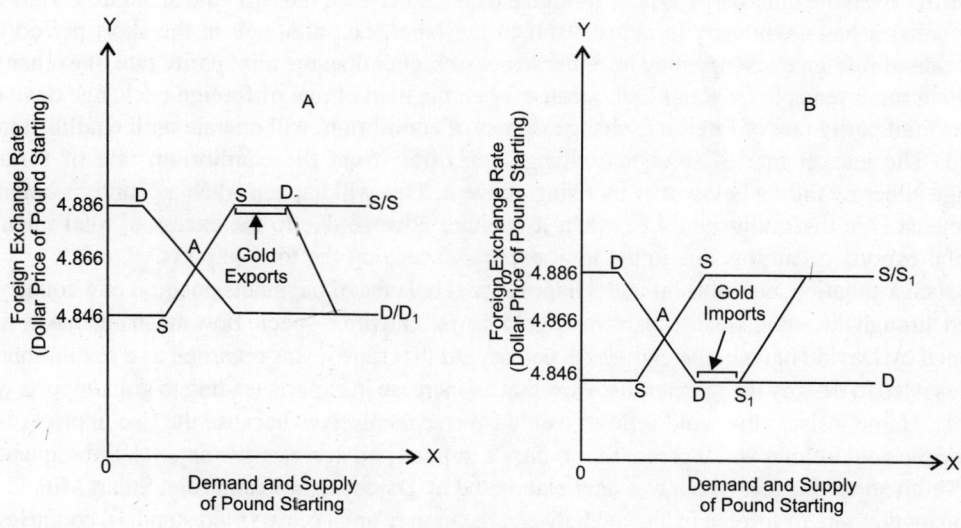

Figure 13.6

Under a stable international gold standard system, changes in the foreign exchange rate between the gold points had great significance between distant money markets where the cost incurred in the shipping of gold separated the two gold points by a considerable margin. Some of these changes were regular and stabilising speculation benefited considerably by these changes. For example, before World War I due to heavy imports in the spring, the American dollar used to remain weak becoming strong in the fall when it was in large demand due to wheat and cotton exports. Consequently, it used to move near the gold-export point in the spring while it hovered around the gold-import point in the fall. Speculators conducted safe transactions by buying the dollars against the pound-sterling in the spring with the object to sell these in the fall. As a result of the activities of speculators, the movements of foreign exchange rate were prevented from touching the extremes of the gold points.

When there occurs a wide change either in the demand for or in the supply of foreign exchange the gold flows generated as a consequence of change in the foreign exchange rate will balance the demand for and supply of foreign exchange in the short period. When the demand for pound-sterling exceeds the supply available at foreign exchange rates (prices) between the gold points, the excess demand will be met through the gold exports which convert dollars into sterling as shown in Figure 13.6A. When the demand curve for the pound-sterling shifts from DD to D_1D_1, the original equilibrium between the demand and supply curves at point A is no more relevant. The new equilibrium takes place at point D_1 where the excess demand for foreign exchange (pound-sterlings) amounting to SD_1 is met through the gold exports. Conversely, as shown in Figure 13.6B, a rightward shift in the supply curve of pound-sterlings from SS to S_1S_1 will push down the foreign exchange rate to the gold-import point where the excess supply of sterling foreign exchange will be partially met through the gold imports amounting to DS_1 which represents the excess supply.

Price-Specie Flow Adjustment Mechanism

Notwithstanding that under the international gold standard the market rate of foreign exchange might differ from the mint parity rate of foreign exchange between the two gold standard countries' money units, it has a tendency to adjust itself to the latter, i.e., although in the short period the market rate of foreign exchange may be either lower or higher than the mint parity rate of exchange, in the long run it tends to be equal to it because when the market rate of foreign exchange deviates from the mint parity rate of foreign exchange, forces of equilibrium will operate until equilibrium is restored. The market rate of foreign exchange can differ from the equilibrium rate of foreign exchange either by falling below it or by rising above it. This will happen when a country's balance of payments is in disequilibrium, i.e., when it is either adverse due to the excess of total imports over total exports or surplus due to the total exports exceeding the total imports.

In such a situation, equilibrium in the international balance of payments position of a country is restored through the automatic mechanism of gold flows. The price-specie flow mechanism was first developed by David Hume in the eighteenth century. At that time it was regarded as a revolutionary notion as it led to destroy the mercantilist view that an increase in exports leading to gold imports was desirable. Hume insisted that gold inflows would reverse themselves because the rise in prices that followed the gold inflows will increase the country's imports and decrease her exports. Subsequently, Adam Smith noted this fact which was later elaborated by David Ricardo and John Stuart Mill.

The market rate of foreign exchange between the money units of two gold standard countries is anchored to the mint parity rate and the factors underlying the balance of payments adjust themselves to the mint parity foreign exchange rate. When a country is faced with an adverse balance of payments, equilibrium in the balance of payments of that country is brought about through a long chain of causation and counter-causation. The excess of imports over exports reflected in the adverse balance of trade of the country causes gold outflows from that country and gold inflows in the country enjoying the favourable balance of trade. The exodus of gold from the former country causes contraction in the money supply leading to deflation due to which prices and incomes fall in the country. Since the fall in prices in the gold-losing country is general, prices of export goods also fall making that country's goods cheaper for the other country. Similarly, some of the goods which the nationals of the gold-losing country imported earlier from the other country would no longer be imported because in the new changed situation these goods become costlier. Consequently, the gold-losing country's exports are stimulated and her imports are reduced.[3] The opposite chain of events operates in the country having a favourable balance of trade which experiences gold inflows and where the money supply expands. Following this, incomes and prices rise in the country. As result of the price and income effects of gold inflows in the country, the demand curves for all the goods and services shift upward to the right. Consequently, the imports expand while the exports contract due to the income and price effects of gold inflows in the gold-receiving country. In other words, through the inflationary process in the gold-receiving country and deflationary process in the gold-losing country equilibrium in the balance of payments of both the gold standard countries is restored. However, the balance of payments equilibrium with fixed foreign exchange rates is only possible at the heavy cost of sacrificing the domestic cost-price stability in the two countries. This classical analysis is known as the price-specie flow adjustment mechanism. Thus, the automatic gold flows initiated under the gold standard between the two gold standard countries bring the market rate of foreign exchange in line with the mint parity rate of foreign exchange whenever it deviates from this norm.

The price-specie flow mechanism under the international gold standard occupied a prominent place in the classical explanation of the determination of foreign exchange rate. Since gold standard no longer exists, the scope for the application of the mint parity theory of foreign exchange rate does not exist today. Under the old gold standard, the foreign exchange rate stability was achieved at the heavy cost of causing instability in the domestic costs, prices and incomes. In other words, the exchange rate stability was achieved by subjecting the economies of the gold standard countries periodically to the frequent bouts of either inflation or deflation. Moreover, the much-publicised gold-flow mechanism could at best correct only a mild disequilibrium in the balance of payments position. If the balance of payments disequilibrium was of a serious and chronic nature, the gold flows even along with the helpful changes in the rate of interest could not restore equilibrium in the balance of payments of the concerned countries.

While it is true that disequilibrium in the balance of payments reacts upon the foreign exchange rate through the gold flows and the consequent changes in the gold-losing and gold-receiving countries' exports (supply of foreign exchange) and imports (demand for foreign exchange) it is not clear that the mechanism of adjustment takes place in the manner stated by the classical economists. Firstly, commodities are not the only important constituent of the balance of payments of a country. Capital movements and services are also important in the total transactions entering into the external balance of payments of a country. Secondly, gold is not the first item to move. Short-term funds, securities etc. move earlier than gold. Consequently, if equilibrium in the balance of payments of the concerned countries is brought about through capital flows, commodity prices may not be affected. In fact, even when gold flows in the country it goes to the central bank where it may be locked in the central bank cellars. Consequently, it may not cause any increase in the money supply in the country. It had happened in America when the massive gold inflows in the thirties were prevented from expanding the money supply in the country by the Federal Reserve Authorities through 'sterilisation' of the gold receipts.

Moreover, the theory of the gold movements places unduly great reliance on the flexibility of the cost-price structures in the gold standard countries. It assumes that prices promptly fall in the gold-losing country in response to gold outflows and rise in the gold-receiving country. This

3. This tendency is reinforced by a rise in the rate of interest leading to capital inflows in the gold-losing country.

involves tight money and credit conditions in the gold-losing country and expansion in bank credit in the gold-receiving country. It must happen because gold is the basis of the issue of currency. The quantity theory of money applies in the system inducing fall in wages and prices in the gold-losing country to whatever level is necessary to restore equilibrium in the country's balance of payments. For this to be possible, an institutional arrangement—*laissez–faire*—conducive to such change must exist. When England returned to the gold standard in 1925 at the old exchange rate parity this was done with the pious hope that England's economy was flexible and wages and prices would fall. But strong trade unions resisted the wage-cuts that were necessary to correct the heavy balance of payments deficit. Strikes paralysed the British economy. In the present era of welfare state, with minimum wage legislations and minimum price support programmes, the theory of gold movements is an anachronism.

Moreover, the theory of the gold movements is based on the heroic assumption that the burden of correcting the external balance of payments disequilibrium is shared evenly by both the countries. This can happen only when the two countries are of equal economic size. Where trade takes place between a small country like England and a large country like the United States of America, prices may fall substantially in England when gold is exported out of the country while there will be only a marginal rise in prices in the USA following the gold inflows since the total gold received is a very small part of the massive US monetary gold reserves.

2. Purchasing Power Parity Theory[4]

The purchasing power parity theory of foreign exchange rate determination was stated by Richard Wheatley in 1802 in his famous work *Remarks on Currency and Commerce.*[5] The theory was later developed by Gustav Cassel at the end of World War I as a means of measuring the departure or deviation from 'equilibrium'. During the war, trade had been interrupted, international gold standard had been abandoned and monetary conditions in various countries were in a bad shape. Consequently, when trade was resumed at the end of war, the problem before the countries was to choose an equilibrium rate of foreign exchange that would ensure equilibrium in their external balance of payments.

Gustav Cassel suggested the purchasing power parity as the appropriate level at which the foreign exchange rate should be set. This rate was calculated by measuring the relative departures or deviations of price levels from some chosen base period in which the balance of payments of concerned countries had been in equilibrium. To illustrate, if the external balance of payments of the two countries X and Y were in reasonable adjustment in time period 0, then these countries should choose a foreign exchange rate (R_1) in time period 1 which reflects the changes in their prices between time period 0 and time period 1. Accordingly,

---

4. According to John Maynard Keynes, the term *purchasing power parity* was first introduced in the economic literature by the Swedish economist Gustav Cassel in an article published in *The Economic Journal* in 1918. For Cassel's views on the theory, see his *Money and Foreign Exchange After* 1914 published in 1922.

5. In his book *The Theory of International Prices*, Angell has shown that this theory which is usually associated with the name of Gustav Cassel is much older. In the famous Bullion Report in 1810 the theory was stated thus: "In the event of the prices of commodities being raised in one country by an augmentation of its circulating medium while no similar augmentation in the circulating medium of the neighbouring country has led to a similar rise in prices; the two currencies will no longer continue to bear the same relative value to each other as before. The exchange will be computed between two countries to the disadvantage of the former."

$$R_1 : \frac{P_{X_1}}{P_{Y_1}} \div \frac{P_{X_0}}{P_{Y_0}}$$

$$R_1 : R_0 \; \frac{P_{X_1}}{P_{X_0}} \cdot \frac{P_{Y_1}}{P_{Y_0}}$$

It says that if the prices (P) in country X doubled relative to the prices in country B from time period 0 to time period 1, the rate of foreign exchange (R_1) in time period 1 should fall to half relative to the exchange rate (R_0) in time period 0; in other words, the price of the domestic currency in terms of the foreign currency should fall to half or that of the foreign currency should be doubled in terms of the home currency.

According to the theory, the equilibrium rate of foreign exchange between two inconvertible currencies is determined by the ratio of their respective purchasing powers. The rate of exchange tends to be stabilised at the point of equality between the purchasing powers of the two currencies. For example, a given amount of certain goods and services demanded in the two countries, say in the United States of America and England, can be purchased for US $ 800 in America and for 200 Euro in England. Accordingly, the purchasing power of 800 US dollars is equal to that of 200 Euros. In other words, US $ 800 = Euro 200, or US $ 4 = EU 1. If the foreign exchange rate deviates from this norm, forces of equilibrium will come into operation and will bring back the foreign exchange rate to this norm. Thus, the general price level in the two countries remaining unchanged, if the foreign exchange rate rises to EU 1 = US $ 4.5 it shows that the purchasing power of the Euro in terms of the US dollar has risen. People owning Euro will convert them into dollars at this rate of exchange, purchase goods in the United States of America for 4 dollars which in England cost one Euro and earn profit of 0.5 dollar. This tendency on the part of British people to convert their Euro holdings into the US dollars will increase the demand for dollars in England while the supply of dollars in England will decrease because British exports to the United States of America will fall. Consequently, the Euro price of the dollar will rise until it reaches the purchasing power parity rate of foreign exchange of EU 1 = US $ 4. Conversely, if prices in England rose by 100 per cent, those in the United States of America remaining unaltered, the dollar value of the Euro will be halved. Consequently, one Euro would be equal to two US dollars because in the new situation two units of the British currency will purchase the same amount of goods and services in England as was bought by one unit before. On the other hand, if prices doubled in both the countries the purchasing power parity rate of foreign exchange would remain unchanged. Gustav Cassel stated the theory in the context of paper money with special reference to the effects of currency inflation on foreign exchange rates. The following quotation represents the early formulation of the theory by Gustav Cassel.

"Given a normal freedom of trade between two countries, *A* and *B*, a rate of exchange will establish itself between them and this rate will, smaller fluctuations apart, remain unaltered as no alteration in the purchasing power of either currency is made and no special hindrances are imposed upon the trade. But as soon as an inflation takes place in the money of *A*, and the purchasing power of this money is, therefore, diminished, the value of the *A*-money in *B* must necessarily be reduced in the same proportion. Hence the following rule: when two currencies have been inflated, the new normal rate of exchange will be equal to the old rate multiplied by the quotient between the degree of inflation of both countries. There will, of course, always be fluctuations from this new normal rate, and in a period of transition these fluctuations are apt to be rather wide. But the rate calculated in the way indicated must be regarded as the new parity between the currencies. This parity may be called *the purchasing power parity,* as it is determined by the quotients of the purchasing powers of the different currencies."[6]

The purchasing power parity theory applies to those countries which have inconvertible paper currencies. For those countries which are on the gold standard, the mint parity suffices to determine the equilibrium rate of foreign exchange between two such countries' currencies. The difference between the mint parity and the purchasing power parity is that while the former is a fixed parity, the purchasing power parity is a moving parity which moves with the price fluctuations in the countries concerned. Changes in the market rate of exchange arising on account of changes in the demand for and the supply of a particular country's currency around this parity go on as before. The limits to these changes are determined by the cost of transporting goods including tariff, insurance, banking and interest charges, costs of uninsurable risks, advertisement of goods in the foreign market etc. from one country to another. These limits are not fixed like the specie points in the case of the gold standard countries. Figure 13.7 illustrates the moving parity. The upper and the lower limits to fluctuations of this moving parity are determined by the commodity import and export points. The market rate of exchange is determined by the intersection of the demand and supply curves of foreign exchange. With S_1S_1 supply curve and D_1D_1 demand curve the market rate of foreign exchange is $0R_1$ which lies within the range of the commodity import and commodity export points. When the demand and supply curves change their positions as shown by D_2D_2 and S_2S_2 curves the market rate of foreign exchange rises from $0R_1$ to $0R_2$.

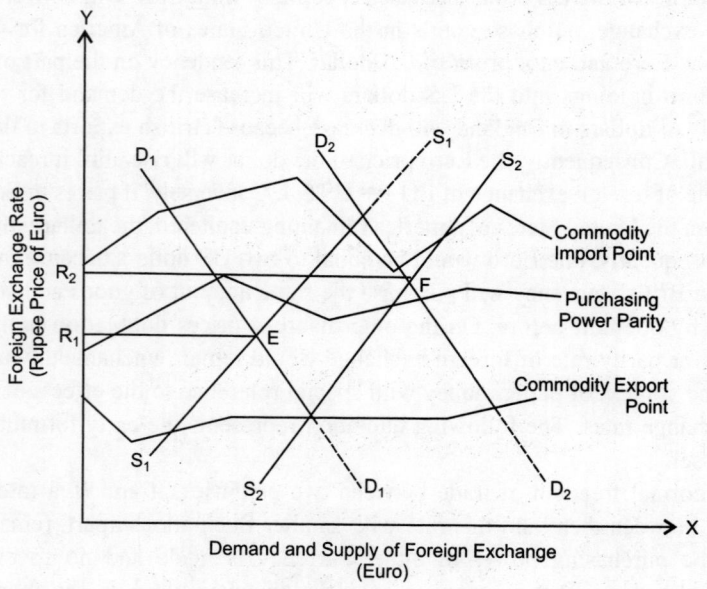

Figure 13.7

6. Gustav Cassel, "Memorandum on the World's Monetary Problems", International Financial Conference, Brussels, 1920, *Documents of the Conference,* pp. 44–45.

Criticisms

1. According to the theory, there is a direct relationship between the purchasing powers of the currency units and the rate of foreign exchange. In practice, however, this is not always so. Foreign exchange rate is frequently influenced by good deal more than the purchasing powers of the currencies of the concerned countries. For example, the foreign exchange rate can be influenced by tariffs. If a country imposes tariffs on her imports while other countries do not levy them, the foreign exchange value of the currency of the tariff-levying country will rise although there is no change in the domestic price level in the country.

2. The theory compares the general price level in the two countries without distinguishing between the prices of domestic goods and the prices of international goods. The prices of international goods are the same in different countries when the costs of transport, import duties etc. are taken into account. When it is said that the rate of exchange depends upon the relative prices in the two countries, the price level includes the prices of all the commodities many of which do not enter the area of international trade.

Eli Heckscher and others have criticised the theory on the ground that the purchasing power parity theory holds only if the analysis is confined to the prices of only those goods and services which enter into international trade and if the relative changes in transport costs and tariffs are ignored. The theory is not valid when applied to the general price levels. Criticising the theory, Heckscher has stated that "the conception that the exchanges represent relative price levels, or what is the same thing, that the monetary unit of a country has the same purchasing power both within the country and outside it is correct only upon the never existing assumption that all goods and services can be transferred from one country to another without cost. In this case, the agreement between the prices of different countries is even greater than that which is covered by the conception of an identical purchasing power of the monetary unit, for not only average price levels but also the price of each articular commodity or service will then be same in both countries if computed on the basis of the exchanges."[7]

The problem is that by considering the purchasing power as consisting of the prices of internationally-traded goods, the theory is reduced to a mere tautology. On removing this restriction although the concept of purchasing power becomes meaningful but no reliance can be placed on it as a determinant of the rate of foreign exchange. In short, the theory does not furnish us with any simple measure of the "true" value of the foreign exchange rate.

Changes in prices can be measured by different index numbers. The choice of an appropriate price index number for calculating the purchasing power parities has always posed a formidable problem. Taking the index numbers of the wholesale prices and calculating the purchasing power parities on their basis involves ignoring the manufactured goods and a vast range of services and other invisible exports. On the other hand, if the cost of living index numbers are used for calculating the purchasing power parities, many goods and services having no relevance to international trade will be included in the computation of purchasing power of the money unit. Consumer price indices contain many personal services, which apart from not being traded, whose prices diverge with economic growth. To the extent the prices of personal services—barbers', beauticians', lawyers', doctors' fees—rise, the consumer prices index is affected without making any direct impingement on the balance of payments. If we take wage rate indices, the position will be more satisfactory because wages figure in all goods' cost of production. But here also the difficulty is that labour efficiency differs from country to country. In the short period, prices of all

7. Eli F. Heckscher (and others), *Sweden, Norway, Denmark and Iceland in the World War,* 1930, p. 151.

domestic goods do not change in the same direction. The theory, therefore, holds good only in the long period. But for all problems of life it is the short rather than the long period which is important because 'in the long run we are all dead' and after death there is no problem.

Moreover, many goods and services which are bought and sold in one country do not compete with their counterparts in the other country. For example, the demand for dry-cleaning of woollens and hair-cuts in India does not compete with the demand for dry-cleaning of woollens and hair-cuts in England. Moreover, due to shipping costs and import restrictions, the prices of even international commodities, such as wheat, coffee and rubber will not be same in different countries expressed in terms of domestic currencies. A price comparison of this kind reveals that internationally-traded goods do not sell at the same prices in the export and import markets. These goods sell at a premium in the importing country and at a discount in the exporting country.

3. According to the theory, changes in the prices must be reflected in changes in the foreign exchange rate. But we may have changing prices in the country co-existing with a fixed foreign exchange rate because "only internationally-traded goods will influence the demand for and the supply of foreign currency and consequently the rate of foreign exchange. Commodities which are only domestically traded, on the other hand, have direct bearing on the exchange value of the currency and their prices may, therefore, fluctuate without affecting directly and immediately the foreign exchange rate. Confined to internationally-traded commodities, the purchasing power parity theory becomes an empty truism."[8]

4. The theory makes us believe that changes in the prices induce changes in the foreign exchange rate but changes in the foreign exchange rate exercise no influence on the prices. According to the theory, the chain of causation is one-sided. But the assertion that changes in foreign exchange rate do not influence prices is not always true. For example, if massive speculative capital outflows from England reduce the foreign exchange rate of the Euro and if the Euro remains depressed for several months, imports in England will become costly. Consequently, more Euro will be required to pay for the imports of the American wheat and other goods. Many British industries use imported raw materials. Since prices of the imported raw materials have risen, prices of the manufactured goods will also rise. Similarly, British exports will become cheaper in terms of the US dollar and the demand for them will increase. Consequently, British exporters may raise prices of export goods, particularly in the short period when their supply is inelastic. If they raise prices of export goods by less than the amount of depreciation of the Euro, the exporters will only increase their profits without reducing their exports. But in spite of the prices of exports having risen they are cheaper in foreign countries. Thus, prices in England rise as a consequence of the fall in the dollar value of the Euro. But this is not all. The Euro having depreciated, the competitive strength of British export industries increases in the foreign markets. Consequently, other countries will also reduce their export prices in order to retain their hold on the foreign markets in the face of growing competition from British exports. In short, depreciation of the Euro not only raises the prices in England but also lowers prices in the other countries. The sequence of events is often the opposite of that predicted by the theory. The experience of some countries leads to the conclusion that the foreign exchange rate governs prices rather than the latter governing the former. According to Geroge N. Halm, domestic prices follow rather than precede the movement of foreign exchange rates. According to him "A process of equalisation through arbitrage takes place so automatically that the national prices of commodities seem to follow rather than to determine the movement of exchange rates."[9]

9. George N. Halm, *Monetary Theory,* p. 524.
8. J.M. Keynes, *A Tract on Monetary Reform,* 1923, p. 101.

5. Changes in the foreign exchange rate affect prices differently in different countries. If a country imports consumer goods in bulk and devotes its resources to producing mostly export goods the effect will be large. Changes in the foreign exchange rate will affect both the prices of imports and exports and when the imports and exports form a large part of the total goods, changes in the general price level will be larger. Thus, a given change in the foreign exchange rate will affect prices more in a country like Holland where foreign trade is an important sector of the national economy than in a country like the USA where foreign trade is a small part of the national economy. Thus, although changes in the foreign exchange rate may exercise different effects on prices in different countries, they always exercise some effect. It is, therefore, wrong to say that changes in prices cause changes in the foreign exchange rate and the changes in the foreign exchange rate do not cause changes in prices.

6. The theory ignores the demand for and the supply of foreign exchange arising from sources other than international trade in merchandise. The theory overlooks the fact that the demand for and supply of foreign exchange also arises on account of the imports and exports of invisible services like insurance, shipping, banking etc. and from other unilateral transfers involving reparation payments, making of gifts, payment of loans and interest by one nation to another. Long-term investments, transfer of capital on account of interest and dividend payments and short-term speculative capital movements also create debits and credits in the international balance of payments of a nation influencing the foreign exchange rate. The foreign exchange rate is the result of the working of a vast complex of forces summarised by the balance of payments and not simply by the balance of trade. In short, the purchasing power parity theory is incomplete. It does not offer a complete explanation of the determination of equilibrium rate of foreign exchange between inconvertible paper currencies.

7. The theory is based on the unrealistic assumption that (a) the balance of payments in the base period was in equilibrium, and (b) no structural changes have taken place in the factors underlying this equilibrium, i.e., technology, resources and people's tastes including their attitudes toward foreign investment influencing the portfolio decisions are all assumed as given.

8. The theory does not consider the fact that economic relationship between countries may change following which the equilibrium rate of foreign exchange may also change although the prices in the concerned countries may not change. For example, when some third country appears on the horizon of international trade as competitor either as a buyer or as a seller of goods and services in the market the flow of trade between the two original countries will be affected, affecting also the foreign exchange rate between their currencies. Ragnar Nurkse has criticised the theory on the ground that it considers only the price movements as the main criterion and ignores other important factors influencing the demand for foreign exchange. According to him, the theory "treats demand simply as a function of price leaving out of account the wide shifts in the aggregate income and expenditure which occur in the business cycle (as a result of market forces or governments' policies), and which lead to wide fluctuations in the volume and hence the value of foreign trade even if prices or price relationships remain the same."[10]

Conclusion

Notwithstanding its serveral criticisms, the purchasing power parity theory is not useless as an explanation of the forces determining the foreign exchange rate and it should not be discarded.

10. Ragnar Nurkse, *International Currency Experience*, p. 126.

Despite its weaknesses, the theory has a practical significance from the point of view of monetary policy. It warns the countries that there are certain things which they cannot do. In 1925, England had fixed a higher foreign exchange rate of the pound-sterling than was justified by the prices prevailing at that time in the country. According to the theory, the high exchange value of the Euro could be maintained only if the prices in England were low relatively to world prices. Failing that, the foreign exchange rate of the Euro would have to fall. Government, however, tried to maintain the external value of the Euro by raising the bank rate and attracting foreign capital. A rise in the bank rate causes fall in prices, wages and employment in the economy. But wages could not fall in England due to the genered strike threat given by the strong trade unions. Consequently, prices remained high and the disequilibrium between the domestic price level and the foreign exchange rate persisted. When costs and prices could not fall, it was inevitable that sooner or later the external value of the Euro must fall. This happened in 1931 when England abandoned the gold standard causing confusion in the world. In a period of fluctuating foreign exchange rates, the theory provides a rough measure of the degree by which the market rate of foreign exchange differs from the equilibrium rate of foreign exchange.

When the calculation of the foreign exchange rate stabilisation is undertaken, the purchasing power parity provides a most important factor for the determination of proper foreign exchange rate that should be adopted. This is particularly true when inflation has been a dominant cause of the foreign exchange rate instability. Although such calculations are crude indicators yet they provide a method to arrive at specific conclusions as to the general location of the appropriate foreign exchange rate. According to John Maynard Keynes, although the theory cannot provide any accurate measure of the true foreign exchange value of the currency units, even so the theory "tells us an important fact about the relative changes in the purchasing powers of money in (e.g.) England and the United States or Germany between 1813 and, say, 1923, but it does not necessarily settle what the equilibrium exchange rate in 1923 between sterling and dollars or marks ought to be."[11]

3. Balance of Payments Theory

The balance of payments theory holds that the foreign exchange rate is determined by autonomous factors which are unrelated to the internal prices and the money supply. According to the theory, a deficit balance of payments leads to the fall or depreciation of the rate of foreign exchange while a surplus balance of payments, by strengthening the foreign exchange, causes an appreciation of the rate of foreign exchange. An adverse international balance of payments of a country shows a situation in which the demand for foreign exchange (currency) exceeds its supply at a given rate of foreign exchange. Consequently, its price in terms of the domestic currency must rise, i.e., the external value of the domestic currency must depreciate. Conversely, if the international balance of payments of a country is favourable it means that there is greater demand for the domestic currency in the foreign exchange market than can be met by the available supply at any given rate of foreign exchange. Consequently, the price of the unit of home currency in terms of foreign currency rises, i.e., the rate of foreign exchange moves in favour of the home currency. Consequently, a unit of home currency begins to command more units of the foreign currency.

The naive balance of payments theory asserts that the foreign exchange rate is determined by the internatic nal balance of payments position of a country in the sense of supply and demand like the determination of any other price. While it cannot be denied that the foreign exchange rate is a

11. J.M. Keynes, *Op. cit.,* p. 58.

function of the supply of and the demand for foreign exchange the important question is: what is it that determines the supply of and the demand for foreign exchange? The theory answers the question by asserting that the balance of payments position (and, therefore, the supply of and the demand for foreign exchange) is determined mainly by factors that are independent of changes in the rate of foreign exchange. Some of these independent factors that influence the foreign exchange rate consist of autonomous payments like the reparation payments and payment of interest on foreign loans etc. Besides these autonomous payments, the demand for imports, particularly for essential raw materials, may be perfectly price-inelastic and, therefore, insensitive to changes in the foreign exchange rate.

The theory has certain strong points as an explanation of the determination of foreign exchange rate. Firstly, it is compatible with the general theory of value to state that like the price of a commodity the rate of foreign exchange should also be determined by the supply of and the demand for foreign exchange. Secondly, it brings the determination of the foreign exchange rate within the purview of the general equilibrium theory. Thirdly, the theory emphasises that there are many significant forces, besides those of exports of goods and services (which are influenced by prices), included in the balance of payments which by influencing the supply of and the demand for foreign exchange affect the equilibrium foreign exchange rate. Fourthly, the theory explains that disequilibrium in the balance of payments of a country can be corrected by making appropriate adjustment in the foreign exchange rate, i.e., either by devaluing or by revaluing the currency unit of the country. It is significant because under the mint parity theory of foreign exchange rate, disequilibrium in the balance of payments could be corrected only through inflation or deflation causing price instability in the country.

The main defect of the theory is that it assumes the balance of payments to be a fixed quantity. In the Keynesian metaphor it applies the theory of solids where the theory of liquids would be more appropriate. The balance of trade depends upon the relationship between the domestic and foreign prices. What the prices expressed in terms of the foreign currency are in the foreign country also depends on the rate of foreign exchange between the two currency units. A rise in the foreign exchange rate is equivalent to a rise in prices abroad in relation to prices at home while a fall in the foreign exchange rate indicates a fall in the foreign prices, assuming no change in the prices in the foreign country.

The second criticism of the theory is that it treats the demand for many imported raw materials as perfectly price-inelastic and consequently independent of changes in prices and foreign exchange rate. It is worthwhile to remember that perfectly inelastic demand for a good is a very rare phenomenon. Even the most essential commodity has some cross elasticity of demand. Let alone the demand for raw materials. Even the demand for imported food grains (food is most essential for life, consequently its demand is price-inelastic) has some cross elasticity because the same physiological needs can be satisfied either by cheap commodities like ordinary bread and vegetables or by expensive ones like meat and fruits. In other words, even in the case of essential-to-life goods, some substitution is possible. Consequently, the demand for them is influenced by price changes which are affected by changes in the foreign exchange rate. The balance of payments is not, therefore, completely independent of foreign exchange rate. It is partly dependent on the foreign exchange rate and changes in it. Thus, the balance of payments, being itself a function of the rate of foreign exchange, cannot explain the determination of the rate of foreign exchange.

Fixed (Stable) Foreign Exchange Rate

The case for fixed or stable exchange rate system is frequently stated in rather practical terms. It is built upon an analogy with the case for a national currency. The following arguments have been advanced for a system of fixed foreign exchange rate.

1. Fixed foreign exchange rates encourage the development of international trade by enabling the importers and exporters to know in advance how much they will have to pay and how much they will receive in terms of the domestic currency. Conversely, freely fluctuating foreign exchange rates add to the hazards of international trade by making the calculation of prices in terms of the domestic currency extremely uncertain and difficult.

2. In an era of extensive overseas investment and economic development, the progress of development plans in the underdeveloped countries may be jeopardised by instability in the foreign exchange rates because the flow of foreign capital is hampered by the speculation about currency appreciation or depreciation.

3. Fixed foreign exchange rates are more suited to the development of regional arrangements like the sterling area or dollar area where a fluctuating foreign exchange rate for the reserve currency can become a bone of contention between the members of the area.

4. For small nations like Denmark, Belgium and England in whose gross national product the contribution of foreign trade is significant, foreign exchange rate stability is the only correct policy for the government to pursue and less the country deviates from this policy the better for it. If the country fails to stabilise the foreign exchange rate for her currency unit, her foreign trade will be dislocated by fluctuations in the foreign exchange rate jeopardising her economic prosperity. Thus, those small countries in whose national product foreign trade plays a significant role can ill-afford the costly experiment of fluctuating foreign exchange rates. If they depend on foreign capital for their economic development they must maintain the foreign exchange rate stability at any cost so long as there are important countries abroad with a fairly stable monetary system to which they can peg their money units.

5. Fixed foreign exchange rates are essential for the orderly growth of international money and capital markets which are vital for international capital flows. When the foreign exchange rates fluctuate widely, the foreign investors become chary of lending to the foreign nationals. Monetary history of the *thirties* presents a most convincing case to prove it. So long as there existed a system of stable foreign exchange rate under the gold standard, smooth flow of international lending continued to the great advantage of orderly growth of world trade and international money markets. International lending ended abruptly when after the breakdown of the international gold standard an era of competitive exchange depreciation began disrupting the international lending.

6. Fixed foreign exchange rates promote international economic integration by establishing a stable link between the world's currencies. Under a system of fixed foreign exchange rate, markets become less fragmented and trade is facilitated. A system of fixed foreign exchange rates reduces or removes uncertainty resulting from speculation about the depreciation or appreciation of currencies.

7. A nation derives immense benefits from having a common currency resulting from economic integration of different regions. Likewise, the world would benefit by having a single currency as it would promote economic integration of world's economies. An approximation of this state is a world economy characterised by fixed foreign exchange rates.

8. Another argument for fixed exchange rate system is the so-called 'anchor' argument. Democratic governments are frequently tempted to overstimulate their economies. In the short run,

expansionary economic policies may reduce unemployment and create boom conditions at the expense of a stable price level. While inflation may create temporary euphoria it will have harmful effects in the long run. It will adversely affect the distribution of income and wealth in the economy. Fixed foreign exchange rate can serve as an anchor imposing a discipline on the governments not to pursue the inflationary policies which are inconsistent with the economic policies pursued by rest of the world.

9. A stable foreign exchange rate system discourages speculation in the foreign exchange market while it ensures international credit operations of normal kind. In the absence of foreign exchange rate stability commercial and financial relations with foreign countries are hampered.

Flexible Foreign Exchange Rate

The case for flexible foreign exchange rate is supported by countering the arguments advocated for fixed foreign exchange rate. It is claimed that fixed foreign exchange rates encourage the smooth flow of free multilateral trading system. This argument is not, however, strengthened by the history of the post-war years. Even when the foreign exchange rate fluctuates from day-to-day, the trend of the currency's value in the foreign exchange market can usually be assessed. When the balance of payments of the country concerned is in equilibrium, the rate of foreign exchange is not likely to alter significantly in such period because the average trade transaction takes time to complete while in disequilibrium the trend of the market may be judged. In either case, the forward foreign exchange market provides protection to the importers and exporters.

Secondly, it is argued by the advocates of flexible foreign exchange rate system that it allows the foreign exchange rates to find their natural level where there is no disequilibrium in the balance of payments. It has been said that it is a very elegant system which solves the problem of external disequilibrium almost without effort.

Thirdly, flexible foreign exchange rates adjust the external balances and prevent the recurring balance of payments crises more effectively compared with the fixed foreign exchange rates. Consequently, their effect on the international lending is likely to be beneficial.

Fourthly, fixed foreign exchange rates are not essential for any system of currency areas. The sterling area is typical of a regional payments group whose framework was cast in the *thirties* when the exchange rate against the pound-sterling was free to fluctuate in the free foreign exchange market. Certain economic, political and social ties have brought together the different countries to form the sterling block and these would not be weakened if the pound-sterling has a flexible exchange rate system after making consultations with the members of the area.

Fifthly, central to the argument in favour of flexible rates is the view that government intervention in the economy should be discouraged. All attempts to fix or manipulate exchange rates harm the economy because the control of exchange rates is entrusted to experts who frequently fail to handle the problems in the national interests. According to the supporters of flexible foreign exchange rate, the great merit of this system is that it frees the government from considerations about the balance of payments. Under the flexible foreign exchange rate system, the burden of the balance of payments adjustment is borne by the foreign exchange rates rather than by the domestic incomes and prices. To ease the problem of maintaining the external equilibrium without burdensome constraints on domestic economic policies, a country should allow its foreign exchange rate to respond freely to the demand for and the supply of foreign exchange. Flexible foreign exchange rates allow the country freedom of monetary management.

Thus, some of the strong arguments against flexible foreign exchange rates have lost their force. Furthermore, the fixed foreign exchange rates suffer from certain other defects. A fixed foreign exchange rate encourages destabilising currency speculation which, despite foreign exchange control, can endanger the stability of a currency and make it impossible for an exchange rate, once the currency falls under the suspicion of impending devaluation, to be maintained. Speculation regarding the devaluation of the pound-sterling played a vital role in bringing about that currency's devaluation in September 1949, of the Indian rupee in June 1966 and July 1991 and of the US dollar in December 1971 and again in February 1973.

Moreover, fixed foreign exchange rates do not reflect the existing and true cost-price relationship between the two currencies; they represent the rates which existed formerly. When countries follow different economic policies, the cost-price relationship alters frequently. The theoretical arguments against fixed foreign exchange rates and the experience of the past few decades point to the desirability of abandoning the fixed foreign exchange rates in favour of flexible foreign exchange rates.

It must be remembered that except for a brief interval of time, no country can afford to allow its foreign exchange rate to drift and follow the day-to-day developments in international and external economic conditions for any length of time. The instability resulting from such policy will have serious repercussions not only on the foreign trade but also on the entire economy of the country. Fluctuating foreign exchange rates are incompatible with the stability of the domestic economy. Unbridled fluctuations in foreign exchange rate cause unwarranted fluctuations in the prices of exports and imports. Consequently, goods which were not formerly internationally-traded enter the arena of international trade while some others that were formerly so traded are excluded from international trade. Consequently, not only the smooth flow and development of international trade but also the inter-industry allocation of resources and factors of production in the economy is disrupted. Fluctuating foreign exchange rates cause "constant shifts of domestic factors of production between export and home market industries; shifts which may be disturbing and wasteful."[12]

Moreover, freely fluctuating foreign exchange rates induce unwarranted capital movements. Speculation in foreign exchange becomes an attractive pastime. The massive capital outflows that take place in anticipation of currency devaluation are dangerous because such anticipation is apt to bring its own realisation. Anticipatory purchases of foreign exchange produce or at any rate hasten the anticipated fall in the foreign exchange rate of national currency and the fall in the foreign exchange rate when it comes about strengthens speculation regarding further fall in the foreign exchange rate. The danger of such cumulative self-aggravating movements under freely fluctuating exchange rate is demonstrated by the French experience of 1922-26 and the American experience leading to the devaluation of the US dollar for the second time on February 13, 1973 in less than 14 months. Foreign exchange rate in such circumstances is bound to become highly unstable. The problem of speculative capital movements under fluctuating foreign exchange rate in turn creates the more difficult problem of an unusually high liquidity preference. "The gains and losses which can be made from unexpected depreciation of one currency or another have come to be realised more and more widely. People, therefore, try to invest their money in as liquid a form as possible, in order to be able to convert it at the first sight of danger into some other currency which appears at the moment to offer greater security."[13] This phenomenon of very high liquidity

---

12. Ragnar Nurkse, *Conditions of International Monetary Equilibrium*, Essays in International Finance, No. 4, Princeton University Press, Spring 1946, p. 3.
13. Gottfried Von Haberler, *The Theory of International Trade*, 1950, p. 45.

preference encourages hoarding and causes the rate of interest to rise leading eventually to fall in the aggregate investment, employment and income in the economy.

Fluctuating foreign exchange rates produce very unpleasant long-period effects. The argument that it is necessary to allow the foreign exchange rate to fluctuate to enable it to find its 'natural' level does not have much force in it. There is no such thing as the natural level for the rate of foreign exchange of a currency. The proper rate, in each case, depends on the economic, financial and monetary policies that are pursued by the country concerned and by the other countries with whom it has important economic relationship. Moreover, whether or not a given foreign exchange rate is at the 'correct' level can be determined only after there has been sufficient time to observe the course of the international balance of payments of a country in response to that rate of foreign exchange.

In a period of continuing inflation, fluctuating foreign exchange rates cause a vicious circle of devaluation and inflation to emerge in the economy. On the one hand as long as internal prices are not stabilised flexible foreign exchange rates are essential to prevent over-valuation of the currency. Yet under conditions of domestic price instability, the foreign exchange rate depreciates regularly contributing to the greater inflation. Moreover, if a country does not promptly adopt a monetary policy aiming at economic stability, the movements of its foreign exchange rate are likely to be oscillations not around a stable value but around a declining trend.

Uncontrolled fluctuations in foreign exchange rate are harmful because such fluctuations seriously hamper the flow of long-term foreign investment. Where the foreign loans are contracted in terms of the borrower country's currency, the creditor faces the hazard resulting from exchange rate changes causing depreciation of the borrower country's currency in relation to lender country's currency. Even when the loans are contracted in the lender country's currency, fluctuating foreign exchange rate increases hazard for the debtor country in so far as a depreciation of her currency increases the burden of country's total debt servicing and repayment of debt.

To conclude, some degree of exchange rate stability is essential for the smooth working of world's economy. Perhaps the best position is neither a continuously fluctuating foreign exchange rate nor a rigidly pegged foreign exchange rate but an exchange rate system which gives freedom to a country to alter her foreign exchange rate when needed within certain well-defined limits such that it promotes the development of stable international economic relations. This has been made possible under the IMF scheme in which a system of flexible exchange rate stability has been promoted. .

Fluctuating Foreign Exchange Rate and the IMF

Although the IMF was created five decades ago in 1946 to promote the foreign exchange rate stability without subjecting the members to the rigid discipline of the gold standard in the matter of national monetary management, fluctuating foreign exchange rates are not uncommon. The Fund has often been criticised for its approach to the "fluctuating foreign exchange rates" or "floating rates".

Although until before the Second Amendment of Articles in 1978, the Fund had supported the system of institutionally agreed par values of members' currencies, pragmatic considerations have, however, induced the Fund to allow certain members in exceptional circumstances to practise fluctuating exchange rate system. Such a situation was faced by the Fund first in July 1948 when Mexico suspended transactions at the par value of its currency and allowed the market to set the exchange rate for her currency. The Fund and the Mexican Government had held continuous consultations with the object of establishing a new par value for the Mexican *peso*. One year after the suspension of the initial parity a new par value was fixed. Consequently, for one full year the Mexican *peso* had a floating exchange rate.

In November 1949, Fund's attitude towards the problem of fluctuating foreign exchange rates changed when Peru had suspended transactions at the initial par value for the Peruvian *sol* fixed in 1946. Upto March 1, 1970, Peru had not set a new par value for her currency and there was a dual exchange market with two fluctuating exchange rates. Meanwhile, during the two decades that Peru had maintained a system of floating rates, the Fund had allowed Peru to draw upon its resources and to assume the obligations of Article VIII, that is, to be regarded as a country which is relatively free of restrictions and which has a convertible currency. The Fund tolerated this action of the Peruvian authorities as it was convinced that the purpose of Peru's action was the establishment of a unitary exchange rate system on a more appropriate level.

The second example of Fund's acceptance of floating foreign exchange rate in exceptional circumstances is that of Canada. The Canadian dollar floated in the foreign exchange market for about 12 years from September 1950 to May 1962. The fluctuating exchange rate was introduced to discourage the heavy inflow of capital, mainly from America, which had become indeed massive in 1950 inflating the money supply, depressing interest rates and augmenting the inflationary pressures in the economy. The object of this floating rate was, therefore, in marked contrast to most other exchange rate changes which were intended to correct an unfavourable balance of trade and to check capital outflows. On account of the speculative nature of bulk of inflowing capital, the Canadian Government was unable to foresee the end of the capital inflows as long as the foreign exchange rate was fixed. Consequently, to discourage the speculative capital movements which were inflationary in nature for the country's economy, the Canadian Government left the Canadian dollar free in the foreign exchange market for its exchange rate to be determined by the market forces of the supply of and the demand for the Canadian dollar. Recognising the gravity of the situation, the Fund allowed Canada to let the exchange rate of the Canadian dollar remain floating in the foreign exchange market.

Besides Peru and Canada, the other Fund members which have had floating foreign exchange rates for relatively long periods included Lebanon, the Syrian Arab Republic and Thailand in 1954. To facilitate the Fund transactions in currencies with fluctuating exchange rates—such as Fund drawings in these currencies or payment of additional local currency in order to preserve the value of Fund assets—the Fund set up special rules for computing the exchange rates of those currencies with fluctuating exchange rates. Computations were based on the mid-point between the highest and lowest rates for the US dollar quoted for cable transfers for spot delivery in the main financial centre of the country of the fluctuating currency on specified days. Computations under this decision had been made only for those countries' currencies for which transactions did not take place at their par values.

In the second set of circumstances, the Fund had gone somewhat further. In conjunction with exchange reforms and stabilisation, the Fund had supported programmes which had included a fluctuating exchange rate. To some extent, Peru's fluctuating exchange rate had been introduced as a part of an exchange reform. In those countries where a flexible foreign exchange rate policy is deemed essential because of continuing inflation, a test has been set up as a means of assuring that a rate will be maintained which conforms to the basic market trends. Such a test consists of a prescribed minimum level at which the foreign exchange reserves of the country are to be maintained during a stated period with exchange rate action taken whenever there develops a threat to that level. Even so since most countries have been reluctant to excessive depreciation of the exchange rate of their currencies, the Fund had to tell the countries with flexible exchange rate not to stabilise the exchange rate of their currency units prematurely or in a hurry.

The ultimate purpose of the Fund, even in cases of exchange reform, has been to create the conditions for the restoration of a fixed and unified foreign exchange rate system. The fluctuating exchange rate has been regarded as a temporary means to an end. The Fund has regarded a general system of fixed foreign exchange rates and institutionally agreed par values superior to a system of fluctuating foreign exchange rates piously hoping that a few exceptions would not cause a return to the exchange rate instability of the pre-Fund era. The Fund era has recognised the possibility that there may develop certain unusual circumstances forcing the country to adopt the fluctuating foreign exchange rate.

Adjustable Peg (AP) and Crawling Peg (CP) Foreign Exchange Rate

In recent years, the debate between the supporters of the 'fixed' and 'fluctuating' exchange rate systems has become increasingly intense. As a consequence of this renewed controversy, various new proposals aiming at achieving greater degree of exchange rate flexibility while still retaining the important aspects of the 'Bretton Woods' adjustable peg system have been advanced. One such proposal advanced as an alternative to the adjustable peg exchange rate envisaged under the IMF scheme is the system of crawling peg exchange rate. The crawling peg exchange rate system, espoused in the mid-sixties by economists of the eminence of William Fellner, John H. Williamson, J. Black, James E. Meade and Carter J. Murphy, is based on the argument that the adjustable peg exchange rate system in the IMF scheme under which the exchange rates could diverge from their par values within a margin of 1 per cent on either side of the exchange rate parity by not allowing a wide-enough band for the exchange rate to move encouraged disequilibraing speculation about the ability of a member to maintain the exchange value of her currency at 1 per cent above or below the exchange parity.

It was argued that a major determinant of the effectiveness of any exchange rate system is the effect it has on the speculative movements of international money. Speculative and destabilising capital flows are stimulated if and when the financial community feels that an exchange rate is likely to be changed so that it can make profit by shifting funds from weak currencies to strong currencies. Greater the scope for the exchange rate to fluctuate, smaller is the reason for speculators to entertain fears about any impending devaluation of the currency. In fact, a perfectly freely fluctuating foreign exchange rate is on the other extreme of the rigid foreign exchange rate stability which was an important feature of the gold standard. The crawling peg exchange rate system is a compromise between these two extremes. The foreign exchange rate is allowed to crawl within the band on either side of the par value. In December 1971, the margin by which the foreign exchange rate could diverge from its par value had been widened from 1 per cent to 2.25 per cent on either side of the exchange parity. This in effect meant that in place of the old system of adjustable peg, a new system of crawling peg had been introduced in the IMF scheme. Under this new scheme, the exchange rate could comfortably crawl within the overall wider band-range of 4.5 per cent in place of the earlier narrow 2 per cent band-range. The introduction of a wider band-encouraged the equilibrating capital flaws while discouraging the disequilibrating speculation. It also enabled the Fund members to. exercise a greater degree of freedom in the use of monetary policy than was possible in the past while still preserving some of the anti-inflationary effects of the fixed foreign exchange rate.

Spot and Forward Foreign Exchange Rates

There are the spot rate of foreign exchange and the forward rate of foreign exchange ruling in the foreign exchange markets. The spot rate of foreign exchange is the rate of price expressed in terms of the home currency which is payable for the spot delivery of specified type of foreign exchange. The forward rate of foreign exchange is the rate of price at which a transaction will be consummated at some specified time in future. In actual practice, there is not one but many spot rates of foreign exchange; the spot rate for cables (telegraphic transfers) being different (usually higher) from the one applicable to cheques and commercial bills.

The spot transactions in foreign exchange in which the buyer buys and seller sells 'on the spot' are conducted in the spot foreign exchange market. In the spot transaction, the seller of foreign exchange has to deliver the foreign exchange sold by him 'on the spot' (within two days). Likewise, a buyer of spot foreign exchange will immediately receive the foreign exchange bought by him. In the forward foreign exchange market the seller agrees to sell and the buyer agrees to buy a certain stated amount of foreign exchange at some specified future date at a price agreed upon in advance. Usually the forward contracts are made on the three-month basis. The spot and forward foreign exchange markets are intimately linked through interest arbitrage, hedging and speculation.

In modern times, the forward foreign exchange rate has assumed great importance in affecting the international capital movements. The foreign exchange banks play an important role in this respect by matching the purchases and sales of forward foreign exchange on the part of prospective importers and exporters respectively. The system of forward foreign exchange rate has actually been developed to minimise the risks resulting from fluctuations over time in the spot foreign exchange rate to the importers and the exporters. An example may be given to illustrate this point. Suppose that an Indian television dealer wants to import the television sets from Japan. Suppose further that the foreign exchange rate at the moment is Rs. 2 per Japanese yen and at this rate the Indian television dealer calculates that he could import the television sets, pay the customs duty on them, sell them in India, pay the price of the television sets and make a profit on them. But by the time the television sets have been shipped, exhibited in Bombay, sold and paid for, several months will have elapsed and the foreign exchange rate may now be Rs. 3 for a yen in which case he has to pay one rupee more for each yen of the price of the television set and in place of the expected profit he may realise actual losses. In other words, the transaction will be profitable only if the Indian importer can import television sets from Japan at an exchange rate of two Indian rupees for one Japanese yen.

There is often a time difference between the entering into a contract to import in the present and the actual arrival of goods at some future date when he will have to pay for the imported goods in foreign currency. However, the rate of foreign exchange at that future date may have changed to the disadvantage of the importer rendering the potentially profitable transaction a loss-making transaction. He, therefore, runs the risk of making a loss. The forward foreign exchange rate market gives him this assurance. His bank will sell him "three months forward" yens at Rs. 2 per yen by charging a small premium. It means that the bank undertakes to sell the named quantity of the yens at an exchange rate of Rs. 2 per yen in three months' time whatever the rate of exchange in the foreign exchange market may be when that time comes. Similarly, persons who expect to receive sums in foreign currency at some future date (i.e., exporters or investors expecting interest or dividend income from abroad on their investments) are able to sell "forward exchange" to the banks in order to be sure in advance exactly how much they will receive in terms of the home currency. This basic importance of the forward foreign exchange rate flows from the fact that the actual rate of exchange is liable to fluctuate from time to time rendering the purchase and sale of goods abroad risky. Forward foreign exchange rate enables the exporters and importers of goods to know the prices of their goods which they are about to export or import. The forward foreign exchange market by providing *hedging* or *covering* protects both the importers and exporters against the risks of foreign exchange rate fluctuations thereby enabling them to concentrate on their pure trading activities.

Arbitrage

Arbitrage is the act of simultaneously buying a currency in one market and selling it in another at a small margin of profit. It is the mechanism by which the two markets, physically separate, are unified into a single market in economic sense. A single market is a place where the buyers and sellers of a commodity trade it at an identical price. Where the same price exists continuously for the same commodity, there exists one market for that commodity. Arbitrage produces one price and one market where there are two markets and the costs of buying in one and selling in the other market are small. Where arbitrage takes place in the foreign exchange market it is among the most perfect markets of the world because money is the most homogeneous of all goods and it can be transferred instantaneously. An arbitrageur is one who generally buys at a lower price and sells at a higher price in a situation where both prices are known to him. In the spot foreign exchange market, for example, the US dollar may be slightly cheaper in Frankfurt than it is in Paris. Foreign exchange arbitrageurs will take advantage of this differential by buying the US dollars in Frankfurt and selling these in Paris.

Another form of arbitrage, mentioned in the literature since 1958, is the 'interest arbitrage' as contrasted with the currency arbitrage. Basically, interest arbitrage is based on the same principle as the currency arbitrage except that it involves interest rate differentials as well as spot and forward foreign exchange rates because the arbitrageur's profit depends not only on the interest rate differential but also on the exchange rates. In this manoeuvre, an operator buys a short-term security in one market at a higher yield than can be had in the country of the currency employed; at the same time he covers the possible foreign exchange risks by immediately selling forward the proceeds of the short-term security at rates that will increase or decrease the return from the security purchased. At times, the cost of covering the exchange risk may be quite heavy and might absorb all or part of the interest rate differential precluding the person from entering into the deal. An illustration would make it clear what interest arbitrage stands for. Suppose that a Mumbai trader purchases a 90-day Treasury Bill in London on which he realises 5 per cent return. In order to protect himself against the foreign exchange risk, he sells the proceeds of the UK Treasury Bill in the forward foreign exchange market at a discount of, say, 2 per cent, which when substracted from 5 per cent leaves him with 3 per cent net return. If the rate of return on the Indian Government Treasury Bill is 2.5 per cent, the trader makes 0.5 per cent higher return on the amount over what he could get on a similar investment made in the government security in India.

Arbitrage is conducted by a few large institutions which are mostly banks and possess the necessary capital and connections for prompt purchase and sale of foreign exchange on a large scale and which can afford to obtain a very small rate of profit. It has been said that 1/32 of one per cent is enough to get the professional arbitrage under way. The economic importance of arbitrage operations flows from the fact that such operations remove the discrepancies in the rate of foreign exchange as between the different international financial centres.

When two or more foreign exchange markets are linked together, the prices of a standard form of exchange in the several markets so linked will tend to be uniform. Thus, Euro cables in New York should normally sell at the same rate at which dollar cables are available in London at that time. This relationship should hold good even if more countries are added to the list. But how do arbitrage operations bring about the equality of rates between different countries? The strategic role of arbitrage operations in achieving this can be illustrated by the following illustration. Suppose that the Euro is quoted at US $ 2.80 in New York; it must also stand at US $ 2.80 in London. But why? If the dealers were free to buy and sell dollars and Euro in any quantity, if the dollar/ Euro exchange rate was $ 2.80 in New York and $ 2.85 in London, they would purchase Euro with dollars @ $ 2.80 in New York, sell them @ $ 2.85 in London and earn profit of 0.05 Euro on each Euro bought and sold. What will be the result of this activity on the part of banks? The purchase of the Euro will push up its dollar price in the New York market while the pressure of sale of Euro in London would, by increasing the supply of Euro, lower its price in terms of dollars. This process of purchase of Euro in New York and sale in London will continue so long as the price of the two currencies (exchange rate) ruling at both the centres does not become equal. The foreign exchange rates between two or more centres closely linked with each other must be uniform, subject only to very narrow fluctuations equivalent to the cost of carrying out the arbitrage operations which is as low as 1/32 of one per cent.

In the real world, however, arbitrage operations are not limited to two currencies or two centres only. In fact, there may be more than two currencies and more than two financial centres involved in an arbitrage operation. Three point (and even wider) arbitrage is also common in case of free currencies. Three point arbitrage involves the consideration of six different rates of exchange since each financial centre has a rate on each of the other two centres. Similarly, the four point arbitrage involves the consideration of 12 different exchange rates. There may be three point arbitrage, say between the US dollar, Japanese Yen and British Euro. In such a case there should be uniformity between the three rates in London, New York and Tokyo foreign exchange markets. Suppose that the dollar-Euro rate is $ 2.80 per Euro, the Yen/dollar rate is 350 Yen per dollar and the Yen-sterling rate is 1,020 Yen per Euro. In this situation, arbitragers will buy Euro against dollars, convert each Euro into 1,020 Yen in London and sell the Yen so obtained for the dollars in New York and earn profit because for an investment of $ 2.80 they will obtain 2.9143 (= 1020/350) dollars. The arbitrage operations will continue until the value of the Yen in terms of the Euro is in line with its value in terms of the dollar. This will be when the Euro Yen rate is 980 Yen per Euro. Continued purchases of Yen in London would raise the value of the Yen lowering the amount of Yen obtainable for the Euro.

The real usefulness of the arbitrage operations consists in linking the three foreign exchange (more number of exchange rates in the event of four, five, six point arbitrage) markets virtually into one and in bringing into being consistent cross rates. When the foreign exchange markets are free, arbitraging unifies all the different foreign exchange markets of the world.

The spot and forward foreign exchange rates are closely linked with each other through interest arbitrage. If the speculators expect the spot rate to rise, they will buy forward putting pressure on the forward rate causing it to rise. Conversely, they expect the spot exchange rate to fall, they will sell forward, forcing it to come down. Consequently, speculation tends to make the spot and forward foreign exchange rates to move together.

Speculation is of crucial importance in determining the size of activities in the forward foreign exchange market. Under a system of flexible exchange rate speculation is an important source for the supply of and the demand for forward exchange. A speculator expecting the spot rate to rise in future buys forward in order to sell spot when he receives his delivery of the currency which he has bought forward. On the other hand, a speculator expecting the spot rate to fall sells forward with the intention of buying spot when he needs currency for delivery. Thus, the activities of the speculators tend to narrow down the fluctuations in the spot and forward exchange rates bringing them closer.

When dealings in the foreign exchange are not free, arbitrage operations become impossible. Foreign exchange controls requiring the residents of a country to surrender all their foreign exchange earnings to the authorities completely hinder the activities of the arbitragers which consist of the buying and selling of foreign currencies. In the illustration cited, if Japan had foreign exchange control, arbitragers would not be able to buy Yen against the Euro at the cheap Yen/Euro exchange rate. This would block the operations of selling Yen at the relatively higher Yen/dollar exchange rate. Thus, the foreign exchange rate between the Yen and the Euro would become inconsistent with the dollar/Euro and dollar/Yen exchange rates. Such inconsistent cross rates are the usual accompaniment of foreign exchange controls.

Suggested Readings

Milton Friedman, *Essays in Positive Economics,* pp. 157–204.

Gottfried Von Haberler, *The Theory of International Trade,* 1950, Chapter IV.

Charles P. Kindleberger, *International Economics,* Fifth Edition, 1973, Chapter 17.

Fritz Machlup, "The Theory of Foreign Exchange", *Economica,* New Series, Volume VI, 1939–40, reprinted in *Readings in the Theory of International Trade,* Chapter V.

James E. Meade, *The Balance of Payments,* 1951, Chapter 1.

Delbert A. Snider, *Introduction to International Economics,* Fourth Edition, Chapter 16.

Bo Sodersten, *International Economics,* Second Edition, Reprinted 1981, Chapter 27.

Jacob Viner, *Studies in the Theory of International Trade,* 1937, pp. 377–387.

Margaret G. De Vries, "Fluctuating Exchange Rates: The Fund's Approach", *Finance and Development,* Volume 6, Number 2, June 1969.

Questions

1. Explain the merits and demerits of the fixed and flexible foreign exchange rates.
2. Show how the rate of foreign exchange is determined under paper currency standard? Would you favour a fixed or a flexible foreign exchange rate system? Discuss.
3. Critically examine the purchasing power parity theory of foreign exchange rate. What special significance does it have in the present-day global inflationary conditions?
4. What is meant by foreign exchange rate? How is the equilibrium rate of foreign exchange determined under the gold standard?
5. What is exchange arbitrage? Distinguish between spot and forward foreign exchange rates.
6. Examine critically the purchasing power parity theory of foreign exchange. How is this theory an improvement on the classical mint parity theory of foreign exchange? Discuss.

CHAPTER 14

Balance of Payments

Introduction

The study of the balance of payments was of great interest for the mercantilists. In fact, the eighteenth-century mercantilist commercial policy was directed towards creating a favourable balance of payments by creating an export surplus. It was so because more gold was associated with more state power and for a country having no gold mines a favourable balance of payments was the only means to procure gold bullion. The term 'balance of payments' made its entry into English economic literature during the mercantilist period.

Originally, by the balance of payments was meant an excess of payments over the receipts. Under the gold standard, this excess meant an outflow of gold from the country. The term, however, soon began to be used in the neutral sense of the 'state of the balance of international accounts' whether negative or positive. Consequently, we speak of the term 'balance of payments problem' whether there is an outflow or inflow of gold and the term 'balance of payments theory' covers the entire subject, being no more restricted to cover the situation of excess of payments over receipts. The mercantilists prefixed the adjectives 'favourable' and 'unfavourable' to denote respectively the gold inflows and outflows.

Both these adjectives were, however, rejected during the process of classical reaction to mercantilism on the plea that it were the commodities and not gold that constituted the real wealth and that there was nothing favourable about a surplus export of commodities in exchange for gold. This led to the use of such prefixes as 'active' and 'passive' and 'positive' and 'negative'. These terms were also found equally unsatisfactory. For example, it was said that there was nothing 'passive' about a gold outflow or 'active' about a gold inflow. The terms in current usage are *surplus* and *deficit* as these do not suffer from any important ambiguity and are also in harmony with the current accounting practice.

Concept of the Balance of Payments

The international balance of payments of a country is a statistical record in the form of a balance sheet comprising all its transactions made with rest-of-the-world or with another country during any given period of time. It presents a summary account of all international transactions of a country during any given period of time. In the language of the International Monetary Fund, "the balance of payments for a given period is defined . . . as a systematic record of all economic transactions during the period between residents of the reporting countries . . ."[1] According to the US Department of Commerce, "the balance of payments of a country consists of the payments made, within a stated period of time between the residents of that country and the residents of foreign countries. It may be defined in a statistical sense as an itemised account of transactions involving receipts from foreigners on the one hand and payments to foreigners on the other. Since the former relate to the international income of a country, they are called "credits", and, since the latter relate to the international outgo, they are called "debits".[2] According to Kindleberger, "the balance of payments of a country is a systematic record of all economic transactions between the residents of the reporting country and residents of foreign countries during a given period of time."[3] It comprehends all payments made by a country and all receipts which accrue to a country from abroad.

Since the balance of payments is a systematic record of a country's total money receipts received from and payments made to abroad, the difference between receipts and payments is the surplus or deficit. A country's total money receipts are the receipts or payments that accrue to its residents from abroad while the total payments refer to the payments made by the residents of a country. Dividing residents' total receipts R, and their payments P, into their domestic and foreign components and if the domestic receipts and domestic payments are identical then the balance of payments B of a country can be expressed as

$$B = R - P = (R_d + R_f) - (P_d + P_f)$$
$$= R_f - P_f \qquad \text{(since } R_d = P_d\text{)}$$

For an open economy, the total receipts may differ from the total payments and their difference represents the difference between *foreign* receipts (R_f) and *foreign* payments (P_f). The positive difference is termed as a *surplus* while the negative difference is termed as a *deficit* in the international balance of payments of a country.

A country's external balance of payments is not a balance sheet showing her total foreign assets and liabilities at any given point of time. It shows for any given period of time the flow of a nation's total foreign receipts and total foreign payments. Following the conventional rules of double-entry accounting, a nation's total payments and total receipts for any given period of time must always be in balance or equilibrium in the accounting sense. Moreover, one nation's receipts are payments for others while the receipts of other nations are payments for that nation.

A country's balance of payments is merely a way of listing international receipts and payments. In this sense, it is an application of the double-entry book keeping. Usually, a country's external balance of payments distinguishes between items on the current and capital accounts. In the current account are included the exports and imports of all goods and services, interest and dividend payments, private gifts, and so on. The capital account, sub-divided into short-term and long-term capital transfers, lists the imports and exports of all debt instruments and corporate stocks as well as imports and exports of monetary gold. Reparations and other unilateral transfers are generally listed separately. Like the domestic transactions, the international transactions recorded in a nation's balance of payments comprise the total purchases (imports) and total sales (exports) of goods and services, purchases and sales of claims and unilateral transfers. The table on next page explains the different items included under various subheads in the international balance of payments of a country.

1. International Monetary Fund, *Balance of Payments Manual,* January 1950, p. 1.
2. US Department of Commerce, *The Balance of Payments of the United States,* Washington, D.C., US Government Printing Office, 1937, p. 1.
3. Charles P. Kindleberger, *International Economics,* Fifth Edition, 1973, p. 304.

A nation's total exports of goods and services are a source of her foreign exchange earnings or receipts. These represent that part of a nation's gross rational product (GNP) which is purchased by rest of the world during the period to which the international balance of payments relates. Apart from the exports of goods, these include exports of invisible services like tourism, shipping, air transport, banking and insurance services rendered to foreign tourists in the country etc. The counterpart of such receipts consists of the payments a nation makes to rest of the world on account of imports of similar goods and services. Receipts of a nation from foreign countries also arise on account of interest income, repayment of loans and liquidation of their investments made by the nationals of a country abroad. Similarly, payments are made for the interest payment on foreign loans, repayment of the foreign loans, etc.

Components of Balance of Payments

Receipts (Credits)	Payments (Debits)

1. Exports of goods and services
    (a) Merchandise
    (b) Services
    (c) Income from foreign investments
	1. Imports of goods and services
	    (a) Merchandise
	    (b) Services
	    (c) Foreign income from investments
	        made at home

2. Purchases of long-term claims
    (a) Equity claims
    (b) Debt claims
	2. Sales of long-term claims
	    (a) Equity claims
	    (b) Debt claims

3. Sales of short-term claims
    (a) Against deposits
    (b) Others
	3. Purchases of short-term claims
	    (a) Against deposits
	    (b) Others

4. Sales of gold	4. Purchases of gold
5. Unilateral receipts	5. Unilateral payments
6. Errors and omissions	6. Errors and omissions

Exports and imports are the only items included in the balance of payments which enter into the income and product accounts of every nation. A nation's total exports (X) expressed in the money terms is the value of the national product demanded by the rest of the world and a nation's total imports (M) is the value of the national product spent on the purchase of the output of rest of the world. The difference of these two values, i.e., X – M, is termed as the *balance of goods and services* or the "balance on current account". The balance on current account is a very important concept in as much as it shows the flow aspect of a country's international transactions. All international transactions entering into a country's system of national accounting should be mentioned in the country's balance of current account.

The claims arising from the sales and purchases of securities are included in the capital account and the difference is termed as the balance on capital account. Unilateral transfers, which are usually not very large, generally consist of the gifts and grants made to and received from rest of the world. Gold sales or purchases also appear as receipts and payments in the international balance of payments of a country. The last item of errors and omissions, sometimes also called *unrecorded transactions,* arises from the possibility that certain transactions may escape identification. For example, payments or receipts arising largely from unrecorded movements of short-term claims may be listed under unrecorded transactions.

The international balance of payments of a country may be in balance in the sense of equality between total payments and total receipts.[4] More generally, however, it shows either a *surplus* or a *deficit*. By a deficit or surplus in the balance of payments is usually meant gold movements plus "accommodating" capital movements—those capital movements which are induced by the conditions of balance of payments and by loans that are given or taken for the explicit purpose of equalising the payments balance. The accommodating capital movements are a direct consequence of the balance of payments situation. These capital flows are unforeseen and take place to bring the country's balance of payments into equilibrium. If a country has a deficit in her balance of current account, there will always be an offsetting transaction on the capital account to bring the balance of payments into equilibrium. For example, if a country had imported more than it had exported it will have to borrow abroad to pay for its excess imports and this will be registered as an inflow of capital on the capital account. In other words, the magnitude of accommodating capital flows represents the extent or size of deficit in the country's balance of payments. Conversely, the autonomous capital flows are ordinary capital flows which take place regardless of the other items in the balance of payments. The autonomous capital flows have no connection with the country's balance of payments situation.

The autonomous capital flows may take many forms. For example, an autonomous capital inflow may be due to a foreign corporation buying a domestic company thereby acquiring the assets equal to the capital inflow in the country. It may also be due to a foreign firm or a foreign resident repaying an old loan to a firm or person in the country. In all these cases private firms or persons have capital transactions conducted with the foreigners. Although these capital transactions have their effect on the country's balance of payments but these are in no way caused by the balance of payments situation. On the other hand, accommodating capital flows arise due to either a surplus or a deficit in a country's balance of payments. These accommodating capital flows are the direct consequence of the balance of payments disequilibrium. A deficit in the country's balance of payments will cause accommodating capital outflows while a surplus in the balance of payments will cause accommodating capital inflows.

While the autonomous capital movements can be regarded as planned capital movements emerging from the decisions of individuals, firms or the government to engage in capital transactions with rest of the world, accommodating capital flows are *ex post* in nature which are discovered only at the end of the period whether such capital movements have occurred. Autonomous capital flows should be regarded as planned or *ex ante* flows as they result from the different decision-making units—individuals, firms or government for that matter—plan to engage in these capital movements at the beginning of the relevant planning period.

Accommodating capital movements are politically of great significance. If a country has a deficit in her balance of payments which is covered by an accommodating capital inflow, such inflow should serve as a warning signal for the country. The deficit can be settled either by a short-term loan or a depletion of the reserves. In either case, this situation cannot continue for long because the foreign lenders will seldom be willing to extend short-term loans for ever while the foreign exchange or gold reserves of the country will become depleted after some time. Thus, the government will have to change its economic policy so as to eliminate causes of the balance of payments deficit causing the accommodating capital inflow. The government will have to adopt

4. In the *ex-post* or accounting sense the balance of payments must always balance. This only means that as a matter of double-entry book-keeping convention, the two sides of the balance sheet are always made equal by putting the difference, under a suitable heading, on the smaller side of the balance sheet.

appropriate economic or trade policy which may increase foreign exchange earnings and/or reduce the foreign exchange payments.

Importance of Balance of Payments

A nation's international balance of payments is a quantitative summary of a country's inter-national transactions over a given period of time. It reveals the various aspects of a country's international economic position. The international balance of payments of a country informs the government about the international financial position of the country. It also helps the government in taking decisions on monetary and fiscal policies on the one hand and on the external trade and payments issues on the other. The balance of payments is now also used to determine the influence of foreign transactions on the level of national income. In the case of underdeveloped country, the balance of payments shows the extent of dependence of the country's economic development on the financial assistance given by the developed capital lending countries. In the case of an old country, financially well-off having far-flung foreign investments and receiving a large income flow in the form of dividend and interest income, the balance of payments can show the extent to which its citizens are living on their past exports. However, the greatest importance of the study of international balance of payments lies in its serving as an indicator of the changing international economic position of a country. The balance of payments is the economic barometer which, if properly handled by economic analysts, can be used to appraise a nation's short-term international economic prospects, to evaluate the degree of its international solvency, and to determine the appropriateness of the foreign exchange rate of country's currency.

Notwithstanding that a country's international balance of payments serves as its economic barometer, there are many things which cannot be known by a mere study of the balance of payments. Thus, a country's favourable balance of payments is not an infallible indicator of the economic prosperity of the country nor is the balance of payments adverseness always an indicator of country's economic bankruptcy. A balance of payments deficit *per se* is not the proof of the competitive weakness of a nation in foreign markets. We should know a good deal more about the causes and prospects of the deficit. The longer a deficit continues, however, the more it would seem to point to some fundamental difficulty. Similarly, a favourable balance of payments should not always make the government of a country complacent. A poor country may have a favourable balance of payments due to massive inflows of foreign loans and equity capital. Similarly, an economically strong nation like the United States of America will have an adverse international balance of payments due to the massive assistance she may have given to world's poor debtor countries. While a debtor and economically backward nation at the time of receiving foreign equity capital and loans will enjoy a favourable balance of payments, a creditor country may have unfavourable balance of payments at the time of giving loans or making massive equity investments abroad.

Thus, a deficit or surplus in the international balance of payments of a country *per se* should not be taken as an index of economic bankruptcy or prosperity of the country because a deficit (in the case of a creditor country) is compatible with economic prosperity and provides no cause for national alarm while a surplus (in case of a debtor nation receiving loans) does not always indicate sound economic condition of the country. Moreover, the balance of payments does not give information in sufficient detail to be useful. Insufficient material is made available about the short-term capital movements of the period and in some cases it may be very important to distinguish one kind of the capital movement from the other. Apart from this, the balance of payments, as it is generally prepared by some countries, does not give the breakdown of transactions by reference to the countries with which these have taken place. In the modern world, it would make significant difference whether a country's exports are sold to England instead of to China and *vice versa.*

The main difficulty is that the balance of payments deals only with the transactions of the period under review. It does not provide data about the assets and liabilities that relate one country to others. It is not possible to determine from the balance of payments the debtor creditor status of one country in relation to another or others although it is possible to know the changes in that status during the period in question. Consequently, the balance of payments of a country does not tell the whole story; at the most it tells only a part of it. It is essential to have an analytical mind to get facts out of the bald statistics contained in the balance of payments. To get an idea of the true picture of a country's economic situation it is not enough to know about the international balance of payments, i.e., whether it has deficit or surplus. All these criticisms notwithstanding, the balance of payments provides us with some information vital to an understanding of a country's economic dealings with the other countries. It shows us, for instance, the composition of these dealings, the flows of goods and services and changes, if any, in her status as a debtor or creditor in relation to other countries.

Balance of Payments and Balance of Trade

The meaning of the concepts of balance of trade and balance of payments are often misunderstood. The statement that while the balance of payments includes the balance of trade, the balance of trade does not include the balance of payments makes confusion worse confounded. A country exports and imports a vast assortment of tangible goods and invisible services. Invisible service items include shipping, banking and insurance services for imports and exports of which payments are made and received by a country.

A country's balance of trade refers to the value of imports and exports of merchandise—goods—only. The exports and imports of services—transport services, shipping freights, passenger fares, harbour and canal dues, postal, telephone and telegraph fees, banking and insurance services etc.—comprise the balance of services but the balance of payments is more comprehensive including as it does the total debits and credits relating all the items for which a country makes payments to and receives payments from rest of the world. In short, the balance of trade is only a part of the balance of payments.

The balance of trade is simply the difference between the value of commodity exports and commodity imports and it has no analytical significance *per se.* It is, however, the largest component of a country's balance of payments. The other major components of a country's balance of payments are payments made to and received from rest of the world on account of the imports of shipping, tourists' travel, banking and insurance services, interest and loans, government expenditure, gold and capital movements, gifts, reparation payments etc.

A favourable balance of trade may coexist with an adverse balance of payments and *vice versa.* For example, until recently although England's balance of trade was unfavourable but her balance of payments was favourable because due to the exports of invisible services and interest earnings on her foreign investments she received more in payments from rest of the world than she paid to rest of the world on account of the imports of goods. To conclude, although the balance of payments does not tell the complete truth about a nation's economic situation, the balance of trade does not reveal even that much.

Adjustment Mechanism of Balance of Payments Disequilibrium

A nation's international balance of payments is in equilibrium when the autonomous[5] supply of and the *autonomous* demand for foreign exchange are equal. This is an equilibrium situation, whether there is a monetary authority committed to maintain the exchange rate stability without having to interfere as a residuary buyer or seller of gold and/or foreign currencies in the foreign exchange market to achieve this or whether exchange rates are flexible and their movement assures the equality. A disequilibrium in a country's external balance of payments appears either as a *surplus* or a *deficit*. The balance of payments disequilibrium is favourable (surplus) when the difference between the autonomous supply of and the autonomous demand for foreign exchange is positive. When this difference is negative, the disequilibrium is unfavourable (deficit). While a disequilibrium in the balance of payments of a country is a cause for worry from the point of view of stable international economic relations and balanced growth of world trade, for a nation a *deficit* disequilibrium and not a *surplus* disequilibrium in the international balance of payments is worrisome because the burden of adjustment falls heavily on the deficit and not on the surplus balance of payments country. However, since one nation's deficit is necessarily other nation's or nations' surplus, these deficit and surplus disequilibria cannot be sustained indefinitely.

A country's external balance of payments may suffer either from temporary disequilibrium of a passing nature that is likely to disappear after a short period of one or two years or from fundamental disequilibrium which persists unabated and plagues the economy year after year and requires fundamental structural changes in the economy for its removal. For example, a deficit in a country's balance of payments arising from increased flow of imports and a sudden fall in exports due to famine in any particular year is a case of temporary disequilibrium in her balance of payments as famine being temporary in the subsequent year or two the country should be able to correct her balance of payments deficit by reducing her imports and increasing her exports requiring no structural adjustments in the economy. Quite different from this is the balance of payments deficit which emerges due to persisting inflation in the country and which for its removal requires the fundamental correction in country's cost-price structure. The disequilibrium in this case is a fundamental disequilibrium and the country must take all necessary corrective steps aiming at raising labour productivity, credit squeeze, devaluation etc.

Causes of Disequilibrium

Disequilibrium in a country's external balance of payments may be caused by various factors. It may be due to those factors which may simultaneously worsen a country's balance of payments position and reduce income in the country and *vice versa*. It may also be due to such factors which while raise the level of national income in the country deteriorate the balance of payments position of the country and *vice versa*. There may be some other factors which while cause disequilibrium in the balance of payments position of a country leave the level of income unaffected.

In the first group is included the balance of payments disturbance arising from a shift in the demand from one country's output to another's because such a shift causes similar changes in a

5. As distinct from the *autonomous* supply of and the demand for foreign exchange there is the accommodating supply of and the demand for foreign exchange arising from monetary authority's obligation to keep the foreign exchange rate stable. The monetary authority buys and sells gold and/or foreign currencies at an officially stated prices or rates in the foreign exchange market. It does it by holding gold and/or foreign currency reserves.

country's balance of payments position and the level of income. Similar disequilibrium in the balance of payments of a country will also be caused by international capital transfers. The second group comprises the payments disequilibrium caused by differences between different countries' cost or price rise in the country. In the third group are those disturbances in the balance of payments which while leaving a country's current account unaffected, affect only its liquidity position. These disturbances arise from the domestic or foreign wealth-holders' desire to change the composition of their asset portfolio.

Removal of Disequilibrium

Disequilibrium in a country's international balance of payments shows the imbalance between the autonomous international payments (demand for foreign currencies) and receipts (supply of foreign currencies). Removal of this imbalance means a change in the relationship between the payments and receipts sides of the ledger such that the two sides become equal. To remove a deficit in the balance of payments of a country, autonomous receipts must expand relatively to payments while to eliminate a surplus payments must expand relatively to receipts. In short, adjustment between receipts and payments in the balance of payments requires necessary changes in those variables to which the payments and receipts are functionally related. A complete list of all these variables embraces the economic universe since all the parts of an economic system are interrelated. Attention may, however, be confined to only the key variables which in the case of a country's balance of payments are the foreign exchange rates, prices of internationally-traded goods and services, levels of income and government controls over international transactions. Any change in any one or more of these variables will affect the balance of payments of a country on account of their strong functional relationship with autonomous international payments and receipts. Assuming that disequilibrium in the balance of payments of a country is of a deficit variety we can discuss the adjustment of the balance of payments disequilibrium through the different methods.

1. Change in the Price-Specie Flow Mechanism

Under the international gold standard monetary system, the balance of payments disequilibrium is supposed to be automatically corrected through the automatic price-specie flow adjustment mechanism generated through the gold inflows and outflows. This celebrated mechanism was first sighted by David Hume in the eighteenth century. Following David Hume, Adam Smith, David Ricardo and John Stuart Mill elaborated this in their writings. A country faced with the balance of payments deficit automatically achieves equilibrium through the gold outflows by causing money, price and income deflation in the country.

However, the theory of gold movements and the accompanying process of automatic equilibrium restoration in the balance of payments of the gold standard countries is based on the heroic assumptions that the economies of the concerned gold standard countries are perfectly competitive, flexible and of similar economic size such that the burden of the balance of payments adjustment is evenly borne by both the countries.

2. Change in the Foreign Exchange Rate (Devaluation)

Disequilibrium in the balance of payments is a state of imbalance between the total demand for and the total supply of foreign exchange revealed by the existence of either excess supply of or excess demand for foreign exchange at any given rate of foreign exchange. *Ceteris paribus,* if at any given rate of foreign exchange the balance of payments of a country is in disequilibrium, equilibrium can be brought about through making appropriate changes in the foreign exchange rate. *Ceteris paribus,* if the foreign exchange rate is higher in a country, the prices of her exports will be lower in the foreign markets while the prices of her imports will be higher. It will strengthen the competitive position of the country as a seller of her goods in the international markets. Although autonomous capital movements and unilateral transfers are not directly influenced by changes in the foreign exchange rate, nevertheless the aggregate payments and receipts certainly are, unless the foreign demand for a country's exports and country's own demand for imports are perfectly price-inelastic. The degree to which the balance of payments deficit will be reduced due to a given change in the rate of foreign exchange will depend on the price elasticity of domestic demand for imports and on the price elasticity of foreign demand for a country's exports. Higher the price elasticities of these demands greater will be the effect of any given change in the foreign exchange rate in reducing the deficit of a country's international balance of payments. On the contrary, less elastic these demands are, smaller will be the impact on the balance of payments of any given change in the foreign exchange rate, i.e., greater will be the change which will have to be made in the foreign exchange rate to remove a given deficit in the balance of payments of a country.

The usual effect of a given rise in the foreign exchange rate is to reduce the balance of payments deficit by decreasing the autonomous international payments and increasing the autonomous receipts. Consequently, the demand for foreign exchange decreases and the supply of foreign exchange increases. A fall in the rate of foreign exchange has the opposite effects. A rise in the foreign exchange rate is known as *exchange depreciation* or *devaluation*[6] of currency because the international value of currency unit in terms of other currencies depreciates as the rate on the latter increases. Conversely, a fall in the foreign exchange rate is called *exchange appreciation* or *revaluation* of currency involving an increase in the value of domestic currency in terms of other currencies.

Exchange depreciation or appreciation will remove a country's balance of payments deficit or surplus only if certain minimum elasticity conditions are fulfilled[7]. This can be best understood in terms of the demand and supply curves of foreign exchange. If the demand curve for foreign exchange is negatively sloping while the supply curve is positively sloping, a rise in the foreign exchange rate will decrease the total amount of foreign exchange demanded and will increase the total quantity of the foreign exchange supplied. A fall in the foreign exchange rate has the opposite effect of increasing the total quantity demanded while decreasing the total quantity supplied of foreign exchange. While the demand curve for foreign exchange will almost surely have a negative slope showing that a given change in the foreign exchange rate will contribute to the balance of payments equilibrium adjustment, we cannot be so sure about the supply curve which over a certain range of the foreign exchange rate may be backward bending, i.e., negatively sloping.

Where the supply curve of foreign exchange is perverse or backward bending, a rise in the foreign exchange rate will decrease rather than increase the quantity of foreign exchange supplied. Consequently, so far as the supply of foreign exchange is concerned, devaluation will increase rather than decrease the deficit in the balance of payments while revaluation of the currency will increase rather than decrease the surplus in the external balance of payments of a country.

6. The terms *devaluation* is more properly used only when the country's currency is valued in terms of gold and the gold contents of the currency unit are reduced.
7. For details see Chapter 13.

The supply curve of foreign exchange will be backward bending if the foreign demand for a country's exports is inelastic. When the demand for a country's exports happens to be inelastic, devaluation will not mitigate the deficit of the country's balance of payments unless the demand curve for foreign exchange is so highly elastic as to offset the negative effects produced on the supply side. Nor would the currency revaluation reduce surplus in the balance of payments of a country. In other words, the supply curve of foreign exchange should be so situated that it lies to the right of the demand curve for foreign exchange above the intersection point and to the left of the demand curve below the intersection point. It means that where both the demand curve and the supply curve have negative slopes, the slope of the demand curve must be smaller than the slope of the supply curve. Figures 14.1A and 14.1B illustrate the stable and unstable foreign exchange rate situations when both

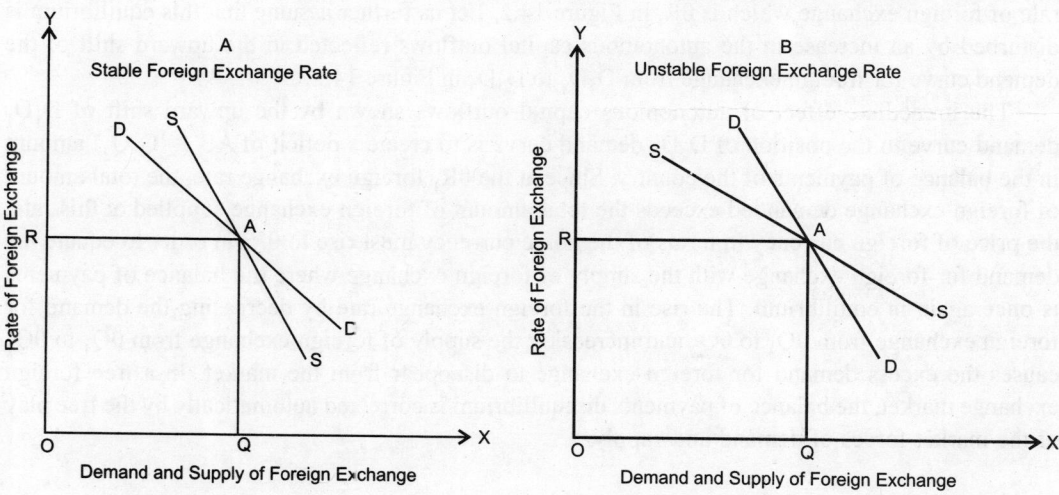

Figure 14.1

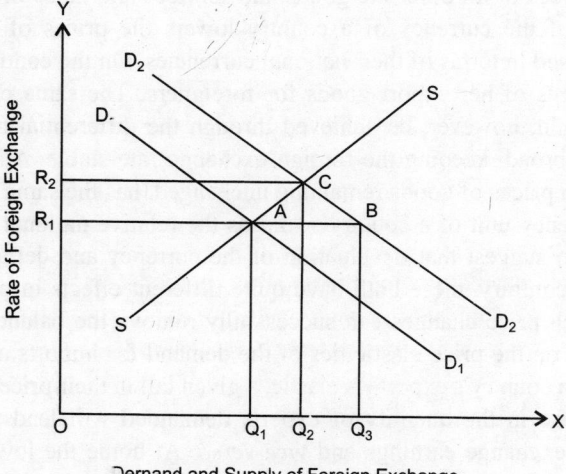

Figure 14.2

the demand curve and the supply curve have negative slopes, i.e., when the demand curve for foreign exchange is normal but the supply curve of foreign exchange is abnormal or unusual.

Figure 14.1B shows that any deviation of the rate of foreign exchange from the equilibrium foreign exchange rate 0R will add to the already existing instability in the balance of payments of the country. Figure 14.1A shows the stable equilibrium where a change—fall or rise—in the rate of foreign exchange helps in the equilibrium adjustment process of the balance of payments of the country. To illustrate how in a perfectly free foreign exchange market a change in the rate of foreign exchange corrects the balance of payments disequilibrium let us assume that initially a country's balance of payments is in equilibrium, i.e., her total payments are equal to her total receipts. In other words, the demand for foreign exchange is equal to the supply of foreign exchange at some given rate of foreign exchange which is $0R_1$ in Figure 14.2. Let us further assume that this equilibrium is disturbed by an increase in the autonomous capital outflows reflected in the upward shift of the demand curve for foreign exchange from D_1D_1 to D_2D_2 in Figure 14.2.

The immediate effect of autonomous capital outflows shown by the upward shift of D_1D_1 demand curve to the position of D_2D_2 demand curve is to create a deficit of AB (= Q_1Q_3) amount in the balance of payments of the country. Since at the $0R_1$ foreign exchange rate, the total amount of foreign exchange demanded exceeds the total amount of foreign exchange supplied at this rate, the price of foreign currency in terms of the home currency must rise to $0R_2$ in order to equate the demand for foreign exchange with the supply of foreign exchange where the balance of payments is once again in equilibrium. The rise in the foreign exchange rate by decreasing the demand for foreign exchange from $0Q_3$ to $0Q_2$ and increasing the supply of foreign exchange from $0Q_1$ to $0Q_2$ causes the excess demand for foreign exchange to disappear from the market. In a free foreign exchange market, the balance of payments disequilibrium is corrected automatically by the free play of the market forces of demand and supply.

3. Change in Price (Price Effect)

The rate of foreign exchange is a special price because changes in it change the relationship between the relative prices of all domestic goods and services and those of all foreign goods and services. Devaluation of the currency of a country lowers the prices of her export goods for foreigners when expressed in terms of their national currencies. On the contrary, revaluation of the currency raises the prices of her export goods for foreigners. The same changes in the relative national price level could, however, be achieved through the differential rates of absolute price changes at home and abroad, keeping the foreign exchange rate stable. A 20 per cent fall in the domestic prices, foreign prices of goods remaining unchanged, has the same effect as a 20 per cent devaluation of the currency unit of a country so far as the relative national prices are concerned. This is not, however, to suggest that devaluation of the currency and deflation are similar in all other respects. On the contrary, these both have quite different effects in other respects.

The extent to which price changes can successfully remove the balance of payments deficit disequilibrium depends on the price elasticities of the demand for imports at home and abroad. If the foreign demand for a country's exports is elastic, a given fall in their prices by causing relatively large percentage increase in the quantity of exports demanded will lead to an increase in that country's total foreign exchange earnings and *vice versa*. At home the lower domestic prices of home-made goods competing with higher-priced imports will nearly always lead to a greater fall in the imports and consequently to a fall in the total amount of foreign exchange demanded, the fall in the demand for imports and consequently in the demand for foreign exchange being greater, higher the cross-price elasticity of demand between the imported goods and domestic goods.[8]

The balance of payments deficit of a country will be eliminated as a consequence of devaluation, i.e., the position of the balance of payments will improve if the price elasticities of demand for the currency-devaluing country's imports and exports are high and at any rate the combined absolute values of the two demand elasticities must be greater than unity,[9] i.e.,

$$|\eta_x| + |\eta_m| > 1$$

where the terms η_x and η_m respectively denote the price elasticity of demand for exports and imports of the currency-devaluing country.

4. Change in Income (Income Effect)

Apart from changes in the rate of foreign exchange and prices, autonomous balance of payments transactions are also functionally related to income. Given the prices and foreign exchange rate, higher income at home leads to higher imports while higher income abroad will cause an increase in the exports of the country.[10] Consequently, a deficit in the balance of payments of a country may be corrected through either a fall in the national income at home or a rise in the national income abroad.

The degree of responsiveness of imports and exports to changes in income at home and abroad is measured by the *marginal propensity to import* or by the *income elasticity of demand for imports*. The marginal propensity to import is the ratio of change in imports in response to a unit change in income and may be symbolically written as m = dM/dY, where m stands for marginal propensity to import, d stands for an infinitesimal change, M stands for total imports and Y stands for national income. To measure the ratio of *relative* change in imports to changes in income, the income elasticity of demand for imports (δ) is the most appropriate concept. It may be defined as the percentage change in the total imports divided by the percentage change in the aggregate income, i.e.,

$$\delta = \frac{dM}{M} + \frac{dY}{Y}$$

$$= \frac{dM}{dY} + \frac{Y}{M}$$

= Marginal propensity to import

Let us assume that there is deficit of Rs. 200 crores in the balance of payments of country A while there is surplus of an equal amount in country B's balance of payments. Consequently, this amount is to be transferred from A to B. Country A raises taxes and restricts credit sufficiently to reduce the aggregate disposable income by Rs. 200 crores. As a consequence of the fall in the aggregate disposable income in country A of Rs. 200 crores, her imports will decrease by an amount which will

---

8. The cross price elasticity of demand between the imported and home-made goods can be defined as the percentage change in the quantity of import goods demanded divided by the percentage change in the price of home-made goods.

9. This is, however, a *sufficient* and not a *necessary* condition. In other words, the balance of payments will certainly always improve when this condition holds. It may, however, improve even in the absence of this condition provided the supply elasticities are sufficiently small. It can be easily shown that provided the elasticity of supply of exports as well of imports is zero (i.e., $e_m = e_x = 0$), the balance of payments will improve even if the sum of the price elasticities of demand for exports and imports was less than one.

10. Income levels at home and abroad possibly also affect the autonomous capital movements.

depend on A's marginal propensity to import (M_A). Assuming the value of $M_A = 0.4$, country A's expenditure on imports will decrease by Rs. 80 crores and her balance of payments will improve to this extent. The transfer of Rs. 200 crores in the form of increase in country A's exports to B will raise income and ease credit conditions in country B. Consequently, country B's imports will increase by an amount which will depend on her marginal propensity to import (M_B). Since country B's imports are country A's exports, country A's balance of payments will also improve on this score. Suppose that the value of $M_B = 0.6$. In this case, A's exports (B's imports) will increase by Rs. 120 crores while her imports (B's exports) will decrease by Rs. 80 crores, improving country A's balance of payments by Rs. 200 crores. Country A's balance of payments will now become in equilibrium. The income effects ($M_A + M_B = 1$) are just sufficient to restore equilibrium in the balance of payments of country A as well as of country B. Since for most countries $M_A + M_B < 1$, the direct income effects will not be strong enough to restore equilibrium in the country's external balance of payments. If we, however, assume the extremely rare case in which $M_A + M_B > 1$, the direct income effects will be so strong that the balance of payments deficit will be more than removed and the country will experience the happy situation of favourable balance of payments.

A country's national income and her balance of payments are intimately interrelated such that changes in the one cause changes in the other. In real terms, national income is the total domestic production of final goods and services, broadly classified into consumption and investment or capital goods, during any given period of time which is ordinarily taken as one year. In money terms, it is simply the aggregate money expenditure incurred on the production of domestically produced goods and services. This definition of national income which is identical with the net domestic product is, however, appropriate only for a closed economy. For an open economy trading—exporting to and importing from—with the rest of the world the correct definition of national income must take into account on the plus side the exports of domestically produced goods and services and on the minus side the imports of goods and services which are available for domestic consumption but have not been produced at home. Thus, in defining national income as the aggregate expenditure incurred on the home produced goods and services, the foreign expenditure incurred on a country's exports should be included while the domestic expenditure incurred on the imports should be excluded. This yields the following identity equation.

$$Y \equiv C + I + X - M \tag{1}$$

where

    Y is national income,

    C is the aggregate consumption outlay,

    I is the aggregate investment outlay,

    M is the aggregate import outlay, and

    X is the aggregate export outlay.

The difference of exports and imports, $X - M$, appearing as a component of national income in the equation relates to the current account of the balance of payments. When the term $X - M$ is positive, the effect of foreign trade on national income is to expand it. Conversely, when the term $X - M$ is negative, the effect of foreign trade on the national income is to reduce it. When the term $X - M = 0$, i.e., when $X = M$, the effect of foreign trade on national income is neutral.

Looked at differently, exports are similar to domestic investment outlay in affecting the national income since each one of these two is like an injection into the aggregate income expenditure flow expanding the stream of aggregate income generated by the purchase of goods and services. Both are largely determined independently of the level of domestic income. Imports affect national income in a manner similar to saving. Both represent 'leakages' in the aggregate income-expenditure flow since both represent disposal of income in such a manner that it does not flow back into the domestic income-expenditure stream. They are also similar in another respect; they are primarily functionally related to the level of domestic income. National income will, therefore, change in the same direction in which the exports change and if income changes—because of changes in exports or for the other reasons—imports will change in the same direction. In other words, the aggregate income is an increasing function of exports while imports are an increasing function of aggregate income. Starting from a situation in which both national income and the balance of payments of a country are in equilibrium, if the balance of payments of a country is disturbed by an autonomous change in her exports, the balance of payments equilibrium adjustment will take place through a change in the imports induced by the income effects of the given autonomous change in exports. On other hand, due to a change in the national income, exports remaining unchanged, the consequent induced change in the imports will create disequilibrium in the balance of payments of the country.

The relationship between a country's national income and her balance of payments can be more precisely stated through equations by stating the national income equilibrium for an open economy. The aggregate income-expenditure flow identity equation was stated as

$$Y \equiv C + I + X - M \tag{2}$$

Deducting the import component of consumption and investment outlays and of exports separately the income-expenditure identity equation (2) can be rewritten as

$$Y \equiv C_d + I_d + X_d \tag{3}$$

where subscript d stands for the production out of the domestic resources. National income also equals the sum of the many uses to which it is put, i.e., the sum of expenditure incurred on the purchase of goods and services produced at home and on imports plus the unspent part of income, if any, i.e., saving. Consequently, we have the following identity equation

$$Y \equiv C_d + M + S \tag{4}$$

From equations (3) and (4), we obtain

$$C_d + I_d + X_d \equiv C_d + M + S \tag{5}$$

or

$$I_d + X_d \equiv M + S$$

Rearranging equation (5), we obtain

$$S \equiv I_d + X_d - M \tag{6}$$

The relationship expressed in the identity equation (6) is extremely useful and revealing. It shows that when the exports and imports of a country are equal, i.e., when the term $X_d - M$ in the equation is equal to zero, domestic investment is equal to domestic saving. When, however, exports exceed imports ($X_d > M$), domestic investment is less than domestic saving to the extent of export surplus and *vice versa*. This shows the potential role of international trade in providing extra resources for the economic growth of a country when adequate amount of capital formation can not be financed out of the meagre domestic savings. When a country's exports exceed her imports, the total domestic investment is less than the total domestic savings to the extent of total foreign investment made by the country.

The above identity relationships are *ex post* relating to the realised events of the past time period. They are not necessarily equilibrium relationships which will prevail in the context of the balance of payments when *ex ante* or planned decisions of different groups are mutually consistent so that the decisions can be carried out harmoniously and with no built-in tendency for change to take place. In the context of national income, equilibrium will occur when planned or *ex ante* 'injections' into the aggregate income-expenditure flow equal the planned or *ex ante* 'leakages' from the flow. Since net investment—addition to economy's total capital stock—and exports represent injections and savings and imports represent leakages, when all these are *ex ante* terms the income equilibrium equation is shown by

$$I_d + X_d = M + S \tag{7}$$

Rearranging equation (7), we obtain

$$S = I_d + X_d - M \tag{8}$$

Now relationship between the current account of the balance of payments and the national income can be expressed thus: Assuming that the current account equilibrium ($X_d = M$) is a necessary and sufficient condition for the balance of payments equilibrium (it means that the net balance on all other autonomous balance of payments transactions is zero), it is evident either from equation (7) or from equation (8) that the national income equilibrium and the balance of payments equilibrium do not necessarily coexist. Let us, however, assume that initially the balance of payments and national income equilibria coexist. Equation (8) will, therefore, be satisfied under the special assumption that $X_d = M$. Suppose that exports of the country now fall either due to a fall in the level of national income abroad or due to the imposition of an import tariff or quota by foreign countries or due to changes in the tastes of foreign consumers disfavouring country's exports etc. The fall in country's exports will create a deficit in the country's balance of payments. It will also immediately cause national income of the country to fall since the exports are a part of national income. The question to be answered is: how will a given fall in the national income affect the balance of payments of a country?

It was stated earlier that imports are a function of the level of national income. The marginal propensity to import *(dM/dY)* stated the precise relationship between the infinitesimal change in the national income and the resulting change in the country's imports. The total amount by which imports will change (ΔM) in response to a given change in the national income (ΔY) can be found out by multiplying the change in the national income by the marginal propensity to import, i.e.,

$$\Delta M = \Delta Y \cdot \frac{dM}{dY}$$

Given the value of the marginal propensity to import, the total induced change in the imports of a country can be known if we know the change in the national income. Therefore, the important question to be answered is: by how much does the national income decrease in consequence of a given autonomous fall in the exports of a country? Let us suppose that country's exports fall by Rs. 200 crores. As an immediate consequence of this fall in the exports, national income will also fall by 200 crores. This, however, does not end here because the fall in the national income will cause a further fall in all those variables—consumption, saving, imports—which are functionally related to national income with their "feedback" effects on the national income. The consequence of a given autonomous fall in the exports is manifested in a series of repercussions by generating a series of diminishing income-expenditure flow rounds until a new equilibrium level of national income is established. This phenomenon is the familiar *foreign trade multiplier effect*. The operation of the foreign trade multiplier can be illustrated by assuming the following values for those variables which are functionally related to national income, assuming an initial autonomous fall of Rs. 200 crores in the exports of a country.

We assume the following values for the marginal propensity to save ($\Delta S/\Delta Y$), the marginal propensity to import ($\Delta M/\Delta Y$), the domestic marginal propensity to consume ($\Delta C_d/\Delta Y$) and the marginal propensity to invest ($\Delta I_d/\Delta Y$) respectively.

$$\frac{\Delta S}{\Delta Y} = 0.2; \frac{\Delta M}{\Delta Y} = 0.3$$

$$\frac{\Delta C_d}{\Delta Y} 0.5; \frac{\Delta I_d}{\Delta Y} = 0$$

The following table shows the series of successive changes which will occur in the successive income spending time periods, starting with a given initial fall of Rs. 200 crores in the country's exports in time period 1.

Time Period	ΔS	ΔM	ΔC_d	ΔI_d	ΔX_d	ΔY
1	–	–	–	–	– 200	– 200
2	– 40	– 60	– 100	0	0	– 100
3	– 20	– 30	– 50	0	0	– 50
4	– 10	– 15	– 25	0	0	– 25
–	–	–	–	–	–	–
–	–	–	–	–	–	–
Equilibrium values	– 80	– 120	– 200	0	– 200	– 400

In the table at the end of the equilibrium adjustment process, $\Delta S + \Delta M = \Delta I_d + \Delta X_d$. Consequently, the condition of equilibrium is satisfied. The equilibrium adjustment process will operate if changes in the national income generated by the balance of payments deficit are free to work themselves out in the absence of countervailing government policies.

5. Direct Controls

The balance of payments deficit may also be corrected through direct controls enforced by government over the international transactions of a country. Direct controls may be divided into physical controls and financial controls. In order to improve the balance of payments position, the physical controls may be used to increase the exports and decrease imports. Since exports cannot be increased substantially through enforcing physical controls, say export quota, such controls are usually applied to restrict imports. By using financial controls, the government can successfully restrict imports by enforcing foreign exchange restrictions. It can permit only those imports which are considered essential by releasing foreign exchange for importing only such goods and none others. Correction of the disequilibrium through the physical and financial controls rules out completely the free operation of the market forces which establish a genuine equilibrium. Import quotas and exchange controls correct the balance of payments disequilibrium by establishing an effective control of government over either the foreign trade or over the allocation of foreign exchange or both. In practice, the physical controls on exports and imports and control over the foreign exchange are simultaneously imposed. At present, under the arrangement evolved by the International Monetary Fund the permanent deficit in a country's external balance of payments is removed by pursuing orderly changes in the par values of the member countries' currency units as agreed to between the countries concerned and the IMF while a temporary deficit in the balance of payments of a member is tided over by drawing upon the Fund resources under the various schemes initiated by the IMF. The IMF scheme is based on the concept of flexible exchange rate stability.

Suggested Readings

Poul Host-Madsen, "Balance of Payments Problems of Developing Countries", *Finance and Development,* Volume IV, Nos. 2 and 4.

IMF, *Balance of Payments: Its Meaning and Uses,* Pamphlet Series No. 1.

IMF, *Balance of Payments: Concepts and Definitions,* Pamphlet Series No. 1.

Harry G. Johnson, "Towards a General Theory of the Balance of Payments", published in *International Trade and Economic Growth,* Allen and Unwin, 1958.

Charles P. Kindleberger, *International Economics,* Fifth Edition, 1973, Chapter 19.

James E. Meade, *The Balance of Payments,* Parts IV and V.

Delbert A. Snider, *Introduction to International Economics,* Fourth Edition, 1967, Chapter 17.

Bo Sodersten, *International Economics,* Second Edition, Reprinted 1981, Chapters 25 and 26.

Lorie Tarshis, *Introduction to International Trade and Finance,* Third Printing, 1962, Chapter 20.

Questions

1. Distinguish between balance of trade and balance of payments, Explain the various sources of disequilibrium in a country's balance of payments. How can adverseness in the balance of payments be corrected?
2. Distinguish between current account and capital account in the balance of payments of a country. Discuss the impact of (a) inflation in the country, (b) over-valuation of the currency, and (c) foreign aid on the balance of payments of a country.
3. Discuss the price-specie flow mechanism of the balance of payments equilibrium adjustment under the gold standard. Under what conditions will this mechanism work successfully? Explain.
4. What do you understand by 'fundamental disequilibrium' in the balance of payments? What remedies would you suggest to correct it?

CHAPTER 15

Devaluation

By devaluation of a country's currency unit is meant the decrease in the external value of a unit of that currency expressed in terms of gold, SDR or foreign currency by government edict. A fall in the gold, SDR or foreign currency value of the currency unit amounts to an increase in the number of units of that currency per ounce of gold, per SDR or per unit of foreign currency. A country may reduce the foreign exchange value of her currency unit for more than one reason, e.g., to create a surplus in her international balance of payments by dumping her goods abroad or to remove the deficit in her external balance of payments. In short, devaluation means an act of officially reducing the external value of the currency unit of the country and an appreciation to the extent of devaluation in the external value of the currency unit of the country in whose relationship the country has devalued her currency. For example, when the pound-sterling was devalued on September 20, 1949 in terms of the US dollar, the old sterling-dollar exchange rate of £ 1 = US $ 4.03 was altered to the new exchange parity of £ 1 = US $ 2.80. Consequently, the dollar-price of the pound-sterling had depreciated by about 30 per cent.

Generally, a country devalues her currency in order to correct the persisting deficit in her international balance of payments. When a country is faced with the chronic deficit in her balance of payments which cannot be removed through other methods at the pre-devaluation foreign exchange rate, e.g., through deflation or export subsidies, she devalues her currency. The desired effects of devaluation follow from the fact that while the external value of the currency is reduced there is no change immediately in the internal value of the currency. In other words, the purchasing power of the currency in terms of domestic goods and services remains unaltered, i.e., the prices of goods and services in the country do not rise immediately following the devaluation of the currency. Devaluation restores the balance of payments equilibrium primarily through an improvement in the domestic versus international cost-price ratio.[1]

The immediate effect of devaluation is to cause a change in the relative prices at home and abroad. If a country devalues her currency by, say, 30 per cent, it means that prices of her imports counted in the home currency will rise by 30 per cent. It will cause a fall in the demand for imports while the import-competing industries will be in a stronger competitive situation. On the exports side, the exporters will receive 30 per cent more in the home currency for each unit of foreign currency earned by them. Consequently, they can lower their prices counted in the foreign currency and exports will become more competitive. Thus, exports will expand while imports will

1. Conceptually, a given percentage of currency devaluation is equivalent to a like amount of export subsidy and import duty. Consequently, the problems that arise in the wake of devaluation are similar to the problems that would arise when a policy of uniform export subsidy and import tariff is adopted.

contract causing improvement in the balance of payments. But by how much imports (foreign exchange payments) will fall and exports (foreign exchange earnings) will rise will depend on the demand elasticities for imports and exports of the country.

Calculating Percentage Change in Currency Value

Devaluation of a currency represents the decrease in the gold, SDR or foreign currency value of a unit of national currency. An increase in the gold, SDR or foreign currency value expressed in terms of the devalued currency, can be measured either by stating the change in percentage terms expressed in terms of the rise in the currency price of gold or foreign currency or in terms of the fall in the value of the currency unit expressed in terms of gold or foreign currency. The percentage figure would not, however, be identical in the two methods of stating the change. For example, a 10 per cent fall in the gold value of the US dollar is equivalent to a 11.1 per cent (100/90) rise in the dollar price of gold.

For those currencies which are traditionally quoted in terms of the value that one unit has in terms of gold, SDR or another currency, calculating the percentage of devaluation involves subtracting the old rate from the new rate and expressing the difference as percentage of the old rate. For example, when the gold content of the Finnish markka was reduced from about 0.1996 gramme to about 0.1889 gramme, this meant a devaluation of

$$\frac{0.1889 - 0.1996}{0.1996} = -\frac{0.0107}{0.1996} = -5.4$$

The minus sign indicates devaluation.

To obtain the percentage of devaluation or revaluation for currencies traditionally quoted the other way round, i.e., as so many units of the national currency per unit of gold or per US dollar, one can interchange the new and old rates and proceed as in the above example. Thus, when the rate of the Finnish markka was revised from 4.1 to 3.9 markka to the US dollar, the value of one markka in terms of the US dollar appreciated by 5.1 per cent, which is calculated as

$$\frac{4.1 - 3.9}{3.9} = \frac{0.2}{3.9} = +5.1 \text{ per cent}$$

This result is reconciled with the preceding example by noting that if the markka devalued by 5.4 per cent in terms of gold, its value fell to 94.6 per cent of the previous level while the 10 per cent devaluation of the US dollar reduced it to 90 per cent of its previous value. The markka in relation to the US dollar then became 94.6/90 = 105.1 representing a relative appreciation of the markka by 5.1 per cent against the US dollar.

Methods of Correcting Balance of Payments Deficit

Apart from the devaluation of currency there are other methods of removing the balance of payments deficit. As a method of correcting the balance of payments deficit, devaluation was abhorred before World War I. Even after the War, nations did not take recourse to devaluation although they were faced with the chronic deficit in their balance of payments. Until the thirties, devaluation was regarded as a sign of national bankruptcy. Consequently, nations avoided devaluation of their currencies preferring deflation to devaluation. Moreover, under the international gold standard, which was in vogue upto 1936, devaluation was not feasible and the enlightened public opinion abhorred devaluation as a method of correcting the balance of payments deficit. This is borne out by the views expressed by the Macmillan Committee appointed by the British government in November 1929 "to enquire into banking, finance and credit conditions paying regard to the factors—both internal and international—which govern their operation, and to make recommendations to enable these agencies to promote the development of trade and commerce and employment of labour." Rejecting devaluation as a method of correcting the balance of payments deficit, the Committee wrote: "In our opinion the devaluation by any government of a currency standing at its par value suddenly and without notice (as must be the case to prevent foreign creditors from removing their property) is emphatically one of those things which are not expedient." In Committee's view, international trade, commerce and finance were based on confidence. One of the foundation-stones on which international confidence rests is prevalence of the belief that all countries would maintain the external value of their currency units as it has been fixed by law and will only give legal recognition to its depreciation when that depreciation has already come about *de facto*.

The Committee knew of many cases which arose either through exigencies of war or mistakes of policy or the collapse of prices when the foreign exchange rates of currency units had fallen so far below the mint parity that their restoration involved either great social injustice or national efforts and sacrifices. The British case in 1925 was precisely of this nature. The pound-sterling had depreciated for some years before 1925 and it was an open question whether the restoration of the gold standard in 1925 at the old parity was in the national interests. England should have returned to the gold standard by fixing the mint parity value of the pound-sterling at a lower level. However, if England would have decided to devalue her currency in 1931 "it would be to adopt an entirely new policy, and one which would undoubtedly be an immense shock to the international financial world, if the government of the greatest creditor nation was deliberately and by an act of positive policy to announce one morning that it had reduced by law the value of its currency from the par at which it was standing to some lower value."

The disastrous effects of a chronic deficit in the external balance of payments of a country are too serious to be ignored for an orderly growth of international trade and for the domestic economy of the country. Consequently, the government looks upon such deficit with serious concern and takes appropriate measures to restore equilibrium in her external balance of payments. A deficit disequilibrium in the balance of payments may be corrected through one or more of the following methods.

Firstly, if the balance of payments deficit is due to fall in the export earnings, exports may be increased by making them cheaper by lowering their prices. This involves reducing the unit factor cost of production by lowering wages, interest etc. If the increase in exports by this method is inadequate to remove the deficit in the balance of payments, efforts may be made to cover the deficit by directly restricting imports through import quotas or indirectly by levying import tariffs. Such measures are the non-monetary and direct measures of correcting the deficit in the balance of payments of a country. These methods were employed by many European countries, particularly by France and Germany, which imposed import embargoes, import quotas, exchange clearing, dumping, multiple exchange rates etc. to correct the deficit in their international balance of payments.

Secondly, deficit in the balance of payments may be removed through monetary squeeze by decreasing the money supply in circulation in the country. Deflation means contraction of currency and fall in the domestic prices of goods and services. Deflation also involves a fall in the total consumption of people in the country consequent upon decrease in their incomes. When the total consumption falls, the consumption of all goods and services, including the imported goods and services, will decrease. To the extent that consumption of imported goods and services in the country falls, the total import-payments bill of the country would fall. Conversely, the total value of the country's exports would rise due to fall in their prices at home. Consequently, exports become cheaper for the foreigners and the total amount of exports demanded will increase more than proportionately in response to a given proportionate fall in their prices unless their demand is inelastic. Thus, the effect of price deflation on the balance of payments of the country is to remove the deficit by increasing the exports and decreasing the imports. However, since deflation involves adjustment of the balance of payments through causing general wage-cut and unemployment in the economy, it is not generally favoured.

Thirdly, deficit in the balance of payments may be corrected by means of devaluation of the currency. The success of devaluation, however, mainly depends on the reaction of those foreign countries in relation to whose currencies the country devalues her currency. Those countries must cooperate with the country devaluing her currency by not raising import duties or by not subsidising their exports in order to nullify the effects of devaluation. Moreover, the degree to which the country devaluing her currency will succeed in removing the deficit in her balance of payments also depends on the price elasticity or foreign demand for her exports. If the price elasticity of demand for the exports of the country devaluing her currency is high, there will be net gain to the country because the loss of foreign exchange earnings resulting from a given percentage fall in the foreign prices of exports consequent upon devaluation will be more than made good by relatively larger percentage increase in the total quantity of exports demanded so that there will be a net addition to the total foreign exchange earnings of the country devaluing her currency. However, if exports of the country consist mostly of those goods whose foreign demand is inelastic no gain will accrue to the country by devaluing her currency. In fact, the situation might rather worsen as a consequence of devaluation. In such a situation, revaluation, and not devaluation, will improve the country's balance of trade.

This is in so far as the contribution of the demand for exports is concerned. Before taking recourse to devaluation, the nature of the elasticity of demand for imports of the country devaluing her currency should also be carefully studied. If the elasticity of demand for the country's imports is low, the total import-payments bill in terms of the demand for foreign exchange will increase rather than decrease as result of devaluation. To the extent the foreign exchange value of imports increases, the gain made from the increased exports will be lost. Consequently, from the demand side it is desirable that the price elasticities of demand for a country's exports and imports should be high if devaluation has to remove the deficit of the balance of payments of a country. Furthermore, the elasticities of supply of the country's exports and imports must also be carefully studied before resorting to devaluation. In short, devaluation will not always help the country in removing the deficit of her balance of payments. Its effectiveness depends on many factors, including the demand and supply elasticities of exports and imports which must be carefully studied before devaluing the currency.

Fourthly, the balance of payments deficit may be removed through exchange controls and import quotas. Deflation is harmful for a country's economy while devaluation can have only temporary effects and it may provoke the other countries to retaliate. Moreover, devaluation imperils the international financial soundness of a country. Consequently, deflation and devaluation are avoided and a country takes recourse to the surer method of quantitative controls—direct and indirect—to correct the deficit in her balance of payments. Under the foreign exchange control, exporters are directed by the exchange control authority to surrender to it their entire foreign exchange earnings and the foreign exchange so received is allocated between the importers according to certain criteria. None else is allowed to import the goods. The balance of payments deficit is rectified by keeping the imports well within the limits of total foreign exchange earnings. Apart from exchange control, the government may also impose complete physical controls over her imports through a comprehensive licensing and quota system according to which, in addition to determining the total quantity of imports, even the countries from which goods can be imported are determined by the licensing authority. Nowadays, the balance of payments deficit of a country is corrected by the government by approaching the International Monetary Fund which advises the concerned member country to adopt certain corrective measures. For taking the various measures, the balance of payments disequilibrium has been divided into (a) fundamental disequilibrium, and (b) temporary disequilibrium. Measures considered necessary to correct the disequilibrium of the first type are different from those that are needed to correct the second type of disequilibrium. Before a member can practise any measure in this regard, consent of the International Monetary Fund must be sought.

Devaluation and the Balance of Payments Deficit

The effects of devaluation are not permanent. The benefits of devaluation continue only for a limited time so long as the cost-price structures abroad and at home do not adjust to the new exchange parity. Normally, the effects of devaluation last for two to three years. Consequently, devaluation provides a breathing time during which the country devaluing her currency must correct her cost-price structure vis-a-vis the rest of the world so as to eliminate the possibilities of recurring deficit in her external balance of payments. The immediate effects of devaluation will be reflected in the increase in exports and the decrease in imports of the country devaluing her currency. The domestic prices of goods and services in the country devaluing her currency remaining unchanged, devaluation will make the country's exports cheaper in terms of the foreign currency because a unit of foreign currency will now exchange for more units of the home currency than could be had at the pre-devaluation exchange parity. Similarly, the cost-price structures in the foreign countries remaining unchanged, devaluation will make the foreign currency and consequently the imports costlier in the country and will thereby reduce imports. In short, devaluation will cause expansion of the export trade and contraction of the import trade of the country.

Devaluation of a country's currency sets in motion two divergent processes in the balance of payments adjustment. While the one adjustment process helps to improve the balance of payments situation the other hinders it. Firstly, the 'impact effect' of devaluation is to improve the domestic versus international cost-price ratio in favour of the country which devalues her currency although not necessarily to the extent of devaluation. This improvement in turn alters the relative prices in the country in favour of the exports and import-substitutes which begin to attract resources for increasing the output of export goods and import-substitutes. Secondly, as the realignment of the cost-price ratio, given the appropriate demand and supply elasticities, begins to improve the balance of trade, the multiplier process of income generation within the Keynesian theory sets in. The 'income effect' of the multiplier process tends to offset the favourable price effect of devaluation on the balance of payments—the degree of adverse effect depending on the level of employment prevailing at the time of devaluation.[2] Devaluation requires for its success certain

2. Whether or not the adverse income effect of the multiplier process under conditions of unemployment would nullify the favourable price effect, would depend on the marginal propensity to save. If, however, devaluation is done under conditions of full employment, the multiplier would raise prices rather than raise real income in the country and would consequently offset the favourable price effect of devaluation on the country's external balance of payments. In such a situation, an appropriate policy of disabsorption should be enforced in order to reap the benefits of devaluation.

prerequisites. Firstly, the cost-price structure in the country devaluing her currency should not react unfavourably following the devaluation. The domestic currency prices, particularly of the export goods, in the country in the wake of devaluation should not rise because to the extent these will rise the country's competitive strength in the foreign markets will suffer impairment and the beneficial effects of devaluation will be lost. It is essential to enforce an effective government control on the activities of the speculators who would like to benefit from devaluation at the cost of the community. The domestic prices of exports in the country should be kept stable at the pre-devaluation level.

Secondly, the foreign countries should accept devaluation in order to make it successful. It implies that the foreign countries should not retaliate by enforcing counter-devaluation measures. In the first place, they should not raise tariffs on their imports from the country devaluing her currency because to the extent they will do the beneficial effects of devaluation through cheaper imports will be lost for the country. Secondly, the domestic cost-price structures of foreign countries should not be readjusted at a lower level by means of export subsidies otherwise the imports of the currency-devaluing country will not decrease.

The Elasticity Approach

The conditions under which devaluation would cause an improvement in the external balance of payments of the currency-devaluing country have been derived and discussed by Abba P. Lerner, Mrs. Joan Robinson, Lloyed A. Metzler, James E. Meade, H. Von Stackelberg and others. The extent to which deficit of the external balance of payments of the currency-devaluing country can be corrected through devaluation depends on the price elasticities of demand and the price elasticities of supply of her exports and imports. Considering first the price elasticity of demand for the exports of the country, if the price elasticity[3] of demand for the exports of the currency-devaluing country is less than unity in absolute value ($|\eta_x| < 1$ where the term η_x denotes the price elasticity of demand for the exports), devaluation will not reduce deficit of the international balance of payments of the country; the balance of payments deficit of the country would rather increase as a consequence of devaluation. This is due to two factors that operate consequent upon devaluation. This first effect of devaluation by reducing the prices of export goods expressed in terms of the foreign currency is to cause an increase in the total amount demanded of the export goods. As a consequence of devaluation, the physical volume of a country's exports will increase. The important question, however, is: by what percentage will exports of the country increase as a consequence of devaluation? It will depend on the price elasticity of demand for exports of the country devaluing her currency.

If the price elasticity of demand for the exports of the country is less than unity, a given percentage fall in the rate of exchange, say of 20 per cent, will cause less than 20 per cent (say, only of 10 per cent) increase in the total amount of exports of the country. This is the positive physical quantity effect of devaluation. But while the total amount of exports increases by 10 per cent the total amount of foreign exchange earnings or the total value of exports of the country will be less than the

3. The term 'elasticity of demand' has been used here in the Marshallian sense. The formula for the price elasticity of demand $\dfrac{dD}{dP} \cdot \dfrac{P}{Q}$ has a negative sign showing that the demand curve for a country's exports is negatively sloping while the formula for the price elasticity of supply $\dfrac{dS}{dP} \cdot \dfrac{P}{Q}$ has a positive sign showing that the supply curve for exports has a positive slope.

pre-devaluation total value of exports because while the total amount of exports has increased by 10 per cent, the foreign currency price per unit of export goods has fallen by 20 per cent. The negative price factor more than offsets the positive quantity factor reflected in the increase of the total quantity of exports, making the net effect on the total amount of the foreign exchange earnings of the country made through exports negative. It has been shown in Figure 15.1A where the two countries taken are India and the United States of America with India devaluing her currency rupee in order to correct her external balance of payments deficit in relation to the USA. In Figure 15.1A, the demand and supply of India's exports have been shown on the X-axis and the price expressed in the US dollars has been shown on the Y-axis. $D_x D_x$ is the inelastic foreign (American) demand curve for India's exports; AS_x is the perfectly elastic supply curve of the Indian exports. Before devaluation of the Indian rupee, the dollar price of exports was 0A, the total amount exported was $0Q_1$ and the total foreign exchange (American dollars) earned by the country through exports was $0Q_1 \times 0A$ (= rectangle $A0Q_1F$).

Consequent upon devaluation of the Indian rupee, the dollar (foreign currency) price of Indian exports falls by AB from 0A to 0B showing that the dollar price of India's exports in percentage term has fallen by the full extent of devaluation. Consequently, the supply curve of India's exports AS_x shifts downward to the position of BS'_x while the total amount of exports increases by $Q_1 Q_2$ (= ΔQ) from $0Q_1$ to $0Q_2$. The new total foreign exchange earnings (total US dollar earnings) are $0Q_2 \times 0B$ (= rectangle $B0Q_2G$). The net change in the total foreign exchange earnings of the country due to devaluation will be positive (increase) or negative (decrease) according as rectangle ABEF is smaller or greater than rectangle $Q_1 Q_2 GE$. In this case, the foreign demand for India's exports being inelastic, rectangle ABEF > rectangle $Q_1 Q_2 GE$. Consequently, devaluation increases rather than decreases the deficit of India's balance of payments. It, therefore, follows that although devaluation will always lead to an increase in the total quantity of exports of the currency-devaluing country (unless the price elasticity of demand for her exports is zero), it does not necessarily always follow from this that it will always also lead to an increase in the total foreign exchange earnings of the country from exports.

Figure 15.1B shows the position of total export earnings in terms of rupees (domestic currency). It is obvious that the total rupee earnings will always increase as a result of devaluation

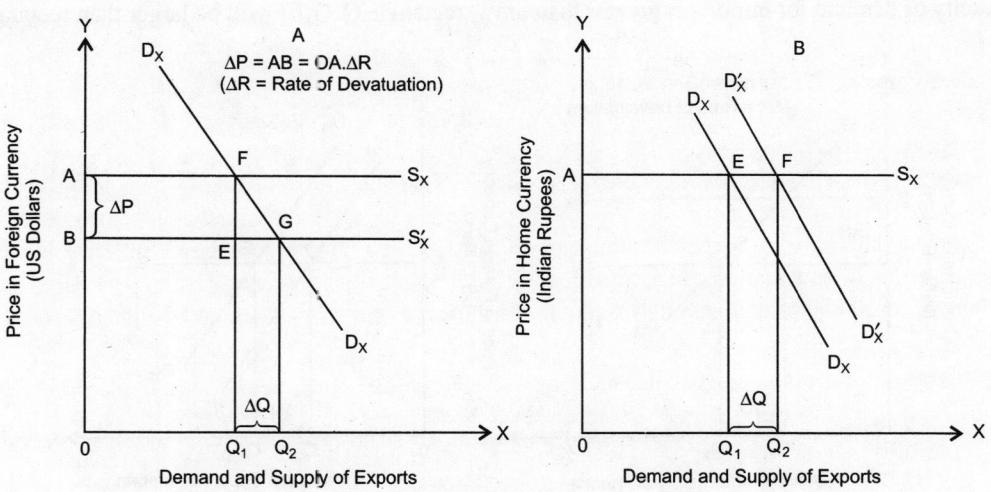

Figure 15.1

even when the foreign demand for the country's exports is inelastic. In this case, instead of the country's perfectly elastic supply curve shifting downward the demand curve shifts upward from $D_x D_x$ to $D'_x D'_x$ showing higher demand for exports at each different rupee price since the same rupee price now means lower dollar price for the Americans inducing them to buy $Q_1 Q_2$ higher amount of exports. However, although the total rupee earnings from exports increase by $E Q_1 Q_2 F$ amount but it does not necessarily follow from this that the total dollar earnings also increase. It is so because the rupee is now worth less in terms of the dollar by the extent of rate of devaluation ($-\Delta R$ the). Since the rate of exchange defined as the dollar price of the rupee has fallen, by multiplying the total rupee earnings by the new foreign exchange rate we will obtain the total dollar earnings which may be greater than, equal to, or less than the pre-devaluation total dollar earnings according as the elasticity of demand for India's exports is greater than, equal to, or less than unity.

Coming to imports, devaluation will help in reducing the deficit of country's external balance of payments only if the price elasticity of domestic demand for imports is greater than unity in absolute value ($|\eta_m| > 1$) because in that situation a given percentage increase in the price of imports measured in terms of the domestic currency would cause more than proportionate fall in the quantity of imports leading to a net fall in the total value of imports as is shown in Figure 15.2A where the demand and supply of imports have been shown on the X-axis and the price of imports expressed in the domestic currency has been shown on the Y-axis. $D_M D_M$ is the demand curve for imports while AS_M is the perfectly elastic supply curve for imports. Before devaluation, the rupee price of imports was 0A and the total imports were QQ_1 giving the total rupee value of imports represented by rectangle $0Q_1 EA$ ($= 0A \times 0Q_1$). As a consequence of devaluation, although the price of imports in foreign currency (US dollars) remains unchanged but as the supply curve for imports has been expressed in terms of the domestic currency, it shifts upward by the full amount of the difference between the pre-devaluation and the post-devaluation rupee price. Consequently, the new supply curve is BS_M. In the post-devaluation situation, the total amount of imports falls by $Q_1 Q_2$ ($= -\Delta Q$) from $0Q_1$ to $0Q_2$. However, after devaluation India pays the higher price 0B for her imports. The total rupee cost of imports is now rectangle $0Q_2 GB$. The improvement in India's external balance of payments resulting from decrease in the total amount of imports consequent on devaluation of the rupee would depend on the relative sizes of rectangle $Q_2 Q_1 EF$ and ABGF. If the elasticity of demand for imports is greater than unity, rectangle $Q_2 Q_1 EF$ will be larger than rectangle

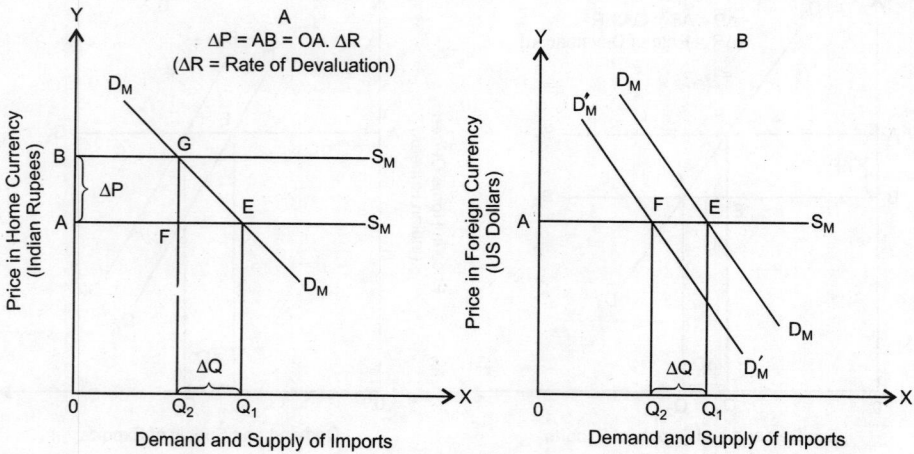

Figure 15.2

ABGF. Consequently, the balance of payments position of the country, in so far as imports are concerned, will improve due to devaluation.

Figure 15.2B shows the effect of devaluation on the total imports and total foreign exchange bill which must fall unless the demand for imports is perfectly inelastic. However, although the total dollar payments obligation is reduced from $0Q_1EA$ to $0Q_2FA$ but it does not necessarily follow from this that the total rupee payments burden of the country will also fall. It will fall only if the demand for imports is elastic, i.e., if the elasticity of demand for imports is greater than unity. There will be no change in the total rupee payments burden if the elasticity of demand for imports is unity and the burden will increase in the event of the elasticity of demand for imports being less than unity.

It can be proved by means of a formula[4] that under conditions of perfectly elastic supply of imports and exports if a country devalues her currency in order to remove her external balance of payments deficit, the sum of the price elasticities of demand for the currency-devaluing country's exports and imports should be greater than unity in absolute value, i.e., $|\eta_x| + |\eta_m| > 1$. This is the well-known Marshall-Lerner condition which is broadly correct if the price elasticities of supply are very large (approaching infinity) and the balance of trade is initially in equilibrium when currency's devaluation is done. If $|\eta_x| + |\eta_m| = 1$, no improvement in the external balance of payments deficit of the country will result from devaluation. If $|\eta_x| + |\eta_m| < 1$, devaluation will increase rather than decrease the balance of payments deficit of the country. Where the sum of these two price elasticities of demand is less than unity there revaluation, not devaluation, of the currency will improve the balance of trade of the country.

The core of the traditional approach contained in the so-called Marshall-Lerner condition stating that the sum of the demand elasticities for a country's exports and imports has to be greater than unity for devaluation to have a positive effect on the country's balance of trade can be expressed in terms of the following formula:

$$dB = KX_f(\eta_x + \eta_m - 1)$$

where the terms dB, K, X_f, η_x, η_m denote respectively the change in currency-devaluing country's balance of trade, currency devaluation in percentage terms, the value of country's exports expressed in terms of foreign currency, foreign country's (rest of the world's) demand elasticity for the currency-devaluing country's exports and currency-devaluing country's demand elasticity for imports.

It is obvious from the above expression that the sum of the two demand elasticities has to be greater than unity in order for devaluation to cause an improvement in the country's balance of trade. If this sum is less than unity, a revaluation or appreciation of currency instead of devaluation should be used to offset deficit in the country's trade balance. The conclusion that devaluation will improve the balance of payments of a country only if the absolute value of $|\eta_x| + |\eta_m|$ is greater than one holds good on the assumption that the price elasticities of supply of a country's exports as well as of imports are infinite, i.e., $e_x = \infty$ and $e_m = \infty$ or are at any rate very large. If the price elasticities of supply of a country's exports as well as of her imports are zero or very low[5] (approaching zero) devaluation places the country in a favourable situation and the price

4. Relative change in the value of exports is $\Delta\eta_x$ where the term Δ stands for the currency devaluation. Following devaluation, the term $\Delta\eta_x$ will be positive. Relative change in the value of imports will be $\Delta(1-\Delta\eta_m)$. Consequently, the change in the balance of payments will be $\Delta(\eta_x - 1 + \Delta\eta_m)$ which should be positive if an improvement in the balance of payments is to result from devaluation. This can happen when the term $\Delta(\eta_x + \eta_m - 1)$ is positive, i.e., when $|\eta_x| + |\eta_m| > 1$.

5. The supply elasticities of certain agricultural products may be relatively low or even zero. Furthermore, under full employment conditions the supply will be inelastic.

inelasticities of supply of exports and imports will cause improvement in the balance of payments of the country regardless of the demand elasticities. In other words, the two will act as exchange stabilisers. In this case, although the Marshall-Lerner condition is sufficient for the balance of payments improvement but it is not necessary, i.e., the price elasticities of the two demands for exports and imports will not matter. In other words, the sum of the two demand elasticities can be less than unity and still the balance of payments will improve if the supply elasticities are low.

If the supply of exports and the supply of imports is perfectly inelastic, i.e., if $e_x = e_m = 0$ and if India devalues her currency, in this unusual case the balance of payments as well as the commodity or net barter terms of trade P_x/P_M will improve in favour of India. The commodity terms of trade move in favour of the country devaluing her currency because (1) the prices of her exports expressed in terms of the domestic currency rise by the full amount of devaluation, and (2) the prices of her imports expressed in terms of the domestic currency do not rise. It means that by exporting the same amount of goods and services as she did before devaluation, the country can obtain a larger amount of imports after devaluation or she can get the same quantity of imports by exporting less amount of goods and services as compared with the pre-devaluation amount.

Figures 15.3A and 15.3B show the effect of devaluation on a country's external balance of payments when the price elasticity of supply of exports and the price elasticity of supply of imports of the currency-devaluing country are zero. In the figures, the exports and imports have been shown on the X -axis while the price, in terms of the home currency, has been shown on the Y-axis. In the country devaluing her currency (India), the supply of exports being perfectly price inelastic ($e_x = 0$), the rupee price of exports will rise by the full percentage amount of devaluation. The exports of the country will, however, not fall because although the price of exports in terms of the home currency has risen by the full amount of devaluation but for the foreigners the foreign currency (dollar) price of exports remains unchanged. Consequently, even when the rupee price of exports increases from 0B to 0A ($\Delta P = AB = 0B. \Delta R$,) where ΔR is the rate of devaluation), the demand for the country's exports remains unchanged, i.e., even at the 0A higher rupee price of exports the foreigners demand the same amount of exports as they demanded before devaluation when the rupee price of exports was 0B, i.e., the foreigners demand $0Q_1$ amount of export goods both before and after the devaluation. This has been shown in Figure 15.3A by an upward shift of the original demand curve for exports $D_x D_x$ to the position of $D'_x D'_x$ demand curve. The increase in the total rupee earnings is $0Q_1 \times AB$ ($= \Delta P \times 0Q_1$) shown by rectangle ABEF in the figure.

Figure 15.3D shows the effect of devaluation on the total foreign exchange earnings through exports. It shows that even after devaluation the total foreign exchange (dollar) earnings of the country remain unchanged at AQ_1AB. It is so because the post-devaluation dollar price of the country's exports remains unchanged at 0B since the rupee price of her exports increases by the full extent of devaluation.

On the side of imports, the price in terms of the home currency will not increase although devaluation should have raised it. The rupee price of imports does not rise because the supply of imports being perfectly price-inelastic ($e_m = 0$), the American exporters reduce the dollar price of their exports to the full extent of devaluation in order to prevent the demand for their export goods from falling. Consequently, the rupee price of imports in India does not increase as result of devaluation and the same amount of imports at the same pre-devaluation rupee price continues to be imported. The net gain from devaluation would be measured by the difference between the change in the total rupee value of exports measured by the rectangle ABEF in Figure 15.3A and the change in the total rupee value of imports which is zero (as shown in Figure 15.3B) since the total amount of imports as well as the price of imports both before and after devaluation of the rupee remain unchanged. Consequently, the net gain resulting from currency devaluation to the country is equal to $AB \times 0Q_1 = ABEF$.

Figure 15.3C shows the effect of devaluation on the total foreign exchange (dollar) bill or payments the country has to make on account of imports. It is evident that the country gains to the extent of ABEF. Since there is no change in the total foreign exchange earnings while the total foreign exchange payments of the country have fallen by ABEF amount, there is net gain to the country and her balance of payments deficit has been reduced as a consequence of devaluation.

When the price elasticity of supply of exports and the price elasticity of supply of imports are neither infinite nor zero but are somewhere between these two limiting situations, i.e., when

$$\infty > e_x > 0, \text{ and}$$

$$\infty > e_m > 0$$

we cannot say off-hand whether the balance of payments of the country will improve or deteriorate as result of devaluation. The net change in the balance of payments due to a given percentage rise

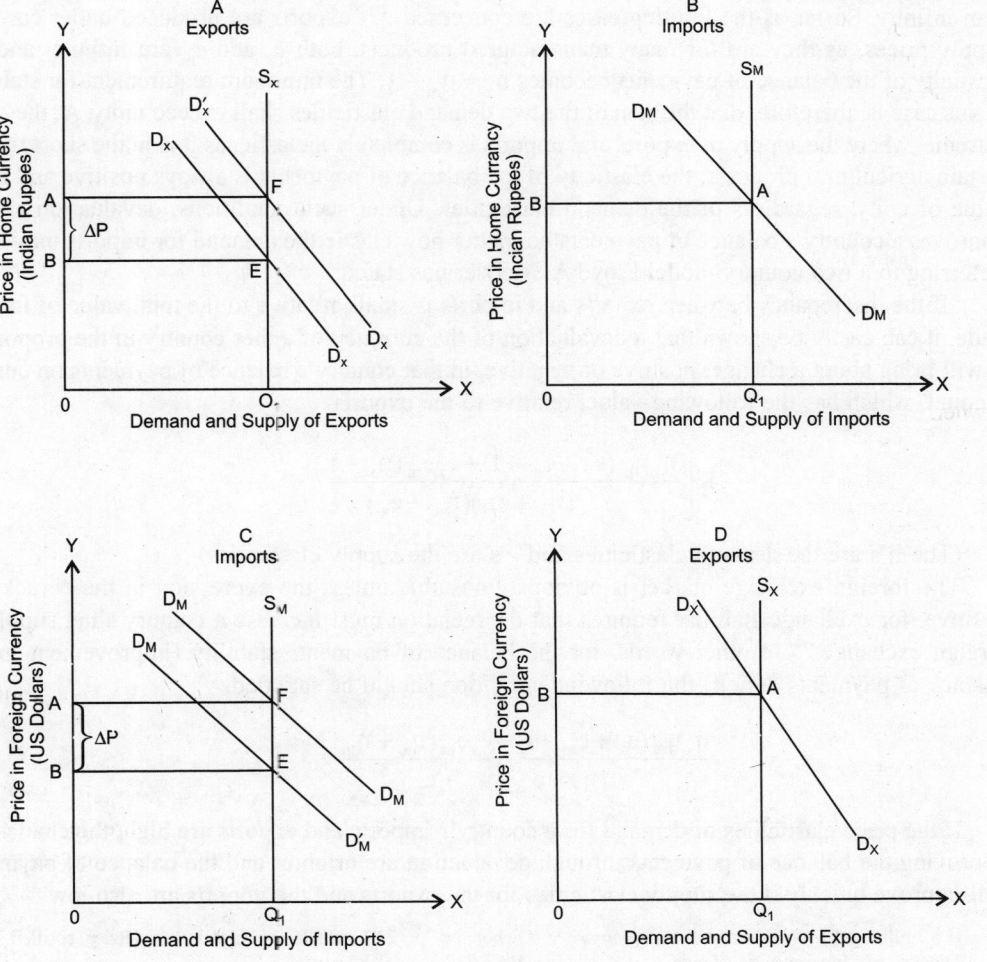

Figure 15.3

in the rate of foreign exchange (devaluation of the currency) can be found out by employing the following formula:

$$\frac{dB}{d\pi} = K \left[\frac{\eta_x \eta_m (e_x + e_m + 1) + e_x e_m (}{(\eta_x + \eta_m)(\eta_m + \epsilon} \right.$$

where

K is devaluation of the currency in percentage terms,

η_x and η_m are the price elasticities of demand for exports and imports respectively of the currency-devaluing country.

e_x and e_m are the price elasticities of supply of exports and imports respectively, and $d\pi$ and dB denote the change in the rate of foreign exchange (devaluation) and the resulting change in the external balance of payments of the country respectively.

To sum up, we have studied devaluation as leading to exchange stability firstly, in the two limiting cases and secondly, in the intermediate case which does not fit in either of the limiting cases, where the supply elasticities of exports and imports are finite being greater than zero but less than infinity. So far as the limiting cases are concerned, "if exports are produced under constant supply prices, as they are for many manufactured products, both e_x and e_m are infinite, and the elasticity of the balance of payments becomes $\eta_x + \eta_m - 1$. The minimum requirement for stability in this case is, therefore, that the sum of the two demand elasticities shall exceed unity. At the other extreme, where the supply of exports and imports is completely inelastic, as it is in the short run for certain agricultural products, the elasticity of the balance of payments is always positive and has a value of unity, regardless of the demand elasticities. Under such conditions, devaluation always improves a country's balance of payments no matter how elastic the demand for imports may be."[6] Referring to a two-country model Lloyd A. Metzler has stated:

"If the discrepancy between exports and imports is small, relative to the total value of foreign trade, it can easily be shown that a devaluation of the currency of either country in the proportion K will bring about a change, positive or negative, in that country's balance of payments on current account, which has the following value, relative to the exports

$$K \left[\frac{\eta_x \eta_m (e_x + e_m + 1) + e_x e_m (\eta_x + \tau}{(\eta_x + e_m)(\eta_m + e_x)} \right.$$

(The η's are the demand elasticities and e's are the supply elasticities)

The foreign exchange market is obviously unstable unless the expression in the "bracket is positive, for exchange stability requires that depreciation must increase a country's net supply of foreign exchange."[7] In other words, for the balance of payments stability (improvement in the balance of payments deficit) the following condition should be satisfied.

$$\frac{\eta_x \eta_m (e_x + e_m + 1) + e_x e_m (\eta_x + \eta_m -}{(\eta_x + e_m)(\eta_m + e_x)}$$

If the price elasticities of demand for a country's imports and exports are high, the chances of improving the balance of payments through devaluation are brighter and the balance of payments will improve quickly if the supply elasticities for the exports and the imports are also low.

---

6. H.S. Ellis (ed.), *A Survey of Contemporary Economics*, p. 227, article by Lloyd A. Metzler entitled "The Theory of International Trade."
7. *Ibid.*

To conclude, the result to be expected from a given change in the foreign exchange rate depends on the price elasticities of demand and supply in each country for the export goods of the other country. Generally speaking, devaluation has much to commend as a remedy for removing the balance of payments deficit of a single country. But since it adversely affects the trade of other countries there is a danger, particularly when not one but many countries suffer from the balance of payments deficit, of its becoming competitive in which case it may fail to serve its purpose. There is, therefore, need for international consultations and agreement, before devaluation of the currency should be done by any country.

The elasticity analysis of the effects of currency devaluation on the balance of payments of a country is a partial equilibrium analysis and the supply and demand curves employed are the partial curves. Devaluation causes changes in the exports and imports of a country which can be calculated from the elasticities because the other things are assumed constant. Some writers, particularly Jacob Viner, have raised fundamental objections against the use of partial demand and supply curves and also the partial elasticities of demand and supply on the ground that it is unscientific and misleading to take recourse to the partial equilibrium analysis in order to study a problem which is essentially of a general equilibrium nature. In other words, it has been forcefully argued by Jacob Viner that it is illegitimate to assume that the demand for imports is "independent of what happens to exports" and the supply of exports is "independent of what happens to imports". A change in the rate of foreign exchange produces changes that reverberate throughout the economy changing incomes and goods' prices over a wide range with the result that other things cannot be taken as equal. Consequently, what is needed are not the partial elasticities which assume other things equal, but the total elasticities which consider the total effects of devaluation on the economy of the country.

The Absorption Approach

So far we have discussed currency devaluation as a method of correcting the external balance of payments deficit disequilibrium of a country in terms of the elasticity approach which works directly on the balance of payments equation

$$B = X - M$$

where the terms B, X and M denote the balance of trade, the value of exports and the value of imports respectively. By differentiating the above equation with respect to the rate of foreign exchange we can find out the effects of a given change in the rate of foreign exchange on the balance of payments. This analysis, however, runs in terms of the partial equilibrium analysis and suffers from all the shortcomings inherent in this type of analysis.

An alternative approach to study the effects of devaluation formulated in macroeconomic terms, as against the elasticity approach formulated in microeconomic terms, is the absorption approach. This approach was first developed by Sidney Alexander in his famous paper published in 1952.[8] According to the absorption approach, the balance of trade can be treated as the difference between the national income (Y) and the total expenditure (E) or absorption such that

$$B = Y - E \qquad (1)$$

8. Sidney S. Alexander, "The Effects of a Devaluation on the Trade Balance", *International Monetary Fund Staff Papers*, Volume 2, No. 2, 1952, p. 263ff.

Designating the total expenditure (E) or total demand as total absorption (A), the above equation can be rewritten as

$$B = Y - A \tag{2}$$

Since the total absorption is composed of the demand created for all purposes, it includes the demand for consumption and investment purposes. Thus, for a three-sector economy

$$A = C + I + G \tag{3}$$

Devaluation will affect the balance of trade either by affecting the real national income Y or by affecting the total absorption A in the economy. Thus, the change in the balance of trade may be expressed as

$$dB = dY - dA \tag{4}$$

The total absorption can be separated in two parts. Firstly, any change in the real income (Y) will cause a change in the total absorption (consumption demand) which will be equivalent to the change in real income (dY) multiplied by the propensity to absorb (c). Secondly, devaluation will directly affect the absorption which will depend, among other things, on the level of real income at which devaluation takes place. Let us call this direct effect on the absorption as D. Consequently,

$$dA = cdY + dD \tag{5}$$

By substituting equation (5) for the term dA in equation (4) we obtain

$$dB = (1 - c)dY + dD \tag{6}$$

Equation (6) is useful as it focuses our attention on the three important factors which determine the outcome of currency devaluation. It states that the effects of devaluation on the balance of trade depend on (i) the effect of devaluation on the real income (Y), (ii) the propensity to absorb (c), and (iii) the effect on direct absorption (D).

In order to deal with the effects of devaluation under the absorption approach, distinction should be made between two main situations of an economy where idle resources exist and of an economy where resources are fully employed. We may start our discussion with the first case of our economy with less than fully employed resources.

If there are enough unemployed resources in the currency-devaluing country (as is generally true of the most developing countries) production can increase in the short period following the devaluation. The expansionary process will start from an increase in the country's exports which will raise the real national income through the multiplier process. The extent of increase in the exports will, however, depend on the extent to which exports' prices in the currency-devaluing country rise and the extent to which the rest-of-the-world is willing or has the capacity to absorb the exports from the currency-devaluing country.

The net effect of the increase in real national income on the balance of trade does not, however, comprise the total increase in the real income or production; it comprises the difference between the increase in real income and the induced increase in total absorption. This difference between the increase in real production and real absorption may be called real hoarding (saving). The effect of devaluation on the balance of trade is, therefore, equal to the amount of real hoarding which takes place in the economy and which is determined by the propensity to hoard ($1 - c$). As long as c, the propensity to absorb, is positive and less than unity, the propensity to hoard will also be positive and less than unity. Consequently, some hoarding will take place and to this extent devaluation will exert a positive effect on the currency-devaluing country's balance of trade. The effect on the country's balance of trade is, therefore, equal to the amount of real hoarding (saving) which takes place in the economy consequent upon devaluation.

Keeping the effects on direct absorption aside, in analysing the effects of devaluation on the balance of trade, the propensity to absorb (*c*) or the propensity to hoard (1 − c), is the crucial factor in this case. So long as c is less than unity, hoarding will be positive and devaluation will have a positive effect on the balance of trade.

If the propensity to absorb is greater than unity, then devaluation will have a negative effect on the country's balance of trade since the induced increase in absorption will be greater than the original increase in the real national income. It cannot be ignored. It is thus concluded that under conditions of less than full employment, devaluation will have a positive effect on the national income. If the propensity to absorb is less than unity, devaluation is an attractive policy for a country to pursue in depression as it will have both a positive effect on the national income and also improve the balance of trade. However, in inflation workers who are employed will have a high propensity to consume. Moreover, the expansion in income will boost investment. Together these factors can make the propensity to absorb larger than unity. Consequently, devaluation will have a negative effect on the country's balance of trade.

If the propensity to absorb is less than unity or as result of policy measures taken by the government it is made less than unity, devaluation will prove a very attractive policy for a country to adopt in a depression because it will both help raise the level of national income as also improve the balance of trade. The positive impact of devaluation on national income was recognised even in the pre-Keynesian analysis. It was probably for this reason that a wave of devaluations was witnessed during the great depression of the 1930s.

So far we have discussed the effects of devaluation when there exist idle resources—unemployment in the economy. Under conditions of full employment in the economy or the marginal propensity being greater than unity, the principal favourable effect of devaluation on the balance of trade is through the direct effect on absorption which is not connected with any change in the real national income. Following the devaluation, its direct effect on absorption will depend on several factors such as the extent of change in the price level, money supply, the *real balance effect* etc. In the full employment situation, the policy measures have to be more straightward and should aim at depressing the absorption. The decrease in absorption will permit the necessary reallocation of resources leading to an increase in exports and decrease in imports eventually improving the balance of trade.

Since successful devaluation by one country will adversely hit the exports of the other non-devaluing countries, devaluations tend to be competitive inducing the other countries to devalue their currencies. The International Monetary Fund has attempted to avoid such situations by providing effective safeguards against the competitive devaluations of their currencies by its members.

Devaluation Deluge

During the post six decades of the IMF era, there has been a deluge of currency devaluations. During the past more than forty years beginning from 1948 up to 1995 several countries including the United States of America have devalued their money units. Many countries have devalued their currencies by more than 75 per cent and more than once. In fact, there are many countries which have devalued their currencies more than a dozen times. For example, Brazil devalued her currency for the fourteenth time in 1975, changing the exchange rate for buyers from 8.85 cruzeiros to 9.02 cruzeiros for the US dollar. The maximum number of countries devaluing their currencies have belonged to Africa with Latin America coming next.

During the *seventies,* the international monetary system had been subjected to chaos and disorder. The leading world currencies had been under heavy pressure of conversion on the part of speculators seeking to convert their foreign currency holdings, particularly the US dollars, into other safer currencies like the Japanese yens, German marks, Swiss francs in anticipation of the impending devaluation of the dollar. So unprecedentedly heavy had been the pressure on the US dollar that within less than fourteen months it was twice devalued raising the price of gold from $ 35 per ounce before December 1971, when it was first devalued, to $ 42.22 per ounce on February 12, 1973. In the wake of the dollar devaluation leading international foreign exchange markets had remained closed for 17 days from March 2, 1973 following an extensive run on the US dollar. During this period, the yen was allowed to float in the foreign exchange markets. Following the devaluation of the dollar, a new international foreign exchange rate structure had emerged in a pattern of roughly four groupings of member countries: those holding their gold parities, in effect revaluing their currencies upward by 11.1 per cent against the dollar; those following the dollar down with a 10 per cent devaluation; those devaluing by less than 10 per cent; and those floating their currencies. Following the dollar devaluation, world's financial leaders assessed the conditions which had led to the dollar devaluation and many of them placed renewed emphasis on the urgent need for the reform of the international monetary system which has now largely taken place.

Devaluation of the Indian Rupee in 1949

After the World War was over in 1945, England faced acute dollar shortage. Although dollar shortage had been there throughout the inter-war period, it had worsened in the post-war period. The U.S.A. had emerged financially powerful after 1945. Large debts had been contracted by England to finance the war. The United States of America had supplied England large amount of defence goods on credit and after the war these debts had to be repaid. While there was the problem of repaying the war loans to the United States, the British economy was to be reconstructed. A country can pay her debt only by increasing her exports and decreasing her imports. It was, therefore, necessary that British exports to the U.S.A. should have considerably increased after the war. But this was not possible in the context of the shattered British economy. Not only was it difficult for England to increase her exports but imports had to be increased to feed her population and to reconstruct the war-ravaged economy of the country. These imports further increased the debt burden of England.

India, Burma and other countries had made large supplies of war materials needed by the British forces stationed on the eastern front. England paid for these supplies by crediting the pound-sterling accounts of these countries. Consequently, at the end of the war India and other countries had accumulated a huge amount of pound-sterlings in the form of Sterling Balances. These countries asked for conversion of their sterling balances into the US dollars to enable them to purchase capital goods which they badly needed for their economic development. Indeed, there was some justification in their demand. They wanted machines and other capital goods not necessarily from the U.S.A. In fact, they would have accepted such capital goods from England even at higher prices. But England had no machines and other capital equipment needed by India and other countries available with her. The point was either England should herself supply the goods or else she should supply dollars in exchange for sterling balances to enable the countries of East Asia to purchase the capital goods from the U.S.A. Thus, there was a great scramble for dollars from every quarter—domestic and foreign.

England met the situation by liquidating her investments in the U.S.A. But this did not suffice. The import surplus was too great. Some relief was provided by the U.S.A. through the Marshall Aid. However, all these palliatives were insufficient to remove the fundamental deficit in the balance of payments. There was such a great scramble for dollars that in the second half of 1947 there was "dollar crisis" and the dollar was declared hard currency. The dollar crisis ensued in the second half of 1947 and after it subsequent trends showed that the situation was improving. After the second half of 1947, the export surplus of the U.S.A. decreased and that of the sterling area increased. While in the second quarter of 1947 the US exports were 263 per cent of the 1937 level, in the second half of 1948 these fell to 183 per cent of the 1937 level. Similarly, her imports rose from 96 per cent of the pre-1937 level in the first half of 1947 to 106 per cent of the pre-1937 level in the first half of 1948. This was a healthy trend. But things did not move smoothly. Imports from the United States of America increased subsequently while the exports from England lagged behind. What was the reason?

In order to increase the exports to the United States of America, Canada and other hard currency areas it was necessary to reduce the prices of British export goods. But this did not happen. England was suffering from inflation. Wages, interest and other components of cost of production did not fall. The Labour Government headed by Prime Minister Attlee did not face the problem boldly at the cost of displeasing the trade unions. British economy became rigid and failed to readjust itself to the changing international situation. Neither the exports increased to the extent these should have increased nor did the imports fall to the extent they should have fallen. Consequently, the balance of payments deficit persisted. Superimposed on this basic problem was the speculation regarding the impending devaluation of the pound-sterling entertained by the speculators. There was a tendency for the capital to take flight to the other countries, particularly to the United States of America, in anticipation of an impending devaluation of the pound-sterling. For some time things were manoeuvered through exchange control but it could be imposed only temporarily because England was a member of the International Monetary Fund. The only remedy open to England was devaluation of the pound-sterling to correct the mounting deficit in her external balance of payments and it was done in September 1949. To summarise, the main causes that forced England to devalue the pound-sterling in September 1949 were

(i) Increased dependence of the sterling area on the US imports.

(ii) War-tisme destruction of the productive capacity of England.

(iii) Higher cost of production of the British goods.

(iv) Currency speculation.

Decision to Devalue the Indian Rupee

India devalued the rupee with the devaluation of the pound-sterling. The dollar value of the pound-sterling was devalued to the extent of 30 per cent from $ 4.03 to $ 2.80. The Indian rupee was also devalued in terms of the dollar to the same extent on 29th September, 1949. In the matter of devaluation India had to act "not on conviction necessarily born out of logic, but so to speak by the compulsion of events". Consequent upon devaluation, the value of the rupee in terms of the US dollar fell from 30.225 cents to 21 cents and the gold parity fell from 0.268691 gramme to 0.186621 gramme of fine gold.

India had to devalue to the same extent as the pound-sterling because 75 per cent of India's export trade was with the sterling area countries and in the post-war period next to Britain, India made the biggest demand on the Central Dollar Reserve Pool established by the sterling area countries. As long as this position continued, severance of the ties with sterling area was harmful for India's interests. The link of the rupee with the pound-sterling was vital for India's interests. India's dependence to the extent of 75 per cent of her export trade on the countries of sterling area made it necessary for India to devalue the rupee simultaneously with the devaluation of pound-sterling. Had it not been done, India's export trade with the sterling area would have suffered a great setback.

Another reason why India decided to devalue was her unfavourable balance of trade. It was necessary to increase exports and reduce imports in order to improve the balance of payments position. It is worthwhile to quote late Dr. John Mathai, the then Union Finance Minister, in the defence of Government's decision to devalue. The Finance Minister stated: "Because of our dependence on sterling area for exports, because of our unfavourable balance of trade, because of our high level of prices, everybody who understood finance and business in any country should have presumed that a high rate for the rupee would be unmaintainable. The result would have been that whole of our trade would have become disorganised. It would, in course of time, have come to a standstill." There was thus ample justification for devaluation of the rupee to an equal extent with the pound-sterling. Failure to do so would have adversely affected India's exports. Had the Indian rupee not been devalued to the same extent as the pound-sterling, India's competitive strength in the export markets would have been undermined. For instance, the prices of Indian textile goods were higher than those in England and had India not devalued the rupee the Indian textiles could not have competed with the Lancashire textile goods in export markets. Thus, devaluation of the rupee to the same extent as that of the pound-sterling was inevitable because otherwise India's export trade with the sterling area countries would not only have greatly diminished but would have been wiped out completely in certain directions. Non-devaluation of the rupee would have meant loss of India's exports of tea, manganese, groundnut, jute, textiles etc. Thus, as the Reserve Bank of India remarked "devaluation became a defensive necessity in India and its choice became a Hobson's choice". The decision to devalue the Indian rupee and to the same extent as the pound-sterling was proper in view of the circumstances faced by the country.

The devaluation of the rupee made the Indian exports cheaper in the United States of America and the other hard currency countries. Consequently, India's exports increased. The immediate effect of devaluation was seen in the improvement of India's balance of payments. In 1948-50. India's balance of trade showed a deficit of Rs. 183.42 crores, The magnitude of this deficit was reduced to Rs. 118.89 crores in 1949–50 and to Rs. 22.01 crores in 1950–51. Thus, it can be said that devaluation was instrumental in substantially reducing the deficit of the balance of payments of the country.

Devaluations of the Indian Rupee in 1966 and 1991

India devalued the rupee by 36.5 per cent on June 6, 1966 fixing the new par value of the rupee at 0.118516 gramme of gold in place of the old par value of 0.186621 gramme of gold before devaluation. The decision to devalue the rupee was preceded by a strong controversy regarding the efficacy of the measure. The rupee was devalued when the domestic price level was under heavy pressure arising out of acute food shortage following the 1965 drought. In other words, the rupee devaluation took place when the Indian economy was faced with a kind of cost-push inflation. Furthermore, the decision in favour of devaluation was taken at a time when there was considerable idle industrial capacity particularly in the engineering goods sector. After devaluation, in terms of the foreign exchange rate, one US dollar which formerly fetched Rs. 4.76 fetched Rs. 7.50, while pound-sterling which formerly was equivalent to Rs. 13.33 fetched Rs. 18. Similarly, a rouble commanded Rs. 8.33 as against Rs. 5.21 before June 6, 1966. Thus, the rupee was devalued for the second time since independence, the first devaluation having taken place in September 1949. Devaluation was done to increase India's exports and foreign exchange earnings as the various export subsidy schemes which aimed at promoting exports—import entitlements to exporters, direct subsidies and tax credit certificates—did not have the desired results. Devaluation was devised as a more enduring and reliable way of restoring and indeed of increasing the competitive power in foreign markets for India's exports. Apart from restoring the competitiveness of India's exports, devaluation was also expected to provide strong inducement for the flow of investment into export goods industries and thereby to strengthen India's exports position. Devaluation was also expected to quicken the pace of import substitution and to that extent help reduce India's dependence on imports. It was argued that since as a result of the devaluation the rupee-cost of imports will rise, this will make it worthwhile and attractive to invest in those industries which produce those goods which are still being imported. Devaluation, it was argued, will provide enduring encouragement to both exports and import-saving activities and it will enable India to move faster towards all-round self-reliance. It would also strengthen the foreign exchange reserves of the country by increasing tourist trade of the country. Since in the traditional items of export no stimulus was needed to be given to the exporters, export duties were levied on certain items to moderate the stimulus leaving sufficient incentive for growth.

Along with the devaluation, the existing special export promotion schemes providing import entitlements against exports and the scheme for tax credit certificates were abolished. Moreover, in order to protect the unit values of exports in terms of foreign exchange, export duties were levied on several commodities, mostly agricultural commodities and agricultural-based manufactures. A variety of additional measures were taken to promote the exports. A liberal import policy was announced for 59 priority industries, including a number of export-oriented industries. A new import replenishment scheme enabled the registered exporters to obtain raw materials, components and spares against the exports of specified products. It was decided to provide cash assistance for exports of selected products with a good export potential. A scheme for the supply of steel at international prices to exporters of engineering goods was announced. Imports of some raw materials were placed under an Open General Licence.

Faced with the chronic balance of trade deficit and the fast dwindling foreign exchange reserves the rupee was devalued twice between July 1 and July 5, 1991 in terms of the US dollar, pound sterling, German mark, Japanese yen and French franc. While the July 1, 1991 devaluation of the rupee was of the order of 9.7 per cent the subsequent one done on July 4, 1991 was of the order of 12.3 per cent. Thus, the rupee was devalued to the extent of 22 per cent. The balance of payments deficit, however, still persists and the government has been making frantic efforts to manage it by seeking the financial assistance from the International Monetary Fund in order to carry forward the structural reforms in the economy consequent upon pursuing the policy of economic liberalisation.

Suggested Readings

Sidney S. Alexander, "The Effects of Devaluation on a Trade Balance," *IMF Staff Papers,* Volume V, April 1952, pp. 263–378.

Sidney S. Alexander, "Effects of a Devaluation: A Simplified Synthesis of Elasticities and Absorption Approaches," *The American Economic Review,* 1959, pp. 22–42.

P.T. Ellsworth and J. Clark Leith, *The International Economy,* Fifth Edition, 1975, Chapter 19.

Harry G. Johnson, *International Trade and Economic Growth,* 1958, Chapter VI.

Charles P. Kindleberger, *International Economics,* Fifth Edition, 1973, Chapters 19 and 22.

Fritz Machlup, "The Analysis of Devaluation," *The American Economic Review,* Volume XIV, June 1955.

Robert A. Mundell, *International Economics,* 1968, Chapter 10.

Joan Robinson, "The Foreign Exchange," *Readings in the Theory of International Trade,* pp. 83–103.

V.C. Shah, "Devaluation—The Indian Case," *Indian Economic Journal,* Volume XVIII, No. 1, July September, 1970.

Bo Sodersten, *International Economics,* Second Edition, Reprinted 1981, Chapter 25.

Questions

1. Analyse the effects of devaluation on a country's balance of payments. Explain in this context (a) the Marshall-Lerner condition, and (b) the absorption approach.
2. What is meant by devaluation of currency? Under what conditions will devaluation help in removing the adverseness of a country's balance of payments?
3. Explain the elasticity approach and the absorption approach to devaluation as the means to remove a country's balance of payments deficit.
4. Discuss the objectives of devaluation. Under what conditions can devaluation bring about an improvement in a country's balance of trade?

CHAPTER 16

Internal and External Balance

The chapters 14 and 15 have been devoted to the examination of the adjustment process in the balance of payments and we have seen how external balance is maintained through price and income changes and changes in the foreign exchange rates. So far, however, we have concerned ourselves entirely with the mechanism for maintaining or achieving external balance giving little consideration to the question of internal balance (full employment with price stability) but governments are concerned not only with the balance of payments, i.e. external balance but also with the internal balance or with the problem of full employment with stable prices. Now to satisfy the requirements, external and internal balance often give rise to serious policy problems. When a government uses the fiscal or monetary policy to achieve internal balance it has serious repercussions on the external balance also. The modern view is to regard economic policy as having the twin objectives of internal and external balance simultaneously by applying the fiscal, monetary and the foreign exchange rate policies.

The earliest systematic account of the difficulties in reconciling the policy objectives is found in James E. Meade 'Balance of payments' Basically the argument is that if two policy targets are to be hit simultaneously two policy weapons or variables are needed to be employed, one to hit full employment target, the other to achieve the balance of payments equilibrium. Meade assumes that there are two types of policy variables (a) income instrument working through changes in the domestic expenditure or absorption and (b) price adjustment which can be brought about in various ways but particularly by means of operating on the rate of exchange. According to Meade, the appropriate technique is to use one set of policies, namely financial (fiscal and monetary) policy to look after the internal balance while employing the price policy as exemplified by exchange rate adjustment to achieve the external balance.

In his generalized scheme of balance of payments adjustments Harry G. Johnson divided the two policies into expenditure switching and expenditure changing. Expenditure switching devices direct expenditure from one field to another. Such devices include depreciation, direct control on exports and imports, subsidies and tariffs etc. Expenditure changes vary the total absorption (C = consumption + I = investment + G = government expenditure), where C is concerned with households, I is concerned with business and G is concerned with government expenditure.

Internal and External Balance Curves

In Figure 16.1 Y-axis measures the competitive cost ratio or exchange rate, X-axis measures the real domestic absorption (C + I + G). The level of domestic expenditure and the state of the balance of payments both depend upon the level of real domestic expenditure and competitive cost ratios or cost of the foreign currency in terms of the domestic currency. A movement along the X-axis is expenditure changing while movement along the Y-axis is expenditure switching policy. Let us take first, the level of domestic employment. If competitive ratio is favourable, i.e. if the cost-ratio of foreign prices to domestic prices is high, then a given level of employment can be supported with low level of real domestic expenditure. Exports will be higher in relation to imports and additional employment will be generated in the exports sector. On the other hand, if the competitive cost ratio is unfavourable, exports will diminish and employment generation will be small in the export sector; Consequently, a high level of domestic expenditure is necessary to maintain the given level of employment. The internal balance curve will be moving downward from north-east to south-west. The relationship between competitive cost-ratio and domestic expenditure has been shown in Figure 16.1 by I family of curves (I, standing for internal balance, the IF curve represents full employment, IO represents overfull employment, and IU stands for under employment.

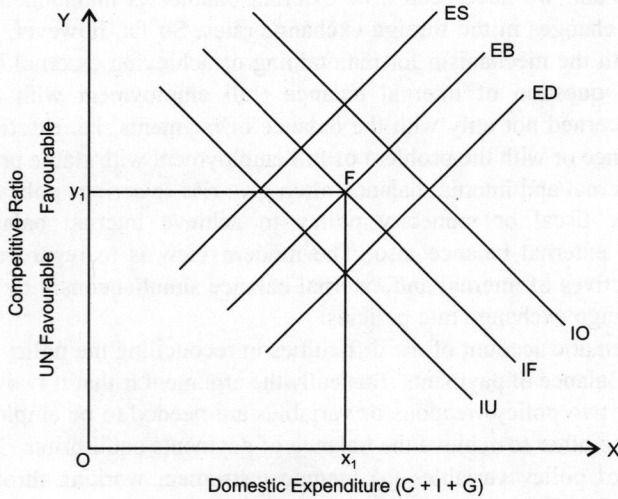

Figure 16.1 Determinants of External and Internal Balance.

Now let us turn to external balance curve. Here we ignore capital movement and let us define external balance occurring whin net exports $(X - M)$ is equal to zero. In the figure we represent external balance curves by E family of curves (EB symbolizing external balance). A given degree of external balance requires various combinations of competitive ratios and domestic absorptions. If the competitive ratio is high, there will be export surplus and more domestic absorption is needed to maintain external equilibrium. On the other, if the competitive ratio is very low imports will be much higher than exports and domestic expenditure will be reduced to maintain external balance. Consequently E curves will slope upward from south-west to north-east. In Figure 16.1 only one of this (EB) represents equilibrium in external balance. The curve ES represents the balance of payments surplus and ED represents the deficit. The object of policy makers is to keep on EB curve. Clearly there is only one point (F) where internal and external balance curves intersect each other, left of this curve is surplus and right to it is deficit. At the point of equilibrium domestic expenditure is ox_1 and competitive ratio is oy_1.

Swan Model of Internal and External Balance

The impact of expenditure switching and expenditure changing policies on the balance of payments and the level of employment can be demonstrated with the aid of a diagram worked out by Trevor Swan. It is given in Figure 16.2 where the X-axis measures domestic absorption and the Y-axis measures the competitive ratio. Figure 16.2 consists essentially of two curves IF and FB respectively for the internal balance and the external balance and they intersect each other at point E. By their intersection four zones (I, II, III and IV) are created. These four zones are known as 'zones of economic unhappiness'. In each zone there is some sort of unhappiness. For instance, in zone I there is surplus-inflation, in zone II there is surplus-recession, in zone III there is deficit-recession and in zone IV there is deficit-inflation. In zones I and II there is surplus in the external balance but in zone I surplus is accompanied by overfull employment whereas in zone II it is accompanied by under employment. In zones III and IV there is deficiet in the external balance while in zone IV there is overfull employment in zone III there is under employment.

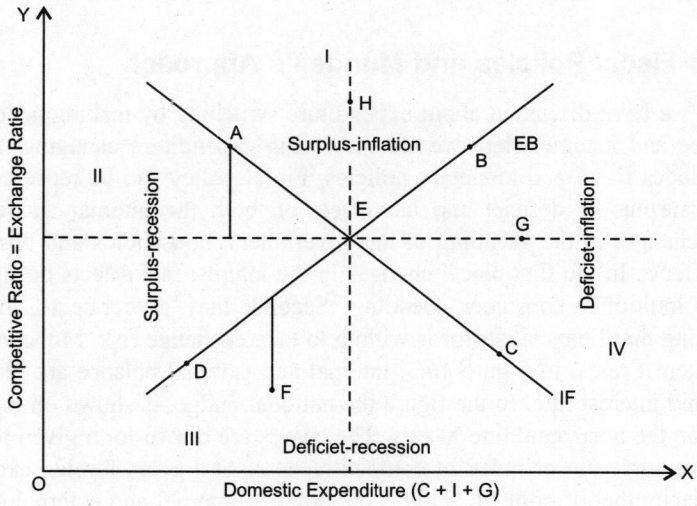

Figure 16.2 Policy Criteria for Internal and External Balance.

If we draw two straight lines (shown by dotted lines in the figure), one parallel to Y-axis and passing through equilibrium point E, and the other line parallel to X-axis and passing through E, the four zone are divided into eight sub-zones. If we are at any point on these dotted lines a single policy variable will be sufficient to bring the external and internal equilibrium. For instance, if we are at H on the vertical dotted line, expenditure switching policy-appreciation of domestic currency is sufficient to bring us to equilibrium point E and if we are below E on the same dotted line in zone III, depreciation will be sufficient. Similarly, if we are at point G on the horizontal dotted line, ccntraction (expenditure changing) in domestic absorption will bring equilibrium in the economy. If we are at any point left to point E on the same horizontal dotted line in zone II, expansion in domestic expenditure is sufficient to bring to equilibrium point E . However, if we are in any one of the sub zones, two policy variables are needed to bring internal and external balance. For instance, if we are at point A in sub-zone I which enjoys full employment and surplus; in this case both expenditure switching and expenditure changing policies are needed to be in operation for achieving the equilibrium. Appreciation of domestic currency accompanied by expansion in the domestic expenditure will bring us from A to E equilibrium point. If we are at point C we should pursue the opposite set of policy. If we are at point B on EB the economy should pursue the restrictive fiscal policy and appreciate the exchange rate. The opposite set of policies should be followed when we are at point D.

If both deficit and recession exist simultaneously as at point F, it is obvious that there is no recourse for the economy but to become more competitive. Expansionary policy that raises the internal demand increase the domestic employment but this causes the deficit to widen. This was the unfortunate situation in which the US economy found itself during 1970-71. It is perhaps the worst of all states and has often been described as (dilemma zone) even though all zones give rise to one dilemma or another. While going through the above discussion we should keep in mind that the internal balance includes only the general unemployment and not the structural unemployment. Similarly, external balance does not include capital destruction and increase in the labour force and productivity.

Monetary and Fiscal Policies and Mundell's Approach

In previous page we have discussed about expenditure switching by making adequate changes in the levels of price and income. Here we are discussing expenditure changing through financial policy which includes fiscal and monetary policies. Fiscal policy can be represented by national budget, whether surplus or deficiet and has effect on both the internal and external balance through causing changes in the spending of the government, households and business. Monetary policy has two effects. In the first place, changes in the interest rate affects bushiness investment and through the multiplier consumer spending. Second, they give rise to short-term capital movement assuming that financial sector is willing to take exchange risk. Monetary policy mainly works through interest rates. In Figure 16.3, internal and external balance are plotted against the national budget and interest rate. In the figure the national budget is shown on the vertical Y-axis and interest rate on the horizontal line X-axis. The curves are drawn for a given foreign exchange rate or competitive cost ratio or index of competitiveness. At a given foreign exchange rate both curves intersect each other at point M. This is the point of internal and external equilibrium. The steeper slope of external balance curve indicates that it is more responsive to monetary policy while the internal balance curve is more responsive to fiscal policy. R. A. Mundell has suggested that if the foreign exchange rate can not be altered and if the other price adjustment are impossible or likely to be inadequate, monetary policy should be given the task of ensuring external balance while to fiscal policy is assigned the work of maintaining the internal equilibrium. If the task is assigned otherwise the actions of the two authorities may move the economy further away from equilibrium. Mundell had referred to it as 'assignment problems'.

In Figure 16.3 in zones II and IV, no assignment problems exist. At a given competitive cost ratio, monetary and fiscal policy should both be expansionary in zone II and contractionary in zone IV. The problems arises in zones I and III. In the former with recession and deficit, the monetary policy should be assigned external balance while tightened fiscal policy should be assigned internal balance and shifted to the less surplus or more deficit since the slope of the external balance curves is steeper than that of internal balance curve, monetary policy has a comparative advantage in working on external balance and fiscal policy on internal balance.

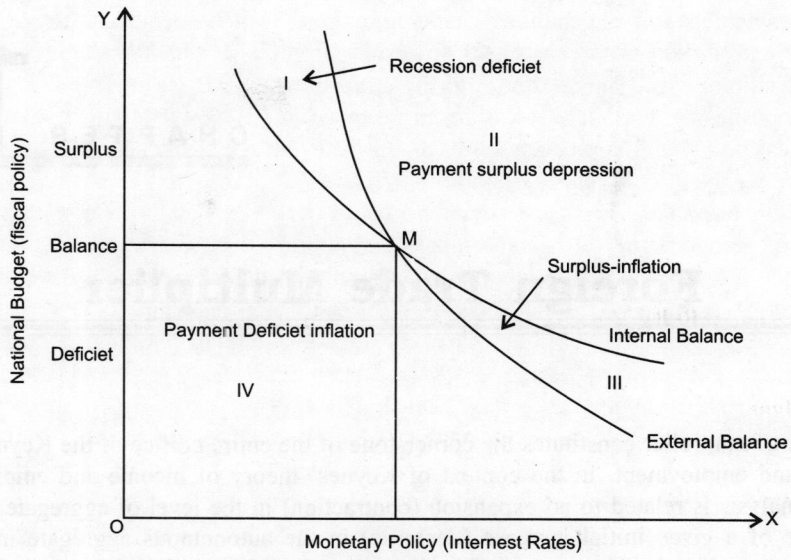

Figure 16.3 Monetary and Fiscal Policy used in Pursuit of Internal and External Balance.

Objections have been raised against Mundell's view of this analysis. Economists are of the opinion that this raises a 'welfare problems' along with the assignment problems and thus detracts from the merit of the analysis, except for the short run.

Suggested Readings

Sidney J. Wells, *International Economics*, Third Impression, 1973, Chapter 16. pp. 212-221.

C. P. Kindleberger, *International Economics*, Fifth Edition, Chapter 26. pp. 496-501.

H. G. Johnson, Theoretical Problems of the International Monetary System, *The Journal of Economic Studies,* Volume 7. No. 2, January 1968.

R. A. Mundell, The Appropriate use of Monetary and Fiscal Policy for Internal and External Stability, IMF Staff Papers Volume 9 No. 1, March 1962.

T. W. Swan, 'Long-Run Problems of the Balance of Payment', Readings in *International Economics.*

Questions

1. Distinguish between the internal balance and external balance. How equilibrium is achieved between the two ?
2. Discuss Swan Model of internal and external equilibrium with the aid of suitable diagrams.
3. Discuss Mundell's approach to equilibrium in internal and external balance.
4. Write short notes on (a) zones of economic unhappiness, and (b) dilemma zone.

CHAPTER 17

Foreign Trade Multiplier

The concept of multiplier constitutes the cornerstone of the entire edifice of the Keynesian theory of income and employment. In the context of Keynes' theory of income and employment, the multiplier analysis is related to an expansion (contraction) in the level of aggregate income as a consequence of a *given* initial increase (decrease) in the autonomous aggregate investment or consumption spending in the economy. The importance of the multiplier analysis for economic policy purposes stems from the assumption that a *given initial* change in the total autonomous investment spending causes a magnified change (this change is multiplier times the change in the total autonomous investment spending) in the aggregate income in the economy.

John Maynard Keynes has discussed the investment multiplier in Chapter 10 of his scholarly work entitled *The General Theory of Employment, Interest and Money* published in 1936. His discussion is largely based on his student Richard F. Kahn's work who discussed the concept of employment multiplier in his classic article entitled "The Relation of Home Investment to Unemployment" published in *The Economic Journal* in June 1931. The multiplier theory states that in the absence of offsetting factors the total expansion in the economy's income occasioned by a given initial increase in the autonomous aggregate investment (or consumption) outlay will be a certain multiple of the increase in investment spending and the magnitude of income expansion will depend upon the value of the marginal propensity to consume (MPC). The investment multiplier (K) and the marginal propensity to consume (MPC) are related in such a way that higher the marginal propensity to consume, higher will be the investment multiplier and *vice versa*.

Simple Investment Multiplier

The simple investment multiplier is the direct function of the marginal propensity to consume ($\Delta C/\Delta Y = b$) such that given the value of the latter the value of the former can be easily derived by using the formula

$$K = \frac{1}{1-b}$$

where K is the investment multiplier and b is the marginal propensity to consume (MPC). The above formula for deriving the investment multiplier has been derived in the following manner.

$$Y = C + I \tag{1}$$

$$C = a + bY \tag{2}$$

$$I = I \tag{3}$$

Substituting for C and I in the equilibrium income equation (1) we get

$$Y = a + bY + I \tag{4}$$

$$Y - bY = a + I \tag{5}$$

$$Y(1 - b) = a + I \tag{6}$$

$$Y = \frac{a+1}{1-b} \tag{7}$$

When the aggregate, autonomous investment I increases by ΔI, the total higher investment becomes $I + \Delta I$. Consequently, the new higher equilibrium aggregate income will be $Y + \Delta Y$ which will be equal to

$$Y + \Delta Y = \frac{a+1+\Delta l}{1-b} = \frac{a+1}{1-b} + \frac{\Delta l}{1-b} \tag{8}$$

Subtracting equation (7) from equation (8) we get

$$Y + \Delta Y - Y = \left[\frac{a+1}{1-b} + \frac{\Delta l}{1-b}\right] - \left[\frac{a+}{1-}\right] \tag{9}$$

$$\Delta Y = \frac{\Delta l}{1-b} \tag{10}$$

It, therefore, follows that the total change (increase) in the equilibrium aggregate income (ΔY) equals the given initial change in the autonomous investment (ΔI) times $1/1-b$. Thus, $1/1-b$ is the investment multiplier. Dividing both sides of equation (10) by ΔI we get

$$\frac{\Delta Y}{\Delta l} = \frac{1}{1-b} = K \tag{11}$$

The ratio $\Delta Y/\Delta I$ is the ratio of change in the aggregate income to change in the aggregate investment spending which is the definition of investment multiplier. The investment multiplier is the inverse of one minus the MPC. For example, if the marginal propensity to consume (b) is 0.8, the investment multiplier (K) will be $1/1-0.8$ or 5.

The investment multiplier is also related to the marginal propensity to save (MPS = $\Delta S/\Delta Y$). We know that the sum of the marginal propensity to consume and the marginal propensity to save should be equal to one. In other words, the marginal propensity to save is equal to one minus the marginal propensity to consume, i.e., MPS equals $1-b$ since b is the marginal propensity to consume. Thus, we can substitute MPS (s) for $1-b$ in equation (11). Accordingly, the investment multiplier $K = \dfrac{1}{\text{MPS}}$. Writing s for the MPS we can write $K = 1/s$. From this it follows that the investment multiplier and the marginal propensity to save are inversely related such that a high marginal propensity to save (MPS) denotes a low investment multiplier (K) and *vice versa*.

The investment multiplier analysis reveals that the total increase in aggregate income equals the given initial increase in the autonomous investment times the investment multiplier K, i.e., $\Delta Y = \Delta I$. K. The only assumption made here is that the marginal propensity to consume (and, therefore, also the marginal propensity to save) is positive and constant but less than one, i.e., $0 < b < 1$. This assumption is necessary for the stability of the system. Since b is a *positive fraction*, it follows from equation (11) that the investment multiplier K, which is reciprocal of $1 - b$, will become higher and higher as the value of b approaches one. In the extreme case of the value of b being either zero or one, the investment multiplier would be either one or infinite. Since b is greater than zero but less than one, it follows that the value of the investment multiplier is finite being greater than one and less than infinite. Consequently, the change in income is greater than the initial given change in the aggregate autonomous investment outlay.

The multiplier analysis has so far been based on the assumption of a closed two-sector economy, i.e., it is assumed that there are only two sectors comprising the household units and business firms in the economy. In a three-sector economy comprising the household, business and government sectors, an additional term G will appear in the above equations with no change in the value of the simple investment multiplier. In an open four-sector economy, we have imports and exports as additional sector to be included in the national income equations. In the four-sector economy, the equilibrium national income will be that corresponding to which the aggregate demand—the sum of consumption, investment, government and net foreign spending—is equal to the aggregate income. This gives us the following equation for the equilibrium aggregate income in a four-sector economy comprising of the domestic household, business, government and foreign trade sectors

$$Y = C + I + G + (X - M) \tag{12}$$

Expressed differently, since $C + S + T = C + I + G + (X - M)$, the equilibrium aggregate income would be that corresponding to which

$$S + T + M = I + G + X \tag{13}$$

Where T = taxes and G = government expenditure, S = saving

Import and Export Functions

For an open economy, the income and output will increase from one time period to the next as its total exports increase or its total imports decrease because as a result of both these changes the economy's net exports expand. Conversely, the domestic economy's income and output will fall over time as its total exports decline or total imports rise as both these changes will tend to cause a fall in its net exports. It follows from this that the effect of imports and exports on the economy's equilibrium aggregate income and output inheres in the factors that determine the economy's imports and exports.

Generally, a country's total volume of exports depends on the prices of goods in the country relative to their prices in other countries, on the tariff and trade policies prevalent in the country and other countries, on the shortage or surfeit of foreign currencies in the foreign exchange markets, on income in other countries, on the own imports of the country, etc. Some of the more important factors that determine a country's exports are not directly related to the conditions within that country. Consequently, it is assumed that the gross exports of a country are entirely autonomously determined, i.e., exports are determined by the external factors.[1]

The volume of imports of a country is determined by similar factors. However, many of these factors are influenced by conditions within the country. *Ceteris paribus*, a country's total imports are determined by the level of national income. In other words, assuming given international price differences and unchanging tariff, trade and exchange restrictions, a country's imports are functionally related to her national income. This functional relationship is such that, *other things being equal*, imports will increase with the rise in the level of the economy's income. This will

---

1. This assumption will be subsequently relaxed when it will be shown that the domestic economy's imports may influence the domestic economy's exports.

happen because as the level of income in the economy will rise the aggregate consumption spending and possibly the investment spending will also increase. Normally, a part of the increased consumption spending will be made on the purchase of the imported goods and services. Consequently, in an open economy imports can be expressed as an increasing or a positive function of national income and the import function may be written as

$$M = M(Y), \text{ and } \frac{dM}{dY} > 0$$

For simplicity, we may assume a linear relationship between income and imports of the following form which defines the import function

$$M = M_a + mY \tag{14}$$

where M_a is the autonomous spending on imports and m is the *marginal propensity to import* (MPM). It shows that part of the change in national income (ΔY) is spent on imports. This marginal propensity to import ($\Delta M/\Delta Y$) is the slope of the linear import function shown in Figure 14.1. In the figure, we have shown the export function $X = X_1$ which is parallel to the X-axis showing that exports of the economy are autonomously determined quite independently of the level of income in the economy. The positively sloping import function $M = M_a + mY$ shows that while the M_a amount of total expenditure incurred on the imports is independent of the level of income, the balance of the total expenditure incurred on imports is m times of the level of income; m is the slope of the import function, i.e., it is the fraction of any given change in income which will be spent on the imports. Expressed differently, m = $\Delta M/\Delta Y$ and its value will be positive but less than one, i.e., $0 < m < 1$.

The upward or positively sloping linear import function $M = M_a + mY$ in Figure 17.1 shows the import-income relationship. At $0Y_1$ aggregate income, the imports are equal to exports. Consequently, $0Y_1$ is the equilibrium aggregate income because at this level of national income the total supply equals the total demand. Any upward or downward shift in either the export function or the import function or a change in the slope of the import function will cause a change in the level of the economy's equilibrium income as defined by the equality between the gross exports and gross imports.

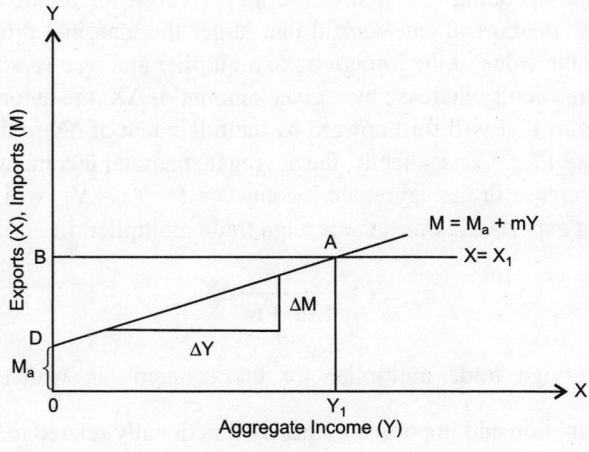

Figure 17.1

We have seen earlier that in a four-sector open economy the equilibrium income is given either by the equation $Y = C + I + G + (X - M)$ or by the equation $S + T + M = I + G + X$. Since our consumption is defined by the equation of the consumption function $C = a + b (Y - T)$ and imports are defined by the import function equation $M = M_a + mY$, by substituting for C and M the above equilibrium aggregate income can be written as

$$Y = a + b (Y - T) + G + X - (M_a + mY) \tag{15}$$

This may be rewritten as

$$Y = \frac{1}{1 - b + m}(a - bT + 1 + G + X - M \tag{16}$$

where the term $1/1 - b + m$ is the foreign trade multiplier for the economy in which exports are wholly autonomously determined while both consumption spending and import expenditure are linear functions of the level of national income. If taxes are assumed to be functionally related to the level of income so that the total tax function is $T = d + t\,Y$, the equilibrium national income would then be

$$Y = \frac{1}{1 - b + bt + m}(a - bT + 1 + G + X \tag{17}$$

where the term $1/1 - b + bt + m$ is the foreign trade multiplier for the system in which consumption, imports and taxes are all linear functions of the level of national income. Furthermore, if we treat investment spending also linearly related to the level of income so that the investment demand function can be written as $I = I_A + eY$, the equation for the equilibrium income would become

$$Y = \frac{1}{1 - b - e + bt + m}(a - bT + 1_A + \text{(} \tag{18}$$

in which the term $1/1 - b - e + bt + m$ is the foreign trade multiplier for the system in which consumption, investment, imports and taxes are all linear functions of the level of national income.

The foreign trade multiplier, sometimes also called the export multiplier, operates in exactly the same manner as does the ordinary investment multiplier. An autonomous increase in the exports (ΔX) causes an increase in the incomes of the exporters and the factors employed in the export industries who in turn spend a part of their increased incomes on domestic goods. The exact part of their increased incomes spent on the purchase of domestic goods will depend on the leakages in the form of savings and spending on imports. Neither savings nor import spending create new incomes in the country. In short, it can be said that larger the marginal propensities to save and import, smaller will be the value of the foreign trade multiplier and *vice versa*.

If the exports of the country increase by a given amount of ΔX, the autonomous export supply function $X = X_1$ in Figure 17.1 will shift upward by the full extent of ΔX to the position of $X' = X_1 + \Delta X$ as shown in Figure 17.2. Consequently, the aggregate national income will increase from $0Y_1$ to $0Y_2$ and the total increase in the aggregate income $\Delta Y\ (= Y_2 - Y_1)$ will be equal to the total autonomous increase in exports ΔX times the foreign trade multiplier, i.e.

$$\Delta Y = \frac{1}{1 - b + m}\Delta X \tag{19}$$

where $\dfrac{1}{1 - b + m}$ is foreign trade multiplier for the economy in which exports are wholly autonomous and consumption and imports are linearly functionally related to national income. The total increase in national income may also be written as

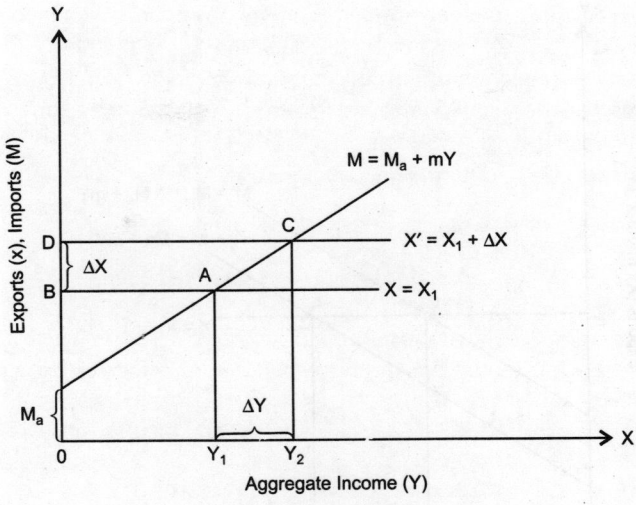

Figure 17.2

$$\Delta Y = \frac{1}{1-b-e+bt+m} \Delta X \qquad (20)$$

where $\dfrac{1}{1-b-e+bt+m}$ is the foreign trade multiplier for the system in which consumption, investment, imports and taxes are all linear functions of the level of national income.

A look at the foreign trade multiplier $1/1-b-e+bt+m$ brings home the fact that *ceteris paribus* the value of the foreign trade multiplier is inversely related to the value of the marginal propensity to import m such that higher m is associated with lower foreign trade multiplier and *vice versa*. Like the savings, imports also constitute a leakage from the aggregate spending stream in the domestic economy. Consequently, like the high marginal propensity to save (s), a high marginal propensity to import (m) tends to reduce the value of the foreign trade multiplier. In an open economy the leakages schedule is represented by S + T + M.

In the same way as an autonomous increase in the country's exports causes an expansion of the aggregate (national) income, an autonomous increase of ΔM_a in the country's imports will have a contractionary effect on the aggregate or national income as shown in Figure 17.3.

When due to an autonomous increase of ΔM_a the import function $M = M_a + mY$ shifts upward to the position of new import function $M' = M_a + \Delta M_a + mY$ the aggregate income contracts in the process falling from $0Y_2$ to $0Y_1$. The reason for this contraction in the aggregate income is explained by the fact that while the aggregate consumption outlay remains unchanged, the demand for the home-made goods declines leading to fall in the equilibrium aggregate income from $0Y_2$ to $0Y_1$.

Exports as a Function of Imports

The analysis has so far been carried on the assumption that exports are determined by external factors. Although this assumption is a valid approximation for small economies in which even a large percentage change in income will be relatively small in absolute terms, the assumption is nevertheless not valid in the case of a large economy like the United States of America, China or Russia where

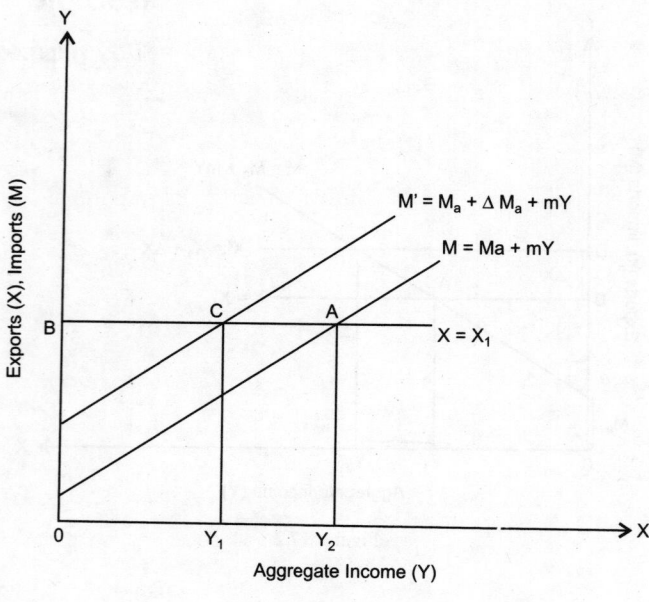

Figure 17.3

changes in the level of domestic income are a factor determining changes in the US, Chinese or Russian exports. In any country an increase in consumption, investment or government spending will lead, in the absence of any offsetting decrease in her net exports or increase in her net imports, to an expansion of domestic income in the economy. However, since the marginal propensity to import is positive, income expansion in the economy will cause an expansion in her imports. The increase in the gross imports of the country will mean an equal increase in the gross exports of rest-of-the-world with no simultaneous increase in the gross imports (since exports of the particular country, say Russia, did not increase initially in the first instance). This being the case, one or more countries in rest-of-the-world must experience an increase in their net exports and consequently in their domestic income. Since these other countries have also a positive marginal propensity to import, their imports too will increase. It is natural to expect that the particular single country (Russia, China or the U.S.A.) will secure a part of the increase in the purchases made by the rest-of-the-world countries, causing the Russian or the US gross exports to rise. In this analysis, changes in a country's gross exports are influenced by changes in her imports which are in turn influenced by changes in the level of domestic income.

This goes to suggest that world countries are interdependent such that rising or falling income levels in a large country like the U.S.A. will tend to raise or depress the income levels in the other countries. With the growing interdependence among the world's nations as a result of ever-growing foreign trade, countries experience international propagation of the trade cycle, i.e., the trade cycle generated in one country tends to spread elsewhere in the world. A depression in one or more large countries soon spreads its rot in the other countries. Conversely, prosperity in one or more countries also benefits the others in the world community of nations. In short, the foreign trade multiplier approach helps us in understanding the determination of national income in an open economy and the important fact that fluctuations in one economy spread to others. A serious study of foreign trade multipliers brings home the fact of interrelationships between world's nations and hanging together of national incomes.

Suggested Readings

Edward Shapiro, *Macroeconomic Analysis,* Fifth Edition, 1984, Chapter 7, pp. 119–123.
Bo Sodersten, *International Economics,* Second Edition, Reprinted 1981, Chapter 24.

Questions

1. Explain the concept of foreign trade multiplier. What is the importance of this concept in the theory of balance of payments adjustment?
2. Explain the concept of foreign trade multiplier with suitable examples.

CHAPTER 18

Exchange Control

Introduction

During and after the First World War, the world had witnessed a massive trend towards increasing state control over the economic affairs. This trend became more pronounced in the immediate postwar period. According to Paul Einzig, exchange control was "the dream of Socialists and Fascists in various countries to secure complete control over the international movement of funds in the interest of their political and economic plans."[1] The belligerent and neutral countries exercised strict blanket control over foreign exchange transactions during the war period of 1914–1918 and it remained in operation even after the war until 1926 although its extent was modified because many countries had stabilised their currencies. From 1926 to 1931 although the controls over foreign exchange transactions were largely removed but the prewar freedom of exchange was never restored. The crisis of 1931 stimulated resort to exchange control spurred by the great depression of the *thirties.* Subsequent events culminating in World War II once again intensified the movement towards foreign exchange control which developed into an elaborate world-wide institution with many forms and varieties.

The origin of exchange restrictions can be traced back to 1931.[2] Forced by the exigencies of the Great Depression of the *thirties,* several debtor countries of Europe imposed payments restrictions as they lacked adequate foreign exchange and gold reserves to fend off sudden and massive capital withdrawals. Currency depreciation was not resorted to due to the fear of the loss of confidence of foreigners and their own nationals in the stability of the domestic economic and monetary system and the painful memories of the concomitant inflation and depreciation after World War I. Besides, neither deflation nor exchange depreciation would have stimulated the exports of these countries to a level which was necessary to sustain the large scale capital withdrawals. Consequently, exchange restriction was the only way out of the impasse.

Full-fledged exchange controls were initiated in Germany in 1934 by Hjalmar Schacht in order to prepare Germany for war and to facilitate the expropriation of the Jews and to prevent them from taking their capital out of the country. The era of payments restrictions started at this time can be sub-divided into four periods. During the first period, extending over about eight years beginning

---

1. Paul Einzig, *Exchange Control,* p. 8.
2. In the immediate post-War I period only partial and incidental restrictions were imposed by certain countries to restrict free capital movements in order to avoid exchange depreciation. The general trend, however, was toward relaxations of exchange restrictions until 1931 when the restrictions were reintroduced more rigorously.

from 1931 to 1939, certain debtor countries of Europe and Latin America enforced exchange payments restrictions of which they had no previous knowledge. Since during this period in most countries international payments were free, application of exchange restrictions by some countries strained the harmonious relations between these countries and those others which did not apply such payments restrictions. Some countries used the device of exchange payments restrictions to carry out the contra-cyclical policies. Germany fruitfully exploited the technique of foreign exchange control to prepare herself for the global war.

The second period extending over six years beginning from 1939 to 1945 brought the multilateral payments system to a complete end. During this period, international economic relations were completely disrupted. The justification for government control was based partly on political and partly on war exigencies grounds. During the war it was essential to maintain strict control over the international balance of payments situation of a country. Payments restrictions became ancillary to economic warfare. After the cessation of hostilities it was considered both necessary and desirable to resume the severed international economic relations. However, there were countries which took to the policy of government controlled economic reconstruction of their economies. Many underdeveloped countries eager to build their economies through the process of forced capital formation imposed restrictions on foreign exchange payments. Inflation in the immediate postwar period prompted the governments of many countries to enforce foreign exchange restrictions on consideration of difficult external balance of payments situation. Consequently, during the third period beginning from 1945 to 1950 exchange restrictions became common.

The fourth period which began from 1950, witnessed a gradual relaxation of exchange restrictions. During the 1950s currency inconvertibility crumbled in the world and restrictions in industrial countries on imports from the dollar area dwindled. However, notwithstanding that exchange restrictions have been removed by many developed countries, we are still far from an era of a free international monetary system comparable to the pre-1914 gold standard. This is amply borne out by the findings of the IMF in its annual reports on exchange restrictions. The Introductory Note to the 1964 Report begins by observing that "in the past year, as in the immediately preceding years, the intensity of exchange restrictions differed broadly between the developed and the developing countries. The industrial and high-income countries were able to maintain their external economic relations with few significant limitations on the acquisition or use of foreign exchange for current transactions, while very many, though by no means all, low-income countries in the process of development found it necessary to continue to rely on such restrictions." This situation has not materially changed.

Meaning and Essential Features

By exchange control is meant any form of interference by the state, central bank or any other agency specially created for the purpose with the free play of the forces that affect the foreign exchange rate. It involves restrictions on international payments, exchange restrictions, and transfers of domestic balances among non-residents and the enforced surrender of exporters' earnings of foreign exchange. As a technique of intervention, exchange control, as it was enforced during the 1930s, had two special characteristics. Firstly, it enabled governments to influence not merely trade flows but also capital flows. Secondly, it supplied a financial ambience which provided an incentive for discrimination and for the balancing of international transactions on a bilateral basis. By means of exchange control, residents could be effectively prevented from exporting and foreigners from withdrawing their capital out of the country. An exchange control is designed to suppress or circumvent the market forces in the foreign exchange market. In a thorough-going exchange control system, all international economic and financial transactions of the country are legally subject to the prior control of the exchange control authority.

The aim of a full-fledged exchange control system is to establish complete control of the government over the foreign exchange market. Exporters must surrender all their foreign exchange earnings from exports to the government. To prevent evasion, exports are subjected to export licences which must be presented to customs officials before making shipment. All current foreign exchange receipts must be sold to the exchange control authority and all foreign exchange is purchased from it. Imports are subjected to import licences and foreign exchange can be obtained only from the central bank or other authorised agencies up to an amount stated in the import licence and for importing only those goods and only from those countries which are mentioned in the import licence. According to Haberler, exchange control is "the state regulation excluding the free play of economic forces from the foreign exchange market."[3]

The steps which a government may take to control foreign exchange transactions include the establishment of official rates of exchange which the residents of the country must pay to obtain foreign currencies, the rates at which they must sell whatever foreign currencies comes into their hands, the rates at which foreigners may be permitted to acquire domestic currency besides setting official rates influencing the same, the enforcement of regulation compelling the residents of the country to surrender to the government whatever foreign currencies they hold, allocation of limited supplies of foreign currencies between the applicants according to a certain pre-determined system of priorities, enforcement of restrictions on the use of domestic currency by foreigners, and the negotiation of agreements with other governments by which payments among them are arranged according to specific procedures, e.g., exchange clearing and payments agreements.

A thorough-going foreign exchange control establishes complete government control over the foreign exchange market of the country. Foreign exchange earned from exports and other sources must be surrendered to the exchange control authorities. Consequently, the making and receiving of all international payments become concentrated in government hands. To prevent the evasion of exchange control provisions export licenses, requiring delivery of foreign exchange earned through exports to the authorities, must be presented to the customs officials before shipments of exports are permitted. The available supply of foreign exchange is then allocated among the various buyers (importers) according to the criterion of relative national importance. Capital exports are frequently banned while interest and amortisation payments to foreigners are severely limited. Imports of only those goods which are essential for the functioning of the economy, e.g., foodgrains, petroleum products, capital goods and industrial materials are permitted. Luxury and non-essential goods do not figure in the imports of the country. Violators of foreign exchange control rules are punished with heavy penalties in the form of heavy fines and confiscation of shipments.

Objects or Purposes

Historically, foreign exchange controls have been instituted by the state for various purposes. One important reason accounting for the presence of exchange controls is the presence of centrally controlled economies and exchange control is merely an extension of central control to the international sector of the economy.

3. Gottfried Von Haberler, *The Theory of International Trade,* English Translation, Third Impression, 1950, p. 83.

Secondly, a country may resort to exchange control when she is faced with a massive deficit in her external balance of payments and does not want to leave the process of adjustment either at the mercy of automatic mechanism or fluctuating foreign exchange rates or on deflation. In the interwar period, exchange restrictions were mainly motivated by considerations of deficit in the external balance of payments. Since the highly disturbed conditions of international trade and capital movements coupled with the growth of rigidities in the internal cost-price structure impeded the correction of international balance of payments disequilibrium, countries resorted to foreign exchange control as the only easy way out of the impasse.

Thirdly, many countries were forced to seek recourse to exchange control in order to prevent the "flight of capital" from the country. A flight of capital is a concerted action by the owners of short-term securities and bank deposits in a country to convert their cash holdings into foreign currencies. A massive flight of capital if allowed to proceed unchecked, may completely exhaust a country's limited foreign exchange reserves. Destabilising and disequilibrating massive 'hot money' movements exert avoidable unduly heavy pressure on the country's limited foreign exchange reserves forcing the country to safeguard her national interests against such movements through imposing exchange control.

Fourthly, countries adopted exchange control to ensure availability of sufficient foreign exchange to enable the government to import essential commodities from abroad. After the outbreak of Second World War in 1939, the belligerent countries adopted exchange control to conserve their limited foreign exchange reserves because they wanted to restrict the use of these limited foreign exchange reserves for importing essential war materials. At present the developing countries are practising exchange control to conserve their limited foreign exchange reserves to import those capital goods which are essential for their accelerated planned economic development.

During the war exchange control had been used by the governments to prevent the use of purchasing power by the enemy nations and their subjects and agents residing in either the controlling or neutral countries. Such regulations were characterised by "freezing of assets" held in the controlling or neutral country by the residents of the enemy countries. Governments have also instituted foreign exchange control to acquire foreign exchange for foreign debt servicing and loan repayments. This was the object of many debtor countries in the thirties. Unable to float new foreign loans to refund their outstanding bonds these countries were compelled instead to rely upon exports for creating the necessary surplus in their external balance of payments in order to repay their external debt.

Yet another reason for employing exchange control has been the desire on the part of some governments to stabilise the foreign exchange rate between their currencies and the currencies of those other countries with which important economic and trade relations exist. For example, the members of the sterling area took action to maintain stability of their exchange rates in relation to the pound-sterling after the abandonment of the gold standard by England in 1931.

A country may resort to exchange control for many other reasons too. It may impose exchange control to overvalue its currency in order to obtain cheap imports of essential raw materials. Another reason for overvaluation of currency through exchange control may be the liquidation of country's external debt more cheaply in terms of the home currency. For example, the countries of central Europe after World War I had contracted large debts in the pound-sterling and the US dollars. They had to make large payments for payment of interest and principal in pound-sterling and dollars. They could pay their foreign debts easily if the exchange value of their currencies was higher than it would be under the free exchange market conditions. Consequently, Germany,

Hungary, Austria, etc. overvalued their currencies through exchange control to reduce their foreign debt burden.

A country may also take recourse to exchange control through undervaluation of her currency to increase her exports and decrease her imports. Here the object of instituting foreign exchange control is to reduce the foreign exchange value of the currency so that country's exports become cheaper and her imports are rendered costlier. Increase in exports followed by a fall in imports will have a favourable influence on the country's external balance of payments position. During the thirties many countries of Europe, including Germany, resorted to exchange dumping and multiple exchange rate practices to increase their exports and decrease their imports by depreciating the foreign exchange value of their currencies. A country may also employ foreign exchange control to avoid temporary fluctuations in her foreign exchange rate although in practice it is difficult to define and carry it out. In fact, only those fluctuations in foreign exchange rate need to be prevented which are purely temporary and are due to the speculative activities of the speculators in the country. It is, however, very difficult to say in advance what is a temporary fluctuation although looking back it may be easy to see what was and what was not a temporary fluctuation. This policy was pursued by England in collaboration with the Exchange Equalisation Fund between 1932 and 1939. A country may also institute foreign exchange control as part of the national programme for the country's economic development with the object of direct rationing of the scarce foreign exchange reserves.

METHODS OF EXCHANGE CONTROL

The different methods which a country may adopt to employ foreign exchange control may be broadly classified into direct and indirect methods. Among the direct methods mention may be made of intervention which may be employed by a government to hold the foreign exchange value of its currency either up or down (it takes the form of either 'pegging up' or 'pegging down' of the foreign exchange rate), exchange restrictions (e.g., blocked accounts, multiple exchange rates etc.), exchange clearing, transfer moratoria etc. The indirect methods of foreign exchange control include export subsidies, import tariffs, changes in the rate of interest when these are practised to influence the foreign exchange rate. The various forms of direct and indirect methods of foreign exchange control may be discussed below.

Intervention

The government may intervene in the foreign exchange market for either holding up or down the foreign exchange rate of its currency. In practice, however, recourse to intervention has generally been taken by the government of a country with the clear purpose to raise the exchange rate of its currency. This interference takes the form of bulk buying and selling of home currency either by the government or by the central banking authority acting under its direction in exchange for foreign currency in the foreign exchange market. The act of fixing the exchange value of the currency to some chosen rate of exchange is known as 'pegging'. 'Pegging' is the most frequent form of intervention. The importance of exchange pegging follows from the fact that in the absence of such pegging the free market rate of foreign exchange would be different from the pegged one. During 1914–18, the government pegged the dollar value of the pound-sterling at 4.765 US dollars. That at this pegged rate the pound-sterling was overvalued was proved in March 1919 when in less than an year of the withdrawal of government support the pound-sterling-dollar exchange rate fell from the pegged rate of £ 1 = $ 4.765 to £ 1 = $ 3.40. The New Zealand Government in 1933 by fixing the exchange rate between its currency and the British pound at £ N.Z. 125 = £ 100 had pegged her currency down.

While both 'pegging up' and 'pegging down' involve the maintenance of fixed foreign exchange rate, intervention does not necessarily mean fixed rate. Intervention may be practised by a government without bothering for pegging up or down the exchange value of its currency. Consequently, government might intervene through the purchase or sale of the home currency against the foreign currency in the foreign exchange market in order to support or depress the exchange value of the home currency without trying to evolve any fixed rate of exchange for it. Thus, intervention is not restricted merely to 'pegging' or 'pegging down' although these two are its most typical examples. When the government wants to 'peg' its currency to a higher exchange rate than the one which would prevail in the absence of intervention in the free foreign exchange market on the basis of the demand for and supply of foreign exchange it must reduce the supply of the home currency in the foreign exchange market by purchasing the home currency for foreign currency. "A government that is 'pegging'[4] its currency must be in a position to payout foreign currencies and receive its own currency; a government that is 'pegging down' its currency must be in a position to payout its own currency and receive foreign currencies; and both must be prepared to go on indefinitely unless they want either to resort to restriction or to fail in their purpose of controlling the rate of exchange. The ability of the government to control the exchange rate by intervention thus depends entirely on the resources which it can dispose for the purpose."[5]

The Exchange Stabilisation Funds established by England, America and France in 1932, 1934 and 1936 respectively, represented attempts by these countries to peg the foreign exchange rate of their respective currencies at a fixed rate through purchase and sale operations of home and foreign currencies. The ability of government or the exchange stabilisation fund to control the foreign exchange rate depended on its resources needed for the purpose. Since the 'pegging' of the currency required the government to sell the foreign currency at the pegged rate, the ability with which the government could achieve its purpose depended on its foreign currency reserves which were strictly limited at any given time. On the contrary, the government was better placed with regard to the sale of its own currency since it could print more currency notes in case of emergency but it could not do so with respect to foreign currency. Consequently, it was far more easy for a government or the exchange stabilisation fund to 'peg down' than to 'peg' the foreign exchange rate of the home currency.

Exchange Restriction

An exchange restriction refers to the policy whereby the government reduces the supply of home currency in the foreign exchange market. The exchange value of the home currency is maintained by reducing its supply. There are three features of an exchange restriction. Firstly, the government centralises all trading in foreign exchange either with itself or with its agent. Secondly, people have to obtain government's permission before offering the home currency in exchange for foreign currency. Thirdly, nobody can make foreign exchange transactions without obtaining prior permission of the government and the transactions can be done only through the government

4. "Pegging" is used for 'pegging up'. Since 'pegging up' is most common, in the textbooks the writers prefer to use the word 'pegging' instead of repeatedly using the relatively inconvenient term 'pegging up'.

5. Geoffrey Crowther, *An Outline of Money,* Revised Edition, 1948, Reprinted 1958, p. 249.

agency. In 1931 Germany and Austria had employed the exchange restrictions. In Germany non-compliance of the currency regulations was declared a crime punishable with death. During World War II, exchange restrictions were imposed by other countries including England and France. On September 21, 1939 England subjected its foreign exchange market to full state control and according to the government order gold holdings and certain other foreign currency assets acquired by the British residents were to be surrendered to the government in exchange for the home currency at the prevailing exchange rate. Exemption from these exchange control provisions was granted only after a careful review of each individual case. For every conversion of the pound-sterling into foreign currency official permission was required. A limit was also placed on the currency—domestic or foreign—which could be brought in or taken out of the country without government permission. Residents were required to sell to the Bank of England through the authorised agents all foreign currencies obtained by them from foreign countries.

There are different kinds of exchange restrictions with each differing from the other only in the degree of severity of its application. Thus, there are blocked accounts, multiple exchange rates and other types of exchange restrictions. The practice of blocking the accounts of creditor countries was adopted by the debtor countries, particularly by Germany, during the financial crisis of 1931.

The governments of the Central European countries which were heavily in debt, were faced with the problem of sudden repayment of short-term foreign loans. For some time they faced it at the cost of putting heavy pressure on their weak foreign exchange reserves but eventually restrictions had to be imposed on the withdrawal of foreign capital. The debtors in these countries paid their foreign debts into an account opened in the name of a foreign country with the country's central bank. The money paid in this account was blocked and could not be converted into creditor countries' currencies. The practice of 'blocked' accounts caused great hardships to the foreign creditors who could not use their money for any purpose in any other country. Many refugee Jews who had migrated to London to escape Hitler's prosecutions found themselves on the verge of starvation in London although they owned millions of marks that had been 'blocked' by the Nazi German government. The creditors of 'blocked' accounts consequently were forced to sell their blocked foreign currency holdings at very low price. Similarly, foreign exporters of Germany, Argentina etc. suffered much as a result of the resort to blocked accounts by these countries.

Multiple Exchange Rates

The multiple exchange rates were employed first by Germany followed by Latin American countries of Chile, Argentina, Brazil, Peru, Ecuador and others. Different exchange rates were fixed for the imports and exports of different goods. Even for the different categories of imports different exchange rates were fixed. The object of fixing different rates of foreign exchange for imports of different goods was to earn the maximum possible amount of foreign exchange by increasing the exports and reducing imports to the maximum possible extent. Different kinds of German marks—Travelmark, Askimark, Registermark, Blockmark, Sondermark, Handelmark, Degomark, Effektensperrmark—were sold in London at different prices ranging between 2 d. and 1 sh. 19d. per German mark. Until recently, Argentina was efficiently maintaining the highly complicated system of multiple exchange rates. Since she was not a member of the International Monetary Fund, she was free to adjust its different exchange rates for different purposes in such a way as to secure the maximum return for her exports and pay a relatively lower price for her imports. In the immediate postwar period nearly more than 50 countries used multiple exchange rates in one or the other form.

The advantage of multiple exchange rates to the country practising these rates lies in the fact that it abandons the need to employ quantitative restrictions and export-import licensing. The system of multiple exchange rates substitutes a system of rationing the demand by cost or price for rationing of imports by quantitative restrictions on supply. Multiple exchange rates generally encourage exports and discourage imports in the country and consequently promise a favourable balance of payments for the country adopting the policy of multiple exchange rates.

In spite of their weighty advantages, multiple exchange rates have many shortcomings. Multiple exchange rates introduce complexity and if either the exchange rates themselves or the commodities which are allowed to be imported or exported at each exchange rate are frequently changed, as is often the case, further uncertainty is created. Moreover, unless some positive preventive action is taken inconsistent cross rates are likely to emerge. From the point of view of international economic prosperity multiple exchange rates result in uneconomic utilisation of scarce domestic and world resources. In so far as multiple exchange rates tend to overvalue the country's currency, they amount to a tax on exports and subsidy on imports while in the opposite situation they have the effect of a subsidy on exports and a tax on imports. In either case, as a consequence of practising multiple exchange rates international trade is diverted towards unnatural channels. Multiple exchange rates obstruct the optimum utilisation of a country's resources and hamper the planned development of a country's economy. They increase the dependence of the country on foreign goods which could have been produced at home. The examples of Chile and Peru— countries which practised multiple exchange rates—are there to show the injury that multiple exchange rates did to the economy of these two countries. The scheduling of imports of essential foodstuffs at low exchange rates under multiple exchange rates greatly handicapped the development of Chile's agriculture. From being a net exporter of agricultural products in the 1930s, within a decade by the late 1940s Chile had become a net importer of these products. Although this shift was purely due to the growing domestic demand for agricultural products which could not be grown within the country, it was aggravated by the constantly rising imports of wheat and meat whose production at home was made unprofitable by the low import exchange rates applied to these products under the system of multiple exchange rates. Similarly, before practising multiple exchange rates Peru's domestic production of meat was sufficient for home consumption. When meat was permitted to be imported at low exchange rates the domestic production fell to less than half of the total domestic consumption. Ecuador had a similar experience with wheat flour.

Apart from causing the inefficient use of the domestic resources, the system of multiple exchange rates has a discriminatory effect on competitors. When a subsidy rate is introduced for exports, it applies only to a limited number of exports. Consequently, only a few competing countries suffer from its jolt. This amounts to a form of unfair competition and discrimination against the few selected countries.

Exchange Clearing Agreement

Exchange clearing agreements were practised by many countries in the interwar period, particularly during the "Great Depression" of the thirties in order to effect an arrangement, in the face of restrictive trade practices like exchange control and import quota, to boost the exports of a country. An exchange clearing agreement was adopted to increase the volume of trade between two countries by being not unduly worried about the scarce foreign exchange—a problem which was at the root of all restrictive practices in the interwar period. Through the fruitful device of exchange clearing agreement two countries regulate the making of payments between them. In both countries the importers pay the purchase price of the merchandise imported by them into a central account the proceeds of which are utilised to pay exporters of goods in the country.

In the 1930s in spite of the imposition of quota restrictions, exchange controls, export subsidies and multiple exchange rates many European countries found that the chronic deficit in their external balance of payments was so serious that it could not be removed by these restrictive measures alone. Many nations had become economically weak and financially bankrupt. The adverseness of their balance of payments could not be liquidated. The system of foreign lending had been abruptly stopped subjecting the balance of payments of debtor nations to unduly heavy strains. In such a grave situation, Germany evolved the novel device of exchange clearing agreement. Thus, one of the earliest forms of foreign exchange control developed into a system of bilateral clearing. When a country's foreign exchange reserves had been totally exhausted and she did not want to change the foreign exchange rate of her currency, her foreign trade shrank to a stage where it could be expanded through barter agreements between the governments. From this barter, there evolved a kind of clearing arrangement which did away with the necessity of dealing in foreign exchange.

Method of Operation—Under an exchange clearing agreement the balancing of the payments due to and the payments due from the traders in a country becomes the exclusive function of the central banks in the two countries. The central bank in country A opens an account with itself in the name of the central bank of country B. In country A there will be some persons who are creditors of country B and certain other persons who are debtors to country B. The debtors (those who have imported goods from B) deposit with the central bank of country A the sum due to them in the home currency in discharge of their foreign payments obligations. Similarly, the creditors (those who have exported goods to country B) receive payments from the central bank of their country in the home currency out of the proceeds collected by the central bank from the debtors (importers). Similar operations take place in country B in respect of the import and export transactions relating to country A.

By this arrangement the foreign trade between the two countries is made possible without the concerned countries being worried about the problem of finding scarce foreign exchange. In the countries which are party to an exchange clearing agreement the importers pay for their imports into a central account, these proceeds are then used to pay to the domestic exporters. Consequently, the conduct of foreign trade is left entirely in the hands of a central banking agency strictly controlled by the government. Moreover, an exchange clearing agreement works on the assumption that the countries concerned will do everything in their power to conduct their foreign trade in such a way that in the final striking of the payments balance sheet an equilibrium is maintained, i.e., the total value of a country's imports must be equal to the total value of her exports, leaving no necessity for their making net payments to or receiving net payments from the other country. To the extent this fact is ignored the purpose of the exchange clearing agreement is defeated.

The method of operation of an exchange clearing agreement may be explained by means of a concrete example. In 1929 Germany exported goods worth 627 million Reichsmarks to Switzerland while she imported goods worth 318 million Reichsmarks from Switzerland. Consequently, the balance of trade between the two countries was unfavourable to Switzerland. On invisible items and on capital account, however, there was a net balance of payments due to Switzerland from Germany to pay for the German tourists in the Alps and interest payments on Swiss loans made to Germany. But when all items, visible and invisible, were included, it was almost certain that

Switzerland made more payments to Germany than she received from her. Germany 'blocked' the payments of interest on Swiss loans. Switzerland gave a very effective reply in return. The Swiss Government passed a law compelling Swiss citizens who had payments to make to Germans to make those payments to the Swiss National Bank instead of paying to their German creditors. With these payments in its hand, the Swiss Government proceeded to threaten Germany that the payments would not be forwarded unless the payment of interest due to Switzerland by Germany was made. And since the payments that were due to Germany which Switzerland had seized were more valuable than the payments that were due to Switzerland on account of interest payments on old Swiss loans the threat proved most effective.

An agreement was concluded between the two countries by which all payments arising between them were to be offset. A Swiss owing money to a German was required to pay the money into the Swiss National Bank which paid it to another Swiss who was owed money by Germany, whether for tourists' expenses or for goods sold or for interest on past loans. Similarly, a German debtor to Switzerland instead of remitting the money to his Swiss creditor paid it to the Reichsbank which utilised it to pay to those Germans who had money owed to them by Swiss debtors (importers). All that passed between the two countries was the mere notification from one bank to the other that such payment had been made. Similarly, under the Hungarian-French exchange clearing agreement, the French importers paid francs to the Bank of France for the goods imported from Hungary and the Hungarian importers paid for the goods which they imported from France by

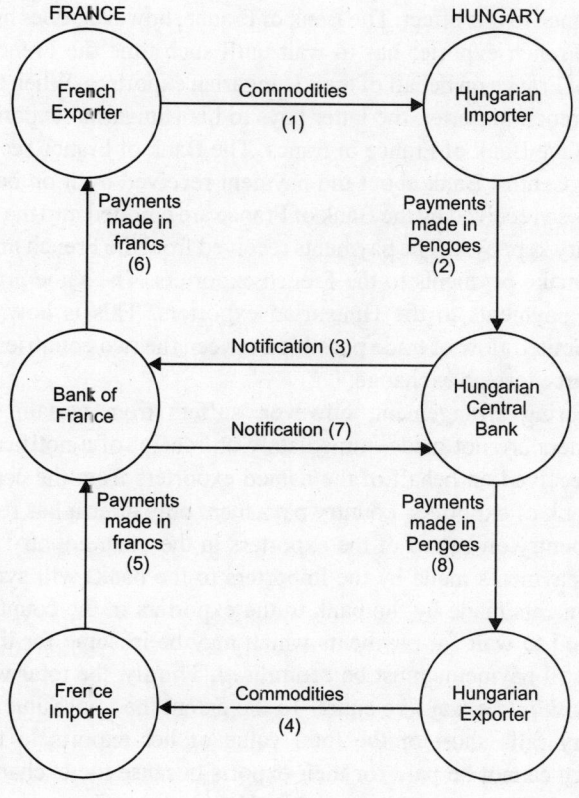

Figure 18.1

making payments in pengoes to the Hungarian Central Bank. Exchange clearing agreements in this way provided scope for carrying on foreign trade between the two countries without involving the use of foreign exchange. Many exchange clearing agreements also provided for the repayment of old debts and interest on these debts. Figure 18.1 illustrates the procedure of effecting the payments under an exchange clearing agreement.

When France and Hungary enter into an exchange clearing agreement, the payments made by the residents of one country favouring the residents of the other country are effected in the domestic currencies through the agency of the Bank of France and the Hungarian Central Bank. The central bank in each country opens a blocked account with itself in the name of its counterpart in the other country to which all money paid into it by the country's importers is credited. The central bank in the other country is notified about each payment which the central bank in the country receives on behalf of exporters of the other country. The exporters in each country receive the payments from the central bank in their country which the bank pays out of the payments made to it by their country's importers on behalf of their exporters of the other country.

In Figure 18.1, the chain starts when the French exporter, exports commodities to his Hungarian importer. The Hungarian importer does not make the payment for the imported goods to his French counterpart directly. He pays on his French exporter's behalf to the Hungarian Central Bank in pengoes. The Hungarian Central Bank utilises this money to make the payments to those Hungarian exporters who have exported goods to France. The Hungarian Central Bank on receipt of the payment from the Hungarian importer credits the same to the blocked account of the Bank of France and notifies the Bank of France to this effect. The Bank of France, however, does not immediately pay the French exporter. The French exporter has to wait until such time the French importers make the payments to the Bank of France on behalf of their Hungarian exporters. When the Hungarian exporter exports goods to his French importer, the latter pays to his Hungarian exporter not directly, but by making the payment to the Bank of France in francs. The Bank of France receives the payment and notifies the Hungarian Central Bank about the payment received by it on behalf of the Hungarian exporter. The payments so received by the Bank of France are credited into the blocked account of the Central Bank of Hungary kept by it. The payments received from the French importers by the Bank of France are utilised to make payments to the French exporters. The same procedure is followed in Hungary to make the payments to the Hungarian exporters. This is how an exchange clearing agreement makes the limited flow of trade possible between the two countries without involving the difficult problem of scarce foreign exchange.

An exchange clearing arrangement, however, suffers from certain shortcomings. Firstly, payments to the exporters are not made immediately on receipt of a notification to the effect that payments have been received on behalf of the named exporters from the central bank of the other country. The central bank of exporters' country pays them only after it has received payments from the importers in the country on behalf of the exporters in the other country. There is no guarantee that the in-payments (payments made by the importers to the bank) will synchronise in time with the out-payments (payments made by the bank to the exporters in the country). Consequently, the exporters may be forced to wait for payments which may be irksome for them. Secondly, for the success of the scheme all payments must be centralised. Thirdly, the total value of each country's exports and of imports should always be equal. To the extent the total value of a country's imports from the other country falls short of the total value of her exports to the other country, the exporters in the country cannot be paid for their exports because the exchange clearing institution receives less in the form of payments received from the importers in the country than it has to pay to the exporters in the country. At the other end, in the country with the total value of her imports more than the total value of her exports the exchange clearing institution will acquire surplus cash which will remain unutilised as the total in-payments received by the exchange clearing institution exceed the total out-payments made by it.

Exchange Clearing Agreements in Practice

Many countries in Europe resorted to exchange clearing agreements during the 1930s. While the basic principle in all these agreements remained the same, details varied from agreement to agreement. Sometimes the payments from country A to country B were so much larger than those from country B to country A that country A, after making all payments due to its citizens from country B, placed the balance at the free disposal of country B. In some cases, the amount of this free balance was specified from the beginning in the agreement. The underlying idea of exchange clearing agreements, however, was always the offsetting of payments so that the transactions would not have to pass through the foreign exchange markets.

During the 1930s, under the leadership of Dr. Colodius and Dr. Schacht, Germany exploited the bilateral exchange clearing agreements to her great advantage to borrow from the poor countries of south-east Europe. She bought essential-for-war materials and food articles from Rumania, Bulgaria, Yugoslavia and Hungary. She paid high prices in lei, leva, pengoes and dinars. The dominant exporter interests in these countries prevailed upon the central banks to finance their export surpluses. Large claims resulting from massive exports from these countries were accumulated in Germany to the credit of these exporting countries. These claims could be used only as Germany reluctantly made goods available to these countries from her rearmament programme at the prices dictated by Germany.

Germany also utilised the exchange clearing agreements to extend her political influence and economic domination over the countries of south-east Europe. These countries were the exporters of agricultural goods and the economies of these countries depended significantly on the exports of cereals. Germany bought large quantities of their food crops and forced these countries to take in return large consignments of the manufactured goods like aspirin and harmonicas at high prices. These German goods formed an important part of the total supply of manufactured goods in these countries and their high prices, combined with the high prices paid by the Germans for the crops they bought, caused the prices in these countries to rise. This cut them off economically from other countries because high prices made it difficult for them to compete in other world markets. So they had to do still more trade with Germany. Why, it may be asked, were these countries prepared to continue with a process that made them, year after year, more dependent on a country they feared? The answer is that for the bulk of their agricultural products Germany was either the only buyer or else offered attractive prices. In the countries of south-east Europe agricultural crops are the principal source of the national income and it was virtually impossible to reject the attractive German offers. The counterpart was the dependence on the German manufactured goods which these countries had to accept even though other countries' goods were of better quality and cheaper too.

Germany placed herself in a dominating position by exploiting her large monopsonistic power. There was virtually no end to the tricks that Germany played through the exchange clearing agreements on these countries. "For example, German firms sold bicycles to Rumanian peasants on hire-purchase terms, giving them years to pay. As soon as the bicycles crossed the frontier, an equivalent amount of Rumanian wheat or oil was released. But the German firms were, of course, paid at once out of the clearing account in Berlin, while the people who actually extended the long-term credit were the Rumanian exporters who were compelled to wait until the payments for the bicycles were paid into the clearing account in Bucharest".[6] Thus, a new way of paying for the oil or wheat was found by the Germans by forcing the Rumanians to find the necessary credit. The case of Germany proves the great extent to which the exchange clearing agreements can be manipulated by an aggressive country into a powerful weapon of economic warfare. Exchange clearing agreements are, however, capable of promoting world trade on a limited scale in the absence of the existence of an orderly international monetary system.

Although the exchange clearing agreement system is better than the pure exchange restrictions in that it permits expansion of world trade within circumscribed channels, it has many disadvantages. Firstly, it leads to the exploitation of the economically weak seller nation by the economically strong buyer nation. A country which sells goods to the exchange clearing agreement country has no other option than to purchase the goods from the same country in exchange of her own goods. It is likely that she might be forced to purchase unwanted goods. For example, Hungary had to purchase Swiss watches from Switzerland of which she had little use. At one time, Hungary was in fact flooded with Swiss watches. Exchange clearing agreements may be an excellent means for a country of purchasing goods to build her economy for war. Through the fruitful device of exchange clearing agreements Germany was able to purchase essential goods for her economy by giving only those goods which she could spare at her own terms with regard to the quality and prices of these goods. The bilateral nature of exchange clearing agreements diverts the world trade into unnatural channels reducing the total welfare from trade. Exchange clearing agreements favour those countries which are party to these agreements at the expense of other countries which are not party to such agreements. Exchange clearing agreements reduce the total volume of world trade and cause interference with the working of the foreign exchange markets and if used extensively they would ultimately suppress the customary institution of a free foreign exchange market. Exchange clearing agreements run counter to the basic principle of unconditional most favoured nation clause which safeguards the interests of free world trade.

Payments Agreement

Exchange clearing agreements suffered from two serious defects. To remove these difficulties, the exchange clearing agreements were replaced from 1945 onward by the *payments agreements*. The payments agreements solved the waiting problem for the exporters and the centralisation of payments which were inherent in the nature and working of the exchange clearing agreements. Under the exchange clearing agreements, payments made by the importers in Hungary did not by themselves guarantee that the French exporters were able to cash their claims immediately because the Bank of France might not hold sufficient cash balances to pay promptly the French exporters. This problem causing hardship to the creditors (exporters) is removed under a payments agreement because the French creditor (exporter) is paid as soon as information is received by the Bank of France from the Hungarian Central Bank that his Hungarian debtor (importer) has made the payment and *vice versa*.

Under a payments agreement delays in the matter of making payments to exporters are eliminated because both the countries establish mutual credit facilities. In the example cited, the French exporter receives prompt payment because the monetary authorities in France extend credit

6. Geoffrey Crowther, *Op. cit.*, pp. 268–269.

to the monetary authorities in Hungary. Since payments to exporters are promptly made hardships resulting from waiting are avoided. There is no longer any need to establish the order of in-payments and out-payments. There is also no longer any need for the centralisation of payments. Under a payments agreement, payments between the two concerned parties are made through the special non-resident accounts opened for that purpose. We can illustrate this by taking the payments agreement concluded between Belgium and the United Kingdom.

According to the Belgium-United Kingdom payments agreement, the National Bank of Belgium and other authorised Belgian banks kept the 'Belgian Accounts' in sterling with their correspondents in the United Kingdom. Conversely, the Bank of England and other authorised British banks kept the 'United Kingdom Accounts' in francs with their correspondents in Belgium. Under the terms of the payments agreement the central banks of the two countries undertook to buy each other's currency at an agreed foreign exchange rate. Consequently, the central bank in each country could obtain a balance of its partner's currency if at the same time it credited the partner's bank for a corresponding amount of its own currency. Each central bank could send a part of its holdings to its own authorised banks to enable them to make payments to the other country. Under the payments agreement, a payment made by a British resident to a Belgian was effected either in pound-sterling through the credit of the Belgian Account maintained by a Belgian authorised bank in England or in francs through the debit of a United Kingdom Account kept by a British bank in Belgium. The opposite happened when a Belgian made payment to a Briton.

The payments agreement nevertheless suffers from two defects. Firstly, the payments agreement accounts could be debited or credited for only licensed or approved payments as agreed upon by the governments of the two countries. Secondly, the balances in the payments agreement accounts could only be used for payments from one partner to another.

Transfer Moratorium

Under transfer moratorium, payments to foreign creditors or exporters are suspended during the period of moratorium. By taking recourse to transfer moratorium a country solves her payments problem temporarily. The importers and the debtors make payments in the home currency and this payment is deposited with certain authorised banks. On the expiry of moratorium period, during which the government puts its house in order by solving the foreign exchange problem, these deposits are released to the foreign exporters and creditors. During the financial crisis of the 1930s many countries in Europe had imposed moratorium on the transfer of funds by the nationals of the country to their foreign creditors and exporters.

Indirect Methods

Among the indirect methods of foreign exchange control mention may be made of import restrictions. Tariffs and import quotas are the chief instruments of indirect methods of exchange control. Although an import tariff seeking to restrict the flow of imports in the country in general and of the goods subjected to import tariff in particular, may have been imposed by considerations of foreign exchange shortage it is not always correct to assert that the duty on imports always constitutes an indirect method of exchange control. Before we can legitimately say so the object of imposing any particular import duty should be carefully studied. An import duty the object of which is to encourage the development of domestic industries in the country by providing them necessary protection against the imports of cheap foreign goods and an import duty imposed on revenue consideration do not fall within the scope of indirect methods of exchange control. Similarly, an export subsidy whose object is to encourage exports is not an indirect method of exchange control although when such a subsidy is allowed on the exports with a view to strengthening the foreign exchange reserves of a country, it will be an indirect method of exchange control. However, when export subsidies are given to stimulate the exports and simultaneously also to support the foreign exchange rate, such subsidies must be regarded as an indirect method of exchange control.

Changes in the interest rate also indirectly affect the foreign exchange rate. A rise in the domestic interest rate, *ceteris paribus,* attracts foreign capital. The nationals of the country also tend to keep their funds in their own country. As a consequence, pressure on the limited foreign exchange reserves of the country is reduced. Foreigners also export their short-term funds for investment in the country where the interest rate has risen. Germany attracted massive amount of foreign capital in this manner between 1928 and 1930. Due to the inflow of foreign capital in the country, the demand for country's currency increases and consequently the foreign exchange rate moves in its favour.

Indirect methods of influencing the foreign exchange rate are also applied generally for reasons other than the foreign exchange control. For example, import duties may be levied to protect home industries against the competition of cheap imports or the rate of interest may be reduced either to boost investment or it may be raised to control inflation in the country. In such situations, changes in the rate of interest cannot be designated as indirect method of exchange control since these are not initiated by foreign exchange considerations. Nevertheless, these indirect methods influence the course of the foreign exchange market. The effect, however, is indirect because they do not interfere directly with the working of the foreign exchange market.

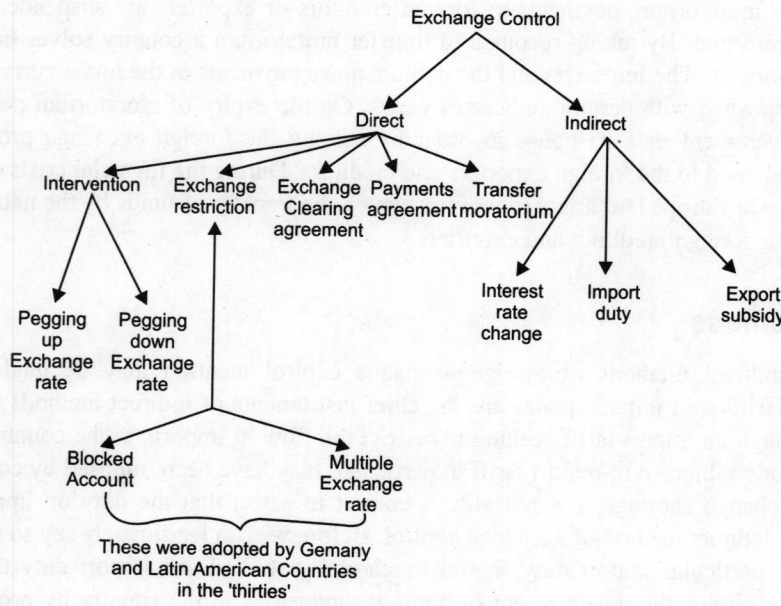

Figure 18.2

Indirect methods, however, have their own limitations. For example, if a country imposes a tariff on her imports and other countries also follow suit, the desired results of imposing an import tariff may not follow. Similarly, manipulation of the foreign exchange rate through changes in the bank rate may not be always effective. Figure 18.2 shows the different methods of foreign exchange control which have been employed by different countries at different times for serving the various purposes.

Progress towards Liberalisation

By 1955 most of the postwar difficulties had faded into the past and with them had gone the acute payments problem that had given rise to tight exchange restrictions. By mid-1956, the IMF observed that "foreign exchange restrictions impose a less serious obstacle to international commerce today than at any time since the outbreak of World War II."[7] The year 1958 witnessed the most signal achievement of the postwar period in the matter of relaxation of exchange restrictions because 14 West European countries agreed to establish external convertibility for their currencies.[8] Thus, by the early 1960s restrictions on trade and payments were considerably reduced for the world as a whole. The European countries which had established the external convertibility and defended it against something of a crisis in 1960-61 were soon joined by others.

Although important obstacles to international trade persisted but relatively few impediments remained to conducting world trade on the basis of a multilateral payments system. The successful maintenance of widespread convertibility also continued to provide the basis and incentive for further reduction in restrictions, especially those still prevailing on imports or on sending remittances abroad by residents of the countries concerned. Such restrictions were also made less discriminatory. Many countries reduced their restrictions considerably. For example, Australia, Austria, Belgium, Denmark, France, Germany, Greece, Iceland, Italy, Japan, Jordan, Luxembourg, Malaysia, the Netherlands, New Zealand, Norway, Portugal, Spain, Sweden, Thailand, Turkey, the United Kingdom and Yugoslavia had either removed completely or partially the payments restrictions.

Many countries even went further in reducing the payments restrictions than they were obliged to reduce under the Fund Articles. Although the Fund's Articles permitted controls on capital movements, liberalisation was gradually extended even to these capital movements. A view which had been dominant before 1930 began to gain ground, namely, that freedom of capital movements was highly desirable in itself and also that the movements of short-term funds might be the equilibrating factor in international payments diminishing the need for keeping the foreign exchange reserves.

Evaluation

Reviewing the exchange control as a method of removing the deficit in the external balance of payments of countries, we are led to the conclusion that exchange control has serious consequences both for the economy of the country practising it as well as for the rest-of-the-world economy. Exchange control is not in the best interest of an expanding world trade. It destroys the working of the

7. IMF, *Annual Report,* 1956, p. 89.
8. The 14 countries were: Austria. Belgium, Finland, France, Germany, Ireland, Italy, Luxembourg, the Netherlands, Norway, Portugal, Sweden and the United Kingdom. Greece did the same in May 1959.

free world markets which have slowly developed during the nineteenth and the twentieth centuries. It replaces the system of multilateral trading and multilateral convertibility of currencies by substituting in their place bilateral trading arrangements and currency inconvertibility and thereby substantially reduces the total gain from trade, both for the controlled as well as for the free regions.

Under exchange control, it is impossible for the buying country to purchase goods in the cheapest market. Similarly, it is difficult to sell them in the best market from the selling countries' point of view. Free access to world markets—an accepted principle of international trade—is denied to the individual countries both as seller and as buyer of goods. Autarky begins to dominate the world's economic affairs. Exchange control checks the natural flow of international investments which are essential for the optimum utilisation of world's searce economic resources.

Apart from reducing the gains from trade, exchange control tends to add directly to the cost burden of a country because a sizeable bureaucracy has to be employed to administer it. In short, exchange controls have hardly anything to commend them and are incompatible with the expanding world trade. They throttle the development of international trade and lending. The system of international private lending proves most susceptible to exchange control and restrictions. To the extent the flow of funds from the developed creditor countries to the developing debtor countries dries up under the impact of exchange restrictions, the process of development of the poor countries which depends on foreign investment funds is adversely affected. Unless exchange restrictions are kept moderate and are practised only as a temporary measure while other fundamental adjustments are made they will pose a serious threat to stable world peace. Consequently, the use of exchange controls must be avoided at all costs.

With the establishment of the World Trade Organisation (WTO) in 1995 and with its membership having risen to over 150, the era of non-discriminatory multilateral trading and payments system is well under way. It is expected that the system of foreign exchange control will become a phenomenon of the past.

Suggested Readings

E.M. Bernstein, "Some Economic Aspects of Multiple Exchange Rates", *IMF Staff Papers,* Volume I, No. 2, 1950.

Geoffrey Crowther, *An Outline of Money,* Revised Edition, Reprinted 1958, Chapter VIII.

Paul Einzig, *Exchange Control.*

H.S. Ellis, *Exchange Control in Central Europe,* Harvard University Press, 1941.

P.T. Ellsworth and J. Clark Leith, *The International Economy,* Fifth Edition, 1975, Chapter 21.

League of Nations, *International Currency Experience,* 1944, Chapter VII.

Margaret G. de Vries, "Exchange Restrictions: Progress Toward Liberalisation," *Finance and Development,* Volume 6, No. 3, September 1969.

Questions

1. Describe the objects and various methods of exchange control.
2. Explain the importance and effects of exchange clearing agreements and multiple exchange rates as methods of exchange control.
3. Explain the meaning of exchange control. Discuss the various objectives and methods of exchange control that were employed by the world countries to achieve these objectives during the thirties.

CHAPTER 19

Free Trade and Protection

FREE TRADE

Concept of Free Trade

A free trade policy is one which does not impose any restriction on the exchange of goods and services between different countries. According to Adam Smith, a free trade policy refers to "a system of commercial policy which draws no distinction between domestic and foreign commodities and thus which neither imposes additional burden on the latter nor grants any special favour to the former." Thus, a free trade policy permits international flow of goods and services without any artificial impediments. In other words, a free trade policy is necessarily a non-discriminatory trade policy. According to Haberler, however, a small extent of state interference is not incompatible with the concept of free trade. Consequently, "free trade is the external trade system of *liberalism,* which opposes every interference by the State with the free play of economic forces. But it by no means follows from this that it is inconsistent to advocate, on the one hand, unrestricted free trade, and on the other hand, certain interference with the free play of economic forces, for example in the labour market."[1]

According to Jagdish Bhagwati, free trade requires complete absence of all restrictions on the free movement of goods and services across national borders. Thus, a free trade policy involves complete "absence of tariffs, quotas, exchange restrictions, taxes and subsidies on production, factor use and consumption."[2] A free trade policy does not, however, require removal of all kinds of restrictions on the imports and exports. Consequently, under free trade revenue import duties may be levied not either to protect the domestic industries or to discriminate against the cheap imports.

In the eighteenth and nineteenth centuries, arguments for free trade were mainly based on the philosophy of individualism which found ample expression in the writings of political philosophers and economists John Locke, David Hume and Adam Smith. Adam Smith had severely criticised the mercantilist foreign trade regulations by showing that free trade between nations enables each nation to increase its wealth by making use of division of labour which opens the flood-gates of the gains from specialisation. Smith assiduously shows that the extent to which the advantage of division of labour can be reaped depends on the extent of the market. A wider market allows the division of labour and specialisation to be carried out to a fuller extent. By increasing the size of the market, free trade enables the world's nations to exploit the advantages of specialisation resulting

1. Gottfried Von Haberler, *The Theory of International Trade,* English Translation, Third Impression, 1950, p. 225.
2. Jagdish Bhagwati, *International Trade,* Penguin Edition, 1969, p. 5.

from division of labour to the fullest extent. Adam Smith has explained the advantages of free trade in the following classic passage of his magnum opus entited *An Inquery into Nature and Consite of the wealth of Nations.*.

"The natural advantages which one country has over another in producing particular commodities are sometimes so great that it is acknowledged by the world to be in vain to struggle with them. By means of glasses, hot beds, and hot walls, very good grapes can be raised in Scotland, and very good wine too can be made of them at about thirty times the expense for which at least equally good can be brought from foreign countries. Would it be a reasonable law to prohibit the importation of all foreign wines merely to encourage the making of claret and burgundy in Scotland?.. As long as one country has those advantages, and the other wants them, it will always be more advantageous for the latter to buy of the former than to make. It is an acquired advantage only, which one artificer has over his neighbour, who exercises another trade; and yet they both find it more advantageous to buy of one another than to make what does not belong to their particular trades."[3]

The advantages of free trade demonstrated by Adam Smith were further highlighted by his worthy followers, particularly by David Ricardo, John Stuart Mill and Charles F. Bastable. The most relentless supporter of the cause of free trade was, however, David Ricardo who propounded the principle of comparative cost advantage to support the cause of free trade. Thus, for about a century free trade enjoyed its unchallenged heydays. It was vehemently argued that free international exchange of goods increases the real incomes of all the countries participating in trade.

This is not, however, to deny the sporadic attacks on the principle and policy of free trade during the nineteenth century. In fact, a German national named Friedrich List had chided Adam Smith for his biases. According to List, free trade was the ideological weapon of the dominant country like England. List was, however, no match for Smith. Consequently, no serious harm was done to the cause of free trade policy.

The great depression of the thirties, however, struck a severe blow to the cause of free trade. People began to entertain serious doubts about free trade serving the best interest of all nations. Economists seriously questioned the assumptions underlying the classical theory of free trade. John MayneridKeynes who was brought up in the classical tradition parted the company of classicists and advocated the cause of development protection for the developing countries. However, even in the midst of confusion and change in ideas caused by the great depression of the thirties the free trade policy was never completely abandoned showing that it had a great vitality to sustain attacks.

Case for Free Trade

The supporters of free trade have advanced the following arguments in favour of free trade.

1. Maximisation of Total Output: Besides equity or maximum social welfare, free trade is defended on the ground of maximisation of social output. It is argued that the Smithian 'invisible hand' operating under free trade, makes it possible to exploit the advantages of division of labour and specialisation thereby enabling the countries to produce the maximum amount of aggregate output by allocating the world's scarce resources most efficiently between their rival uses. In free trade, the price mechanism works as a guide to investment and ensures that each country specialises in the production of those goods which she is best suited to produce and imports those

3. Adam Smith, *An Inquiry into the Nature and Causes of the Wealth of Nations,* Volume One, Everyman's Library Edition, 1910, Reprinted 1937, pp. 402–403.

goods which it can obtain more cheaply from abroad rather than produce herself. This increases the real incomes of all the participating countries.

Under the assumptions of perfect competition and absence of transport costs, free trade will equalise commodity prices between different regions eliminating any possibility of further gain from trading. Furthermore, the prices of different goods and services are equated to their marginal costs ensuring optimum production. Since the factors of production earn equal remuneration in every industry, apart from differences in their productivity, it ensures optimum allocation of scarce resources. "If social and private marginal values are everywhere equal to social and private marginal costs, society has reached an optimum of efficiency in the allocation of resources, the production of goods and the distribution of goods."[4] Thus, if the maximisation of total output is the desired end, free trade policy is scientifically justified.

2. Cheap Imports: The most appealing argument for free trade is the procurement of cheap imports since everyone as a consumer wants goods at lower prices. This argument, however, considers only the interests of consumers and ignores the important social issues of employment and the interest of producers. Free traders, however, answer this objection by pointing out that not only will the prices of goods and services fall causing the demand and employment to increase but, in addition, there will be a movement of factors of production to other parts of the economy inducing them to move into more gainful and specialised branches of production. Consequently, factor rewards—wages, interest, rent and profit—will increase under free trade.

3. Competition: Free trade guarantees against the establishment and development of the exploitative monopolies. By ensuring free competition, free trade safeguards the consumers against the monopolistic exploitation by producers. Experience shows that the restriction on free competition may lead to inefficient conduct of economic affairs. Free trade has an educative effect. Domestic producers are spurred by foreign competition to become more efficient and to adopt promptly any improvement in the technique of production regardless of where it was first adopted. Free trade does not, however, provide complete safeguard against the formation of exploitative monopolies and even under free trade there may emerge international and local monopolies which by curtailing the output and raising prices exploit the consumers. In any case, free trade, however, renders the establishment of injurious monopolies more difficult.

4. International Gold Standard: Free trade is compatible with international gold standard. Any international monetary standard depends upon the free buying and/or selling of the different currencies, i.e., upon the free convertibility of currencies. Multi-convertibility of currencies is not, however, possible in the absence of free trade. Thus, the multilateral convertibility of national currencies is essentially linked with free trade. The free trade system broke down in the *thirties* and this caused the collapse of the international gold standard. England went off the gold standard in 1931 followed by the U.S.A. in 1932 and by France in 1936. The international economic relations were disrupted as soon as the international gold standard had broken down. This led to political rupture culminating in the outbreak of the Second World War in 1939. Free trade is a prerequisite of international brotherhood and peace. Consequently, free trade presents a most convincing case for its restoration in the modem world.

5. Protects Interests of All Nations: Free trade safeguards the economic interests of all the trading countries. During the interwar period the problem of access to raw materials became very

4. Charles P. Kindleberger, *International Economics*, Fifth Edition, 1973, p. 190.

acute. Industrial countries like Italy, Japan, and Germany faced the acute shortage of raw materials. These countries were called "have-nots" contrasted with the "haves" countries which had a hold on the colonies having rich supply reserves of basic raw materials. The reason for this schism between the *haves* and *have-nots* was that during the "thirties" free trade had been disrupted and there were phases of a series of bilateral trade agreements. The world trade pattern was reversed. That was why Germany, Italy, Japan and other "have-not" countries cried for the redistribution of the colonies which were the rich sources of raw materials. Japan attacked China and took away Manchuria which is rich in coal, iron ore deposits, soybean, etc.

6. Optimum Use of World Resources: Free trade offers maximum scope for the optimum utilisation of world's scarce resources. It is compatible with the application of the principle of maximum advantage for every country. Under free trade, every country is enabled to sell her products in those markets where she gets the best prices for them and to purchase raw materials and other goods in the cheapest markets. Thus, under free trade a country enjoys full freedom both as the seller of her exports and the purchaser of her imports. Exploitation of one country by another is difficult since every country has numerous markets where she can sell her surplus goods and purchase those goods which she needs. Thus, free trade guards against the monopolistic and monopsonistic exploitation of one country by another country or other countries.

7. Economic Development of Underdeveloped Countries: Free trade can promote the economic development of world's underdeveloped countries in which world's more than one-half population lives. Gottfried Von Haberler mentions the following four ways in which free trade helps the underdeveloped countries to accelerate their rate of economic development.[5]

1. Free trade enables the underdeveloped countries to import capital goods and essential raw materials which are required for their economic development.
2. Free trade enables these countries to import the necessary capital, technical know-how, managerial talents, entrepreneurship etc. from the developed countries at the most competitive terms.
3. Free trade promotes international capital movements which assist the economic development of the underdeveloped countries.
4. Free trade promotes free competition and throttles the growth of injurious and inefficient monopolies.

These arguments for free trade are not, however, very convincing to the underdeveloped countries. According to these countries, free trade, as it has been in practice, is a colonial pattern of trade which has caused their exploitation by the developed countries. Consequently, it is not surprising that the underdeveloped countries have been raising their voice at various international forums such as the World Trade Organisation (WTO) and the United Nations Conference on Trade and Development (UNCTAD) for a thorough-going change in the present pattern of trade which protects the interests of the developed countries at the cost of the underdeveloped countries.

In addition to the above arguments for free trade, free trade also exerts its beneficial educative impact on nations. It discourages the spreading of the harmful spirit of autarky. Instead, it makes different nations realise that they are interdependent. It encourages international brotherhood and brings different nations together to solve their economic problems through mutual discussions held at international conferences in the give-and-take spirit.

5. Gottfried Von Haberler, *International Trade and Economic Development*, 1959, pp. 4–10.

Despite all its weighty disadvantages, free trade policy has not been abandoned by those who have already adopted it. Economic history shows that during the past two centuries international trade has flourished under the canopy of free trade and not protection. The WTO, which has succeeded the General Agreement on Tariffs and Trade (GATT) established in 1947, is committed to promote the development of nondiscriminatory free trade in merchandise and services through progressive liberalisation and elimination of tariff and non-tariff barriers to trade in goods and services, rejection of all forms of protectionism, elimination of discriminatory treatment in international relations and integration of the developing and least-developed countries and economies in transition into the multilateral trading system.

PROTECTION

The classical economists had championed the cause of economic liberalisation enshrined in the phrase *laissez-faire, laissez-passer.* They argued that trade between countries must be free from all restrictions with each country adhering to the open-door policy. It was argued that if one country follows the restrictionist policy while the rest of the world follows free trade policy, the restrictionist policy of the country will induce other countries also to take to the restrictionist policy. Consequently, tariff barriers would become ubiquitous and the volume of international trade would shrink. This would hinder the optimum allocation of the world's scarce resources leading to their uneconomic utilisation. As a result, welfare of the world community would not be maximised.

The classical argument for free trade leading to maximum total world welfare holds good under the static assumptions of full employment of factors of production and their free mobility within each country, equal economic growth of the countries which are trading partners, perfect competition, etc. Given these assumptions, if the resources are employed rationally in different branches of production, protection would divert the factors of production from more productive to less productive employment. Consequently, there would be a clear loss of the social net product and social welfare. It is, however, an ideal case which is never perfectly realised in practice. "There are many types of frictions and deviations from the ideal conditions (distortions) caused by monopolistic and oligopolistic imperfections of the market, external economies and diseconomies, price and wage rigidities, lack of information, irreversibilities of the various curves involved, etc. Each of these conditions may operate in such a way as to make certain deviations from the free trade policy rational on purely economic grounds."[6] During the great depression of the *thirties,* it was realised that the classical argument for free trade was far removed from the reality of life. The argument is valid only if national resources are fully employed but the case of an economy with its vast resources either underemployed or unemployed challenges the classical argument for free trade. If there are large unemployed factors in the economy and if protection absorbs these factors there will be no loss of total product (output) in the community as a result of giving up the free trade policy and resorting to protection because the total product will increase to the extent the unemployed factors are employed in the country. Consequently, in recent years protection has attained respectability in the underdeveloped countries where surplus labour is a common phenomenon.

---

6. Gottfried Von Haberler, "Some Problems in the Pure Theory of International Trade", *The Economic Journal,* Volume LX, June 1950, pp. 223–240.

Meaning of Protection

In common usage, the term 'protection' means a commercial policy which is adopted by a country to encourage domestic industry by shielding its high-priced products against the competition from cheap imports either by subjecting the imports to import duties so as to bring their prices at par with the domestic prices of import-competing goods or by restricting imports either by banning them altogether or by subjecting them to import quotas. The domestic industry may, however, be protected against the competition of cheap imports by subsidising the domestic producers equivalent to the difference between the domestic price of the product and the price of the imported good. Protection also includes the paying of export subsidies and manipulation of the foreign exchange rates for the purpose of favouring the domestic producers. A protective trade policy pursued by a country seeks to maintain a system of trade restrictions with the object of shielding a large portion of its domestic industry from the competition of cheap foreign goods. On the other hand, a trade policy that does not maintain a system of trade restrictions sufficient to protect the domestic industries and which does not pay export subsidies or does not manipulate the foreign exchange rates for the purpose of favouring the domestic industries is known as a free trade policy. According to Harry G. Johnson, the term 'protection' refers to those "policies that create a divergence between the relative prices of commodities to domestic consumers and producers and their relative prices in world markets."[7]

The earliest explanation of the modern theory of protection can be found in the famous Report on Manufactures submitted to the House of Representatives in 1961 by Alexander Hamilton (1757–1804) who was the first Secretary of the United States Treasury. Hamilton, who had carefully read and thoughtfully considered Adam Smith's book *The Wealth of Nations,* made out a convincing case for protection as a national commercial policy for the United States. Hamilton recognised the advantages of the division of labour and of free trade. He argued in favour of protection on the ground that protection would result in the extension of the division of labour in the United States. Differing from Adam Smith on his doctrine of free trade, Hamilton explained that Smith's doctrine was a theory of universal free trade which favoured England's ends. After Hamilton, came Friedrich List (1789–1846) who, deeply influenced by Hamilton's views, pleaded in very strong terms for protection which he considered necessary for Germany's industrialisation. In the early nineteenth century there developed a strong school of protectionists under the able leadership of Henry Charles Carey in America. Carey defended protection on the plea that it led to "diversification of employment while specialisation under free trade compels the whole population to employ themselves in scratching the earth, in the carriage of merchandise or in the work of exchange". The ideas of Hamilton, List and Carey have spread throughout the world over time and have developed into a mighty creed that has led to the development of a strong protectionist school of economists which champions the cause of protectionist trade policy in the world.

Although all the arguments that have been advanced in support of protection are not so simple and clear-cut as are the arguments for free trade, we may nevertheless examine some of these arguments in detail. These arguments may be grouped under the following three broad categories:

A. Fallacious Arguments,

B. Non-economic Arguments, and

C. Economic Arguments.

7. Harry G. Johnson, "Optimal Trade Intervention in the Presence of Domestic Distortions", published in *Trade, Growth and Balance of Payments,* 1965, pp. 3–34, reprinted in Jagdish Bhagwati (ed.), *International Trade,* Penguin Edition. 1969, p. 187.

A. Fallacious Arguments

Although according to the supporters of free trade all the arguments for protection are fallacious, we will, however, characterise only those arguments as fallacious which being scientifically untenable do not merit serious attention. Consequently, they do not strengthen the case for protection as a scientifically just and analytically sound trade policy.

1. Pauper Labour Argument: One of the most talked-about arguments for protection in the industrially developed countries is based on the fallacious ground of safeguarding the interests of labour in these countries. It is the famous sweated labour argument advanced by the Americans and the Europeans that in the absence of protection the high-paid workers in the industrially advanced countries would face the hazards of competition from cheap foreign labour. Thus, while the Americans cry for protection against the cheap foreign labour of the Continent, one often hears the cry for protection against the superior efficiency of the American labour in the developing countries.

The misleading nature of this contention has been explicitly stated by many economists. It has been included among the particular arguments for protection which according to Haberler do not need any serious thought. The alleged fear of lower foreign wages leads to the foolish idea that tariff should be imposed in order to counteract the lower costs of foreign countries. Taussig has also stated that "perhaps most familiar and most unfounded of all is the belief that complete freedom of trade would bring about an equalisation of money wages the world over .... There is no such tendency for equalisation." Haberler dismisses the argument by stating that it is an argument among the category of those arguments that do not merit serious discussion. "An equalisation of wages comes about only if labour is mobile and can move from districts where wages are lower to districts where they are higher."[8]

As opposed to Taussig and Haberler who deny any pull-down influence of free trade on high wages in a country, there is Bertil Ohlin's view which has aroused renewed interest in the problem. According to Ohlin, it is the differences in the proportion of the productive factors between different countries that give rise to international trade. A country exports those goods which it produces with its relatively abundant factors of production and imports those in the production of which its relatively scarce factors figure prominently. As a consequence of this shift in a country's productive factors toward the increased production of those goods in which the abundant factors dominate, there will be a tendency toward factor price equalisation between the trading countries. As a consequence of trade in such a situation, the prices of scarce factors in one country will fall to the level of prices of abundant factors in the other trading country. Assuming the total amount of productive factors in the country to be fixed, Ohlin's theory suggests that international trade will reduce the relative share of the scarce factor in the real national income. This reinforces the claim of the Americans for protection against the Continental and Asian sweated labour.

The immediate effect of lifting protection from an industry hitherto enjoying it will be against the workers' interest in that industry. The neo-classical economists, however, argued that it had only short-run effect. The long-run effect of free trade on workers' incomes would be beneficial. For example Taussig writes: "The free trader argues that if the duties were given up and the protected industries pushed out of the field by the foreign competition, the workmen engaged in them would find no loss in well paid employment elsewhere. It is possible that as a result of free trade the money wages of the workers might fall but such a fall in their money wages will be more than made up by a consequent greater fall in the prices of goods so that as a result of the removal of tariffs the

8. Gottfried Von Haberler, *The Theory of International Trade*, Third impression, 1950, p. 251.

real wages of the workers will go up. And it is only the real and not money wages that matter". According to Taussig, "the question of wages is at bottom one of productivity. The greater the productivity of the industry at large, the higher will be the general level of wages."

Bertil Ohlin, however, holds that under certain circumstances it is quite possible for free trade to reduce the standard of living of the manufacturing labour class. According to him, "if manufacturing and agricultural labourers form two non-competing groups, high protection of manufacturing industries may raise the real wages of workers in these industries at the expense of the other." Here Haberler agrees with Ohlin and he does not deny the possibility of the painful effects of free trade on the relatively specific factors in the short period. Thus, he states: ". . . in the short run specialised and immobile group of workers like the owners of specific material factors may suffer heavy reduction in incomes, when for one reason or another they are faced with more intense competition."

The question, however, is that if small and specialised factors can be harmed by free trade in the short run then will not the large and mobile factors like labour be harmed in the long run? Haberler is doubtful about this. Holding the contrary view he writes: "We may conclude that in the long run the working class as a whole has nothing to fear from international trade since in the long run labour is the least specific of all factors. It will gain by the general increase in productivity due to the international division of labour and is not likely to lose at all seriously by a change in the functional distribution of the national income." Haberler, however, qualifies his above statement by stating that labour may possibly be harmed by free trade if it enters more intensively in the protected industry.

Jacob Viner is more clear and emphatic than Haberler in this respect. He sees no reason why Haberler should have subjected his statement to a qualification. According to Viner, free trade cannot harm the vital interests of labour. Consequently, he states that "even if labour on the average had low occupational mobility, and were employed relatively heavily in the protected industries, its real income might still rise with the removal of tariff protection if it was an ever important producer of the hitherto protected commodities and if the prices of these commodities fell sufficiently as a result to offset the reduction in money wages in the new situation".

According to Stopler and Samuelson, there is some truth in the pauper labour argument for protection if labour is a scarce factor. According to them, "international trade necessarily lowers the real wage of the scarce factor expressed in terms of any good."[9] Jagdish Bhagwati has pointed out that protection would not necessarily always raise the real wages of the protected factors. Whether or not it would raise them would depend upon the internal relative prices of the goods produced by the scarce factors. Accordingly, he has stated that "protection (prohibitive or otherwise) will raise, reduce or leave unchanged the real wage of the factor intensively employed in the production of a good accordingly as protection raises, lowers or leaves unchanged the internal relative price of that good."[10] He further points out that if tariff is non-prohibitive, the complication arises from the revenue earned by the government. If the revenue is so distributed that it benefits the scarce factors then they will derive income both from the real wage in the employment and the distributional proceeds of the tariff revenue.

It is, therefore, obvious that the pauper labour argument for protection is not based on any sound ground. Rejecting the pauper labour argument, Ellsworth has stated that "it is possible to advance it seriously only if one is completely ignorant of both the principles and the facts of

9. W.F. Stolper and Paul A. Samuelson, "Protection and Real Wages", *The Review of Economic Studies,* Volume 9, 1941, pp. 58–73; reprinted in Jagdish Bhagwati (ed.), *International Trade,* Penguin Edition, 1969, p. 257.
10. Jagdish Bhagwati, *Op. cit.,* p. 281.

international trade. As for the facts, every day, year in and year out, the products of high-wage American labour are sold abroad in competition with goods made by low-paid workers. High wages are clearly no bar to low-cost production, at least in many important lines."[11]

High-wage labour can compete without difficulty with low-wage labour for two reasons. Firstly, labour is not the only factor of production, being always used in combination with capital and natural resources. Consequently, high wages do not necessarily result in high costs of production, particularly if labour is used along with a large amount of capital and natural resources in the production of goods. Thus, high wages will not cause any significant increase in the prices of capital-intensive goods. In other words, low-wage countries command an advantage over the high-wage countries only in those goods whose production is labour-intensive so that wages constitute a large part of per unit cost. The remedy for this is that the high-wage countries should leave the production of such goods for the low-wage countries. A high-wage country can find enough number of other goods whose production being capital-intensive, the country will not suffer from any disadvantage. Secondly, even if labour was the only factor of production, a high-wage country could still compete with a low-wage country in those lines of production where its relative productivity was higher than its relative wages.

2. Expanding the Home Market: It is argued by the protectionists that tariffs cause expansion of the home market by reserving the home market for the domestic goods. If a country's manufacturers are protected they will expand the home market for agricultural products. The rise in agriculturists' incomes will generate additional demand for the manufactured goods. Consequently, the whole market will be expanded and the size of the home market for the products will become larger.

This argument is, however, criticised on the ground that while the home market indeed expands, simultaneously the export market contracts. Thus, the home market is substituted for the world market and protection may not bring any net gain through expansion of the home market. Keynes has settled this problem by stating that "if protection merely means that under this system men will have to sweat and labour more, I grant their case. By cutting our imports we might increase the aggregate of work; but we should be diminishing the aggregate of wages. The protectionist has to prove not merely that he has made work, but that he has increased the national income. Imports are our receipts and exports are payments. How, as a nation, can we expect to better ourselves by diminishing our receipts."[12]

3. Keeping Money at Home: It is fallaciously argued that sending money out of the country, for the purchase of imports will make the country poor. This fallacy is falsely attributed to Abraham Lincoln who had uttered: "I do not know much about the tariff but I know this much that when we buy manufactured goods from abroad we get the goods and the foreigner gets the money, when we buy the manufactured goods at home we get both the goods and the money." This is reminiscent of the old mercantilist philosophy and has been criticised by Beveridge who has said that "it has no merit; the only sensible words in it are the first eight words."[13] In international trade goods pay for goods and money (gold) moves from one country to another only to adjust the balance of payments disequilibrium. The currency of one country is useless for the other unless the country spends it on the purchase of goods in that country.

11. P.T. Ellsworth and J.C. Leith, *The International Economy,* Fifth Edition, 1975, p. 53.
12. J.M. Keynes, *The Nation and Athenaeum,* 1st December, 1923.
13. Sir William Beveridge, *Tariffs: The Case Examined,* 1931, p. 27.

4. Equalising Cost of Production: Another similar argument for tariff is that it is a suitable instrument to even out the existing gap between the domestic cost of production and foreign cost of production. The scientific tariff is one which equalises the cost of production at home and abroad. The argument has an engaging appearance of fairness since it says "no favour, no under rates. Offset the higher expenses of the American producer without being under a disadvantage and then let the best men win."[14] This kind of tariff is, however, necessarily discriminatory. It picks out the least efficient of a country's industry and injures the most efficient one of the other. Rejecting this argument for tariff, Samuelson has stated that ". . . it is widely regarded by economists as a tissue of nonsense which, if taken seriously, would wipe out all trade and all the benefits of trade".

5. Full Employment: Tariff is considered as the most powerful means of solving the problem of unemployment in a country. The existence of unemployment in an industry is regarded as a good ground for the imposition of tariff. Provided that the products of the protected industry compete with similar imported goods and provided also that the demand for such goods is less than perfectly elastic, undoubtedly unemployment in one single industry will be reduced as a result of the imposition of tariff on the competing imports. The short term effects of tariff in the country in the form of reducing unemployment in the country are more likely to prevail than the long term effects. It is likely that in the short-run the imposition of a new tariff may considerably reduce unemployment in the newly protected industry. If goods are imported in large quantity, the imposition of import duty will have an expansionary effect on domestic production and some unemployed workers will be absorbed in the new jobs.

It is, however, argued by free traders that tariff will not reduce total unemployment in the country. Since exports pay for imports, fall in imports as a result of tariff will cause an equal fall in exports. Additional employment created in the economy as a consequence of import substitution will be neutralised by an equal or probably greater unemployment caused in the shrinking export industries.

This argument is, however, criticised on two grounds. Firstly, it is argued that a diminution in imports will not cause a pari passu diminution in exports. Secondly, even if it is granted that exports fall off almost immediately, this does not necessarily mean any fall in employment in the export industries. The purchasing power hitherto spent by the home consumers on the imported goods and subsequently spent by the consumers on the goods produced in the country may in turn be used by the recipients to purchase the goods of their own export industries. This being the probable situation, the new home demand for the products of the export industries may replace the previous foreign demand. The export industry now becomes complementary to the protected industry instead of depending on the foreign markets.

This is, however, over-simplifying the matter. It is not always correct to presume that the demand for the newly employed workers will always consist exclusively of those goods which were formerly exported. If this is correct, then it follows at once that there will be some unemployment in the export industries as a result of tariff although new employment will be created in the protected industry. How then are we to decide the issue? The fact, however, is that the new home demand caused by the creation of fresh employment avenues in the protected industry will reflect itself in one or the other direction— in the increased demand for this or that product. To whatever branches of industry the new demand flows, unemployment will fall. The fall in unemployment will offset the

14. Frank William Taussig, *Free Trade, the Tariff and Reciprocity,* 1920, p. 134.

fall in employment in the export industries. In addition, there will be a net fall in unemployment to the extent of employment created in the protected industry.

It is not always that all the factors which find employment in the protected industry may be previously unemployed. It is likely that some or most of them may have been diverted from other industries and in so far as labour or other factors of production have been diverted from other home industries, there has been no increase or diminution in employment. But notwithstanding all these arguments, it is likely that when due to levying a new protective duty the previously unemployed workers are absorbed in the newly protected industry, the total unemployment in the country might diminish. The extent of diminution in total unemployment in the economy due to tariff will depend upon (a) the extent to which the newly protected industry provides new employment, and (b) the extent of resulting increase in the price of the product.

The long-run beneficial effects of tariff on unemployment are open to more serious challenge. Its favourable effects on employment can at best be expected in the short run. In the long run, a tariff is of little or no practical use. In appraising the long-run effects of tariff on unemployment, it is necessary to distinguish between three broad types of unemployment and the effect of tariff on each type of unemployment, i e., on the (a) frictional unemployment, (b) cyclical unemployment, and (c) permanent or chronic unemployment.

(a) Tariff and Frictional Unemployment: In every country there is always some unemployment due to frictions in the system. For instance, due to inertia or lack of sufficient knowledge on their part about the availability of job opportunities, workers may not be able to locate the industries offering them employment. Thus, while there may be some firms or industries in the economy which may be winding up their business rendering the workers unemployed there may at the same time be other firms and industries offering fresh employment to the unemployed workers. The workers may, however, fail to avail of the employment opportunities due to want of knowledge or lack of mobility on their part. The shift from one job to another seldom takes place without involving some time lag. This is especially true when the change over from one job to another involves the migration of workers from one district or region to another district or region. Although temporary in nature, frictional unemployment is present in every economy. Sometime it is possible to remove the frictional unemployment by means of tariff. When the domestic industry is facing growing pressure of competition from cheap imports and threatens to cause unemployment, it is possible to diminish or avoid such unemployment for the time being by means of an import tariff. But solving the problem of unemployment by means of tariff every time competition from cheap imports appeared, entails great sacrifice on the part of the nation. The nation has to sacrifice the immense gain of international division of labour and of technical progress taking place in other countries for a mess of pottage. The policy of seeking to fill up a gap by means of tariff involves the abandoning of all the gains of technological progress in as much as such gains accrue to a nation through international division of labour.

(b) Tariff and Cyclical Unemployment: Cyclical unemployment accompanies those rhythmic swings of trade which are typical of every free economy. The cyclical unemployment can be reduced by means of tariff. In fact, every depression brings with it a wave of tariff increases. For example, tariffs were imposed by many countries during the great depression of the 'thirties' as a means to reduce the cyclical unemployment. It is argued that by isolating a country from world economy tariffs render it immune from crises. Experience, however, shows that the highly protectionist countries such as the U.S.A., Germany and France are no more immune from depression than are the free trade countries like the United Kingdom.

(c) Tariff and Chronic Unemployment: Tariff can remove chronic unemployment which has been prevailing in England, Austria, Germany and elsewhere since the First War. But such permanent unemployment is always the result of high real wages. Consequently, a very high tariff will have only an adverse effect. This being the case, unemployment will be reduced at a very high sacrifice on the part of the domestic consumers. If permanent unemployment is concentrated in the export industries, tariff will prove of no avail because import restrictions will never help to diminish unemployment in the export industries. In such a case, tariff will accentuate rather than reduce the unemployment. Keynes advocated protection to fight the permanent mass unemployment for an economy which was neither in equilibrium nor in sight of equilibrium. If unemployment in the country is general and permanent, the solution lies in either reducing wages in the country or in adopting the technological progress in order to raise the marginal productivity of labour. If, however, unemployment is cyclical it will disappear on the onset of recovery which must follow the depression. The recovery may, however, be accelerated by pursuing a policy of wage reduction. If unemployment is sectoral–in any particular branch of industry—the country must wait till the transition to full employment comes about in a natural way. Unemployed workers may be trained in job requirements in order to enable them to find jobs in the other sectors of the economy.

6. Retaliation Argument: Tariff is also used as a weapon to beat down foreign duties that affect a country's exports adversely. Such a tariff is called 'retaliatory' tariff. A retaliatory tariff is imposed on the plea that a country which is all around surrounded by highly protectionist states, cannot follow the free trade policy without injuring her economic interests. In a world of mutual give and take, a free trade country is always in a weak bargaining position since it has nothing to offer to its protectionist neighbours to induce them to lower their tariffs on her goods. It is also argued that one-sided free trade is harmful and a solitary free trade country finds herself helplessly delivered into the hands of foreign competitors unless it can successfully retaliate the tariff moves of the other countries on its products.

The mere fact that other countries have imposed tariffs makes, however, no case for tariffs by a particular country. Thus if, for example, one country hinders our export trade then by so doing she not only injures us but injures herself as well, thereby reducing the balance of trade and balance of payments on both sides. In such a situation, if we retaliate by imposing tariff, we only help to accentuate the injury both to ourselves and to the other country without reducing in any manner the injury inflicted on us by the tariff moves of the latter. The validity of the free trade argument does in no way depend upon the absence of foreign tariff and the often repeated contention of the inability of a country surrounded by a ring of hostile tariff barriers to maintain her balance of payments in equilibrium does not deserve serious attention.

B. Non-economic Arguments

It is not possible to draw any sharp dividing line between economic and non-economic arguments for protection. The distinction lies exclusively in the desired ends. The so-called non-economic arguments are best conveyed by phrases like national defence, self-sufficiency, nationalism, etc. These arguments are not based on rational ground of national benefit but are based on political considerations.

I. Defence: The supporters of protection argue that economic independence is of vital importance for the development of a nation. A nation should least depend on others for her defence requirements. The industries producing military equipment must be developed in the country since one never knows who one's allies will be in the next war. It is, therefore, unwise to have sources of supply of essential war materials located in another country. Even Adam Smith, who stood for free trade, had uttered: "Defence is more important than opulence". Thus, industries supplying essential military requirements, unable to face competition in peace time, should be protected even if they involve a heavy economic loss to the nation. The higher price which the consumer pays in peace time is his insurance against a war time shortage.

But in the modern economy, with a high degree of horizontal and vertical linkages it is too difficult to draw any precise dividing line between military and non-military industries. There is hardly any industry which in some way or other does not produce commodity which is directly or indirectly not required to maintain the industry producing military equipments. Thus, every industry, whether it is an inefficient cheese industry, candle industry or radar industry, will exaggerate its own importance to get protection under the national defence label. Politicians who want to "do something" for a given area are there to help them. The whole economy will, therefore, need protection. This will bring nothing but inefficiency since national resources will be diverted from more efficient to less efficient uses. And inefficiency does not serve defence. However, in the matter of defence, economic arguments do not hold good because security of a country is of utmost importance. Freedom should be protected at all costs. To avoid the undue pressure of politicians, a committee of experts might be appointed by the government to draw a logical dividing line between the essential and the non-essential industries.

2. Preservation Argument: Protection had been advocated as a means of promoting the growth of some specific industry or the desire to preserve a certain way of life isolated from foreign influences may be so strong that a country is willing to pay certain economic price for the attainment of this social objective. For example, if India wants to preserve the Indian handloom weavers as a token of her traditional way of life she should impose heavy import duty on the cheap foreign mill-made textiles. In France, protection against wheat imports provided in the 1880s and 1890s was motivated by the urge to preserve the family and family farm. But such protection involves the willingness to forgo potential real income. Thus, this argument for tariff is non-economic and is largely based on the sentimental feelings.

C. Economic Arguments

Following are some of the strong economic arguments for protection.

1. Distortions in the Domestic Market: According to many economists, the ideal conditions for free trade are never realised in real life. Frictions and deviations operate in such a way that the Paretian optimality is not achieved. The existence of domestic distortions such as the externalities in consumption and production, monopolistic and monopsonistic pricing, and disequilibrium in factor markets prevent the attainment of the gains obtained in a competitive free trade equilibrium. The Paretian optimality ensures equality of the marginal rates of transformation in domestic production (MRT_d) and foreign trade (MRT_f) and the marginal rate of substitution in consumption (MRS). Symbolically, $MRT_d = MRT_f = MRS$. In the absence of this equality, there will exist distortion in either foreign or domestic economic system which will prevent market prices from being equal to the marginal rates of substitution and transformation. In this situation, restoration of equilibrium requires imposition of either a suitable tax or a subsidy or a combination of the two at the point where this distortion occurs.

If distortion exists in the foreign market either due to the imperfectly elastic demand or supply, the competitive free trade solution will be characterised by MRS = MRT_d = MRT_f. In other words, foreign prices will not be equal to the marginal rate of transformation. In this case, the first best optimal policy will change away from free trade and the country can achieve the Paretian optimally assuring maximisation of community welfare by imposing a suitable tariff. If instead of imposing a tariff, a subsidy is given on the domestic production of import substitutes it could equalise MRT_d and MRT_f. It is, therefore, obvious that a tax-cum-subsidy on domestic production is necessarily inferior to an optimum tariff.

If the economy is characterised by domestic distortion there will be divergence between the autarky price-ratio and the marginal rate of transformation under free trade. Symbolically, MRS = MRT_f MRT_d. The distortion is the result of several factors like the externalities in consumption and production, monopolistic and oligopolistic imperfections of the market, price and wage rigidities, lack of information, etc. Several studies have been made on this subject by economists including those by Gottfried Von Haberler,[15] Harry G. Johnson,[16] E. Hagen,[17] Jagdish Bhagwati and Ramaswami.[18] Without cataloguing the various types of distortions stated by these writers let us discuss the various methods of elimination and examine the validity of tariff as a means of protection.

It is clear that in the case of domestic distortion the Paretian optimality is disturbed. To restore the Paretian optimality, if we apply taxes or subsidy on foreign trade, one disturbance is removed at the cost of the other. For example, a suitable tariff can equalise MRT_f and MRT_d but would destroy the equality between MRS and MRT_d. Consequently, no tariff may exist that would yield a solution superior to that under free trade. In this case, a suitable tax or subsidy on consumption, production or factor supply would enable the policy makers to secure the Paretian welfare maximisation. This is what the theory of the second best developed by Meade, Lipsey, Lancaster and others suggests. Harry G. Johnson in his classic article 'Optimum Trade Intervention in the Presence of Domestic Distortions' has categorically given the different types of domestic distortions and the remedy as follows:

"Where externalities in consumption make social marginal rates of substitution diverge from private, taxes or subsidies on consumption are required; where external economies in production exist or where monopolistic influences raise prices above marginal costs, marginal subsidies on production are required, and where external diseconomies are present, marginal taxes on production are required; and where the price of factor in particular occupation exceeds its price in other occupations by more than can be accounted, a subsidy on the use of that factor in that occupation is required. The point of central importance is that the correction of domestic distortions requires a tax or subsidy on either domestic consumption or domestic production or domestic factor use, not on international trade."

II. Development of Infant Industries: One of the most intellectually sustainable argument for protection is the infant industry argument. The infant industry argument has been generally accepted

15. Gottfried Von Haberler, "Some Problems in the Pure Theory of International Trade," *The Economic Journal,* Volume LX, June 1950, pp. 223–240.
16. Harry G. Johnson, "Optimum Trade Intervention in the Presence of Domestic Distortions," published in Jagdish Bhagwati (ed.), *International Trade,* 1969, Penguin Series, pp. 184–217.
17. E. Hagen, "An Economic Justification of Protection," *The Quarterly Journal of Economics,* Volume LXXII, November 1958, pp. 496–514.
18. Jagdish Bhagwati and Ramaswami, "Domestic Tariff and the Theory of Optimum Subsidy," *The Journal of Political Economy,* Volume LXXI, No. 1, February 1963, pp. 44–52.

by international trade theorists. An infant industry is one which has latent comparative advantage of growth but whose operating cost is very high in the early stages of growth. If these industries are exposed to the competition of old established foreign suppliers they would be stifled in infancy. Consequently, they legitimately need protection for their survival and growth. The justification for granting protection to certain new industries is based on the plea that those industries which in their infancy are unable to compete in world markets with the longer established and more experienced foreign producers may nevertheless be able to compete very effectively in their maturity. In order to see such industries established, it is necessary to afford them temporary protection against foreign competition. In such a situation, it is worthwhile to abandon free trade and spurn the short-term advantages of a more complete international division of labour in favour of the long-term fruits of a higher rate of technical progress. At the core of the infant industry argument for protection lies the existence of some sort of internal economies which can be exploited only if the size of the firm is large enough to face foreign competition. Consequently, it should be allowed to grow for sometime under the cover of protective umbrella. After the firm has fully developed, the tariff can be removed.

The infant industry argument for protection is the oldest argument in favour of protection. It was first emphatically stressed in 1840s by the German economist and politician Friedrich List who had the underdeveloped economy of the nineteenth-century Germany and the developed economy of Great Britain in his mind. According to List, Germany could never dream of becoming an industrially developed nation unless her infant industries were protected from the competition of the established and old British industries. List was modern in his views and he considered industrial development as a prerequisite of Germany's progress. On this score he had no doubt and he boldly stated that "manufactories and manufactures are the mothers and children of municipal liberty, of intelligence, of the arts and sciences, of internal and external commerce, of navigation and improvements in transport, of civilisation and political power." [19] At the core of the infant industry argument for protection is the existence of some kind of internal economies. A firm cannot compete against the old well-established firm if it is small. It has to grow large enough to exploit all the economies of scale in production and become competitive. It has, therefore, to be protected for some time, and be allowed to grow during this time.

According to Friedrich List, tariffs were necessary for the development of the young emerging German industries. List stated that "the reason for this is the same as that why a child or a boy in wrestling with a strong man can scarcely be victorious or even offer steady resistance." [20] The protection needed for the development of infant industries cannot be justified for ever. List held that it should continue "only until that manufacturing power is strong enough no longer to have any reason to fear foreign competition and henceforth only so far as may be necessary for protecting the inland manufacturing power in its very root." [21] List's infant industry argument for protection soon won acceptance even in England, the home of Smithian *laissez-faire laissez-passer* philosophy and policy and even John Stuart Mill favoured the idea of protecting infant industries in these words: "A protecting duty, continued for a reasonable time, might sometimes be the least inconvenient mode in which a nation can tax itself for support of such an experiment (introducing new industries). But it is essential that protection should be confined to cases in which there is good ground of assurance that the industry which it fosters will after a time be able to dispense with it; nor should the domestic producers ever be allowed to expect that it will be continued to them beyond time for a fair trial of what they are capable of accomplishing." [22] Thus, the infant

19. Friedrich List, *The National System of Political Economy*, 1885, p. 230.
20. Friedrich List, *Op. cit.*, p. 240.
21. *Ibid.*
22. John Stuart Mill, *Principles of Political Economy*, Book V, Chapter 10, Section 1, p. 92.

industry argument is purely dynamic and it urges only a temporary intervention to remedy a transient distortion. Incurring the costs for a limited period in return for further benefits is a type of investment of 'nursing capital'. If this investment is not done it is possible that free competition would produce socially inefficient allocation of resources. In short, according to the "Mill test", protection must be temporary in order to enable the industry to overcome a historical handicap.

Not agreeing with Mill and treating the "Mill test" as rather a soft condition to be satisfied by the industry in order to claim the protective umbrella, Bastable wanted a more rigorous condition to be satisfied by the industry before it could be legitimately protected. For him, the mere prospect of overcoming a historical handicap (as suggested by John Stuart Mill) was not enough to warrant protection. Something more was required. It was also necessary that the ultimate saving in costs should compensate the community for the high costs of the protected learning period. It was necessary, in particular, that after applying a suitable time discount to the early excess costs and to the ultimate cost savings, the commodity should be worthwhile to produce. According to Bastable, in addition to satisfying the "Mill test," it was also necessary on the part of the industry in order to warrant assistance to show that if protected the ultimate saving in costs would be more than enough to compensate the community for the high costs of the protected learning period.[23] Thus, not only the "Mill test" but a harsher "Bastable test" must be passed by the industry before the community should protect it in infancy. In short, the "Mill-Bastable test"[24] flowers itself in the proposition that if an industry successfully passes both the "Mill test" and the "Bastable test", it is the duty of the community to protect it until such time that it can stand on its own feet.

The core of the infant industry argument for protection consists in the existence of some kind of internal economies which can be exploited only when it grows under protection. After a few years of protection, it will enjoy the economies of scale and will be able to compete with the old established industries in the other countries. When the industry has developed and protection is no longer needed it can be dismantled and free trade can be resumed.

Figure 19.1 illustrates the infant industry argument for protection. Initially AB is the country's production possibilities curve and P_1P_1 is the international terms of trade or exchange ratio line. This international terms of trade line is tangent to the country's production possibilities curve at point C. The country produces at point C and through trade moves on to point D in consumption under conditions of free trade.

Now, in order to protect its industry the country imposes a prohibitive tariff raising the price of the protected industrial goods for the domestic consumers and producers. The new price ratio line which emerges under protection is P_dP_d which is tangent to the country's production possibilities curve AB at point E. The country will produce and consume at this point under protection. It is evident that the community welfare has diminished as a result of protecting the infant industry because at point E the community is forced to consume a smaller quantity of both the goods compared with the larger quantity of both these goods which is available at point D under free trade.

However, the loss in the community's welfare is only a short-run phenomenon and the community will be more than compensated for this temporary loss. As a result of protection, in due course of time, the production of industrial goods increases substantially as the protected infant industry grows under the umbrella of protection. Internal economies are exploited and the

23. Charles F. Bastable, *The Commerce of Nations*, Tenth Edition, revised by T.E. Gregory, 1921, pp. 140–143.
24. Murray C. Kemp, "The Mill-Bastable Infant Industry Dogma," *The Journal of Political Economy*, Volume 68, February 1960.

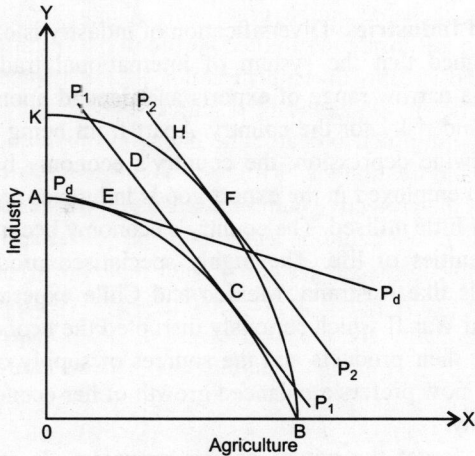

Figure 19.1

country's production possibilities curve AB stretches outward to the position of BK showing considerable expansion of the country's industrial sector. Now free trade may be resumed at the pre-protection international prices as shown by P_2P_2 international price-ratio line. The country will now produce at point F and will reach at point H in consumption showing larger community welfare because at point H the community consumes a larger quantity of both the goods compared with the earlier free trade consumption equilibrium point D. The obvious conclusion following from this is that by nurturing the infant industry for a short period the country reaches a higher level of community welfare than was available to her under free trade. The community, however, pays for this gain in welfare in the form of initial fall in the social welfare.

Notwithstanding the theoretical validity of the argument, serious difficulties arise in practice. Firstly, it is difficult to lay down general rules for ascertaining whether such a possibility is in fact present. Secondly, it is difficult to say that the protected industry would eventually be able to stand on its own feet without protection and face competition. According to Kindleberger, "if foreign competition is kept out by tariffs, domestic industry tends to become sluggish, fat and lazy. Or it may start that way, and resist the change which foreign innovation threatens for it by persuading government to impose tariffs."[25] Experience of the protected industries unmistakably shows that protection lulls the home industries into a sense of complacency. Thirdly, an infant once protected refuses to develop into an adult. Consequently, protection becomes a permanent burden on the consumers without any corresponding gains to the community.

Moreover, the highly persuasive nature of the Mill-Bastable test notwithstanding, neither Mill nor Bastable has explained his assumptions about the learning process in sufficient detail to permit of a judgement on his protectionist conclousious.

However, in spite of these criticisms, the infant industry argument is most prevalent in the underdeveloped countries. It is argued that for rapid industrialisation and sustained economic growth the underdeveloped countries need all-round protection. Gunnar Myrdal has pointed out the validity of 'infant economy' in place of 'infant industry' argument for protection for the underdeveloped poor countries of the world.

25. Charles P. Kindleberger, *Op. cit.,* p. 119.

III. Diversification of Industries: Diversification of industry is often pointed out as a goal of protective tariff. It is argued that the system of international trade where industrial nations increasingly specialise on a narrow range of exports and depend upon others for a wide range of imports, is very unstable and risky for the country. Apart from being subjected to the hazards of recurrent shocks of worldwide depression, the country's economy becomes highly unbalanced. Some factors of production employed in the export goods industries are recklessly exploited while others are not at all or very little utilised. The country's economy becomes dependent on others for the supplies of basic amenities of life. The highly specialised producers of foodgrains, dairy products and raw materials like Australia, Mexico and Chile experienced the disadvantages of specialisation during World War II which seriously disrupted the economies of these countries by cutting off the markets for their products and the sources of supply of essential goods for many years. Thus, every country now prefers a balanced growth of her economy through diversification of industries in the economy.

Let us not, however, forget the nature of this argument. It applies only to very highly specialised economies, in practice to those which export a narrow range of primary products and depend upon imports for most of their supplies of the manufactured goods. And even in such cases which are rare in real world, the diversification of industries can only be justified provided it substantially insulates the economy against the harmful effects of depression and wars without involving heavy costs.

IV. Anti-Dumping: Tariff is favoured to counteract the harmful effects of dumping of goods in the country by the foreigners. In fact, many nations contain special provisions against dumping which arouses the indignation of domestic producers in the market in which dumping takes place. However, certain difficulties arise in imposing anti-dumping tariffs. Firstly, the meaning of dumping should be clearly understood. "Contrary to widespread impression dumping is *not* selling abroad below costs of production. It means instead *sales in a foreign market at a price below that received in the home market,* after allowing for transportation charges, duties, and all other costs of transfer. Discrimination between the home and foreign price is the essential mark of dumping. Thus sales abroad below cost of production would not constitute dumping *unless* the foreign price were lower than the domestic price."[26] Secondly, it is equally important to be certain about the nature of dumping. Dumping may be sporadic, predatory or persistent. Sporadic dumping occurs when a company finds itself only occasionally with more inventory stocks than it can hope to dispose of in an orderly manner through its normal trade outlets. If for this company the demand abroad is more elastic, it will want to cut its losses abroad by selling the goods for any price that can be realised. In this case anti-dumping tariffs are not in the interest of consumers. Predatory dumping is selling goods at loss abroad for the purpose of destroying foreign competition. It can be most disturbing, even ruinous, for the local firms since it is followed by an increase in prices after the market has been established or competition overcome. In this situation, protective tariff becomes inevitable. Persistent dumping, on the other hand, is not different in its effects from sales by a low cost foreign producer. Consequently, this type of dumping may be regarded as providing permanent gain to the consumers. To prohibit such dumping by protective tariff would permit the domestic producers to charge the foreign monopoly price and exploit the consumers. Obviously, an anti-dumping tariff has much to commend provided dumping is defined carefully.

---

26. P.T. Ellsworth and J.C. Leith, *Op. cit.,* p. 250.

V. Economic Recovery: Protective import duties may serve as an instrument of recovery of the economy from depression. In a depressed economy where factors of production are unemployed and where the country's economy has suffered a great setback going completely out of equilibrium, protection will help revive the economy. It will pull the economy out of the rot of depression. But this will be more true if other countries cooperate with such a country by not raising simultaneously the tariff walls within their own political territories. Imposition of protective import duties by a country will have the effect for the time being of reducing imports without having any unfavourable repercussions on the country's exports. Imports having been reduced employment will be created in the country since the goods hitherto imported will be produced in the country. On the basis of the theory of employment multiplier, primary employment will generate secondary, tertiary and quarternary employment in the economy. The expansionary effect of initial employment generated due to the imposition of protective tariff would not, however, continue indefinitely. After some time it will peter out. While serving as a member of the Macmillan Committee in 1931 which was set up to study the condition of depressed British economy and to suggest measures, John Maynard Keynes stated that for an economy "which is neither in equilibrium nor in sight of equilibrium, protection and not free trade is the most rational trade policy." Accordingly, Keynes had made a proposal to levy a general protective tariff with rebates on all imported materials entering into exports. Keynes' proposals were denounced by William Beveridge and Lionel Robbins when they asserted that protective tariffs could only do harm since reduction in imports would cause a *pari passu* reduction in exports. Keynes, however, sanguinely replied to this charge by saying that "there is ... no simple and direct relationship between the volume of exports and the volume of imports." Now there is considerable unanimity among the economists in believing that protective duties can help in initiating recovery in a depressed economy.

Suggested Readings

P.T. Ellsworth and J.C. Leith, *The International Economy,* Fifth Edition, 1975, Chapter 14.

Charles P. Kindleberger, *International Economics,* Fifth Edition, 1973, Chapter 7.

Bo Sodersten, *International Economy,* Second Edition, Reprinted 1981, Chapters 5 and 15.

Jacob Viner, *Studies in the Theory of International Trade,* 1937, Chapters 8 and 9.

Murray C. Kemp, *The Pure Theory of International Trade,* 1964, Chapter 12.

Questions

1. Appraise the infant industry argument for protection as a measure to promote the industrialisation of developing countries.
2. Explain the various grounds on which protection as a trade policy is justified by the developing countries.
3. Examine the external economies argument and market distortions argument for protection and explain the policy implications of these two arguments.
4. Explain the 'Mill-Bastable Test'. Why should an infant industry pass this test before it should be protected by the government?

CHAPTER 20

Protection and Underdeveloped Countries

Protection constitutes an important plank in the commercial policy of underdeveloped countries of the world today. Notwithstanding that the traditional argument for free trade shows that free trade results in the maximisation of world production efficiency under static assumptions, the argument has no direct relevance to the structural and dynamic problems of the world's poor nations. When the conditions of production efficiency are to be considered over *a period of time* and not *at any given point in time* and when our concern is with the consideration of national gains from trade rather than with the mutual gains from trade, protection and not free trade will be the optimal policy for the underdeveloped country in the sense that protection will spur the economic growth of the poor country by enabling it to acquire a large share of the gains from trade by augmenting its rate of capital formation and by promoting its industrial development.

1. Increase in Gains from Trade

The argument that a protective commercial policy can enable a poor nation to change the distribution of gains from trade so as to cause an increase in its share of the total gains from trade, is based on the terms of trade argument for protection—the argument that an import tariff will result in the improvement of the terms of trade for the tariff-imposing poor country. World's poor countries are understandably concerned about the secular deterioration in their terms of trade. It is feared by these countries that in the absence of protection, increase in the production of primary products combined with the low income elasticity of demand for these products will cause a further fall in the prices of poor countries' export products relative to the prices of their imports thereby leading to further deterioration in their already unfavourable terms of trade. It is argued that an export or import tariff that seeks either to raise the prices of a poor country's exports or to reduce the prices of her imports by restricting exports or imports may cause improvement in her terms of trade or may forestall the anticipated deterioration in the terms of trade of a poor less developed country.

Raul Prebisch, a staunch spokesman of the world's poor underdeveloped countries, has pointed out that the terms of trade are systematically evolving against these countries and these countries have to trade in a hostile atmosphere. In their various writings, several economists including Raul Prebisch, Hans W. Singer, Arthur Lewis and Gunnar Myrdal, have successfully argued that the poor countries have suffered a secular deterioration in their commodity terms of trade. Basing their inferences on the commodity terms of trade of the United Kingdom, the supporters of this view argue that "from the latter part of the nineteenth century to the eve of the second World War ...... there was a secular downward trend in the prices of primary goods relative to the prices of manufactured goods. On an average, a given quantity of primary exports would pay, at the end of this period, for only 60 percent of the quantity of manufactured goods which it could buy at the beginning of the period."[1]

According to Raul Prebisch,[2] the less developed countries mainly produce and export primary products such as coffee, rubber, tea, sugar etc. which have not changed in character over the years—these products are the same today as these were 50 years ago. As against this, the quality of manufactured goods such as radios, transistors, automobiles, etc. has changed tremendously over the years. It is evident from the fact that a 1920 model of fiat car is quite different from its 1998 model. Thus, while there is no novelty in the products produced and exported by the underdeveloped countries, the products produced and exported by the developed countries carry novelties year after year giving the developed countries an edge in fixing their prices in the international markets. With almost constant or falling export prices of primary products and rising import prices of manufactured goods, while the terms of trade for the poor developing countries have been deteriorating those for the rich developed countries have been improving. Similar view has been expressed by Singer,[3] according to whom, the fruits of technological advancement reflected in the increase in productivity have been largely reaped by the producers in the form of high factor incomes in the developed countries. In the case of primary products, however, these fruits have not been reaped by the producers in the form of high factor incomes but have been largely passed on to the consumers in the rich countries in the form of falling prices of primary products in the world markets.

The Prebisch-Singer thesis of deteriorating terms of trade for the less developed countries has been supported by Gunnar Myrdal. According to Myrdal, "most of the underdeveloped countries are saddled with basket of traditional export goods ... the prices of which have been lagging behind."[4] Emphasising that underdeveloped countries are untouched by the 'spread effect' of international trade and reap only the 'backwash effect' of such trade, Gunnar Myrdal has suggested that only by pursuing a protective commercial policy can the underdeveloped countries alter the distribution of gains from trade in their favour. The truth, however, is that underdeveloped countries can successfully twist the terms of trade in their favour by pursuing a protectionist trade policy only if they command either a monopoly or a monopsony position in trade. The real situation is that few, if any, of these poor countries are in a position to exercise sufficient monopoly or monopsony power in view of the fact that importing developed nations have alternative sources of supply of foodstuffs. They have also developed synthetics as close substitutes for natural raw materials and the domestic market of anyone poor country for a particular import good is of a relatively small size reducing its bargaining power as a monopsonist.

A tariff can improve the terms of trade for the tariff-imposing country when the foreign offer curve which the country faces is inelastic. Where the foreign offer curve faced by the tariff imposing country is perfectly elastic (i.e., if the elasticity of reciprocal demand of the other country is infinite represented by a linear foreign trade offer curve), imposition of import tariff will not lead to any improvement in the terms of trade for the poor country. Experience shows that the foreign

1. United Nations Department of Economic Affairs, *Relative Prices of Exports and Imports of Underdeveloped Countries*, 1949, p. 72.
2. Raul Prebisch, "Commercial Policy in the Underdeveloped Countries," *The American Economic Review, Papers and Proceedings*, May 1949, pp. 261–264.
3. Hans W. Singer, "The Distribution of Gains between Investing and Borrowing Countries," *The American Economic Review, Papers and Proceedings*, May 1950, pp. 477–479.
4. Gunnar Myrdal, *An International Economy*, 1956, p. 231.

trade offer curve faced by any single poor country is normally elastic showing that less imports will be demanded and less exports will be supplied as the prices of imports increase. The greater is this elasticity, greater will the volume of trade shrink as result of tariff. In such a situation, it becomes necessary to know the optimum tariff, i.e., the optimum import and export duties. An optimum tariff is that amount of duty which maximises the gain resulting from improved terms of trade minus the loss suffered from smaller volume of trade. The optimum tariff will be lower, the higher is the elasticity of the foreign trade offer curve.

Even granting that a single poor country commands sufficient monopoly or monopsony power to alter the terms of trade in her favour by means of a tariff and that an optimum tariff can be easily calculated, the terms of trade improvement argument for protection is dependent upon the price elasticities of demand and supply at only a given moment of time and is consequently static in nature ignoring as it does the possibility that a short-term gain may be easily offset by subsequent changes in the demand and supply elasticities. It is also not unlikely that an initial improvement in the terms of trade brought about by imposing tariff may be later counteracted as a result of the retaliatory moves of the other countries or by an increased demand for imports caused by government's expenditure of the tariff revenue or by an internal redistribution of income. Once these dynamic effects are recognised, the weakness of the terms of trade arguments for protection becomes self-evident.

2. Promotion of Capital Formation

Countries of the world show marked differences in the nature, character and the degree of their economic development. A few countries of the world have achieved the coveted position of developed economies. Consequently, the nationals of these developed countries enjoy the high standard of living with very high per capita real income. On the other hand, there are a large number of countries of the world where due to extremely low per capita real income people live a life of unbelievable misery, being entrapped in the 'vicious circle of poverty' which they find it difficult to break. The low per capita real income is the major impediment to the rapid economic development of these poor countries. This formidable obstacle to economic growth of world's less developed regions has been described by economists in the form of the 'vicious circle of poverty' which justifies the saying that "a poor country is poor because it is poor." The low per capita real income is both a cause and a consequence of the low demand leading to low saving, low investment and low real income in the poor economy.

Besides, the so-called 'vicious circle of poverty', Ragnar Nurkse and James Duesenberry have pointed out the 'demonstration effect' under which underdeveloped countries satisfy their physical appetites and titillate their innate sense of self-expression or exhibition. They frequently want to import Swiss watches, Japanese transistors, American fountain-pens etc. at the most primitive level before they have arranged to earn or borrow the necessary foreign exchange. This kind of psychological behaviour of the consumers discourages capital formation in these countries. Consequently, it is argued that the imports of luxury goods should be prohibited by means of a prohibitive protective tariff in order to promote the capital formation.

But if the expenditure thus saved is diverted to domestic products, the increased home consumption will generate inflation. Furthermore, the domestic factors of production will shift away from capital goods to consumer goods industries and there will be no net gain to capital formation. Moreover, this may have an adverse effect on exports since it is based on the 'beggar my neighbour' policy. It is, therefore, suggested that the purpose of capital formation would be served if consumer goods could be replaced by capital goods through import restrictions. It will not change the volume of imports but will only change the composition of the import goods and there will be no adverse effects on the exports. Gerald M. Meier[5] has, however, pointed out that this procedure will bring at least three adverse effects on trade. Firstly, if there does not exist disguised unemployment in the country, the protected import-replacing industry will expand at the cost of other industries, including export industries. Secondly, if the imported goods happen to be peasants' strong incentive to produce cash crops for exports, the restriction on imports may discourage them to produce for the export markets. The denial of imports may, therefore, reduce farmers' incentive to produce crops for the export market reducing the volume of trade. Thirdly, exports can hardly be maintained if the policies of protection and promotion of the import-competing industries lead to higher internal costs.

3. Increase in Capital Imports

It is argued that tariff attracts foreign investment by making capital scarce in the tariff imposing country. The theoretical validity of this argument has been proved by Robert A. Mundell.[6] Mundell's thesis is based on the well-known general proposition of the neoclassical theory of trade that the commodity movements are a substitute for the factor movements and consequently an increase in trade impediments stimulates factor movements and *vice versa*.

To explain Mundell's thesis, let us assume that a capital-poor country's imports consist of the capital-intensive commodity. A tariff on imports will make the scarce factor relatively more scarce. The relative scarcity of capital raises the marginal product of the scarce factor and consequently its marginal return (interest rate). The increase in the factor returns in the capital-poor country creates an incentive for capital movement from the capital-rich country to the capital-poor country. Equilibrium, in the absence of transport cost, will be achieved when the marginal product and consequently return (interest rate) on capital is equated in both the countries.

Mundell's thesis is, however, based on several restrictive assumptions. Firstly, it assumes that production functions are identical in all the countries. Secondly, that capital is perfectly mobile and responds to changes in the marginal return. Thirdly, it assumes that the terms of trade after tariff remain unchanged. Fourthly, that capital-intensive industries exhibit the economies of scale. But in real life production functions for the commodities differ from country to country. Capital movements take place due to political instability, risk, economic uncertainty, patriotism, etc. and not only due to differences in the marginal return on capital. It is an oversimplification to assume that the terms of trade remain unchanged after the imposition of tariff. Tariff is generally imposed to change the terms of trade. Economies of scale should be sufficient to make up the loss due to transport costs. Mundell himself states that ". . . it would have to be established that the economies of scale are sufficient to make up for transport costs which must be paid on the interest returns. If they are sufficient, the tariff would be unequivocally beneficial." In this case also, the economies of scale will be effective only if the importing country is rest-of-the-world and the exporting country is a small country otherwise the economies of scale in the exporting country will cancel out the effects of the economies of scale in the importing country.

5. Gerald M. Meier, *International Trade and Development*, 1963, p. 121.
6. Robert A. Mundell, "International Trade and Factor Mobility," *The American Economic Review*, Volume XLVII, No. 3, June 1957, pp. 321–335.

4. Improvement in the Balance of Payments

According to many economists and statesmen, economic planning is a remedy for country's all economic ills. It is the grand 'panacea' of modern age. Most underdeveloped countries have launched planning for economic development and transformation of society. The tempo of planned development brings with it disequilibrium in the country's balance of payments since successful execution of plan requires large imports of capital goods, technical know-how and essential raw materials. Until the development projects have materialised and a strong capital base is created, any substantial increase in exports would not be possible. Furthermore, many goods which were previously available for exports may now be consumed by the newly created domestic industries creating deficit in the country's international balance of payments. In addition, the development projects will adversely affect the poor country's external balance of payments through the 'income effect' and the 'price effect.' The development expenditure increases the money incomes of the consumers. The money illusion lures the consumers to import the luxury and superior goods at the initial stage of development. Tastes change with trade, even as trade satisfies the existing wants more fully. Kindleberger has stated that the "demonstration effect is more significant for underdeveloped countries that frequently want to import—Swiss watches, British bicycles, US fountain-pens at the most primitive level—before they have earned or arranged to borrow the necessary exchange."[7] This situation adversely affects the external balance of payments of these poor countries. The external balance of payments of poor countries is adversely affected not only through the increased imports but also through the decreased exports. The introduction of modern methods of production introduces modern methods of consumption also as a result of which commodities which were previously exported are now domestically consumed. Highlighting this fact Kindleberger has stated: "Trade in Europe, in the 19th century continued side by side with different styles of national diet and cooking. The British breakfast was distinct from the Continental breakfast, and on the Continent itself Italian, French, German, and Scandinavian cooking all differed. Today the Indo-Chinese complain that the native Asian breakfast is giving way to European eggs, bacon, and coffee; in Japan rice is increasingly abandoned in favour of wheat; and Coca-Cola is a trademark known round the world."[8]

Moreover, the development projects take time to bear fruits owing to the shortage of complementary factors and bottlenecks. The inelastic supply of goods accompanied with deficit financing results in high prices. A rise in the prices encourages imports and discourages exports adversely affecting the external balance of payments position of the country. Consequently, import restrictions are needed to remove disequilibrium in the balance of payments. Even the IMF and WTO also allow member countries to impose temporary restrictions on the imports during the transition period to solve the problem of deficit in their external balance of payments.

5. Stabilisation of Primary Products' Prices

Exports of primary products constitute by far the most important source of foreign exchange earnings for most developing countries accounting for more than three-fourths of their export earnings. But the prices of primary products have shown wide fluctuations causing instability in export earnings. The economies of the developing countries have been adversely affected both by the secular deterioration and short-run fluctuations in their international terms of trade These countries are

7. Charles P. Kindleberger, *International Economics,* Fifth Edition, 1973, pp. 54–55.
8. Charles P. Kindleberger, *Ibid.,* p. 54

largely the producers of primary products and depend on the developed countries for the exports of these products. The exports of primary products constitute by far the most important source of foreign exchange earnings for the world's most underdeveloped countries accounting for 85 to 90 per cent of their total export earnings. The widely fluctuating prices of the primary products give rise to many problems, including difficulties in planning development and seriously hampers the process of economic development of these countries. With the stabilisation of the prices of primary products which constitute the vast bulk of exports of the developing countries, a major obstacle to the smooth process of development of these countries could be removed. It is, therefore, of utmost importance for the economic prosperity and growth of these poor countries that the international prices of primary products should be (i) stable and free from fluctuations in the short period, and (ii) rising over a longer period. Unfortunately, it is not what has happened over the years. Contrary to this, the prices of the export goods of these poor countries have kept on fluctuating widely in international markets creating instability in these countries' total export earnings which hampers the growth process seriously. Moreover, the prices of primary products vis-a-vis those of the industrial goods have constantly fallen over time. In other words, the deteriorating terms of trade in the long run have added to the economic plight of these countries. It is, therefore, strongly argued that the prices of primary products in the international markets should be stabilised.

It has been suggested that trade should not be free and in private hands. In order to stabilise the prices of primary products, the erratic behaviour of supply and demand, particularly of supply, should be eliminated or at least minimised by undertaking supply stabilisation schemes such as building the buffer stocks, restricting supplies etc. The governments of the developing countries with the cooperation of the developed countries should create buffer stock funds to finance the purchases of surplus production and thereby to arrest price fluctuations. The urgent need for establishing a fund for stabilising primary products' prices has been emphasised in the UNCTAD, WTO and other international economic institutions. Such a scheme is, however, possible for those products only whose storage cost is low. Supply restriction schemes are also difficult to enforce for a long period. The failure of many international cartels proves it.

6. Development of Infant Economy

The 'infant industry' argument has been replaced by the 'infant economy' argument by Gunnar Myrdal. According to Gunnar Myrdal, the economies of the underdeveloped countries are unbalanced. There is a large agricultural sector which is dominated by traditional methods of production and a scantily developed industrial sector using modern techniques of production. Since most underdeveloped countries are the erstwhile colonies of the developed countries, they lack a sound industrial base. Thus, there is a strong and valid reason for pursuing a vigorous policy of industrialisation on the part of these countries. Rapid industrialisation is, however, possible only if they reap the benefits of horizontal and vertical linkages. To reap the fruits of linkages, they should adopt the policy of 'protection all round' until the stage of 'take off' is reached. According to Gunnar Myrdal, "the difficulties of finding the demand to match new supply, the existence of surplus labour, the large rewards of individual investments in creating external economies, and the lopsided internal price structure disfavouring industry"[9] amply justify the imposition of protective tariff by the underdeveloped countries. The infant economy argument for protection is akin to the old infant industry argument for protection. Just as to develop an infant

9. Gunnar Myrdal, *Op. cit.,* p. 279.

industry it needs protection against the competition of fully developed industry, so also in order to develop a less developed economy it should be industrially developed through providing it the development protection.

Keeping in mind the special problems encountered by the developing and the developed countries, even the World Trade Organisation (WTO), an international institution committed to promote nondiscriminatory multilateral trading system in the world, in order to integrate these countries in the multilateral trading system which is important for their economic development as also for the expansion of global trade has made provisions for conferring differential and more favourable treatment for these countries, including special attention to the particular situation of the least-developed countries. The WTO is committed to promote the expansion of trade, sustainable growth and development of the developing countries, including the least-developed countries, through its closer cooperation with the World Bank and the IMF. The WTO's Plan of Action will be applied in respect of the least-developed countries designated as such by the United Nations which are members of the WTO.

In addition to the above weighty arguments for protection, the other arguments such as the diversification of industries, national defence, increase in government revenue and employment in the economy etc. are also given in favour of protection by the underdeveloped countries.

Suggested Readings

P.T. Ellsworth and J. Clark Leith, *International Economy,* Fifth Edition, 1975, Chapter 14.
Charles P. Kindleberger, *International Economics,* Fifth Edition, 1973, Chapter 10.
Gerald M. Meier, *International Trade and Development,* 1963, Chapter 6.
Bo Sodersten, *International Economics,* Second Edition, Reprinted 1981, Chapters 15 and 18.

Questions

1. Make out a case for protection in the less developed countries. Discuss the problem of stabilisation of primary products' prices.
2. Discuss the terms of trade and the infant economy arguments for protection by the less developed countries.

PART FOUR

TRADE POLICY

CHAPTER 21

Tariffs

In many countries of the world, tariff policy has been politically widely discussed because tariff directly affects the welfare of the citizens of a country. A tariff is a type of protectionist device which while restricts the consumers' freedom of choice in the country by forcing them to curtail their consumption of those goods which they prefer much causes the shift in country's economic resources from one productive use to another. Thus, by imposing a tariff, a country is able to change the relative prices of the goods and factors of production. This will result in a different trade pattern than would occur in the absence of tariff. In other words, consequent upon the imposition of tariff, the structure of international trade will materially change. A high rate of tariff will reduce the volume of international trade while a low rate of tariff will only marginally reduce the volume of international trade. A negative tariff or subsidy will lead to an expansion of international trade.

As an important instrument of commercial policy, a tariff is a kind of duty or levy imposed on the commodities when they cross the national border. It is levied both on a country's exports and imports. A tariff is, however, generally levied by a country's government on her imports. Consequently, the terms tariff and import duties are used interchangeably. An import duty is levied on the goods originating abroad and are scheduled for import to the duty-levying country. An export duty, on the other hand, is imposed on the commodities originating in the duty-levying country and are scheduled for abroad. In addition to import and export duties, there is another important duty which is known as the transit duty. It is levied on the goods crossing the national frontier originating in other countries and are scheduled for some other third countries. For example, India might impose a transit duty on the Nepalese imports from or exports to England while crossing the Indian border either at the entry point or at the exit point. Such transit duties are generally of significant concern for the land-locked countries.

Classification of Tariffs

There are several kinds of tariffs which can be placed under the following three broad groups.

Group One consists of those tariffs which are based on the different criteria followed in levying a particular tariff. In this group we have the (i) Specific Duty, (ii) Ad valorem Duty, and (iii) Combined Duty.

Group Two consists of those tariffs the basis of imposition of which is the particular purpose sought to be served by a particular tariff. Included in this group are the (i) Revenue Tariff, and (ii) Protective Tariff.

Group Three relates to those tariffs which are based on the criterion of application between different countries. In this group we have the (i) Single Column Tariff, (ii) Double Column Tariff, and (iii) Triple Column Tariff.

Having classified the different tariffs under the above three groups we may now explain separately the different tariffs mentioned under each one of the above three groups.

Specific Duty—When a fixed sum of money, keeping in view the weight and measurement of the commodity, is charged by government as tariff, it is known as the specific duty. For example, a fixed sum of import duty may be levied on the imports of television sets irrespective of the value of different television sets. Specific import or export duties are easy to administer as they do not involve the problem of determining the value of the export or import goods. The valuation of goods is not so easy as it is determined by the supply and demand prices which vary from place to place and time to time. In addition to the demand and supply prices, there are several other prices, such as invoice price, contract price, market price, f.o.b. (free on board) price, c.i.f. (cost, insurance and freight) price, etc. Which one of these many possible prices should be taken as the basis of commodity valuation for purposes of levying the import duty is a difficult problem which is solved by resorting to specific duty. A specific duty cannot, however, be levied on certain articles like the works of art. For example, a picture cannot be taxed on the basis of its weight or surface area.

Ad valorem Duty—When a fixed percentage of the value of a commodity is taken away as tariff, it is known as an *ad valorem* duty. It ignores the consideration of weight and measurement of the commodity to be taxed. The basis of duty levy is the price or value of the commodity rather than its physical dimensions. An *ad valorem* import duty is equitable because heavier duty burden falls on the costlier goods which are imported and consumed in the country by the rich while the coarse and cheaper goods which are imported and consumed by the poor people bear the smaller tariff burden. The imposition of an *ad valorem* duty is more justified in the case of those goods whose values are not determined on the basis of their physical and chemical characteristics such as the costly paintings, rare manuscripts, etc. An *ad valorem* duty has the advantage of clearly expressing the burden of the duty. Moreover, the duties levied by different countries can be easily compared.

Combined Specific and Ad valorem Duty—Sometimes a combined schedule of specific and *ad valorem* duties is prepared by the legislature and the importers are given the option to choose that duty whose rate is minimum. In case, protective measures are to be tightened the highest rate of import duty is imposed by the customs authorities. Thus, in order to provide effective protection to the home industries this system is an ideal system.

Sliding Scale Duties—Those import duties which vary with prices of the commodities are known as sliding scale duties. These duties may be either specific or *ad valorem*. In practice they are, however, always specific. Historically, these duties have been confined almost entirely to duties levied on foodgrains. Moreover, foodgrains are important stable commodities whose prices can be readily ascertained and governments have attempted to stabilise their prices.[1] These duties have been applied to foodgrains because their prices frequently change owing to harvest variations.

Revenue Tariff—A tariff which is designed to provide only revenue to the home government and not protection to the domestic industry is a duty on imports of a commodity not produced

1. Gottfried Von Harberler, *The Theory of International Trade*, 1950, p. 343.

domestically.[2] The sole consideration in motivating the imposition of a revenue import tariff is to obtain the much-needed revenue for the state. But even in this case, if the consumer shifts his consumption away from the commodity or commodities subjected to revenue tariff in favour of the other commodities there is an unintended protective effect. Generally, a revenue tariff also exerts some protective effect. Consequently, a purely revenue tariff is not possible. Generally, a tariff imposed with the purpose of raising revenue is levied on the consumption goods and particularly on the luxury goods whose demand from the rich may be inelastic. In order to be productive of revenue, the rate of revenue tariff should not be prohibitive. The primary purpose that motivates the imposition of a revenue tariff is to provide the needed revenue to the government. Revenue import duties are a special form of taxation.

Protective Tariff—In order to protect the domestic industries from cut-throat competition of cheap imports, protective tariff is levied on the imports of foreign goods competing with the home-made products. The main motive for imposition of protective duties is not to create a new source of income for the government but to maintain and encourage the development of home industry protected by these duties. Higher the rate of protective tariff, greater will be the protective effect. A completely perfect protective tariff has the effect of stopping completely the imports of foreign goods competing with the home-made goods yielding no revenue to the government. In other words, a purely protective import duty in order to provide absolute or 100 per cent protection to the domestic producers of goods has necessarily to be completely prohibitive of imports, i.e., the import duty must be high enough to eliminate the price advantage of imports over the domestic goods. In practice, however, a protective tariff seldom, if at all, touches the optimum limit of completely banning the imports and yielding no revenue at all. Apart from causing loss of the much-needed revenue to the state, the domestic producers are prone to become lethargic and inefficient if the protective tariff is pushed up to its extreme height. Sometimes, an import duty is levied to serve the dual purpose of collecting revenue and of protecting the domestic producers from foreign competition. Friedrich List (1789–1846) had championed the cause of a high protective tariff against the imports of cheap foreign manufactured goods to protect the infant industries in Germany. At present, the developing countries impose protective tariff on the imports of cheap foreign manufactured goods in order to develop the industries at home.

It should, however, be emphasised here that the simultaneous achievement of the dual aim of obtaining the highest revenue to the government and providing the greatest possible protection to domestic industry are not compatible with one another. The import duty which affords the maximum protection is a prohibitive duty which yields no revenue to government. On the other hand, an import duty which yields maximum revenue to government is one which affords no protection to home industry. An import duty fulfils its protective function if it results in greater expansion of the home production, but if it does so, it yields less revenue. Thus, from purely revenue consideration those import duties are most useful which are imposed on those goods which are not produced at home.

Single Column Tariff—Under the single column tariff system, a list of tariff is prepared without discriminating between the commodities and countries. This tariff list is applicable to all the commodities and countries. Even if the low rates of duty on some commodities appear due to commercial treaties with some country or countries these are also extended to imports from all other countries. This system is very easy and simple to administer. However, the single column import

2. P.T. Ellsworth and J.C. Leith, *The International Economy*, Fifth Edition, 1975, p. 231.

duties are not elastic enough to adjust themselves according to the changing needs of the industries of the country concerned.

Double Column Tariff—The double column tariff provides two rates of import duty on all or some commodities. Both the rates are declared by the government from the very beginning. Sometimes one rate is declared in the beginning and the other rate is declared after some commercial treaties are signed with the other countries. A country adopting this system believes in the commercial treaties and bargaining. This is why a new rate of tariff is declared after entering into commercial treaties which serve the purpose of the country concerned. The double column tariff can be either a general and conventional tariff or a maximum and minimum tariff.

General and Conventional Tariff—This tariff consists of the two separate tariff schedules—general and conventional. The general tariff schedule is determined by the state legislature with the declaration that it would be adjusted if and when needed as a result of commercial treaties. The conventional tariff schedule is the result of commercial treaties entered into by the tariff-imposing country with the other countries. Under the conventional tariff schedule system, regular and gradual change needed according to the changing domestic conditions is not possible. The change is possible only after long negotiations or after the expiry of existing commercial treaty. The main drawback of the conventional tariff schedule is that it is rigid in nature. As against it, the general tariff schedule has the advantage of flexibility. It does not, however, strengthen the bargaining capacity of a country. Sometimes, the general tariff rates have to be maintained even if they are harmful to the economy.

Maximum and Minimum Tariff—Under this tariff two rates of tariff—maximum and minimum tariff rates—are fixed for every commodity by the legislature and the tariff administration is authorised to apply the particular rate of tariff to the goods imported from a particular country. The administration applies the minimum tariff rate on the imports from those countries which are entitled for concession either under the 'most favoured nation clause' or on consideration of cordial relations with the tariff-levying country. Maximum tariff rate is generally applied for purposes of commercial bargaining.

Triple Column Tariff—Under this tariff, three rates of tariff—(1) a general rate, (2) an international rate, and (3) a preferential rate—are fixed. The first two tariff rates are similar to the maximum and minimum tariff rates discussed above. The preferential tariff rate is generally applied by the colony country on the imports made from the mother country. The preferential rate of tariff is certain percentage lower than the general rate of tariff. For example, in India before independence goods imported from England were levied the lower import duty under the policy of discriminating protection and imperial preference while imports from Japan, Germany and other continental countries were subject to a higher rate of import duty. In certain cases, while the imports from the other countries were subject to import duties those from the mother country were completely free from import duty.

Effects of Tariffs

The effects of tariff may be analysed either in terms of economy as a whole or in terms of the market for a particular commodity which is one of the many goods, such as sugar. When the tariff issue is examined in terms of the economy as a whole, the analysis is known as the general equilibrium analysis. When it is examined in terms of the market for a particular commodity it is known as the partial equilibrium analysis. For making the partial equilibrium analysis of the effects of an import tariff we make the following assumptions.

A. Consumers' tastes, their money incomes and prices of all other goods remain constant.
B. Technological improvements in the production processes, the externalities and other changes in cost conditions, are ruled out.
C. Any tariff on any material input used in the production of the commodity is ruled out.
D. Transport costs are ignored.

Effect of Tariff Under Partial Equilibrium

The theory of tariffs is quite complicated. Consequently, initially the effects of tariff may be studied under partial equilibrium analysis assuming that the secondary effects of a tariff are non-existent. The different effects which follow the imposition of an import tariff have been discussed under the (1) Protective or production effect; (2) Consumption effect; (3) Revenue effect; (4) Redistribution effect; (5) Terms of trade effect; (6) Competitive effect; (7) Income effect; and the (8) Balance of payments effect.[3]

1. Protective or Production Effect—In order to discuss the protective effect of an import tariff, let us assume that the world supply of the commodity is perfectly elastic so that it is available at constant price while the domestic supply can be increased only at increasing cost. It means that the foreign suppliers are able to supply at the constant supply price any amount of commodity that the domestic consumers demand. It will be reflected in the perfectly elastic foreign supply curve of the commodity at some given price. In Figure 21.1, SS and DD curves represent the domestic supply and domestic demand curves for the commodity respectively. 0P is the constant price at which foreign producers are able to sell the commodity in the domestic market. The foreign supply

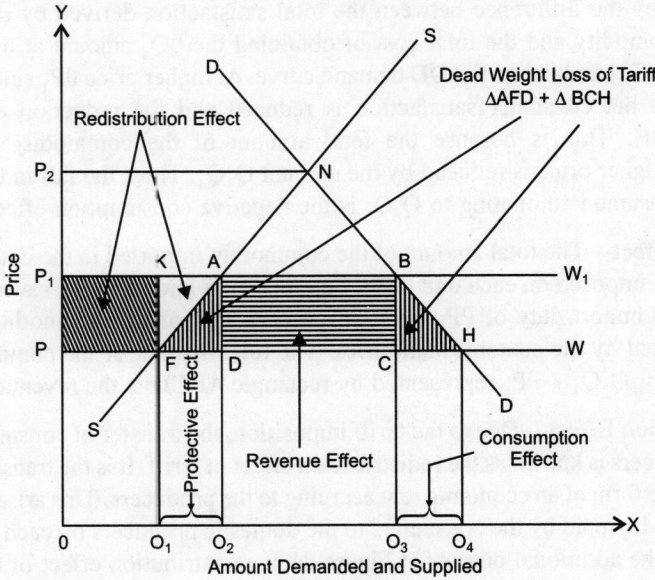

Figure 21.1

3. Charles P. Kindleberger, *International Economics*, Fifth Edition, 1973, pp. 107–128.

of the commodity is, therefore, perfectly elastic at the 0P price and is illustrated by the perfectly elastic foreign supply curve PW. At 0P price, the domestic supply of the commodity is $0Q_1$ while the domestic demand is $0Q_4$. Consequently, there is an excess demand of Q_1Q_4 amount which is met through imports from abroad.

Let us assume that in order to protect the domestic producers, a protective tariff at the per unit rate of PP_1 amount is imposed on the imports of the commodity. The *nominal* or *ad valorem* tariff rate expresses the per unit tariff PP_1 as a percentage of the 0P price. Consequently, the *nominal* or *ad valorem* tariff rate in our example is $PP_1/0P$. Under the assumption of perfectly elastic foreign supply curve of the commodity, the foreign price of the commodity remains unchanged. At the 0P price, the total amount demanded in the country is $0Q_4$ while the total domestic supply is only $0Q_1$. The shortfall of Q_1Q_4 amount in the domestic supply is met by importing the Q_1Q_4 amount. Now PP_1 per unit tariff is imposed. The effect of tariff is to raise the price of the commodity on which import duty is levied. Consequently, the post-tariff price of the commodity in the domestic market increases by the full amount of tariff and becomes $0P_1 (= 0P + PP_1)$. After the imposition of tariff the foreign supply curve shifts upward horizontally from PW to $P_1 W_1$. At $0P_1$ price, imports of the commodity decrease from Q_1Q_4 to Q_2Q_3 while the domestic production of the commodity increases from $0Q_1$ to $0Q_2$. The increase of Q_1Q_2 amount in the domestic production of the commodity is the protective or production effect of import tariff. If PP_2 tariff is levied on imports of the commodity, the domestic producers will be completely protected against the competition of cheap imports because at the $0P_2$ price the total domestic demand for the commodity will be completely met by the total domestic supply. Consequently, there will be no need for importing the commodity.

2. Consumption Effect—If the demand for the commodity is not inelastic, higher the price, smaller will be the total amount demanded of the commodity by the domestic consumers. At 0P price, the total amount of the commodity demanded by the consumers is $0Q_4$. The consumers' surplus measured by the difference between the total satisfaction derived by consuming the $0Q_4$ amount of the commodity and the total cost of obtaining the $0Q_4$ amount at 0P price is the area lying above the line PH and below the DD demand curve. At higher price $0P_1$ since the total amount bought is less, the net consumer satisfaction is reduced and the reduction equals the area of quadrilateral PP_1BH. This is because the total amount of the commodity consumed by the consumers at $0P_1$ higher price is reduced by the amount Q_3Q_4. Thus, the fall in the total amount of the commodity consumed amounting to Q_3Q_4 is the negative consumption effect of tariff.

3. Revenue Effect—The total amount of the commodity imported in the country multiplied by the amount of tariff imposed on each unit of the commodity is known as the revenue effect of tariff. In Figure 21.1, the import duty of PP_1 imposed on each unit of the commodity imported by the importers is collected by the customs authorities. The total amount of the commodity imported is Q_2Q_3. Consequently, $Q_2Q_3 \times PP_1$ represented by rectangle ABCD is the revenue effect of tariff.

4. Redistribution Effect—Due to the tariff imposition, the transfer of consumers' surplus from consumers to producers is known as the redistribution effect of tariff. It is the transfer of income from the consumers in the form of an economic rent accruing to the producers. This arises due to the higher price of the commodity paid by the consumers to the domestic producers on each unit of their entire output and not on the additional output. In Figure 21.1, redistribution effect of the tariff has been shown by the quadrilateral $PP_1 AF$. Of this total quadrilateral area, the area covered by the shaded rectangle PP_1KF is due to the additional economic rent paid to the pre-existing domestic producers and the area measured by the triangle *KAF* is the rent paid to the new producers above their supply price.

In the diagram, the total loss of consumers' surplus due to imposition of import tariff is equal to the area of quadrilateral PP_1BH. Of this total loss of PP_1BH which is suffered by the consumers, PP_1AF amount is transferred to the domestic producers while ABCD amount is taken away by the customs authorities in the form of import duty revenue. The loss of consumers' surplus represented by the shaded triangles AFD and BCH is neither transferred to the domestic producers nor does it accrue to the government. Nevertheless, the consumers suffer this loss. Consequently, the triangles AFD and BCH represent the net loss which is suffered by the society. The loss of shaded triangle AFD is due to the inefficient use of domestic resources. The resources drawn in the new production due to higher domestic price were earning returns equal to $Q_1Q_2 \times 0P$ (= Q_1Q_2DF) in some other industry. But due to protection the expenditure incurred on Q_1Q_2 is Q_1FAQ_2. Consequently, the net loss to the community on the additional domestic production of Q_1Q_2 is AFD. On the other hand, by abstaining from consuming Q_3Q_4 amount of the commodity, the consumers save only Q_3Q_4HC amount of money as against the Q_3Q_4HB amount of loss of satisfaction measured in terms of money. Consequently, the net loss suffered by the consumers due to curtailment in their consumption is shown by the shaded triangle BCH. The total loss of AFD amount suffered on the production and of BCH amount suffered on the consumption is called the 'dead weight loss' or cost of tariff to the community.

5. Terms of Trade Effect—The classical economists believed that the imposition of tariff improved the terms of trade of the tariff-imposing country. Modern economists, however, argue that the specific effects of a tariff depend on the elasticities of demand and supply of the two trading countries. If the foreign supply of the good is perfectly elastic or if the foreign suppliers are ready to supply the effective demand of the importing country at constant price, imposition of tariff will not improve the terms of trade of the tariff-imposing country. However, if the foreign supply f the good is not perfectly elastic, the imposition of tariff will have varying effects as shown in Figure 21.2 depending on the supply and demand elasticities of the two countries.

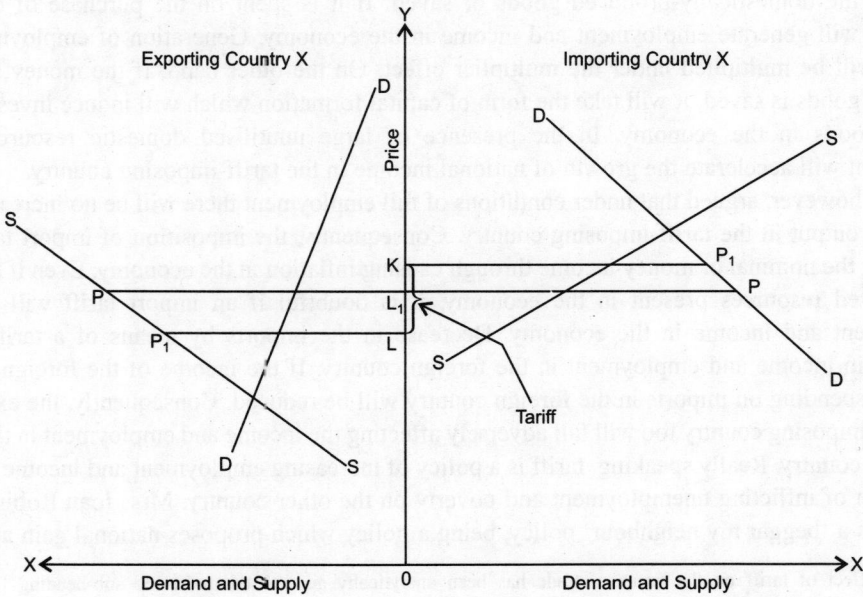

Figure 21.2

In Figure 21.2, Y is the commodity-importing country and X is the commodity-exporting country. The domestic demand and supply curves of country X are inelastic. Country Y imposes a per unit import tariff of KL amount in order to cut down her imports by raising the domestic price of imports. However, a part of the tariff duty is borne by the exporters in country X and the rest is paid by the importers in country Y. As the foreign supply is both large and inelastic and the domestic demand of the country is small, the exporters lower their prices to sell their product to country Y in order to meet the large demand for the imports. Of the total per unit import tariff amount, the exporters in country X bear the LL_1 part of the tariff burden while the L_1K part of the total tariff burden is borne by the importers. Evidently, the tariff burden borne by the importers is less than the tariff burden borne by the exporters, i.e., $L_1K < LL_1$. Consequent upon the imposition of import tariff, the rise in the price of the commodity in the importing country from P $(= 0L_1)$ to P_1 $(= 0K)$ is less than the fall in the price of the commodity in the exporting country from P $(= 0L_1)$ to P_1 $(= 0L)$. In this situation, the terms of trade become favourable for the tariff-imposing country Y.[4]

6. Competitive Effect—It is argued that a growing infant industry which is unable to face the competition of cheap imports produced by the efficient old established industry should be protected by imposing an import tariff. After it has fully developed to face the foreign competition, the import tariff may be removed. After the removal of import tariff, the home industry will be efficient enough to compete with the foreign industry. This is often called the competitive effect of tariff. There are economists who oppose the imposition of import tariff on the so-called infant industry argument. For example, according to Charles P. Kindleberger, "if foreign competition is kept out by tariff the domestic industry tends to become sluggish, fat and lazy."[5] Consequently, an import tariff places a premium on laziness and inefficiency in the protected domestic industry.

7. Income Effect—Tariff reduces the total expenditure incurred on the purchase of imported goods. The money income which is not spent on the purchase of imported goods will either be spent on the domestically-produced goods or saved. If it is spent on the purchase of domestic goods, it will generate employment and income in the economy. Generation of employment and income will be multiplied under the multiplier effect. On the other hand, if the money spent on imported goods is saved, it will take the form of capital formation which will induce investment in capital goods in the economy. In the presence of large unutilised domestic resources, new investment will accelerate the growth of national income in the tariff-imposing country.

It is, however, argued that under conditions of full employment there will be no increase in the total real output in the tariff-imposing country. Consequently, the imposition of import tariff will raise only the nominal or money income through causing inflation in the economy. Even if there are unemployed resources present in the economy, it is doubtful if an import tariff will increase employment and income in the economy. Decrease in the imports by means of a tariff means decrease in income and employment in the foreign country. If the income of the foreign country falls, the spending on imports in the foreign country will be reduced. Consequently, the exports of the tariff-imposing country too will fall adversely affecting the income and employment in the tariff-imposing country. Really speaking, tariff is a policy of increasing employment and income at home at the cost of inflicting unemployment and poverty on the other country. Mrs. Joan Robinson has called this a 'beggar my neighbour' policy, being a policy which proposes national gain at others'

4. The effect of tariff on the terms of trade has been analytically generalised under the sub-heading 'Effects of Tariff under General Equilibrium'.
5. Charles P. Kindleberger, *Op. cit.*, p. 122.

expense. The other country cannot tolerate it and she will retaliate. Consequently, income and employment in both the countries will fall.

8. Balance of Payments Effect—A tariff is a means of correcting the deficit disequilibrium in the balance of payments of the tariff-imposing country through the price mechanism. When the scope for expansion in country's exports is limited, import tariff is imposed to curtail the imports by making these costlier which helps in bringing about a balance in the trade account of the country. As an instrument of correcting the deficit disequilibrium in the balance of payments of the country, an import tariff has been criticised on the following grounds.

1. If the demand for imports in the tariff-levying country is inelastic, tariff cannot reduce the volume of total imports despite the rise in the prices of import goods following the imposition of an import tariff.

2. An import tariff can at best achieve temporary equilibrium in the balance of payments of the tariff-imposing country. It cannot eliminate the fundamental causes of the balance of payments deficit disequilibrium of the tariff-imposing country.

3. If disequilibrium in the balance of payments of the country is due to export surplus, the imposition of tariff will aggravate rather than reduce this disequilibrium. In short, the balance of payments effect of tariff is highly uncertain.

Effects of Tariff Under General Equilibrium

In the general equilibrium analysis, the impact of an import tariff on the entire import competing sector is analysed. Under static conditions, a country gains from international trade through increased specialisation and the exchange of commodities at the more favourable international terms of trade compared with the autarky terms of trade. When tariff is imposed by a country on her imports, import substitution takes place in the tariff-imposing country. Consequently, the export sector of the exporting country contracts. Due to contraction of the export sector, the gains resulting from specialisation (production effect) and exchange of commodity (consumption effect) are reduced. Within the framework of the general equilibrium analysis, these two effects of an import tariff have been explained in Figure 21.3.

Let us assume that the country produces only two goods: cloth and wheat. Let us further assume that under free international trade cloth is an import good and wheat is an export commodity. In the figure, MF curve shows the production frontier of the country. In autarky, the country achieves the simultaneous equilibrium in production and consumption at point U_0 where the country's production possibilities curve MF and the community indifference curve C_0 are tangential (it is the Paretian optimality situation). The international terms of trade have been shown by the slope of line PU_4 showing that cloth is cheaper while wheat is costlier in the international market. At these terms of trade, if the country produces cloth and wheat corresponding to U_0, she will gain by engaging in international exchange of commodities because by giving away a part of wheat in exchange for cloth she can reach point U_1 which is situated on the higher community indifference curve C_1. As the price of wheat in the international market is higher, it will be advantageous for the country to increase the production of wheat and curtail the production of cloth because by exchanging the surplus wheat against cloth she can increase the total community welfare. She produces at point P where the international terms of trade line PU_4 is tangent to country's production possibilities curve MF. In this situation by exporting the DP quantity of wheat in exchange for the DU_4 quantity of cloth, the community will consume the bundle of wheat

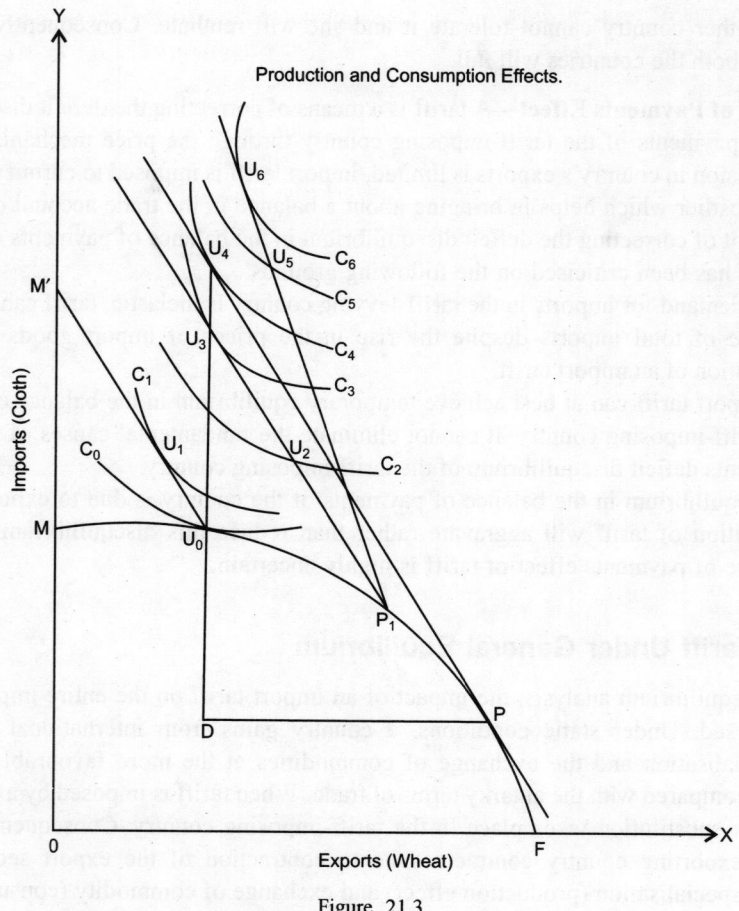

Production and Consumption Effects.

Figure 21.3

and cloth at point U_4 which is situated on a higher community indifference curve C_4. If the country imposes tariff on the imports of cloth, the imported cloth will become costlier in the country. Consequently, resources will be diverted from the production of wheat to the production of cloth. In the changed situation, production will take place at point P_1 which is situated on the country's production possibilities curve MF. If tariff is not imposed and production takes place at P_1 the country will consume through international trade the basket of wheat and cloth at point U_3 which is situated on the community indifference curve C_3. Line P_1U_3 is parallel to the international terms of trade line PU_4. If tariff is imposed, the imports will become costlier in the country. Consequently, the country will curtail the consumption of imports moving down to the lower community indifference curve C_2 which intersects the international terms of trade line P_1U_3 at point U_2. The movement from point U_4 to point U_3 is due to the production effect (due to less specialisation) while the movement from point U_3 to point U_2 is due to the consumption effect (due to smaller exchange of goods) of import tariff.

However, if the country is large enough to influence the international terms of trade by reducing her exports and imports, the tariff will not make her position worse off. If due to smaller exports of wheat the price of wheat becomes higher in the international market the new international terms of trade line becomes P_1U_6. In this situation, after the imposition of tariff the country will consume at U_5 which is situated on a higher community indifference curve C_5 than the pre-tariff equilibrium position U_2 which is situated on the lower community indifference curve C_4.

Terms of Trade Effect—In international trade, a country frequently resorts to import tariff in order to turn the terms of trade in her favour. How far she can succeed in doing so is, however, a different matter and depends upon, *ceteris paribus,* the reaction of the other trading country or countries to the tariff moves of the country concerned. The effect of tariff on the terms of trade can be analysed with the help of Marshallian trade offer curves shown in Figure 21.4.

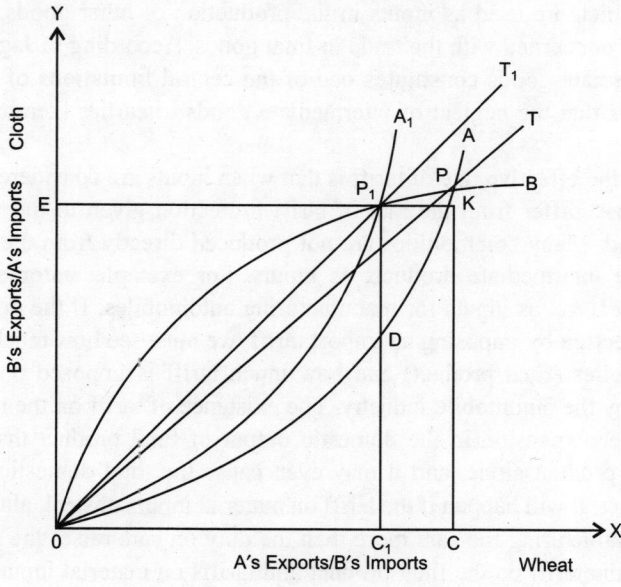

Figure 21.4

In Figure 21.4, the initial pre-tariff trade offer curve of country A is 0*A* and of country B is 0B. Country A exports wheat which has been measured on the X-axis while country B exports cloth which has been measured on the Y-axis. The original trade offer curves of both the countries intersect each other at point P where the balance of trade is in equilibrium. Consequently, the equilibrium terms of trade are indicated by the slope of 0T line. If country A imposes an import tariff, her trade offer curve shifts to the left upwards to the position of the new trade offer curve $0A_1$ which intersects country B's initial trade offer curve at point P_1. Consequently, the terms of trade alter and are represented by the slope of $0T_1$ line. In the new situation, equilibrium takes place at P_1 corresponding to which country A exports $0C_1$ quantity of wheat in exchange for C_1P_1 quantity of cloth. Of the C_1P_1 total amount of cloth imported, DP_1 amount is collected by the customs authorities. If it is an export tariff, the P_1K amount of wheat will be collected by the customs authorities. In both the cases, the terms of trade will move in favour of the tariff-imposing country A. Although before the imposition of tariff, country A was willing to offer 0C amount of wheat in exchange for CP amount of cloth but after the imposition of import tariff she offers only $0C_1$ amount of wheat for C_1P_1(CP–KP) amount of cloth. In other words, country A withholds CC_1 amount of wheat which is the net saving for her. Consequently, there is improvement in the terms of trade of country A after the imposition of import tariff. However, to the extent the terms of trade for country A have improved these have deteriorated for country B after the imposition of import tariff by country A on her imports of cloth from country B. This analysis, however, assumes that country B does not retaliate the tariff moves of country A by imposing retaliatory tariff on her imports of A's wheat.

Effective Rate of Tariff

Empirical studies have shown that a large part of international trade consists of trade in intermediate goods which are used as inputs in the production of other goods. Trade theory has, however, been mostly concerned with the trade in final goods. According to Jagdish Bhagwati, the neglect of the intermediate goods constitutes one of the central limitations of trade theory while Murray C. Kemp feels that the neglect of intermediate goods in earlier literature on trade theory does not invalidate it.

The main idea of the effective rate of tariff is that when inputs are considered, the nominal rate of tariff on a good may differ from the rate of tariff protection given to the value added in the production of the good. Many commodities are not produced directly from the original factors of production but utilise intermediate products as inputs. For example, automobile industry uses tyres, paint, glass, steel, etc. as inputs for manufacturing automobiles. If the domestic automobile industry is given protection by imposing an import tariff, we must see how much tariff is levied on the imported automobiles (final product) and how much tariff is imposed on the imported raw material inputs used by the automobile industry. The existence of tariff on the material inputs will always cause a smaller expansion in the domestic output of final product than if the tariff was imposed on the final product alone, and it may even cause the total domestic production of the final product to contract. It will happen if the tariff on material inputs of steel, aluminium and rubber raises the cost of manufacturing the cars more than the duty on cars raises the price of cars. "The relationship between the tariff on the final product and tariffs on material inputs can be expressed in terms of the *effective rate of protection* enjoyed by producers who *process the final product*."[6]

If the imported automobile parts can enter the country either completely duty free or at low import duty while the automobiles are protected by a high import duty, the effective rate of protection will be very high. On the other hand, if the inputs used in the manufacture of automobiles enter the country under a very high import duty and the automobiles are given a very low protection the domestic price of automobiles will be higher than the world prices and the effective rate of protection will be very low. For example, if the material inputs comprise 50 per cent of the total value of the finished product and if these can be imported duty free and if the rate of import duty on the imported automobile is 50 per cent, the effective rate of tariff will be 100 per cent as against a nominal tariff rate of 50 per cent. If the imported material inputs are levied 50 per cent import tariff, the effective rate of tariff will be equal to the nominal rate of tariff. The effective rate of tariff protection can be calculated by means of the following formula:

$$E_t = \frac{N_t - N_{i-g}}{v}$$

where E_t is the effective rate of tariff on the final product, N_t is the nominal rate of tariff on the final product, N_i is the nominal rate of tariff on the imported material inputs, g is the proportion or

6. P.T. Ellsworth and J.C. Leith, *The International Economy,* Fifth Edition, 1975, p. 232.

percentage of the value of the inputs to the value of the final product and v is the proportion of the value added in the processing.

In general when both the final product and the material inputs are subject to duty, the effective protective rate of tariff enjoyed by the processors increases if the nominal rate of tariff increases and *vice versa;* it decreases if the tariff levied on the material inputs increases and *vice versa.* If the weighted average of tariff on the material inputs exceeds the tariff on the final output, the effective rate of tariff will be negative. It is, therefore, obvious that in calculating the effective rate of protective tariff not only the tariff on the final product should be considered but the tariff on the material inputs should also be taken into account. The tariff on the final product alone is not decisive. It should be observed that, everything else being equal, smaller the value added in an industry, larger will be its effective tariff. For determining the effective rate of tariff, the whole tariff structure plays a significant role in determining the effective protection offered to an industry.

The fact that the effective rate of tariff protection may be different from the nominal rate of tariff protection has interesting implications. It shows that the net protection provided to an industry processing the product can be increased simply by reducing the rates of import duties on the imported material inputs used by the industry rather than by raising the tariff on the imports of final product. The effective protection is frequently substantially greater than the nominal protection on the finished goods. For, example, in 1962 while the nominal tariff rate on the imported textiles in the United States of America was 23.1 per cent, the effective tariff rate was considerably higher, being 50.6 per cent.

Although the notion of effective rate of tariff is interesting it is, nevertheless, a very elusive concept. It is one thing to measure a concept but an entirely different thing to give this measurement a practical meaning and to fit it into a coherent theory. Moreover, the theory of effective rate of tariff differs from the standard tariff theory because it employs the partial equilibrium approach. It is assumed that the general equilibrium repercussions of tariffs are either non-existent or negligible. However, a tariff obviously causes changes in the relative prices. Consequently, the assumption that these tariff-induced price changes will not cause any repercussions on trade volume, factor intensifies, etc. is unrealistic. Secondly, there is also the assumption of fixed factor proportions which is of doubtful validity.

Optimum Tariff[7]

In framing any national tariff policy, the consideration of optimum tariff occupies an important place. From the social welfare point of view, the optimum tariff argument is the only valid argument for a tariff. In the absence of any effective retaliation by the other country, an import tariff under appropriate circumstance—unless the trade offer curve of the other country is perfectly elastic–will improve the terms of trade of the tariff-imposing country enabling that country to obtain her imports cheaper from abroad. If the foreign currency price of the tariff-imposing country's imports falls either by the full or part of the amount of import tariff, in effect the foreign exporters bear either the entire burden or a considerable part of the burden of import duty. The country gains by imposing an import tariff in as much as the terms of trade move in her favour.

7. It is assumed that the tariff proceeds are not redistributed among the people in the country and the government does not spend the tariff proceeds on the imports of goods. It is also assumed in the discussion that the other country does not retaliate the tariff moves of the country. Furthermore, the argument is carried throughout in terms of the two-country and two-commodity exchange model. It is also assumed that the tariff-levying country is not very small because if this was so the improvement in the terms of trade resulting from the imposition of import tariff would be small and insignificant.

There is, however, another weighty aspect of the problem which cannot be ignored while considering the issue of an optimum tariff. As a consequence of levying the import tariff, the post-tariff domestic price of imports will rise in the country unless it is unrealistically assumed that the foreign supply of tariff-levying country's imports in the exporting country is perfectly inelastic. Consequently, unless the very rare and exceptional situation of perfectly inelastic domestic demand for imports prevails in the country, imports will fall following the imposition of tariff. Thus, while an improvement in the terms of trade ensures a gain for the tariff-imposing country, the fall in the volume of her imports and exports which would follow the imposition of tariff causes a loss of satisfaction or welfare to the community in the country.[8] Often the negative effect of misallocation of resources in production and of distortion in consumption consequent upon decrease in the volume of imports and exports would be larger than the positive effect resulting from an improvement in the terms of trade. There is, therefore, a cost involved in imposing a protective tariff measured in the form of loss suffered through decrease in the volume of trade.

It may, however, be that the positive effect of tariff on the terms of trade is larger than its negative effect on the volume of trade. In such a situation, as long as the gain which accrues to the tariff-imposing country due to improvement in her terms of trade exceeds the loss resulting to the country from shrinkage in the volume of her trade as a consequence of tariff, tariff means a net addition or increase in the community welfare of the tariff-imposing country and the country can improve its welfare by applying the "right" tariff. It is evident from the fact that the tariff-imposing country, in the absence of any effective retaliatory action taken by the other country in the form of imposition of similar tariff on her imports, reaches the equilibrium at a point which touches a higher trade indifference curve[9] than the one which was attained at the free trade equilibrium position. There will be a particular rate of import tariff at which the total gain will be optimum, i.e., the difference between the gain accruing to the country from the improvement in her terms of trade and the loss suffered by the country as a consequence of the fall in her total imports (and exports) will be maximum. This tariff is the *optimum tariff* in the sense that the net gain accruing to the tariff imposing country by levying the tariff is maximised. To express in slightly different words, an optimum tariff is reached when the gain due to the movement of the terms of trade in favour of the country imposing the tariff exceeds the loss resulting from the curtailment in the volume of imports (and exports) by the greatest possible margin. In other words, an optimum tariff is the welfare maximising tariff for the country which imposes the tariff. This will be realised at that point where the trade indifference curve of the tariff-imposing country is tangent to the trade offer curve of the other country. Figure 21.5 shows that at point F the highest trade indifference curve TICA$_4$ of country A touches the trade offer curve 0B of country B. Of the total amount of FK imports, FL amount of imports is collected by the government in the form of total tariff proceeds and the *ad valorem* tariff rate is FL/KL × 100 per cent. This tariff rate is called the "optimum tariff" because it maximises the total welfare of country A.

Beyond this optimum tariff rate, at higher rates of tariff although a welfare benefit will still be realised but it will be diminishing until the tariff rate reaches the level of GM/NM × 100 per cent

8. Both the countries simultaneously can never gain by imposing an import tariff, and one country will gain while the other will lose only if the new equilibrium falls within one of the two shaded areas shown in Figure 18.5.
9. A trade indifference curve is the locus or path of the different possible alternative combinations or pairs of imports and exports from which the country derives the same or equal satisfaction. Consequently, the country is indifferent between all the different import-export combinations which are situated on a trade indifference curve.

where, as is indicated by the trade indifference curve TICA$_2$ of country A which passes through points E and G, the gain from imposing the tariff will be nil for the tariff-imposing country A while country B suffers throughout as the result of country A's tariff moves. For still higher rates of tariff, both the countries will suffer a loss. Country A will actually be losing her welfare throughout, the extent of such loss being larger with a higher rate of tariff and with the lower elasticity of its own reciprocal demand. As may be seen from Figure 21.5, beyond point G country A loses from tariff because while improvement in the terms of trade is still possible, such improvement leads to the decline in the volume of trade of such a high magnitude that the loss suffered due to the decline in the volume of trade more than offsets the gain resulting from the improvement in her terms of trade. The 'feasibility region' within which a country will gain by imposing tariff has been shown by the two shaded areas in Figure 21.5 and the optimum tariff must lie somewhere within the shaded area relevant for a particular country.

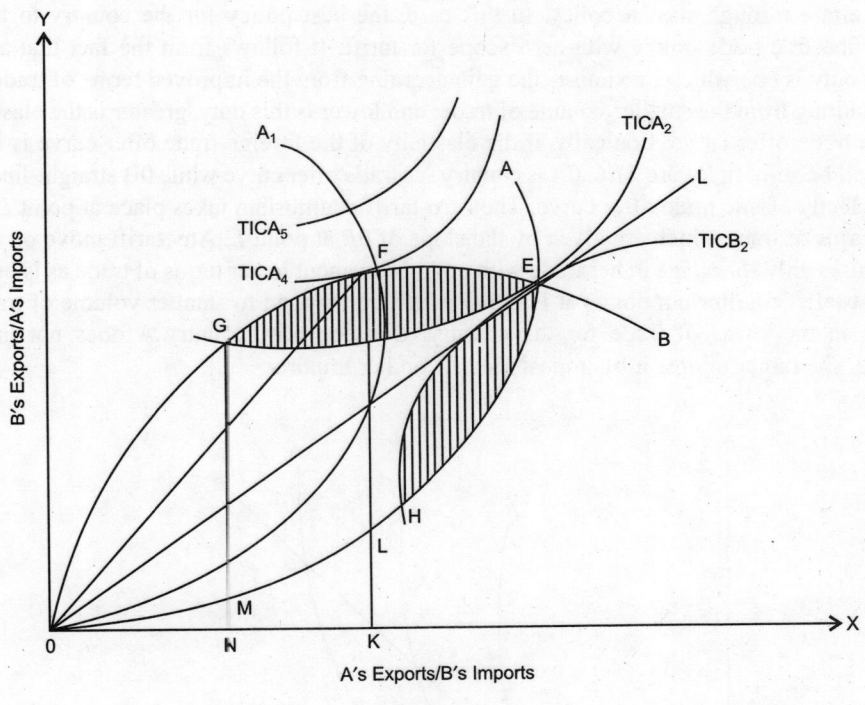

Figure 21.5

We must, however, remember that an optimum tariff only potentially maximises a country's welfare. It will, like any tariff, affect the income distribution in the country by making some people better off and others worse off. Thus, in order that the optimum tariff should result in an unambiguous increase in the community welfare, it should be accompanied by a redistribution policy which seeks to compensate adequately the sufferers of import tariff. Since by definition an optimum tariff increases a country's total welfare, it should be possible to compensate those who have been hurt by the tariff and there will still be something left over for those to enjoy who have been directly benefited by the tariff. Thus, an optimum tariff is an important theoretical theme in welfare economics.

The improvement in the terms of trade of the country resulting from an import tariff will, however, depend on the elasticity of the foreign trade offer curve. In the event of the foreign offer curve being perfectly elastic (i.e., when the elasticity of reciprocal demand of the other country is infinite) represented by the straight line drawn from the point of origin of the axes, the imposition of tariff will only slash the volume of trade while it will cause no improvement in the country's terms of trade. In this case, the optimum solution is a zero-tariff. It is only another way of stating that where the foreign trade offer curve faced by the country is perfectly elastic, the tariff-imposing country cannot improve her welfare through such a policy. In this situation, free trade is the best policy because the positive effect of tariff flowing from improvement in the terms of trade is zero as there is no improvement in country's terms of trade while the country will suffer from the negative effect of tariff due to shrinkage in the volume of her exports and imports.

Figure 21.6 shows that if the tariff-imposing country A faces the perfectly elastic trade offer curve of country B, it cannot gain in any manner by imposing the tariff; she will rather injure her own welfare through such a policy. In this case, the best policy for the country to follow is to pursue the free trade policy with zero scope for tariff. It follows from the fact that an optimum import duty is one which maximises the gain accruing from the improved terms of trade minus the loss resulting from the smaller volume of trade; and lower is this duty, greater is the elasticity of the foreign trade offer curve. Logically, if the elasticity of the foreign trade offer curve is infinite, the tariff will be zero. In Figure 21.6, 0A is country A's trade offer curve while 0B straight-line is country B's perfectly elastic trade offer curve. The zero-tariff equilibrium takes place at point E at the free trade terms of trade which are given by the slope of 0B at point E. Any tariff move on country A's part causes only shrinkage in her trade with no improvement in her terms of trade as is evident from the post-tariff equilibrium points at F and G which correspond to smaller volume of trade with no change in the terms of trade for the country. Consequently, country A does not improve her welfare; she rather injures it by imposing tariff on her imports.

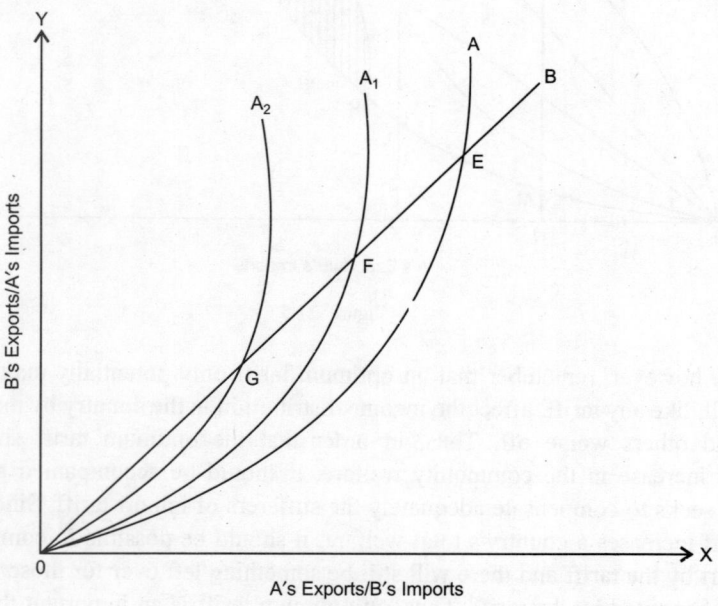

Figure 21.6

The concept of optimum tariff warns the tariff-imposing country against raising the rate of tariff too high in her temptation to improve her terms of trade because if raised beyond a certain limit the loss suffered by the country due to shrinkage in the volume of her trade will exceed the possible gain which accrues to the country due to improvement in her terms of trade.

Tariff Retaliation

The discussion of optimum tariff has been carried on under the assumption that the other country (country B) does not retaliate the tariff moves of country A. To think, however, that country B will not retaliate the tariff moves of country A which injures her welfare is unrealistic. Whenever it is possible to do so, the other country will give a befitting reply by effectively retaliating the tariff moves of the country by imposing a suitable retaliatory tariff on her imports. Thus, if country A optimises her welfare through improving her terms of trade by means of an import tariff, country B may successfully manage to keep the terms of trade unchanged by imposing the appropriate retaliatory tariff on A's exports. This act of imposing import duty by a country in reply to the imposition of tariff on her exports by some other country is known as retaliation or war in tariff. By means of a suitable retaliatory tariff, a country may prevent the other country from improving the terms of trade in her favour by means of imposing an import or export duty.[10] In this tariff war, the volume of trade is, however, reduced substantially reducing the community welfare in both the countries. This has been illustrated in Figure 21.7.

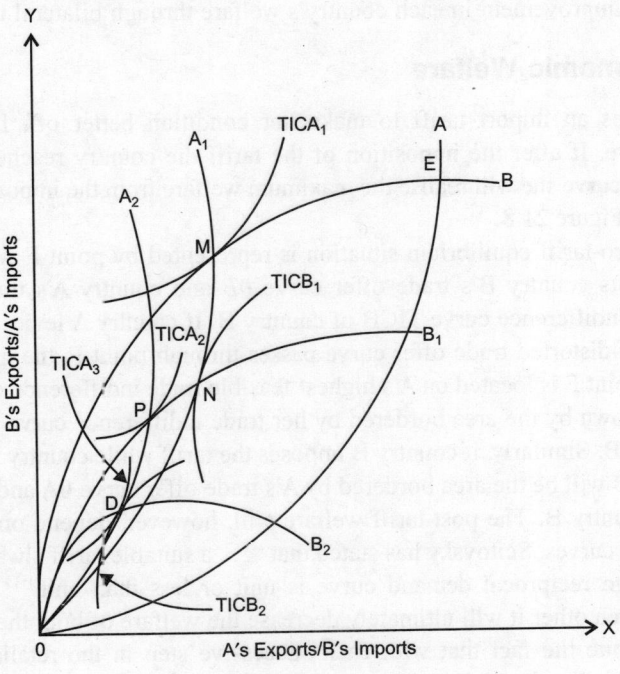

Figure 21.7

10. Harry G. Johnson has, however, asserted that a country *may* gain by imposing an import tariff even if the other country retaliates to the tariff moves of the country. He has determined the conditions under which it *will* gain in a special group of cases. See Harry G. Johnson, *International Trade and Economic Growth*, Fifth Impression, 1970, Chapter II.

In Figure 21.7, the initial trade offer curve of country A is 0*A* and that of country B is 0*B*. Both these trade offer curves intersect at point *E*. Let us assume that country A imposes import tariff in such a way that her tariff-distorted trade offer curve 0*A*. passes through point *M* where country A's trade indifference curve $TICA_1$ is tangent to country B's trade offer curve 0*B*. The terms of trade move in favour of country A. Faced with the new trade offer curve 0*A*$_1$ of country A, country B may impose suitable retaliatory tariff on her imports from country A. She will impose import tariff in such a way that her tariff-distorted trade offer curve 0*B*$_1$ passes through point *N* where her trade indifference curve $TICB_1$ touches country A's trade offer curve 0*A*$_1$. At point *N* country B is on a higher trade indifference curve than she was at point *M*. The process of retaliation and counter-retaliation can continue for many rounds until a point is reached where country A's trade indifference curve is tangent to country B's trade offer curve at a point which lies on country A's trade offer curve. In the figure, this situation is located at point D. After D no further tariff could allow the first country to reach a higher trade indifference curve. It is obvious that each successive round of tariff war will result in a successively smaller volume of trade between the two countries. The volume of trade at the zero-tariff point *E* is greater than that at point M, which in turn is larger than at point N which is larger than at point P which is larger than at point D. In other words, a retaliatory tariff war causes progressively increasing shrinkage in the total exports and imports of the two countries inflicting avoidable misery on both the countries.

An important practical conclusion which follows from this analysis is that whenever both the trading partners (country A and country B) restrict trade by means of an import tariff, it is always possible to cause improvement in each country's welfare through bilateral trade liberalisation.

Tariff and Economic Welfare

A country imposes an import tariff to make her condition better off, i.e., to enjoy a higher community welfare. If after the imposition of the tariff the country reaches the highest possible trade indifference curve she will realise the maximum welfare from the imposition of tariff. This has been explained in Figure 21.8.

The initial zero-tariff equilibrium situation is represented by point E where country A's trade offer curve 0*A* cuts country B's trade offer curve 0*B* and country A's trade indifference TICA touches the trade indifference curve TICB of country B. If country A levies import tariff in such a way that her tariff-distorted trade offer curve passes through point F, the gain from trade will be maximum since point F is located on A's highest feasible trade indifference curve TICA$_1$. The gain of country A is shown by the area bordered by her trade indifference curve TICA and country B's trade offer curve 0B. Similarly, if country B imposes the tariff while country A does not impose any tariff, the gain of B will be the area bordered by A's trade offer curve 0A and the trade indifference curve *TICB* of country B. The post-tariff welfare will, however, depend on the elasticities of the reciprocal demand curves. Scitovsky has stated that ". . . a suitable tariff always gives welfare if the elasticity of foreign reciprocal demand curve is unit or less than unit."[11] If both the countries retaliate against each other it will ultimately decrease the welfare of both the countries involved in the tariff war despite the fact that with each successive step in the retaliation war the country imposing the tariff will gain slightly. Thus, a single country can always gain by levying an import tariff (unless the trade offer curve of the other country is perfectly elastic which is unlikely to be the case) provided that the other country does not retaliate the tariff move of the country. However,

11. Tiber Scitovsky, "A Reconstruction of the Theory of Tariff", *The Review of Economic Studies*, 1942, p. 19.

gain substantially from exploiting their bargaining power by applying high tariffs. Thus optimum tariffs are likely to be high, particularly bargaining with the developed countries.

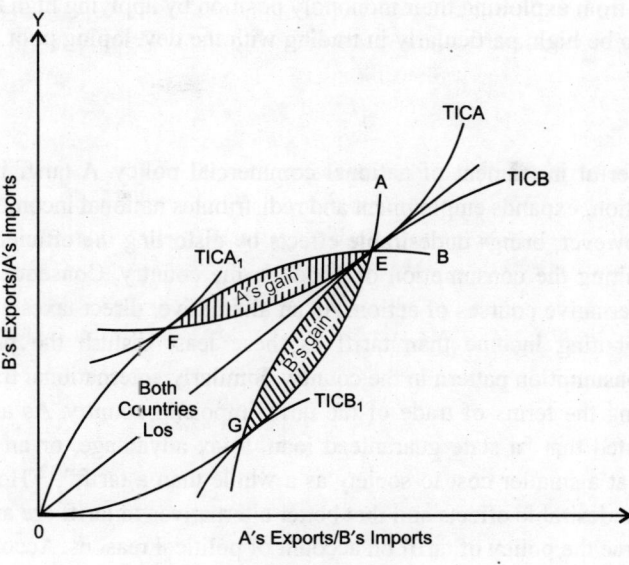

Figure 21.8

taking both the countries as a single unit, free trade is the optimal policy since one country's gain from an optimum tariff will always be smaller than the other country's loss which results from it.

After having explained the meaning of an optimum tariff, the question arises: what determines the height of the optimum tariff? In this connection it may be stated that higher is a country's share of foreign trade, larger is the scope for its optimum tariff. It follows from the fact that larger the share (or percentage) of national income which the country exports, more effective will be a given improvement in the country's terms of trade. If this is not the case, the important factors that determine the height of an optimum tariff stem from the second country—rest of the world. In this connection, the effects produced by a change in the relative prices on the quantities consumed and supplied are of utmost importance. The larger these effects are, the lower is the optimum tariff that a country should levy. A tariff in country A will turn the terms of trade against country B. If country B's price elasticity of supply is high, the country's production structure is flexible enabling producers to shift the factors of production from one line to another line of production. Consequently, there will not be great change in the terms of trade due to tariff and country A will not gain much by levying a tariff. If the elasticity of supply of country B is very large (infinite) there will be no change at all in the relative prices due to the imposition of tariff. In such a situation, the best policy to follow for country A will be that of free trade. This position will also hold good if country B's elasticity of demand for exports (or imports) with respect to a change in relative prices is high. It means that goods are substitutes and consumers can adapt their habits of consumption to a change in relative prices without difficulty.

Treating the issue in the context of the developed and developing countries, it has been observed that usually the developed countries have great flexibility both in their supply and demand, i.e., both in production and consumption while the less developed countries suffer from inflexibility on both these counts. Consequently, the developing countries usually cannot gain much from high tariffs and their optimum tariffs are quite low while the developed countries can gain substantially from exploiting their monopoly position by applying high tariffs. Their optimum tariffs are likely to be high, particularly in trading with the developing poor countries.

Conclusion

Tariffs are a powerful instrument of national commercial policy. A tariff increases the revenue, stimulates production, expands employment and redistributes national income in the tariff-imposing country. Tariff, however, brings undesirable effects by distorting the efficient allocation of scarce resources and limiting the consumption of goods in the country. Consequently, it is desirable to investigate the alternative courses of action. As an alternative, direct taxes and transfers are better means of redistributing income than tariff as these least disturb the pre-existing free trade production and consumption pattern in the country. Similarly, international transfers are superior to tariff for improving the terms of trade of the tariff-imposing country. As an alternative to tariff, Heller has suggested that "a state guaranteed loan, a tax advantage, or an outright subsidy may help the industry at a smaller cost to society as a whole than a tariff".[12] However, even knowing that tariff brings undesirable effects and that better alternatives to tariff are available, we are not in a position to divorce the policy of tariff on account of political reasons. According to Kindleberger, "the movement for tariff is strong because producers' interests in particular are politically more powerful than producer and consumer interests in general".[13] From the international point of view, however, a tariff can only be defended when it is levied to protect the infant industry against the competition of the developed industries abroad.

Tariffs and the WTO

Established on January 1, 1995 as a successor to the erstwhile GATT, the World Trade Organisation (WTO) is the legal and institutional foundation of the non-discriminatory multilateral trading system. For almost fifty years, key provisions of the GATT outlawed discrimination among the members and between imported and domestically produced goods. Following the establishment of the GATT in 1948, average tariff levels have fallen progressively and dramatically from an average of 40 per cent in 1947 on the eve of the GATT establishment to less than 4 per cent in 1994 through a series of seven trade rounds.

Although legal in the WTO framework, tariffs are commonly used by governments to protect the domestic industries and to raise revenues. These are, however, subject to disciplines. For instance, tariffs have to be non-discriminatory among imports and are largely 'bound'. By binding is meant that a tariff level for a particular good becomes a commitment by a WTO member and cannot be increased without comprehensive negotiations with its main trading partners. For example, it can be the case that the extension of a customs union can lead to higher tariffs in some areas for which comprehensive negotiations are necessary. After the establishment of the WTO, the tariffs and quotas have been progressively reduced/eliminated during the past one decade.

---

12. H. Robert Heller, *International Trade*, Second Edition, 1973, p. 164.
13. Charles P. Kindleberger, *Op. cit.*, p. 128.

Suggested Readings

P.T. Ellsworth and J.C. Leith, *The International Economy,* Fifth Edition, 1975, Chapter 13.

Gottfried Von Haberler, *International Trade,* 1950, Chapters XVI and XIX.

H. Robert Heller, *International Trade,* Second Edition, 1973, Chapter 9.

Harry G. Johnson, *International Trade and Economic Growth,* 1958, Chapter II.

Charles P. Kindleberger, *International Economics,* Fifth Edition, 1973, Chapter 7.

Bo Sodersten, *International Economics,* Second Edition, Reprinted 1981, Part III, Chapters 13 and 15.

Questions

1. Explain fully the effective rate of tariff and the optimum tariff. Show the importance of these two concepts in the theory of trade policy.
2. Explain the effects of tariff on the terms of trade and the domestic price ratio.
3. Explain and show in a diagram the effects of an import tariff on production, consumption, government revenue and imports of a commodity in the tariff-imposing country under partial equilibrium. How does tariff affect the distribution of income and community welfare in the tariff-imposing country?

CHAPTER 22

Quotas, Dumping and Cartels

Quotas, dumping and cartels are the different forms of restrictions on the free flows of international goods and services. The purpose of these restrictions is to bring about a change in the volume and direction of free trade. The quotas represent physical quantitative restriction on trade.

1. Quotas

When the free market forces fail to bring the desired result in the pattern of foreign trade of a country several kinds of restrictive measures are practised by the government of a country in order to freeze the volume and direction of trade along the desired channels in order to achieve certain given objectives. Import and export quotas are one of the several important restrictive trade practices which limit either the value or the total quantity of the commodity to be imported into or exported from the country during any specified period of time, the shortest time span being a month and the longest being a year. The use of quotas as a restrictive device was almost ubiquitous in the European and other countries in the interwar period when these countries were under serious deflationary pressure from abroad and in the postwar period when they were forced with the shortage of foreign exchange.

Types of Quota

The system of quota that a country may practise may be divided into the following five main categories:

1. Tariff or custom quota,
2. Unilateral quota,
3. Bilateral quota,
4. Mixing quota, and
5. Licensing of imports.

1. Tariff or Custom Quota—Under the tariff or custom quota system, a specified quantity of imports of a good is allowed by the government of the tariff quota levying country either import duty free or at a low rate of import duty. Additional imports beyond the specified quantity are permitted only at an enhanced rate of import duty. A tariff quota, therefore, combines the features of a tariff with those of an import quota. A tariff quota may either be an *autonomous* quota fixed by law or by decree or an *agreed* quota fixed by the force of some trade agreement made by the tariff quota imposing country with one or more other countries.

The tariff quota system has the advantage insofar as the higher rate of duty imposed on the imported goods discourages their imports and the scarce foreign exchange is saved to be spent on the imports of those goods which are vitally needed for the economic development of country. Furthermore, the home market is not completely insulated from the world market since the imports beyond the import quota limit are not completely prohibited. Consequently, the price of the commodity in the domestic market remains linked with its price in the world market and it cannot exceed the world price by more than the amount of additional import duty.

However, notwithstanding these merits the tariff quota system has the following drawbacks.

(i) When the imports exceed the fixed physical quantity specified under the low rate of import tariff, the entire gains from the low rate are enjoyed by the quota-exporting countries.

(ii) The rush of the imported goods in the importing country may disturb the domestic price level in the importing country.

(iii) It discriminates against the poor consumers who find it difficult to pay the higher rate of import duty. It is generally the rich people who import additional quantity of a commodity at higher rate of import duty.

2. Unilateral or Autonomous Quota—Under unilateral quota system, the total quantity or the total value of the commodity which can be imported in the country is fixed without making prior consultations with the exporting countries. The unilaterally fixed quota may be either global or the quota may be allocated between the different exporting countries according to certain criteria fixed by the quota-fixing country's government. Under the global quota, the entire quota quantity may be imported either from anyone or more countries during any specified period of time. Under the allocated quota system, the total quota quantity is distributed between the different exporting countries on the basis of certain principles.

The global quota system keeps the door of the importing country open for all the exporting countries. The different exporting countries compete among themselves in order to capture the market of the importing country. Consequently, the importing country can take full advantage of the competition prevailing among the exporting countries in the form of low price and other favourable terms offered to her by the rival exporting countries. The global quota system has, however, the following defects.

(i) It favours the nearby sources of supply of imports as against the distant ones. In their anxiety to receive the imported goods quickly, the importers in the quota-fixing country prefer to place orders for the imports with the exporters of those countries which are situated nearby and from where the goods take the least time to reach the port of destination in the importing country. Consequently, the countries situated far away are placed at disadvantage for no fault of theirs.

(ii) It discourages imports from small or less organised countries.

(iii) As soon as the import quota is announced by the government of the import quota-fixing country, in the beginning the economy of the country is flooded with imports and prices fall very rapidly. Subsequently, when due to the limited quota no further imports can be made, the economy is faced with acute shortage of imports and the stockists raise the prices of imports. Consequently, the unilateral quota system encourages speculative stock-piling by the stockists causing wide fluctuations in the prices of imports.

(iv) It does not provide regular protection to the domestic industry.

(v) The importers in the import quota-levying country rush to import the entire quota quantity in the beginning of the period leading to avoidable stock-piling involving substantial storing expenses which are particularly very heavy in the case of perishable goods. This raises the prices of imported goods diminishing in the process the aggregate consumer welfare.

Due to these drawbacks, the unilateral global quota has not found much favour and has given place to the allocated quota which is simply the method of distributing the total quota of imports between the different exporting countries. The unilateral allocated quota system is, however, not free from shortcomings. Following are the main drawbacks of the unilateral allocated quota system.

(i) It imposes avoidable rigidity on the sources of supply of imports and does not take into account the cost conditions abroad.

(ii) On account of the certainty of the share of quota among the exporting countries, the unilateral allocated quota system encourages monopoly practices by those countries which export the allocated quota quantity.

(iii) There is enough scope for practising partiality in the matter of allocation of the quota quantity among the exporting countries.

3. Bilateral Quota—Under the bilateral quota system, quotas for the different exporting countries are fixed after negotiations between the import quota-fixing country and the exporting countries. This system has the following advantages.

(i) Since the quota quantities are determined by mutual negotiations and agreement between the exporting and importing countries, the possibility of discriminating between the different exporting countries is minimised. Consequently, the scope for distrust, heart-burning and mutual bickering between the different exporting countries and the importing country is minimised.

(ii) As the shipment of import goods is spread evenly over the entire quota period, price fluctuations are minimised.

(iii) Since it prevents discrimination as between different exporting countries on the part of the import quota-fixing country, it provokes no retaliatory action on the part of the exporting countries.

(iv) Since producers of both the countries are taken into confidence in determining the quota quantity of the commodity, the bilateral quota system functions more smoothly than does the unilateral quota system.

Following are the principal drawbacks of the bilateral quota system.

(i) The bilateral quota system encourages the formation of international cartels.

(ii) The bilateral quota system opens the way for large-scale competition.

(iii) Since it has a tendency to raise the price level in the exporting country, the importers stand to lose.

(iv) According to some economists, it is an open invitation to monopoly in the exporting country.

4. Mixing Quota—Mixing quotas are also referred to as "indirect quotas" in the sense that quota regulations, instead of being applied to importers, are applied to domestic producers. The domestic producers in the quota-fixing country are required by the quota-fixing authority to use a fixed proportion of the domestic raw materials in combination with the imported raw materials to produce a finished good. This system of quota is advantageous because (i) it provides protection to the domestic producers of raw materials, and (ii) it saves the precious scarce foreign exchange for the country.

The system of mixing quota has been criticised on the following two grounds.

(i) The compulsory utilisation of the domestic raw materials, if these are of low quality, may cause deterioration in the quality of the finished product.

(ii) The compulsory use of the domestic raw materials is not in accordance with the principle of comparative cost advantage theory. The deviation from the principle of comparative cost advantage is an invitation to productive inefficiency resulting in the relatively higher cost of domestic production.

5. Licensing of Imports—Under import licensing, the importers in the import-licensing country are required before importing the goods to obtain an import license from the license issuing authority for a specific quantity of import quota. The system of licensing the imports has the following merits.

(i) The import license-issuing authority (chief controller of imports) in the country while issuing the import licenses keeps a vigilant watch on the existing position of country's total foreign exchange reserves. Since the import licenses are issued keeping due regard to the foreign exchange reserves' position of the country the possibility of the foreign exchange crisis occurring is eliminated.

(ii) Import licenses are generally distributed to the established importers in the country.

(iii) It discourages speculative activities in the foreign exchange dealings and minimises the fluctuations in prices.

The system of import licensing, however, has the following shortcomings.

(a) Since the authorities administering the import licenses are entrusted with wide discretionary powers this system leads to corruption, favouritism and nepotism. It encourages red-tapism and in the hands of corrupt government administration it leads to import license scandals whereby import licenses are issued in the names of bogus importers. This is particularly so in the developing countries where the standard of public morality is low and the administrators are easily corruptible.

(b) This system avoids the entry of new importers in the trade whose entry in business is very essential for improving the efficiency of import trade. Consequently, the import trade in the country tends to be monopolised by a handful of big established importers who thrive on import licenses and make a capital out of their contacts with the government high-ups. Once these established importers obtain import licenses in their names they sell these import licenses to others at high profits in the black market.

(c) Sometime the import licenses are obtained by the unscrupulous importers who sell these for much higher value than the face value of the license in the open market. While this creates inflation in the domestic economy, the unscrupulous license holders gain at the cost of the society.

Objects (Purposes) of Quota

The government of a country resorts to import quota for different reasons. Fundamentally, when the external balance of payments of a country is adverse and this adverseness cannot be eliminated through other easier means, the system of import quota comes in handy for the government. When it is not possible to minimise the import surplus in the balance of trade of the country by other methods, import quotas are fixed to restrict imports effectively and promptly. Import quotas are applied as one of the measures to correct the adverse disequilibrium of the external balance of payments position of the country. Import quotas may also be used to protect the domestic industry against the cut-throat competition from cheap imports by limiting the total quantity of foreign goods which can be imported into the country. It is often useful to use the import quota system for stabilising the internal price level by limiting the total inflow of imports in the country.

A country may also resort to import quota for purposes of commercial bargaining. A country may allot quota of her imports to those exporting countries which in exchange allot to this country quota of their imports. Sometime import quotas are also used as a retaliatory measure. Import quotas are used as an effective instrument for checking the entry of speculative imports in the country which are possible in anticipation of the enactment of tariff legislation to be enforced in future. Import quotas are also used for enforcing equitable distribution of imports among the importers of those goods which are in short supply.

Effects of Quota

The quota system raises problems which are absent in the case of tariffs under which the amount which can be imported and by whom is determined by the market forces of supply and demand. Under tariffs, anyone can import any amount he pleases so long as he pays the duty. But if the total amount which can be imported is restricted to a certain quantity, a new principle of selection of importers divorced from the play of market forces is introduced. The fixing of a quota is an interference with the free market price mechanism.

If the domestic demand and supply curves of a commodity are not inelastic there is little difference whether the country imposes a tariff on her imports or fixes a quota for her imports because given the quantity of import quota, the effective rate of import tariff which reduces the total imports of the commodity to the given quota quantity can be easily calculated. In short, given the form of the demand and supply curves, for every import quota there is an equivalent import tariff as a result of whose imposition the total imports of the country will be equal to the given import quota which the country fixes for her imports. If the quantity of import quota is fixed at that quantity of imports which would result from the imposition of a given import tariff, the protection, consumption, and redistribution effects will be identical in both the cases. This has been illustrated in Figure 22.1 where SS is the domestic supply curve and DD is the domestic demand curve of the country for sugar.

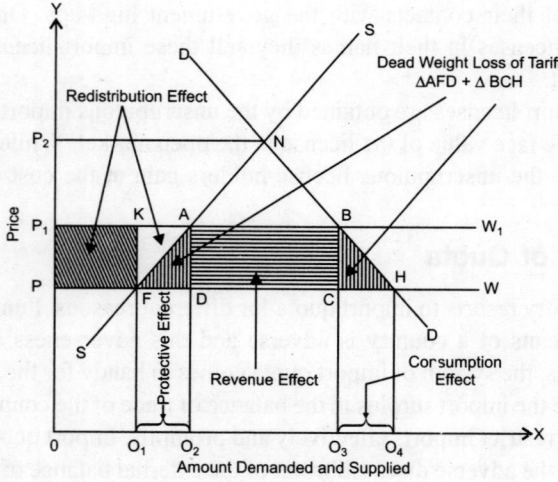

Figure 22.1

Let us assume that at 0P price the foreign supply of sugar is perfectly elastic. At this price the domestic demand for sugar is $0Q_3$ while the supply of the product produced by the domestic producers is only 0Q. Consequently, there is excess demand gap present in the country at this price. The excess demand for sugar amounting to $QQ_3 = (0Q_3 - 0Q)$ will be met through imports from the foreign country if trade is free. If $Q_1 Q_2$ quantity of sugar is fixed as import quota by the government this much quantity of sugar will be imported if instead of fixing the import quota, the government imposes an import tariff equivalent to PP_1. In this situation, there is no difference between an import tariff and an import quota as regards the protective effect QQ_1, the consumption effect Q_2Q_3, and the redistribution effect $PDAP_1$. However, there is difference between an import tariff and an import quota as regards the revenue effect of the two measures. When the PP_1 import tariff is imposed by the government, *ABEF* amount of revenue is collected by the customs authorities of the importing country. On the other hand, if an import quota of Q_1Q_2 imports is fixed by the government the price of sugar in the country will rise to $0P_1$. The question is: to whom will this gain due to the increase in the domestic price of sugar accrue? Nothing can be said *a priori* with certainty. There are three possibilities. Firstly, the government may auction the import licenses in the market and may obtain this extra surplus value due to scarcity. If this happens there is no difference between an import tariff and an import quota. But this device is not widely used by modern welfare states which are averse to the idea of license auctioning since this arouses avoidable public criticism. Secondly, if the exporters abroad are organised and the importers in the country are not, the terms of trade may swing against the importing country and the entire gain may be pocketed by the exporters. Thirdly, if the importers enjoy a monopoly buying power in trade they may succeed in pocketing the entire gain. In practice, however, it is more realistic to assume that the gain will be shared between the exporters and importers with the share of each group depending upon its organisation and bargaining power. Thus, the terms of trade will remain indeterminate. The indeterminate terms of trade effect of an import quota has been explained in Figure 22.2.

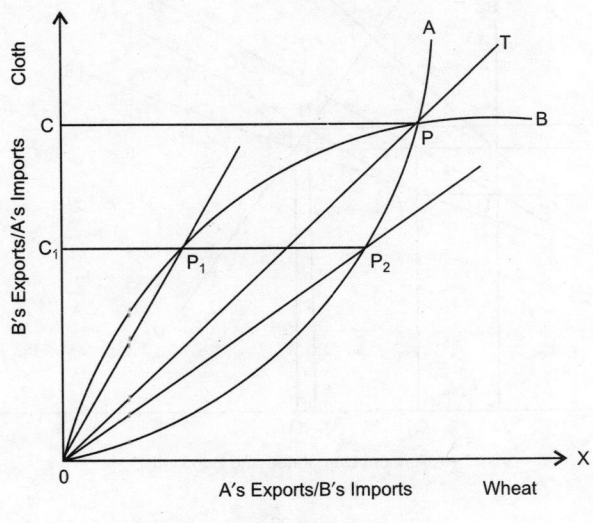

Figure 22.2

Terms of Trade Effect—Let us suppose that there are two countries A and B. In Figure 22.2, $0A$ and $0B$ are respectively the trade offer curves of country A and country B. Country A's import good is cloth and country B's import good is wheat. The equilibrium terms of trade have been shown by the slope of the terms of trade line $0P$. At these terms of trade, country A imports $0C$ amount of cloth. If country A limits her imports of cloth to $0C_1$ quantity, the terms of trade between wheat and cloth may be either as shown by the slope of line $0P_1$ or as shown by the slope of line $0P_2$. If the exporting country B has a monopoly power in trade, the terms of trade will be more favourable to her. Consequently, the new terms of trade will be shown by the slope of the terms of trade line $0P_2$. Conversely, if the importing country A has a monopoly power in trade, the terms of trade will be more favourable to her and the new terms of trade will be represented by the terms of trade line $0P_1$. Thus, after the fixation of import quota the terms of trade will be indeterminate. Since generally traditional importers are more organised the new terms of trade will be more favourable to them.

Price Effect—By limiting the physical amount of goods which can be imported, import quota raises the prices of those commodities which are subjected to quota restrictions. This is also true in the case of import tariff. There is, however, one important difference between the two. The rise in the domestic prices of import goods caused by an import tariff is limited to the amount of import duty less any fall in the prices of import goods that may take place abroad in the country of their supply. But as a result of import quotas prices can rise to any extent since quotas place absolute limit on the physical amount of imports allowing for no adjustment to take place through a fall in the prices abroad. Consequently, the price determination is left purely to the interaction of supply and demand forces. This has been illustrated in Figure 22.3.

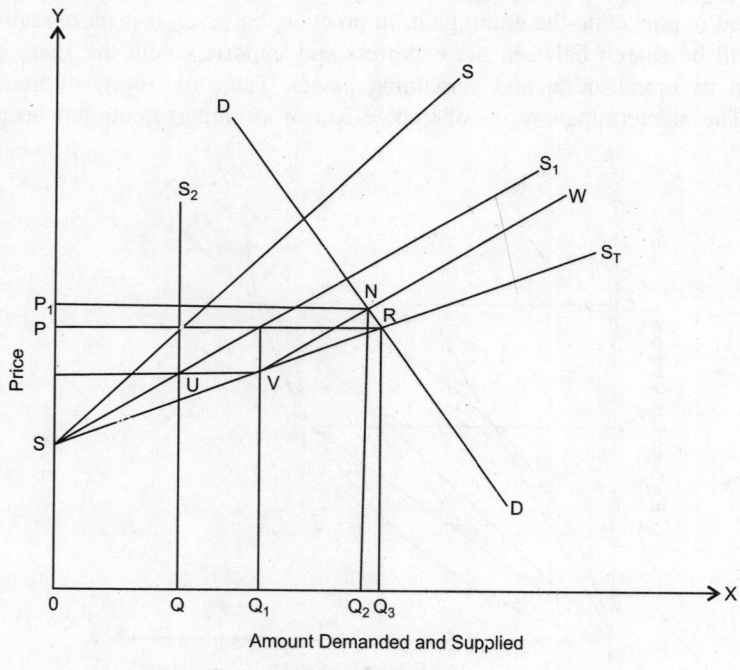

Figure 22.3

In Figure 22.3, DD is the domestic demand curve and SS is the domestic supply curve of the commodity. SS_1 is the supply curve of the exporting country. Consequently, SS_T is the composite supply curve of the commodity showing the horizontal summation of the domestic and foreign supply of the commodity at different prices. Under free trade, the equilibrium price of the commodity will be 0P since at this price the composite supply curve SS_T intersects the domestic demand curve DD at point R. The total amount demanded and supplied at this price is $0Q_3$. Of this $0Q_3$ total amount demanded and supplied, Q_1Q_3 quantity is supplied by the domestic producers while $0Q_1$ ($= 0Q_3 - Q_1Q_3$) amount is imported from abroad. Now, if an import quota of less than $0Q_1$ quantity, say of $0Q$ amount, is fixed by the government, the new foreign supply curve of the commodity will be SUS_2 which is perfectly inelastic beyond the fixed import quota quantity $0Q$. The aggregate supply curve SS_T remains unchanged up to point V. Beyond point V, however, the additional quantity of the commodity will be supplied entirely by the domestic producers. Consequently, VW has been drawn parallel to the domestic supply curve SS. In the new situation when the government has fixed an import quota of $0Q$ quantity, the old composite supply curve SS_T is no longer relevant; the relevant composite supply curve now is SVW which is kinky at point V. The new equilibrium price determined by the intersection of the kinky aggregate supply curve SVW and the domestic demand curve DD at point N is $0P_1$ which is higher than the old equilibrium price $0P$. The quantity supplied by the domestic producers after the imposition of import quota is QQ_2 which is greater than the pre-quota quantity Q_1Q_3 supplied by the domestic producers.

Balance of Payments Effect—The main reason for resorting to import quotas on the part of governments in today's world is the existence of chronic deficit in the external balance of payments of those countries which impose import quotas. It has been argued that import quotas can also serve as a fruitful measure for eliminating deficit in the balance of trade of a country restricting imports to permissible limit dictated by the balance of payments situation. Import quotas can be imposed to cover the whole range of imports in those situations in which the country imports more than it exports. After the imposition of import quota the marginal propensity to import becomes zero when the limit of import quota is reached. This plugs the leakages of national income to foreign countries and the value of the national income is multiplied within the country. This reduces deficit in the external balance of payments of the country. It may, however, be argued that deflation or devaluation are more effective measures to correct the adverseness of the external balance of payments of a country. It should, however, be noted that the application of import quota as a means of correcting the balance of payments deficit is administratively easier to enforce compared with either the devaluation or deflation. Moreover, an import quota is less harmful than the other two measures. The favourable effect of an import quota on the balance of payments position of the country is doubtful only if it provokes retaliation by other countries.

Reasons for Imposing Import Quota

We have seen that the protective, redistribution, and consumption effects of import quota are similar to those of import tariff provided the elasticities of the domestic demand and supply curves are not zero. The question naturally arises that if the two measures are identical in their effects, why did import quotas come to supplant import tariffs on such a large scale in the 1930s that after the introduction of the quota system by France in 1931 the number of countries adopting this system rose to 27 by 1934.

The French Government which first introduced import quota system was neither concerned with the earning of revenue nor with the problem of the balance of payments deficit. In taking resort to import quota, the government was interested in maintaining the high wages of French peasants in the important wheat growing industry. No simple rate of import tariff could have effectively checked the flood of cheap imports of the Australian wheat in the French markets as the supply of the Australian wheat was perfectly inelastic. In the case of an inelastic foreign supply of commodity, an import quota is more effective than an import tariff to maintain the stability of internal prices. This has been illustrated in Figure 22.4.

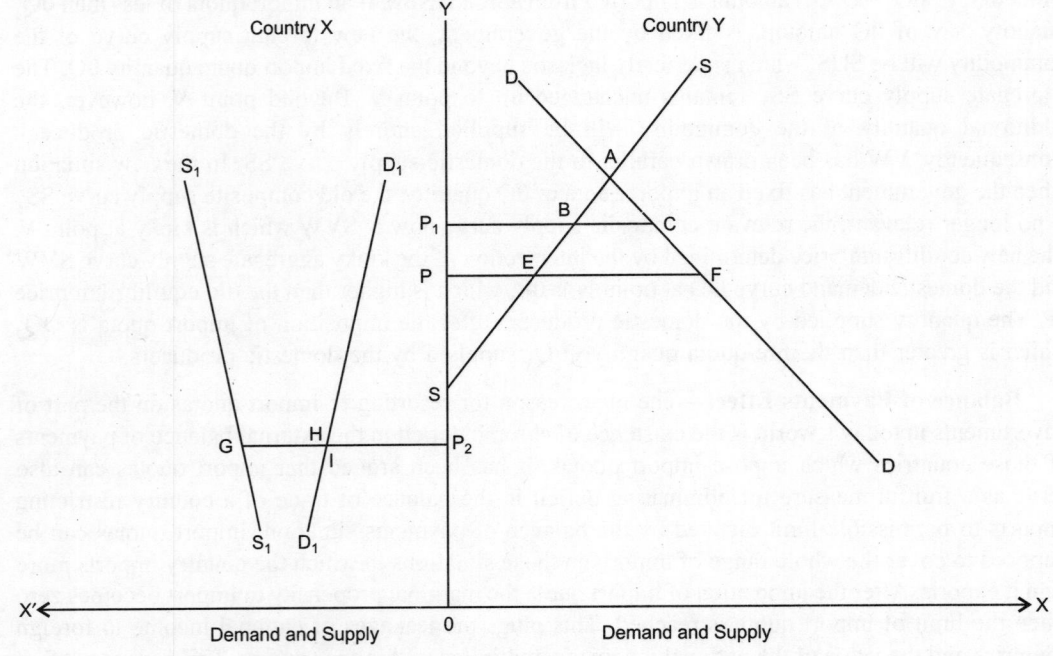

Figure 22.4

In Figure 22.4, the supply curve of the importing country Y has been shown by SS and that of the exporting country X has been shown by S_1S_1. When trade is free, country Y will import EF amount of wheat from country X and the equilibrium price of wheat will be 0P because at this price the domestic demand equals the sum of the domestic supply PE and imports EF. If country Y wants to raise the price of the import good, any reasonable tariff will fail to do this since the excess supply curve of exporting country X is perfectly inelastic. If the import quota of BC quantity is fixed, the domestic price will rise from 0P to $0P_1$. If country X does not retaliate, the export price in country X will fall to $0P_2$.

The second reason for the widespread use of import quota is that the excess supply curve of the exporting country is not known to the importing country. In case of an unknown supply curve no one can say with certainty in advance as to how much amount of the commodity will be imported during any given period of time. It is difficult to know as to how much high an import tariff will effectively prevent a fall in the price in the exporting country from spreading in the importing country. Apart from these reasons, the possibility of dumping also provides a good case for favouring the import quota system.

Finally, administratively it is easier to use import quota than an import tariff as an emergency measure. Kindleberger has rightly stated that "the use of tariffs had been so institutionalised in commercial agreements, with most favoured nation clauses and other restrictions on independent action that it was no longer possible to use tariffs as an emergency measure."[1] Thus, the three

1. Charles P. Kindleberger, *International Economics*, 1968, p. 134.

important reasons that were responsible for the massive use of import quotas in the sphere of foreign trade were the inelasticity of foreign supply, certainty of the outcome and administrative flexibility of the quota system.

Import Quota versus Import Tariff

Although an import quota and import tariff are both important restrictive measures, but the quota system is more harsh and cruel than tariff. Under the tariff system, a commodity may be imported in unlimited quantity provided the duty is paid on the imports. This is, however, not so with the quota system. Under the quota system, the specific quantity of imports is rigidly fixed which cannot be relaxed even if the foreign commodity has become cheaper or the importing capacity of the importers has increased. Thus, an import quota involves a more severe interference with the free working of the price mechanism than does an import tariff.

The effective protection provided by quota is more damaging than tariff because it favours the so-called established importers in spite of the fact that they are frequently less efficient than the new importers. Even if the new importers are competent to convert the raw materials into finished products more cheaply they are not able to secure an import quota. Consequently, they are debarred from undertaking gainful productive activities in the economy. In short, the quota system places a premium on inefficiency and lethargy. The quota system impairs the flexibility of the balance of payments mechanism of the country. If a country with surplus balance of payments position limits the quantity of all imports through quantitative restrictions, it will not be possible for the deficit country to improve her balance of payments position. This is not, however, so with tariff. Under tariff, the volume of imports and exports is always flexible. This flexibility helps in removing the deficit in the balance of payments position of the country.

The quota system gives rise to monopolies not only in the importing countries but also in the exporting countries. These monopolies, being free from the fear of foreign competition, can exploit the consumers by charging higher prices as the demand in the market increases. Under tariff, however, the monopolies cannot charge higher prices above the import price including the tariff levy.

When tariffs are imposed the rise in prices is absorbed either partly or fully by the state in the form of additional revenue earned through the tariff levy. On the other hand, most quotas fail to bring any revenue to the state.

From the above discussion of the quota system, it is clear that quotas are a half-way house to complete prohibition of free trade.

Case for Quota

Although quotas are in no way superior to import tariffs from the point of view of the world as a whole, however, from the point of view of underdeveloped countries quotas are recommended on several grounds. Firstly, in a period of severe scarcity of foreign exchange, import quotas reduce the pressure on the limited foreign exchange reserves of the country by limiting the physical quantity of imports to the desired extent. Secondly, as a means to provide protection to infant industries against foreign competition, quotas are certainly a more effective measure than import tariffs. Thirdly, to check the conspicuous consumption in the underdeveloped countries quotas are more effective restrictive measures than tariffs. Fourthly, for a planned economy, on account of their property of 'certainty', import quotas are superior to tariffs. Lastly, quotas are more effective in stabilising the domestic prices than tariffs if rationing and price control is efficiently enforced.

Both import quotas and tariffs are important restrictive measures. Consequently, these should be employed with great care and only when needed and when the national interests of the country cannot be served without resorting to them.

2. Dumping

Dumping is generally "taken to mean the sale of a good abroad at a price which is lower than the selling price of the same good at the same time and in the same circumstances (that is, under the same condition of payments and so on) at home, taking account of differences in transport costs."[2] Similar views have been expressed by Ellsworth and Leith who have refuted the widespread impression that dumping is selling abroad below the cost of production. According to them, "contrary to a widespread impression, dumping is *not* selling abroad below cost of production. It means instead *sales in a foreign market at a price below that received in the home market,* after allowing for transportation charges, duties and all other costs of transfer. Discrimination between the home and foreign price is the essential mark of dumping. Thus, sales abroad below cost of production would not constitute dumping *unless* the foreign price was lower than the domestic price."[3]

These definitions clearly show that dumping is an act of selling goods abroad at lower prices than those at which goods are sold at home. From an analytical point of view, however, Jacob Viner's definition of dumping is superior to those of others. According to Jacob Viner, "dumping is price discrimination between the two markets." This definition is better than the other definitions for the following three reasons.

1. According to this definition, dumping may occur either between the two independent countries or between two regions of the same country. In other words, a producer or seller, according to Viner's definition, will be said to have indulged in dumping even if he sells his product at different prices in different markets within the country. It is not necessary for a producer to sell his product abroad at a price lower than the domestic price of the product. The *sine qua non* of dumping is the charging of two different prices for the same good irrespective of the locational aspect of the price differential.

2. Price discrimination may occur not only between home and abroad; it may also occur between the two foreign markets.

3. The domestic price may be lower than the foreign price, i.e., reverse dumping is also possible. In other words, although the most common form of dumping has been seen in the selling of his goods by a producer at a lower price in the foreign markets compared to the domestic price of his goods, but it may as well be that he may sell his goods at higher price in the foreign markets compared with the home market. This will happen if the elasticity of foreign demand for his product is particularly low while at home he may have many rival producers.

Dumping presents an interesting case of monopolistic practices in the domain of international trade. This phenomenon has been poorly understood by both the laymen and politicians. Dumping which essentially means selling the same good at different prices in the home and foreign markets evokes the picture of a producer who dumps part of the unsold supply in a foreign market. There is, however, nothing vicious or peculiar about dumping. It is simply an example of price discrimination involving the home and foreign markets. The two essential conditions for it are that the producer should exercise monopolistic control over the domestic market while the demand for

2. Gottfried Von Haberler, *The Theory of International Trade,* 1950, p. 269.
3. P.T. Ellsworth and J.C. Leith, *The International Economy,* Fifth Edition, 1975, p. 250.

his product abroad is more elastic than at home. These two conditions are met quite often in practice. It is not unusual that a seller while he has monopoly at home faces free competition in the international market.

Kinds of Dumping

Economists have distinguished between different kinds of dumping including (*a*) sporadic dumping, (*b*) predatory dumping, and (*c*) persistent dumping.

Sporadic dumping is motivated by the desire to get rid of the inventory stocks which are practically unsaleable in the home market at the end of the season. When the producer is unable to sell off the 'remainder' in an orderly manner through normal trade outlets the policy of price discrimination is adopted. For this, the foreign demand for the product of a producer should be more elastic than the demand at home. Normally, the elasticity of foreign demand for a firm's product is greater than the elasticity of domestic demand for its product.

Predatory dumping is selling at loss (as measured by the average cost and not by the marginal cost of production) abroad in order to get a footing in the foreign market by killing off the rival firms. This aggressive form of dumping is subsequently followed by an increase in prices after the market is firmly captured and the rivals' competition has been overcome. The cost of such a campaign is very high. The seller does not only spend a lot on the advertisement for his product in the foreign markets but he is forced even to pay lower wages to his workers. The danger is always present that the foreign government may resort to defensive measures, such as anti-dumping duty, defensive dumping etc. against predatory dumping.

Persistent dumping occurs when a producer consistently sells his goods at lower prices in one market than in the other. The scope for persistent dumping may arise due to the difference in the demand curve for a particular commodity in different markets. For instance, if the domestic demand for the commodity is inelastic while the foreign demand for the product is highly elastic, the producer can sell a larger quantity of the commodity in the foreign market at lower price and smaller quantity in the home market at higher price. When the advantage of different demand elasticities in different export markets is taken by the seller this is known as "differentiated dumping".

Conditions Essential for Dumping

The first essential condition for the dumping to be successful is that there should be no seepage of the commodity from the foreign market where it is sold at a lower price to the home market where its price is relatively high. In other words, the goods should be prevented from coming back to the exporting country. To prevent the flow-back leakage of the commodity from the foreign market to the home market, it is necessary that the difference between the high domestic price and the foreign price should be less than the cost of transportation involved in re-exporting the commodity to the country of its origin. If the difference in the two prices is greater than the transportation cost, the reselling from the cheaper market should be stopped by imposing an appropriate import duty on the import of the commodity in the exporting country.

The second essential condition for dumping to be successful is the existence of monopoly at home. If there is perfect competition, the average cost of production will be equal to the marginal cost of production and the marginal cost of production will be equal to the selling price. Under this condition, no seller would like to sell abroad at price lower than that at home without the incentive of export bounty. Under monopoly, the average cost and the marginal cost of production are lower than the selling price. Consequently, the monopolist can afford to sell his product abroad at lower price than the price at which he sells at home.

Effects of Dumping on the Exporting Country

It is common belief that foreign selling of the commodity at relatively lower price takes place at the cost of domestic consumers. The loss to the monopolist from foreign market is compensated from the gain he makes by selling his product at higher price in the domestic market. In other words, the belief has been that the domestic consumers subsidise the welfare of the foreign consumers. Schumpeter has expressed emphatically that dumping causes prices to rise in the exporting country. He says that "if the goods now exported could not be dumped, they would not cease to be produced but a large part of them—not the whole—would be sold in the home market, thereby reducing the prices there."[4] The argument that dumping raises the price in the exporting country is, however, true in the case of sporadic dumping only. In permanent or long-term dumping, the quantity produced is not *a datum*. The relevant issue is whether the commodity would, in fact, be produced and marketed at home in the absence of dumping. This would be determined by the demand and cost conditions.

The core of persistent dumping is that domestic price depends upon the laws of returns which operate in production. For instance, if the production of the commodity is subject to the law of decreasing costs, domestic demand remaining constant, the expansion in production resulting from dumping the commodity in the foreign market will lower the cost of production and this will also lower the domestic price of the commodity. Under constant cost conditions, the cost of production and consequently the domestic price of the commodity will remain unchanged. Only in the case of increasing cost conditions, the expansion of output resulting from dumping in the foreign market will raise the cost of production and will, therefore, raise the domestic price of the commodity. Since increasing costs or diminishing returns is a universal law of production, dumping will raise domestic prices. Consequently, dumping is injurious to the exporting country.

The above discussion is, however, one-sided since it does not consider the demand conditions. If the foreign elasticity of demand for the product is greater than the domestic elasticity of demand, discrimination between the two markets must raise the home price of the product whether the marginal cost is falling or is constant. If the two elasticities of demand for the product are identical, there will be no scope for dumping. If the elasticity of demand for the commodity in the foreign market is less than the elasticity of demand in the home market, price in the home market will be lower than the price in the foreign market.

The phenomenon of dumping causes strong emotional reactions for very obvious reasons. The consumers at home get irritated when they come to know that for the same product they are paying higher price than their foreign counterparts. To cite an example of this emotional reaction, in the 1960s when the Japanese found that the Japanese producers of colour television sets were charging them much higher prices than the American consumers, they pressurised the Japanese firms to reduce the prices of the television sets in the domestic market. The behaviour of the Japanese firms is, however, fully understandable when seen against the background of theoretical exposition.

---

4. Joseph A. Schumpeter, "Zur Soziologie der Imperialismen", in *Archiv, fur Soziolwissenchaft*, 1919, volume 46, p. 301, quoted by Gottfried Von Haberler, *Op. cit.,* p. 302.

Effects of Dumping on the Importing Country

If the cheap products continue to come regularly from abroad, the importing country is in no way injured. Haberler has correctly stated: "It makes not the slightest difference from the standpoint of the importing country whether the goods come in cheaply because the exporting country enjoys a natural comparative advantage or because they are dumped, nor does it matter in the least whether the dumping is due to monopoly abroad or to export bounties given by the foreign government or by some other body. Not one of these circumstances disturbs the fundamental free trade argument. They are significant only insofar as they indicate whether or not the cheap imports are likely to continue."[5]

Dumping is injurious to the importing country only if it is intermittent or sporadic. If dumping takes place in long spasm during which production in the importing country is shifted, it is more ruinous for the producers as well as for the consumers in the importing country. Dumping may lead to the establishment of certain industries in the importing country which use cheap raw materials dumped by the exporting country. If dumping ceases, producers in the importing country cannot continue their industries since producers' goods cease to come in. If the dumped goods happen to be consumer goods, consumers will shift their demand which must later be removed. They may, therefore, be injured beyond any remedy.

If dumping is predatory, it is harmful for the importing country. In practice, however, this kind of dumping very seldom occurs. If this happens, the monopolist is robbed of the fruits of his costly victory by legislative intervention.

The phenomenon of dumping causes strong emotional reactions. The consumers in the home country become agitated when they come to know that they are paying higher prices than the foreign consumers. As stated earlier in the 1960's, the Japanese were agitated when they found that they were paying much higher prices for the Japanese colour television sets than the prices in the U.S. markets. Consequently, they pressurised the Japanese producers to reduce the prices of their TV sets in the domestic market.

Anti-dumping Measures

Many countries impose penal import tariff equivalent to the difference between the domestic and export prices of the commodity in the exporting country. The determination of the price difference is, however, very difficult. Consequently, import quotas are fixed for those goods which, the government of the importing country believes, are being dumped in the country from abroad. Quotas are more effective than tariffs in preventing the entry of dumped goods in the importing country. There is, however, always present the fear of retaliation by the exporting country which is harmful from the point of view of community welfare as a whole. Abhorring the retaliatory measures against dumping, Kindleberger has stated that "countervailing measures against alleged dumping are obnoxious because they reduce the flexibility and elasticity of international markets and reduce the potential gain from trade.... With anti-dumping tariffs everywhere adjustment after miscalculations which result in over-production is much less readily effected."[6]

5. Gottfried Von Haberler, *Op. cit.,* p. 314.
6. Charles P. Kindleberger, *International Economics,* Fifth Edition, 1973, p. 156.

3. Cartels

As a kind of trade restriction, international cartels had an important impact on international trade in the inter-war period. In general terms, an international cartel may be defined as an arrangement between the producers of two or more countries for the purpose of regulating competition in the production and selling of an international commodity. Such cartels are sometimes formed with government participation. According to Kindleberger, "cartels are international business agreements to regulate price, division of markets, or other aspects of competition. They occur in industries with less than perfect competition."[7] According to Haberler, an international cartel is an act "of uniting the producers, in a given branch of industry, of as many countries as possible, into an organisation to exercise a single planned control over production and price and possibly to divide markets between the different producing countries."[8] Cartel agreements are not explicit and legal but implicit and habitual. When business enterprises come to understand that their common motive of earning profit is hurt by one another's reaction, they come under the banner of cartel to avoid certain kinds of competition among themselves. Although under cartels their separate identity is maintained but they act together in deciding such matters as the selling price of the product which they are to charge from customers, the amount of the product they are to produce or sell and the share of the market they have to hold so that their motive of earning profit may be fulfilled and the economy may run more smoothly.

International cartels can succeed in achieving their aim of raising the price of the product for any considerable period of time only if it is possible to exercise an effective control on production and enforce discipline among the participants. This is possible only when the output is not too widely distributed among different producers. This fact makes agriculture unsuitable for cartelisation since the production of agricultural goods is distributed among many thousands of producers who are widely spread throughout the world. Furthermore, an industry in which the bulk of output is produced by small and medium-sized firms is unsuitable for cartelisation. Thus, industries producing goods made by individual craftsmen using their skills and art designs are not suitable for cartelisation. The following groups of industries offer great scope for the formation of international cartels.

1. Those industries which are closely based on raw materials and can effectively block the entry of outsiders by exercising strict control over raw material supplies. The international sulphur, zinc, tin, copper, aluminium and lead cartels belong to this group.

2. Cartels are most effective in those industries where products are patented. These are the electrical and chemical industries. Examples are the incandescent lamps and ball-bearings cartels.

3. Cartels have great scope in those branches of production in which the economies of large-scale production are massive and, therefore, there is great concentration of output. The best case example is the iron and steel industry and the Continental Crude Steel Cartel is the best-known example of international cartel in this group.

Advantages and Disadvantages of Cartels

1. The supporters of international cartels hope that the cartel agreements among the producers of different countries might prepare ground for the removal or lowering of tariff barriers which will maximise the welfare of the world community.

2. Cut-throat competition is a common feature of oligopoly or few sellers. They involve themselves in price war to encroach upon one another's share of the market. Consequently, prices

7. Charles P. Kindleberger. *Op. cit.,* p. 157.
8. Gottfried Von Haberler. *Op. cit.,* p. 328.

of goods fluctuate violently. The obnoxious price war comes to an end through some kind of agreement between the rival producers. The agreements manifest themselves in the formation of cartels. Thus, cartels not only establish price stability by putting an end to price war and cut-throat competition among the rival producers of the cartelised product but they also save wasteful expenditure incurred on competitive advertising.

3. According to some thinkers, the agreements between the producers of different countries united in international cartels prepare the ground for lowering the tariff barriers. International cartels can, however, help in reducing the tariffs only when they themselves serve as substitutes for tariffs so that protectionism of cartels replaces protectionism of tariffs.

4. International cartels make possible the pooling of scarce technical knowledge which is very helpful in the expansion of total output at cheaper cost. Consequently, international cartels may be treated as an engine for economic growth.

5. Since the formation of international cartels is based on international understanding and cooperation they relax the severity of economic crises.

There are many economists and businessmen who consider international cartels ruinous for international economic prosperity and harmful for the welfare of the world community. They advance the following arguments for their belief.

1. The classical principle of *laissez-faire, laissez-passer,* has been breached by international cartels. There is always possibility of monopolistic exploitation of consumers by cartelised producers. When the producers are protected against foreign competition they charge very high selling prices. Moreover, since there is no competition among the producers who have formed themselves into a cartel, the commodity supplied is not of superior quality.

2. The argument advanced by the supporters of international cartels that international cartels help in lowering and removing tariffs is not acceptable in general. It is pointed out that big enterprises establish international cartels simply as a show-piece to reduce tariffs and small firms are not in a position to forego their advantages of actual or potential tariffs. According to Haberler, "...international cartels are not a suitable instrument for demolishing tariff walls within any measurable time. Many of the present international cartels *owe their own existence to tariffs.* They are therefore scarcely adapted for destroying tariffs. Unless the industry in question happens to be about equally strong in every participating country, the weaker national groups would not and cannot give up their tariff protection."[9]

3. According to some economists, the members of an international cartel somehow tend to lose patriotism.

4. The members of a cartel are united by loose agreement. If a member country is not satisfied with the allocation of production quota and sales area, it may indirectly dishonour the agreement. This explains why international cartels are short-lived and become ineffective in the long run.

5. Cartelisation is possible only when the total output is not distributed too widely among the producers, i.e., when the industry is either oligopolistic or there is monopolistic competition with bulk of the total output being controlled by only a few large producers. In other words, an effective control over supply is essential for the success of an international cartel. On this criterion, oil industry where the total world oil production is controlled by only few multinational corporations provides the best example of international cartelisation. Oil, however, seems to be a very special product. No other product has the specific characteristics of oil which made the maintenance of drastic price increases by the Organisation of Petroleum Exporting Countries (OPEC) possible. A

9. Gottfried Von Haberler, *Op. cit.,* p. 331.

few minerals such as copper, manganese and bauxite may offer possibilities for practising restrictive policies through cartels. However, even here the scope for progress seems limited since the restrictions on supply would unduly protect the existing producers and bar the entry of new firms leading to stagnation and inefficiency in the long run. Consequently, cartelisation of agricultural products, wood-working industry, etc. is not possible since the output being too widely distributed throughout the world the effective enforcement of cartel agreement by different motley group of members spread far and wide in the world may be very difficult. Consequently, the area of international cartels is very limited.

6. Under the protective umbrella of cartel, established producers are unduly protected. Barriers to new entrants are erected leading in the long run to inefficiency and stagnation. Thus, cartels often thwart growth by inhibiting competition.

From the above discussion it is obvious that the advantages of international cartels are outweighed by their serious disadvantages. Consequently, the informed international opinion is against the formation and growth of international cartels.

Suggested Readings

C. Edwards, *Control of Cartels and Monopolies: An International Comparison,* 1966.
Gottfried Von Haberler, *The Theory of International Trade,* 1950, Chapter XVIII.
Charles P. Kindleberger, *International Economics,* Fifth Edition, 1973, Chapter 9.
Edward S. Mason, *Controlling World Trade,* 1946.
Bo Sodersten, *International Economics,* Second Edition, Reprinted 1981, Chapter 14, pp. 187–188.

Questions

1. Discuss the merits and demerits of quota as a method of protection.
2. What is dumping? What are its effects on the economy of the country in which dumping occurs?
3. What is a cartel? What are its advantages and disadvantages?
4. What is quota? Explain and show with the help of a diagram the various effects of an import quota.

Regional Economic Co-operation

Introduction

After the Second World War, Western Europe presented a pessimistic picture. During the Second World War, the productive capacity of West European countries had been greatly damaged by the war needing their rebuilding and removing their external imbalances. It was felt that in the new set-up rebuilding of Western Europe required cooperation on a wider scale among the neighbouring countries of the region. Soon the idea of integration of the West European countries took the form of compact integration of a small group of countries. At the initiative of France, the six[1] West European countries consisting of Belgium, the Netherlands and Luxembourg ('Benelux' countries), France, Germany and Italy took the first step toward economic unification of their economies by signing a treaty in April 1951 to form the European Coal and Steel Community (ECSC). The aim of the ECSC was to regulate the production of steel and coal in such a manner so as to prove beneficial for the six countries. This partial 'functional integration', however, proved to be inconsistent with the needs of the area. Consequently, the six original member countries assembled in Rome and signed the famous Treaty of Rome on March 25, 1957[2] seeking to create the two new agencies of integration, namely the European Economic Community (EEC) and the European Atomic Energy Community (Euratom). This Treaty came into operation on January 1, 1958. The six EEC countries agreed to the creation of a 'Common Market' which is popularly known as the European Common Market (ECM). The other regional cooperation was effected among a group of countries belonging to the Organisation for European Economic Cooperation (OEEC) in 1948.

Regional economic cooperation can be of several forms with different degrees of integration between the members of the group. Thus, one can think of a free trade area, a customs union, a common market, an economic union and a complete economic integration between the individual member countries of a region. A free trade area consists of a group of small countries generally with geographical vicinity which have abolished all tariffs (and quantitative restrictions) between themselves. Members are, however, free to apply tariff and other quantitative restrictions against outsiders. The European Free Trade Association, commonly known as EFTA, formed in 1959 by the

---

1. The European Economic Community has now twelve members including, in addition to the original six members, England, Ireland and Denmark. England became member of the Community after protracted negotiations that had lasted for more than a decade. With the approval in the second round of meetings concerning membership for Greece, Portugal, and Spain the membership of the community rose to twelve. In 2005 total members were 25.
2. The EEC members celebrated the 20th anniversary of the Treaty of Rome on March 25, 1977.

seven countries outside the European Common Market is an example of the free trade association.

A customs union is one step forward in the direction of integration. Apart from abolishing inter-member tariff and other restrictions on trade, members also agree to impose uniform tariff and other quantitative restrictions against the outsider non-member countries. The Belgium-Luxembourg Economic Union (BLEU) was an example of customs union. A common market represents a still higher form of economic integration. Besides allowing the free trade in goods, it also allows free movement of the factors of production within the common market area. The European Common Market (ECM) is an example of the common market. In the case of economic union, apart from allowing free trade and factor movements between the member countries, the members adopt integrated fiscal and monetary policies. Finally, in the case of economic integration there is complete economic fusion of member countries and there is a supranational authority whose decisions are binding on the members. The monetary, fiscal and social policies of the members are completely unified.

Purpose of the European Common Market

The basic purpose of the European Common Market (ECM) is to foster trade within the community embracing a population of over 180 million people through progressive abolition and regulation of customs duties, quotas, tariff and other exchange control procedures affecting the intra-regional movement of goods, services and capital. However, the immediate purpose of establishing the 'Common Market' was to appropriate the advantages of increased specialisation and division of labour in production by enlarging the size of the markets. "Its sponsors contended that existing national markets were too small for economical operation of certain industries except as monopolies. A large market would make possible mass production without this drawback. Attainment of these ends should, it was argued, make the unified area a more powerful unit, ensure continual expansion, increase economic stability, raise standards of living and develop harmonious relations between its component states."[3]

For achieving the above ends the constituent countries agreed to abolish all tariffs among themselves and to adopt a uniform tariff vis-a-vis other nations. The duties between the members were to be gradually eliminated over a twelve-year period staggered in three stages of four years each. After every four-year period, the tariffs were to be reduced by 30, 30 and 40 per cent in successive stages. The uniform external tariff was to be not higher than the average of the previous tariffs of the members. The overseas territories (or former territories of member countries) were given the option of associating with the Common Market. The European Economic Community has 18 African States as Associate Members. Greece was accepted as an Associate Member of the Community in 1961.

After the formation of the Common Market it was necessary to establish a closer relationship between the member states to harmonise their national policies so that quicker and harmonious economic development of the community could be achieved. To accomplish this difficult task the member countries are committed under the Treaty of Rome to:

I. The abolition, as between member states, of obstacles to the free movement of persons, services and capital.

II. The inauguration of a common agricultural policy.

---

3. P.T. Ellsworth and J. Clark Leith, *The International Economy,* Fifth Edition, 1975, p. 542.

III. The inauguration of a common transport policy.

IV. The establishment of a system ensuring that competition shall not be distorted in the Common Market.

V. The application of procedures which shall make it possible to co-ordinate the economic policies of member states and to remedy disequilibria in their balance of payments.

VI. The approximation of their respective legislations to the extent necessary for the functioning of the Common Market.

Under the above provisions, the European Social Fund was established to ease the readjustment problems of workers experiencing unemployment as a consequence of trade liberalisation within the community. The European Investment Fund was established to help the industrialists to improve workers' conditions in the underdeveloped regions of the constituent states. This fund also helps to finance those projects of European importance that are too large to be handled alone by the individual states.

Organisation

Since the formation of the European Economic Community was a kind of super-government with respect to economic affairs, like any other government it needed agencies to act, to legislate and to settle disputes.

The principal administrative body of the Community is the European Council. It consists of one member from each member state. The Council formulates rules of conduct, prepares new legislation and asks members to carry out the provisions of the treaty. The Council is assisted by the European Commission which studies special problems and makes recommendations to the Council.

There is a Monetary Committee to keep watch over the balance of payments position of the members of the Community. In addition to it, there is the European Economic and Social Committee. It consists of the representatives of industry, workers, farmers, etc. Its area of concern is virtually unlimited.

The legislative branch consists of an Assembly of 106 members which takes final decision on recommendations of the Council. Finally, there is a Court of Justice to adjudicate on disputes between the member countries.

Development Assistance

Since its inception with the Treaty of Rome in 1957, the European Economic Community (EEC) has gradually developed a unique programme of regional cooperation with the less developed countries and today the Community has emerged as an important donor of new forms of multilateral assistance. The aid programme of the Community has grown rapidly in the last 30 years. Net disbursements from 1960 to 1964 were $ 234 million; over the period 1970–80 they exceeded $ 3.7 billion. Today, after the United Nations, the EEC is the second largest source of multilateral grants. It provides about one-third of all multilateral grant capital and one-eighth of total multilateral loans and grants. The 'Nine' members of the community together account for 40 per cent of world trade and are the largest importers of Third World's raw materials. As a result of the Lome Convention in February 1975, the Nine have granted free access for 99.2 per cent of their exports to 46 African, Caribbean, and Pacific States (the so-called ACPS). With the conclusion of the Lome agreement, the

Nine replenished the European Development Fund (EDF) to a level of 3,000 million units of account (about $ 3,700 million). The Community has also established a fund—STABEX—with 375 million units of account (about $ 470 million) to stabilise the developing ACP countries' earnings from agricultural raw materials and iron ore exports to the ECM. If a developing country's exports earnings from these products fall significantly below the average of the four preceding years, the shortfall is met by a loan or grant from STABEX.

Since 1964, over 1,200 projects from the developing countries have been financed by the funds of the EDF. In addition, there is the European Investment Bank (EIB). Although primarily meant to give financial assistance to projects in the economically depressed regions of the EEC, it has got increasingly drawn into development lending outside the territories of the Nine. It had committed over half a billion dollars in commercial loans to ACP nations in the period 1976–80. In addition, EIB's involvement in the Mediterranean region will grow with the development of the EEC's Mediterranean policy. In 1976, the Bank lent large sums for the rehabilitation of Portugal's economy.

Between January 1, 1959 and December 31, 1974 net disbursements from the EEC for the development financing of various sectors aggregated 1,831 million U.a. (about 2,300 million dollars). Of this total assistance, no less than 37 per cent had been given for the development of transport and communications facilities in the developing countries while another 30 per cent had been given for encouraging rural production in these countries. Development of education has also received attention and no less than 12 per cent of the total disbursements made were for the development of education. Industrialisation accounted for 5.2 per cent while urbanisation and rural water supplies received 6.2 per cent of the total aid disbursed by the EEC to the poor nations.

Although the EEC's development assistance policy is often criticised for over-emphasis on Africa and neglect of Latin America and Asia, the criticism is not justified because the Nine have entered into nonpreferential trade agreements with Argentina, Brazil, Uruguay, and Mexico. The Community has also made available technical assistance to the Central American Common Market and to the Latin American Free Trade Association. The EEC has also concluded nonpreferential trade agreements with India, Sri Lanka, Pakistan and Bangladesh. The EEC has also forged closer links with the Association of South East Asian Nations.

In 1971, the Community implemented the Generalised System of Tariff Preferences (GSP) under which developing countries' exports of semi-manufactured and manufactured goods, as well as some processed agricultural products, are imported duty free into the EEC and each year this system has been improved. However, due to lack of information and procedural complexities many developing countries have not been able to avail of the GSP scheme with the result that only half of the granted preferences have been actually taken up. For these reasons, the GSP covers only about 5 per cent of the Community's total imports. The Community had intimated its intention to continue the GSP after 1980. In 1974, the Community pledged $ 500 million in aid to those developing nations which were most seriously affected by the economic upheavals of 1973–74. In short, during the 45 years of its existence the EEC, apart from bringing about the economic and political stability among its member states, has projected enlightened policies in its dealings with the Third World and has played an important role in the community of nations. In the past fifteen years, the Community has not only absorbed three members; it has also built up a politico-economic solidarity that has enabled it to withstand the pressures of the most serious worldwide recession since the Second World War. The Community is now a political and economic reality which cannot be ignored. She Community with 25 members has become European union (EU) wish a single currency known as EURO acid moving fast toward a single constitution.

European Free Trade Association (EFTA)

At the time of formation of the European Economic Community, England did not join the Community. She formed a rival European Free Trade Association in 1959 with six other nations. The other six nations were Denmark, Norway, Sweden, Austria, Switzerland and Portugal. The British motivations, in keeping with the usual motivations of nation-states, were a blending of self-interest and a sense of cosmopolitan responsibility. Although originally intended by England to create a bargaining weapon to force the Six to come to terms with the rest of Western Europe, EFTA was actually never used for this purpose. The countries of European Free Trade Association were spread out in an enormous circle around the EEC. They were known as the 'Outer-Seven'. The main goal of this Association was to reduce the tariffs among the member countries. The gradual reduction of duties on intra-area trade in industrial products culminated in their complete elimination at the end of 1966. However, the member country is free to maintain its own tariffs.

The EFTA had achieved a considerable success. The intra-area imports rose sharply from 7.6 per cent in 1958 to 11.3 per cent in 1961. The share of member country imports also increased from 17.6 per cent to 21.2 per cent. With the exit of the United Kingdom, Ireland and Denmark on their entry into the EEC at the beginning of 1973, the EFTA has become truncated.

Latin American Free Trade Association (LAFTA)

The countries of Latin America started negotiations in 1954 for establishing arrangements to stimulate intra-area trade and contribute to regional economic development. For several years ideas on this subject were exchanged with the Economic Commission for Latin America. Ultimately the Treaty of Montevideo was signed on February 18, 1960 which came into operation on July 1, 1961. The Treaty was signed by the seven Latin American countries of Argentina, Brazil, Chile, Paraguay, Peru and Uraguay in South America and Mexico in North America. Three additional countries Colombia, Bolivia and Equador joined later.

The declared purposes of the Treaty are the liberalisation of intra-Latin American trade, promotion of the complementarity of industrial production and co-ordination of agricultural development and trade among the members. The Treaty of Montevideo provides a framework which can eventually lead to the formation of a true common market for all Latin America. However, many obstacles have to be overcome before the goal can be achieved. The vested interests of national producers and the deep feeling of mutual distrust among the members have to be overcome. However, despite many meetings and frequent discussions it is no exaggeration to say that the LAFTA is still in the stage of talk rather than in active action and 33 years after its inception the LAFTA is virtually dormant.

Among the other regional groupings which have made rapid progress mention may be made of the Central American Common Market (CACM) consisting of Costa Rica, El Salvador, Guatemala, Honduras and Nicaragua. A common tariff has been established and internal trade has been carried on free trade basis since 1966 and as a result intra-regional trade increased ten times during 1963–73. Yet another regional arrangement in the form of Economic Community of West African States (ECOWAS) was formed in 1975 with the aim of establishing a customs union among the Anglophone and Francophone member states. She formation of the North America free Trade Area (NAFTA) comprising the U.S., Canada and Mexico is a challenge to murtslatenal trading system.

ASEAN Free Trade Area (AFTA)

AFTA was established in 1992 to realise on integrated market for 500 million people, the size of original members for AFTA are Brunei Darussalam, Indonesia, Malaysia, Philippines, Singapore and Thailand. They have agreed to bring down tariff on wide range of manufactured and processed agricultural goods traced within Asian content requirement to 5 p.c. within a 15 years period commencing from 1 January, 1993. Vietnam is expected to realise AFTA by 2006, the Lao People's Democratic Republic and Myanmar and Cambodia by 2010. Realisation of AFTA has not presented ASIAN from exploring closer integration with other members in the region.

Conclusion

The successful establishment of the European Economic Community has tempted the statesmen and economists in the other countries to weld themselves into similar regional associations. Consequently, countries which are too small to take advantage of the economies of scale have favoured the idea of associating together into regional associations. However, the growing tendency on the part of small nations in Africa and Asia to be tempted by the success of the EEC should not by any means lead them to conclude that regional associations can and will succeed everywhere. The success of the EEC has been due to several important factors some of which are not at all present in the case of other regional arrangements like LAFTA. The EEC is an association among the equally developed countries having almost similar areas with close and common borders and similar political organs. More important, however, is the impact of these regional arrangements on the free multilateral trading system. It is no exaggeration to say that regional arrangements and free multilateral trading system ill-go together.

Suggested Readings

Victoria Curzon, *The Essentials of International Integration,* 1974, Chapters 1–3.
Sidney S. Dell, *A Latin American Common Market,* 1967.
P.T. Ellsworth and J.C. Leith, *The International Economy,* Fifth Edition, 1975, Chapter 30.
F.B. Jensen and I. Walter, *The Common Market: Economic Integration in Europe,* 1956.
Charles P. Kindleberger, *International Economics,* Fifth Edition, 1973, Chapter 11.
Bo Sodersten, *International Economics,* Second Edition, Reprinted 1981, Chapter 17.

Questions

1. Discuss the objectives of the European Common Market. How far has it achieved these objectives.
2. What is meant by regional economic cooperation? How far are these regional economic groupings compatible with the free trade policy?

Theory of Customs Unions

Discriminatory trade policy has become a common feature with almost all countries of the world today. All countries, however, do not discriminate equally against all the other countries. Consequently, there are several forms of economic integration representing different degrees of trade discrimination against the outside members. Economic integration may be defined both as a process and as a state of affairs. As a process it covers measures aiming at abolishing the discrimination between economic units belonging to different national states. As a state of affairs, it can be treated as an area or region comprising different national states marked by the absence of different forms of discrimination between the member states. Economic integration can take several forms representing different degrees of integration. Thus, one can think of a free trade area, a customs union, a common market, an economic union and complete economic integration. A free trade area comprises a group of countries which have abolished all tariffs (and quantitative restrictions) between themselves but each member is free to apply tariff and other quantitative restrictions against the outside non-member countries.

A customs union involves not only the abolition of tariffs and other trade restrictions between the constituent member countries but also involves the imposition of uniform tariff and other trade restrictions against the non-member outside countries. A common market, which represents a higher form of economic integration, besides allowing the free movement of goods also permits the free movement of factors of production within the common market area. In the case of an economic union not only the free movement of commodities and factors of production is allowed within the union but an integrated fiscal and monetary policy is also adopted at the national level. Finally, economic integration represents the complete economic fusion of member countries in the sense that it presupposes the unification of monetary, fiscal, social and counter-cyclical policies and a supranational authority is set up to take the decisions, which are binding on the members, for the unification of economic and social policies.

This chapter is, however, devoted only to the study of the theory of customs unions whose analysis can also be extended to cover the other forms of economic integration.

Theory of Customs Unions

The tariff system of a country may discriminate between commodities and/or countries. The pure theory of customs unions may be defined as "that branch of tariff theory which deals with the effects of geographically discriminatory changes in trade barriers".[1] According to the definition

1. Richard G. Lipsey, "The Theory of Customs Unions: A General Survey", *The Economic Journal*, Volume 70, September 1960, pp. 496–513.

given in the General Agreement on Tariffs and Trade, a customs union must aim at (*a*) the substantial elimination of all tariffs and other forms of restrictive trade practices among the members of a customs union, and (*b*) the establishment of uniform tariff and other regulations on foreign trade with the non-member countries. Although the main contributors to the pure theory of customs unions are Jacob Viner, Jaroslav Vanek, R.G. Lipsey, James E. Meade and K.J. Lancaster, however, the modern discussion on the theory of customs unions started after the Second World War with the publication in 1950 of Jacob Viner's well-known book *The Customs Union Issue* and James Meade's book *The Theory of Customs Unions* published in 1955. Prior to Jacob Viner, the early discussions on the subject assumed that a customs union was a movement toward free trade since it established a complete freedom to trade among few countries, if not among all the countries of the world. Therefore, a customs union increased international welfare even if it did not lead to the maximisation of world welfare. Jacob Viner objected to this faulty view and developed the concepts of trade creation and trade diversion to examine if a customs union necessarily increases the total world welfare. The subsequent literature on the theory of customs unions has been built on the theoretical framework provided by Jacob Viner who has analysed, under different assumptions, the welfare gains and losses resulting from the formation of a customs union. The main contributions on this theme have been made by James E. Meade, H. Makower and G. Morton, Richard G. Lipsey, Tibor Scitovsky, Harry G. Johnson and Jan Tinbergen.

Trade Creation and the Trade Diversion Effects

Prior to Jacob Viner's pioneering study of the theory of customs unions, it was believed that since customs unions represent a move toward free trade they will increase welfare even though they may not maximise it. Jacob Viner's study showed this conclusion to be incorrect. In order to study the welfare effect of a customs union he introduced the twin concepts of *trade creation* and *trade diversion*. The concepts of trade creation and trade diversion refer to the static effects of the formation of a customs union under the assumption of fixed resources and given technology. Trade creation refers to the expansion of trade between the member countries of a customs union while trade diversion refers to the volume of trade that is diverted due to the formation of a customs union from the foreign country to the union partners due to the elimination of intra-union tariff. Trade diversion occurs when imports from more efficiently producing country are switched over to a less efficiently producing country due to the formation of a customs union resulting in the lowering of welfare as it involves less efficient allocation of resources. Trade creation occurs when as result of the formation of customs union production of goods takes place in the more efficient member country instead of taking place in both the countries. The concepts of trade creation and trade diversion associated with the theory of customs unions have been explained in Figure 24.1.

Let us assume that there are three countries India, Burma and Bangladesh. Let us further assume that India and Bangladesh form a customs union while Burma is a non-member outsider country. The domestic supply and demand curves for rice of India are *SS* and *DD*. The perfectly price-elastic supply curves of Bangladesh and Burma have been shown by S_D and S_B respectively. The perfectly elastic (horizontal) supply curves of Bangladesh and Burma indicate that both these countries produce rice under constant per unit cost or supply price.

Before the formation of customs union and in the absence of imports, the domestic price of rice in India is $0P_3$ since at this price the total domestic demand is just satisfied by the total domestic supply and there is equilibrium in the domestic market. The price of the rice in India would, however, never be as high as $0P_3$ because India can buy rice from the least-cost producer country

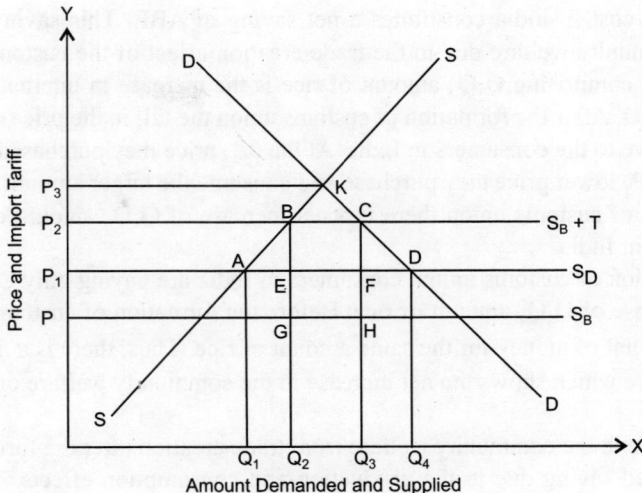

Figure 24.1

Burma and even after paying PP_2 import tariff Indian consumers can obtain Burmese rice at $0P_2$ price which is less than $0P_3$. At $0P_2$ price, the total demand for rice in India is $0Q_3$. Of this total amount demanded, $0Q_2$ amount is supplied by the domestic producers in India while the balance Q_2Q_3 amount is imported from Burma. The effective supply curve of rice after the imposition of import tariff has been shown by the horizontal line $S_B + T$.

Trade Creation—After the formation of customs union between India and Bangladesh, import tariff is abolished between the two countries while the tariff on imports from Burma remains unchanged. Consequently, imports of rice can be made at lower price from Bangladesh. As result of the cheapening of imported rice reflected in the fall in price from $0P_2$ to $0P_1$, larger quantity of rice is imported by the Indians. Consequently, the domestic producers in India will curtail their production of rice from $0Q_2$ to $0Q_1$ and the domestic demand will be met through importing Q_1Q_4 amount of rice from Bangladesh. The increase in the amount of imports of rice amounting to Q_1Q_2 $+ Q_3Q_4 (= Q_1Q_4 - Q_2Q_3)$ is the *trade creation effect* of forming the customs union between India and Bangladesh.

It is obvious that the trade creation effect of a customs union can be divided into two parts comprising of (1) $Q_1 Q_2$, and (2) Q_3Q_4. The first part comprising Q_1Q_2 amount is the increase in the imports of rice in India due to decrease in the domestic production of rice. The Indian producers were supplying $0Q_2$ amount of rice at $0P_2$ price before the formation of custom union between India and Bangladesh. After the formation of customs union, the imported rice from Bangladesh becomes available in India at $0P_1$ price. At this price, the Indian producers can supply only $0Q_1$ amount of rice. Consequently, the increase in international trade represented by additional imports of Q_1Q_2 $(= 0Q_2 - 0Q_1)$ amount of rice is the *trade creation effect* due to the production effect. The total resource cost of producing Q_1Q_2 quantity of rice in India is equal to the area of Q_1ABQ_2 quadrilateral lying under the Indian supply curve. When Q_1Q_2 quantity of rice is imported instead of being produced in India, the consumers in India pay only $0P_1$ price per unit of output and the total payment made is equal to Q_1AEQ_2. The difference between the total domestic resource cost and the total import-cost in India constitutes a net saving of ABE. This saving represents a net increase in the community welfare due to the trade creation effect of the customs union.

The second part comprising Q_3Q_4 amount of rice is the increase in international trade due to the consumption effect. After the formation of customs union the fall in the price of rice from $0P_2$ to $0P_1$ offers an incentive to the consumers in India. At the $0P_2$ price they purchased only $0Q_3$ amount of rice while at the $0P_1$ lower price they purchase and consume the larger amount $0Q_4$ of rice. Thus, due to the formation of customs union there is a net increase of Q_3Q_4 amount of rice in the total consumption of rice in India.

After the formation of customs union, consumers in India are paying only Q_3FDQ_4 amount of money for the purchase of Q_3Q_4 amount of rice. Before the formation of customs union they were paying Q_3CDQ_4 amount of money for the same amount of rice. Thus, there is a net saving of FCD amount of expenditure which shows the net increase in the community welfare on the consumption side.

The total increase in the community welfare from trade creation after the formation of customs union is the combined saving due to the production and consumption effects. The magnitude of the welfare gain depends mainly upon the (1) price elasticities of the supply and demand curves of the importing country, and (2) height of the pre-union tariff. If the supply and demand curves of India are flatter, i.e., if the price elasticities of the supply and demand for rice in India are more than one, the gain due to the trade creation effect will be larger and *vice versa*. If the pre-union tariff is higher, the gain which will accrue from the abolition of tariff on the formation of customs union will be larger and *vice versa*. Thus, the trade creation effect causes an increase in the welfare of India. But there is also the trade diversion effect of the customs union which causes an adverse effect on the welfare of India.

Trade Diversion—Before the formation of customs union, India was importing the Q_2Q_3 amount of rice. The entire imports were made from Burma which is the most efficient producer of rice in the world. Bangladesh which is a member of the customs union was unable to secure the Indian market for her rice on account of her higher unit cost of production of rice. After the formation of customs union, the imports of Burmese rice are discriminated against by imposing the import tariff. Consequently, it becomes possible for Bangladesh to capture the market in India. After the formation of customs union, India no longer buys rice from the least-cost non-member producer country Burma. Consequently, the volume of imports is diverted from Burma to Bangladesh causing trade diversion from the relatively low-cost producer country Burma to the relatively high-cost producer country Bangladesh causing the loss of community welfare in India.

Before the formation of customs union, India was importing Q_2Q_3 quantity of rice from Burma for which she paid the total import bill amounting to $Q_2Q_3 \times 0P_2$ money payment which is represented by the rectangle Q_2BCQ_3 in Figure 24.1. Out of this total money payment made for imports, GBCH amount goes to the customs authorities of India. This tariff revenue represents simply an income redistribution within the country. Crucial for our purpose is, however, the total money payment of Q_2GHQ_3 amount which India has made to Burma. After the formation of customs union, India pays for the same amount of imports of rice from Bangladesh an amount of money equal to Q_2EFQ_3. Thus, the extra payment amounting to GEFH which India makes to Bangladesh is due to trade diversion from Burma to Bangladesh. This represents loss of community welfare in India and due to this loss the country is worse off than before.

The net welfare effect of a customs union is the difference between the welfare gain which accrues to the country due to the trade creation effect and the welfare loss which is inflicted on the community due to the trade diversion effect of customs union. The trade creation effect of a customs union is beneficial while the trade diversion effect is harmful from the community welfare consideration. In order to predict the welfare effect of a customs union we should consider several factors influencing the productive efficiency. In the static analysis framework these factors are as following.

1. If the economies of the countries which are members of the customs union are competitive in the sense that the commodities produced under tariff protection in each country overlap to a large extent, there is greater possibility for substitution of the products of the low-cost efficient customs union member for those of the other high-cost less efficient partner. In this case, the trade creation effect of forming a customs union will be larger than the trade diversion effect. Consequently, the customs union will be more likely to bring greater gain to the country. If, on the other hand, the economies of both the countries are complementary in the sense that the goods produced under tariff protection are dissimilar there will be little reallocation of production between the high-cost and the low-cost sources of supply consequent upon formation of the customs union between the two countries. In this situation, the elimination of tariff would have very little to do with the trade creation while a considerable amount of trade diversion is possible consequent upon the formation of a customs union.

We have seen that the net gain in welfare to the country by forming a customs union depends on the relative strength of the trade creation effect of a customs union which itself depends on the extent to which the high-cost or inefficient domestic production in the country is replaced by the low-cost imports from the member country. It will happen only if the two countries produce the same commodity. If neither of the two countries forming the customs union produces the goods in question, both the countries will continue to import the goods from a third country. Consequently, there will not be any trade creation and trade diversion effects and the customs union will be of no significance. On the other hand, if the goods produced by the member countries are complementary, there will be loss in welfare in as much as the country will import goods from a high-cost member in preference to a low-cost outsider. To the extent trade is diverted from the low-cost to the high-cost channel there is loss in the world's welfare. The obvious conclusion is that a customs union will have detrimental effects if the member countries are producing complementary goods. Conversely, if the two countries produce substitute goods the positive effects on the welfare will be large. Figure 24.2 illustrates these facts.

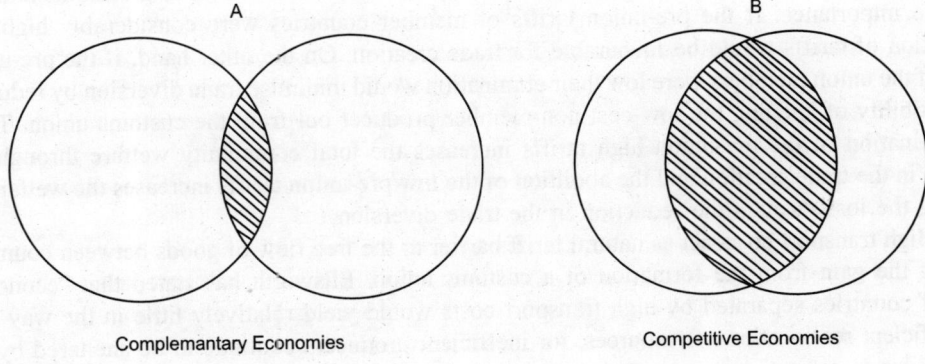

A   B

Complemantary Economies   Competitive Economies

Figure 24.2

Figure 24.2 A shows that the two countries forming customs union are primarily complementary—one being an agricultural economy with the other being an industrial economy. As the two countries produce mostly complementary goods there is very little overlapping and displacement of the inefficient high-cost domestic production. Consequently, there is very little trade creation effect of forming a customs union between the two countries. Figure 24.2 B, however, shows that the two countries forming a customs union are primarily competitive being both primarily agricultural countries, so that there is a very high degree of overlapping in production leading to a very large trade-creation effect of forming the customs union and the resulting increase in community welfare.

2. Jacob Viner has shown that the gain which accrues to a country from forming a customs union will arise if both the countries forming the customs union are producing the same commodity, i.e., the economies of the customs union members are competitive rather than complementary. Makower and Morton[2] have shown that the gain from forming a customs union will be greater if the difference in the relative cost-ratios of the same commodity produced by both the countries is large. If the difference in the relative cost-ratios is greater there would be a substantial readjustment of resources from the inefficient domestic to the more efficient customs union production sources. The lower prices will be a strong incentive for the consumers and there would be a larger gain in the aggregate satisfaction resulting from increased consumption of the commodity.

From this it follows that contrary to the instinctively held belief that an agricultural country should form a customs union with an industrial country, an agricultural country should form a customs union with another agricultural country while the industrial countries should form a customs union with each other because the scope for the trade creation will then be greatest resulting in improved resource-allocation and consequent increase in welfare.

3. Jan Tinbergen and Bela Balassa have emphasised the importance of the size of the customs union. They have shown that larger the size of customs union greater will be the gain from trade due to increased production and productive efficiency. Larger is the size of the customs union, greater is the possibility of world's lowest-cost producer becoming a member of the customs union. The merger of the lowest-cost producer in the customs union will nullify the adverse effect of trade diversion. In the extreme situation, if the customs union is large enough to encompass the whole world, trade diversion will no longer be there and only gain from the trade creation effect of the customs union will be reaped. It follows from this that the non-discriminatory free trade is always far superior to the trade pattern that emerges under a customs union arrangement.

4. The level of the pre-union and the post-union tariffs of the members of a customs union is of prime importance. If the pre-union tariffs of member countries were considerably high, the elimination of tariffs would be favourable for trade creation. On the other hand, if the pre-union tariffs of the union members were low their elimination would minimise trade diversion by reducing the possibility of keeping the low-cost non-member producer out from the customs union. Thus, the elimination of the pre-union high tariffs increases the total community welfare through the increase in the trade creation and the abolition of the low pre-union tariffs increases the welfare by reducing the loss through the reduction in the trade diversion.

5. High transport costs act as natural tariff barrier to the free flow of goods between countries reducing the gain from the formation of a customs union. Ellsworth has stated that "economic union of countries separated by high transport costs would yield relatively little in the way of a more efficient reallocation of resources, for inefficient producers continue to be sheltered by the

2. H. Makower and G. Morton, "A Contribution Towards a Theory of Custom Union," *The Economic Journal*, March 1953, pp. 33–49.

natural protection"[3] provided by the high transport costs. For example, economic union between countries which are situated on the east and on the west coasts of South America and are separated from one another by the lofty Andes mountains and have consequently to ship goods by the long sea voyage all the way around the Cape Horn would fail in bringing about more efficient resource reallocation between the members of the union because inefficient producers would enjoy the natural protection provided by the high transport costs.

In addition to the economic reasons discussed above, there are several other factors, more particularly political, which considerably influence the decision of a country to join a customs union.

Consumption Effect—Jacob Viner's and Modern Views Compared

Jacob Viner's conclusion that the trade diversion which results from the formation of a customs union necessarily reduces the welfare is based on the implicit assumption that the commodities are consumed in fixed proportions irrespective of the price changes caused by changes in the tariff structure after the formation of a customs union. In other words, Viner ruled out intercommodity substitution. In reality, however, changes in the relative prices of commodities lead to some substitution in consumption between commodities. On the supply side, Viner assumed the supply elasticities to be infinite so that all the goods were produced under constant returns to scale. The importance of the substitution effect in consumption in the context of a customs union has been discovered independently by Meade,[4] Gehrels[5] and Lipsey[6]. The approaches of Viner and Lipsey have been explained in Figure 24.3A and Figure 24.3B.

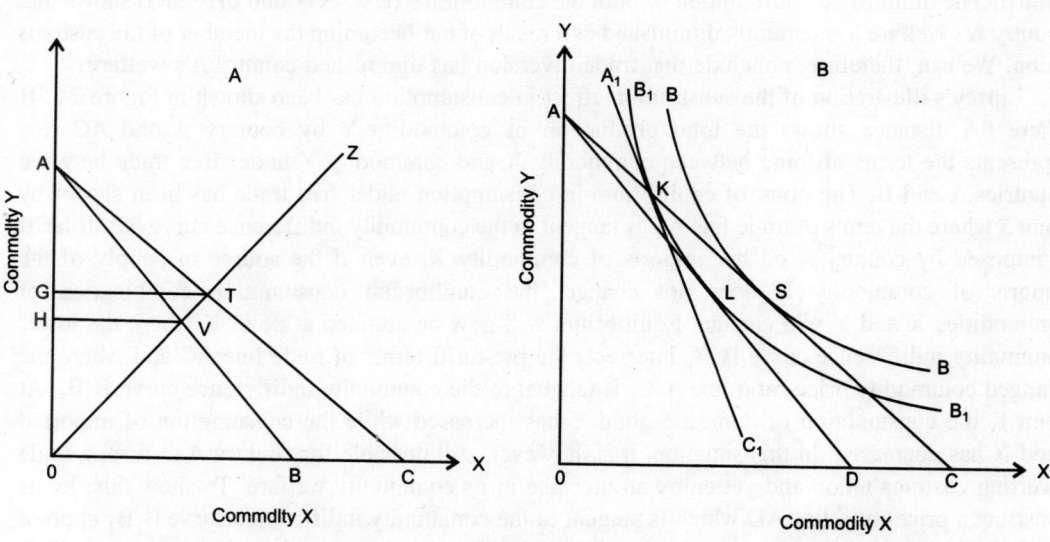

Figure 24.3

3. P.T. Ellsworth and J.C. Leith, *The International Economy*, Fifth Edition, 1975, p. 523.
4. James E. Meade, *The Theory of Customs Unions*, Amsterdam: North-Holland Publishing Company, 1955.
5. F. Gehrels, "Customs Unions from a Single Country Viewpoint", *The Review of Economic Studies*, Volume XXIV, No. 1. 1956–57, pp. 61–64
6. R.G. Lipsey, "The Theory of Customs Unions: Trade Diversion and Welfare", *Economica*, New Series, Volume XXIV, February 1957, pp. 40–46.

Figure 21.3A explains Viner's contention. Country A consumes the two commodities X and Y in fixed proportion indicated by the 0Z straight line drawn from the point of origin 0 which is the price-income consumption line for all (finite) prices and income. 0A shows the total amount of commodity Y produced by country A. Commodity X is imported from the low-cost producer country C. Line AC shows the terms of trade offered by country C. Under free trade, the equilibrium consumption point is T which is the point of intersection of lines 0Z and AC. Country A consumes the 0G amount of commodity Y and GT quantity of commodity X after exporting GA amount of commodity Y in exchange for the imports of GT quantity of commodity X. Now suppose that an import tariff which is not high enough to protect the domestic industry completely is levied on the imports of commodity X. This tariff distorts the domestic prices but keeps the terms of trade unchanged. After the imposition of import tariff, country A's consumption equilibrium takes place at point V. As point V is located on the price-income consumption line 0Z, it shows that although the imposition of import tariff changes the relative prices of the two commodities but the consumers are completely insensitive to this relative price change.

Let us assume that country A forms a trade diverting customs union with country B. This means that country A must buy her imports of commodity X at a price which in terms of commodity Y is higher than that which she was paying before the customs union was formed. In other words, in the new situation country A's terms of trade are shown by AB line. As a consequence, country A's equilibrium is now at point V which is the intersection point of lines 0Z and AB. At point V, although the proportion of the commodity combination remains unchanged (GT/0G = HV/0H) but the quantities consumed of both the goods X and Y are less than those which are consumed at point T. The diminished consumption of both the commodities (HV < GT and 0H < 0G) shows that country A's welfare has certainly diminished as a result of her becoming the member of the customs union. We can, therefore, conclude that trade diversion has diminished country A's welfare.

Lipsey's illustration of the substitution effect in consumption has been shown in Figure 24.3B where 0A distance shows the total production of commodity Y by country A and AC line represents the terms of trade between commodity X and commodity Y under free trade between countries A and B. The point of equilibrium in consumption under free trade has been shown by point S where the terms of trade line AC is tangent to the community indifference curve BB. If tariff is imposed by country A on her imports of commodity X, even if the source of supply of the imports of commodity X does not change, the equilibrium consumption combination of commodities X and Y will change. Equilibrium will now be attained at point K where the lower community indifference curve B_1B_1 intersects the pre-tariff terms of trade line AC and where the changed commodity price-ratio line A_1C_1 is tangent to the community indifference curve B_1B_1. At point K the consumption of domestic good Y has increased while the consumption of imported good X has decreased. In this situation, it is, however, still possible for country A to form a trade diverting customs union and yet enjoy an increase in its community welfare. To show this, let us construct a price-ratio line AD which is tangent to the community indifference curve B_1B_1 at point L. If after the formation of customs union the terms of trade are worse than those represented by the terms of trade line AC but better than those represented by the terms of trade line AD, country's welfare will increase by forming a trade diverting customs union. Conversely, if the terms of trade are, however, worse than those represented by the terms of trade line AD, country's welfare will be reduced as a consequence of the formation of customs union. If the terms of trade after the formation of the customs union are represented by AB, country's welfare will remain unchanged. As long as the final equilibrium point lies within the area lying above the community indifference curve B_1B_1 and below the terms of trade line AC, country's welfare will increase.

To conclude, Lipsey has stated: "In a verbal statement this possibility may be explained by referring to the two opposite effects of trade-diverting customs union. First, A shifts her purchases from a lower to a higher cost sources of supply. It now becomes necessary to export a larger quantity of goods in order to obtain any given quantity of imports. Secondly, the divergence between domestic and international prices is eliminated when the union is formed. The removal of tariff has the effect of allowing... consumers in A to adjust... purchase to a domestic price ratio which now is equal to the rate at which... (Y) can be transformed into... (X) by means of trade. The final welfare effect of the trade diversion customs union must be the net effect of these two opposing tendencies; the first working to lower welfare and the second to raise it"[7].

General Equilibrium Analysis of Customs Union

Jaroslav Vanek has developed the analysis of customs union in terms of the general equilibrium theory by using the technique of the Marshallian trade offer curves as illustrated in Figure 24.4.

Let us assume that two countries India and Bangladesh form a customs union. A third country Burma represents the outside world. The quantities of the two commodities cotton and rice which the union partners are willing to exchange with the outsider country Burma can be summarised in an excess trade offer curve which shows the different cotton-rice combinations which the India-Bangladesh customs union is willing to trade with Burma at different terms of trade. To derive this excess trade offer curve, we should construct the international trade offer curves for the customs union partners.

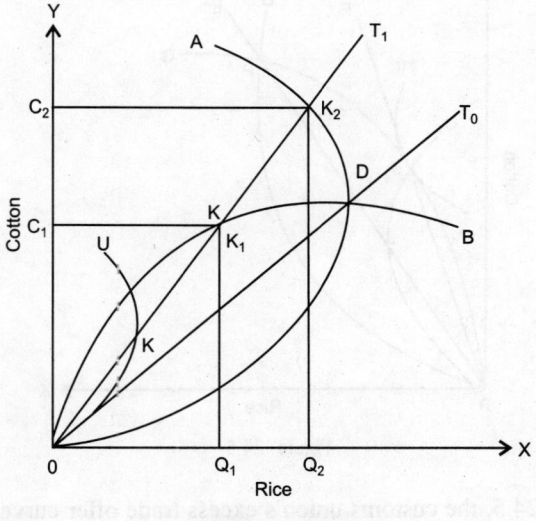

Figure 24.4

In Figure 24.4, the trade offer curve of India is 0A and that of Bangladesh is 0B. We should now determine the net quantities of cotton and rice which the partners are willing to trade with the outsider Burma at all the possible terms of trade. At the terms of trade shown by the slope of $0T_0$ terms of trade line, trade between India and Bangladesh is balanced and there will be no excess demand or offer for either cotton or rice in the customs union. Consequently, the union will not

7. Richared G. Lipsey, *Op. cit.,* pp. 43–44.

trade with the outsider country Burma. Since there is no trade with Burma, the excess trade offer curve will coincide with the point of origin 0. However, at any other terms of trade, the trade between India and Bangladesh will not be balanced. In other words, at any terms of trade other than those represented by the slope of $0T_0$ terms of trade line there will be either excess demand or excess supply for either cotton or rice between the union partners necessitating trade with the outsider country Burma. For example, at the terms of trade represented by the slope of the terms of trade line $0T_1$, trade between the union partners will not be balanced because at these terms of trade India is willing to offer $0Q_2$ quantity of rice to Bangladesh for Q_2K_2 quantity of cotton from Bangladesh while Bangladesh wants to exchange only Q_1K_1 quantity of cotton for $0Q_1$ quantity of rice. Thus, the total offer (supply) of cotton which Bangladesh makes to India does not match the total demand made for it by India at the terms of trade shown by the slope of line $0T_1$. Consequently, there is an excess demand for cotton and an excess supply of rice in India at the current prices, i.e., at the $0T_1$ terms of trade. Such excess demand for cotton (or excess supply of rice) is denoted by the segment KK_1 on line $0T_1$. This is the 'excess offer' of rice for cotton by the customs union and it can be shown separately as the $0K$ distance measured from the point of origin 0 on $0T_1$ line. The collection of all possible excess offers at different terms of trade will generate the customs union's excess trade offer curve $0U$. This offer curve shows the quantity of exports of two commodities which the two countries India and Bangladesh under customs union are willing to offer at different price ratios in exchange for imports from outsider country Burma.

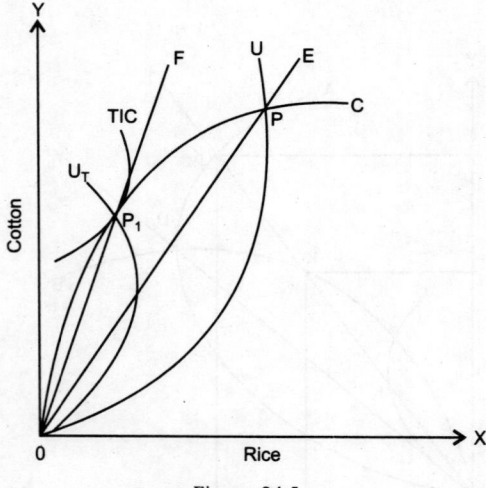

Figure 24.5

As shown in Figure 24.5, the customs union's excess trade offer curve $0U$ and the trade offer curve $0C$ of outsider country Burma intersect each other at point P. Thus, the equilibrium terms of trade represented by the slope of $0E$ line are established showing that at these terms of trade there will be no excess demand or excess supply of the two commodities in the world market. When the customs union imposes an import tariff on the imports from the outsider country Burma, the customs union's distorted trade offer curve $0U_T$ passes through point P_1 where the customs union's trade indifference curve TIC is tangential to the $0C$ trade offer curve of Burma. This is the situation of optimum tariff. The new terms of trade line $0F$ is favourable for the customs union's partners.

The question, however, is: has the formation of customs union between India and Bangladesh made the position of the partners better or worse off? As shown in Figure 24.6, both the situations are possible. The position of the union partners will be better if the pre-tariff excess trade offer curve is $0U$ and after the imposition of a common tariff the excess trade offer curve is $0U_1$. On the other hand, the position of the customs union partners will be worse off if the pre-tariff excess trade offer curve is $0U_1$ while the post-tariff excess trade offer curve is $0U$.

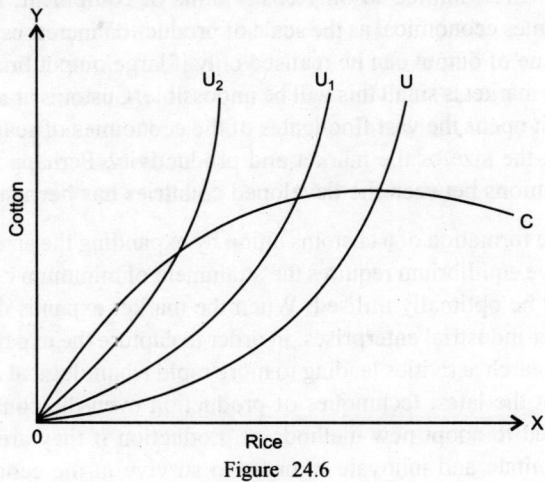

Figure 24.6

It should, however, be recognised that in the determination of the welfare gain it is not only the terms of trade that are important. Several other factors also influence significantly the welfare gain accruing to a country from her membership of a customs union. Thus, the effects of the formation of a customs union cannot be predicted on an *a priori* basis.

Dynamic Effects of Customs Union

The effects of a customs union on the production and consumption of a member country surveyed so far are purely static since it was assumed that the formation of a customs union did not bring any change in the structure of the industry, technology and the pattern of behaviour of the people in the country. In other words, after the formation of a customs union, everything in the country goes on as before. This is not, however, the real situation. After the formation of a customs union, everything does not remain static and the impact of the membership of a customs union is felt on the industrial and technological structure of the member countries as well as the outside world. Even the behavioural pattern of the people in the countries both within the union and outside is affected as a result of the formation of a customs union.

Jaroslav Vanek has rightly pointed out that "the creation of a customs union and the adjustment of the world economies to it is a long-lasting process. During long period of time the structure or structural parameters of the different countries may change substantially. To the extent that such changes are autonomous, they fall within the sphere of the general analysis of international trade. To the extent, however, that they are brought about by the existence and operation of the customs union, they should be examined within the theory of customs union."[8] In

8. Jaroslav Vanek, *International Trade: Theory and Economic Policy*, 1962, p. 366.

fact, the dynamic effects of forming a customs union can be far more important than the static effects of a customs union. The dynamic gains resulting from the formation of a customs union include the economies of scale, the stimulus to competition and the stimulus to invest, etc.

Economies of Scale—The formation of a customs union helps a country in the painless realisation of the gains of the economies of scale in production. The economies of large-scale production are not, however, limited to only costly units of equipment. Even the employment of specialised labour becomes economical as the scale of production increases. Those economies which depend on a large volume of output can be realised only if large output produced on mass scale can be profitably sold. If the market is small this will be impossible. Customs union expands the size of the market. Consequently, it opens the vast floodgates of the economies of scale. A high correlation has been recorded between the size of the market and productivity. Perhaps for this reason alone the formation of customs unions between the developed countries has been strongly advocated.

Competition—The formation of a customs union by expanding the size of the market stimulates competition. Competitive equilibrium requires the attainment of minimum cost which implies that the industrial plants should be optimally utilised. When the market expands due to the formation of a customs union, the larger industrial enterprises, in order to capture the expanded market, spend more freely on conducting research activities leading to more rapid technological advancement. Since their progressive rivals adopt the latest techniques of production even the conservative and traditional industries are also forced to adopt new methods of production if they are to stay in the business. Inefficient firms also imitate and innovate in order to survive in the economic world. Thus, as a consequence of the formation of a customs union the whole economy is dynamised.

Investment—The formation of a customs union creates a better and healthier atmosphere for domestic investment. One important aspect of the change in domestic investment is the establishment of market-oriented industry near the frontier of the partner. "Moreover, if the forces generating increased efficiency operate as expected, aggregate income will rise, saving will increase and aggregate investment will be larger. With expanding markets giving a lift to confidence, the *proportion* of income saved and invested may be raised."[9] In addition to domestic investment, the larger market can also attract foreign investment.

Empirical Validity of the Theory

Several empirical studies have been made to estimate the gains and losses of forming the customs unions. The earliest and the best known of these studies was made by the Dutch economist P.J. Verdoom which was later used by Tibor Scitovsky.[10] Another study was made by Harry G. Johnson.[11] In all these studies efforts were made to measure the gains and losses which accrued from the trade creation and trade diversion effects of customs unions. All these early studies showed that the resulting gains were small. Verdoom's study which related to the original six European Economic Community countries showed that the welfare effect of the union amounted to only 1/20th of 1 per cent of the total national incomes of these countries. Tibor Scitovsky's study

9. P.T. Ellsworth and J. Clark Leith, *Op. cit.,* p. 552.
10. Tibor Scitovsky, *Economic Theory and Western European Integration,* London, Allen & Unwin, 1958.
11. Harry G. Johnson, "The Gains from Free Trade with Europe: An Estimate", *Manchester School of Economic and Social Studies,* Volume 26, September 1958.

also showed that the gain from increased intra-European specialisation was insignificant. Harry G. Johnson's findings confirmed the findings of Verdooms and Scitovsky studies. Johnson found that by joining the EFTA, England could at most expect an insignificant increase of 1 per cent in her national income. Major's study[12] which examined the changes in the market shares during 1958–61 concluded that the European Economic Community did not have any significant effect during this period.

Notwithstanding the pessimistic estimates of the welfare effect of customs unions in these early studies, some more recently made studies have shown substantially higher estimates of the trade-creation effect of the customs unions. In his analysis of the changes in the sources of supply of manufactures to the EEC countries, M. Truman[13] estimated that by 1964 when internal tariffs in the Community were reduced by 60 per cent, the intra-EEC trade was about 30 per cent higher compared with what it could have been in the absence of the EEC. He also found that this increase had been achieved without any net diversion of trade from rest of the world countries. A further study made by J.Williamson and A. Bottrill[14] supported Truman's findings of trade-creation for 1964 and concluded that by 1969 the intra-EEC trade was about 50 per cent higher compared with what it would have been in the absence of the EEC. Their study attributed this increase mainly to the trade-creation effect of customs union. Another study made by M. Kreinin[15] showed that in 1969–70 the trade-creation was eight times of the trade diversion. There are, however, wide variations in the findings of the published studies and it is desirable not to reach any final and settled conclusion on the empirical validity of the gains from customs unions.

It should, however, be kept in mind that the theory of customs union discussed here is of a comparative-static nature starting as it does from an equilibrium with a given tariff structure. Thereafter it makes a discriminatory change in this structure in order to estimate the effects on economic welfare. The above-mentioned estimates are based on this kind of comparative static analysis being blissfully ignorant and unconcerned about any serious defect in them.

It could, however, be argued that this kind of analysis fails to take account of other important dynamic or institutional effects. One such dynamic effect is the presence of unutilised economies of scale. It has, for instance, been argued that due to the market segmentation the European industry has not reaped the fruits of the economies of scale. However, for various reasons—the European industries in which the economies of scale could be expected were the export industries facing already the large world market—the argument for unutilised scale economies does not appear to be too valid.

The other plausible arguments of dynamic nature such as that a customs union will lead to enforced competition although plausible is very difficult to prove empirically. Among the other institutional arguments are those which relate to increased factor mobility. It has been argued that the increased movements of labour and capital will lead to increased factor productivity fostering growth and well-being. In the event of these dynamic effects turning out to be important, these will have to be added to the list of effects of comparative static nature in order to get a complete picture of the effects of formation of a customs unions.

12. R.L Major, "The Common Market: Production and Trade", *National Institute Economic Review*, August 1962.

13. M. Truman, "The European Economic Community: Trade Creation and Trade Diversion", *Yale Economic Essays*, Volume 9, Spring 1969, pp. 210–257.

14. J. Williamson and A. Bottrill. "The Impact of Customs Union on Trade in Manufactures", *Oxford Economic Papers*, November 1971 pp. 323–351.

15. M. Kreinin, "Effects of the E.E.C on Imports of Manufactures", *The Economic Journal*, September 1972, p. 216.

Suggested Readings

Bela Balassa, *The Theory of Economic Integration,* 1961, Chapters 1 to 4.

M.O. Clement, R.L. Pfister and K.J. Rothwell, *Theoretical Issues in International Economics,* 1967,
   Chapter 4.

Victoria Curzon, *The Essentials of Economic Integration,* 1974, Chapters 1, 10 and 11.

Geoffrey Denton (ed.), *Economic and Monetary Union in Europe,* 1974.

P.T. Ellsworth and J.C. Leith, *The International Economy,* Fifth Edition, 1975, Chapter 29.

H. Robert Heller, *International Trade,* Second Edition, 1973, Chapter 10.

Charles P. Kindleberger, *International Economics,* Fifth Edition, 1973, Chapter 11.

Bo Sodersten, *International Economics,* Second Edition, Reprinted 1981, Chapter 20.

Jaroslav Vanek, *International Trade: Theory and Economic Policy,* 1962, Chapter 18.

Jacob Viner, *The Customs Union Issue,* 1953.

James E. Meade, *The Theory of Customs Unions,* 1955.

Questions

1. "All far reaching schemes for a customs union are quite utopian and fantastic. They are
   completely ruled out by the spirit of nationalism and protection which prevails today."
   Discuss.

2. Discuss the effects of a customs union. Is a customs union superior to a system of all-round
   tariffs from the community welfare point of view? Discuss fully.

3. Examine the following statements:
   (i) "A customs union is superior to a system of all-round tariffs because it has fewer tariffs".
   (ii) "A customs union would have more favourable effects if the economies of countries
        forming a customs union are complementary rather than competitive".

CHAPTER 25

Commercial Treaties

Commercial treaties constitute an important economic phenomenon in the sphere of international economic relations, particularly in international trade. These treaties may extend to a wide range of subjects covering trade-marks and copyrights, execution of legal judgments, legal and police protection to foreigners and their property, customs duties, establishment of foreign firms and status of foreign tourists, freight rates, rights and treatment of foreign ships in home ports etc. All these matters forming the subject-matter of commercial treaties may be grouped under the following four broad categories.

1. Consular matters,
2. Rights of foreigners,
3. Transport matters, and
4. Tariffs and trade matters.

With the passage of time as the economic relations between world's governments have tended to become increasingly complex. It has become usual to regulate certain matters such as those relating to copyright and patents, avoidance of double taxation, regulation of foreign direct and portfolio investments etc. by signing special agreements. The agreements relating to tariff matters are usually termed as the commercial treaties.

Coming to the discussion of the form of commercial treaties, it is necessary to distinguish between the *bilateral* and *multilateral* form of commercial treaties. While the bilateral treaties are signed between two countries, multilateral treaties are signed between more than two countries. Multilateral treaties are sometimes also known as collective agreements or international conventions like the Brussels Sugar Convention of 1902 which was an important successful collective agreement on tariff matters. Governments may pledge themselves in commercial treaties to maintain certain relations between themselves either through *direct* or *indirect* method. As an example, under the direct method, it may be provided that an import duty of certain given amount will be levied on a certain stated good coming from the country with which the treaty is made. Under the indirect method, a measure or yardstick is prescribed by which the treatment accorded to the other party is regulated. Three such measures are customary with each one of these three measures having its corresponding *the Parity Clause, the Reciprocity Clause* and *the Most-Favoured-Nation Clause* in the treaty.

Under the *Parity Clause,* the treatment which is given by a State to the citizens and goods of the State with which the treaty is made must not be worse than that accorded to its own citizens and goods. Under the *reciprocity* clause, this treatment must correspond to, or at least must not be worse than, the treatment given to the citizens and goods of the State in question by the other State. Under the *most-favoured-nation clause,* this treatment must not be worse than the treatment accorded to any third country.

In the narrower sense, the commercial treaties are concerned with the imposition and height of import duties. Such treaties can be divided into the *pure most-favoured-nation treaties* and the *tariff treaties.* Under the *pure form of most-favoured-nation treaty,* a state commits itself not to impose upon the goods coming from the other state any duties higher than it imposes upon similar goods coming from any third state or country. Under the *tariff treaty,* particular tariffs' concrete provisions are stated. For example, it may be stated that the import duty on the Australian cheese would not be more than ten rupees per kilogram. Most tariff treaties, however, also include a most-favoured-nation clause. Under a pure most-favoured-nation treaty, while the amounts of import levies are left to the autonomous decisions of the state imposing them, however, any reduction in the import duties granted by the state to any third country must also be granted to any other country with which it has such a treaty.

Most Favoured Nation Clause—Contents and Forms

The early origins of the most-favoured-nation (MFN) clause may be traced to the seventeenth century when the states of Europe began to include it in their commercial treaties. With the passage of time, it became increasingly customary to do this with several variants of the clause being introduced in the commercial treaties. By the second half of the eighteenth century, there was scarcely any commercial treaty which did not include expressly or implicitly in its text the MFN clause. Perhaps the MFN clause is the most hotly debated theme, even of the day, by the politicians, statesmen and economists. While according to some it is the surest guarantee for stable and lasting peace and a symbol of free trade policy, others regard it as an obstacle to reduction in import duties.

According to the unconditional most-favoured-nation clause insertion in the commercial treaty, a state binds itself not to give worse treatment to imports from another state than it gives to imports from any third state immediately and automatically and without receiving any reciprocal concession. Under the conditional MFN clause, however, the other state receives a concession only if it grants to the first state the same, or an equally valuable concession, as the one granted as a *quid pro quo* by the third state to the first state. The conditional variant of the MFN is sometimes also known as the European variant since the European countries adopted it *de facto* during the first half of the nineteenth century and in the post-1890 period adopted it explicitly without any exception. By way of an example of the MFN clause, the following extract from Article 7 of the now-defunct commercial treaty entered into between the United States and Germany on 8th December 1923 and ratified in 1925 may be quoted here.

"Each of the High Contracting Parties binds itself unconditionally to impose no higher or other duties or conditions and no prohibition on the importation of any article,... of the territories of the other than are or shall be imposed on the importation of any like article,... of any other foreign country.

"Each of the High Contracting Parties also binds itself unconditionally to impose no higher or other charges or other restrictions or prohibitions on goods exported to the territories of the other High Contracting Party than are imposed on goods exported to any other foreign country.

"Any advantage of whatever kind which either High Contracting Party extends to any article...., without request and without compensation, be extended to the like article,... of the other High Contracting country....

"With respect to the amount and collection of duties on imports and exports of every kind, each of the two High Contracting Parties binds itself to give to the nationals, vessels and goods of the other the advantage of every favour, privilege or immunity which it shall accord to the nationals, vessels and goods of a third State, and regardless of whether such favoured State shall have been accorded such treatment gratuitously or in return for reciprocal compensatory treatment. Every such favour, privilege or immunity which shall hereafter be granted to the nationals, vessels or goods of a third State shall simultaneously and unconditionally, without request and without compensation, be extended to the other High Contracting Party, for the benefit of itself, its nationals and vessels."

It is, therefore, very obvious from the above unconditional most-favoured-nation treaty cause that every concession in duties which one state grants to another state is immediately extended to all those other states which stand in the most-favoured-nation relationship with the first state. Thus, as it is, the MFN clause establishes a nexus between all the commercial treaties of a country and in the event of the country applying it universally gives the country a single uniform tariff which is applicable to imports from every other country. In sum and substance, the unconditional most-favoured-nation clause when applied universally guarantees purely non-discriminatory international trade policy.

Apart from the conditional and unconditional variants of the MFN treaty clause, a meaningful distinction may as well be drawn between the *unilateral* and *bilateral* most-favoured-nation agreements depending upon whether only one or both the parties to the contract are under obligation to accord the most-favoured-nation treatment. The famous Treaty of Versailles bound Germany unilaterally to give most-favoured-nation treatment to the Allies for five years. Aside this, distinction is also made between the *restricted* and *unrestricted* most-favoured-nation agreements depending upon whether they apply to all matters (like the level of tariffs, rights of foreigners, veterinary regulations etc.) goods and countries. Richard Schueller has made a further distinction between the restrictions of the *content* of the most-favoured-nation agreement in which case the agreement does not cover all matters or topics and the restrictions in the *ambit* of agreement when it does not cover all states.

Advantages of Unconditional Most Favoured Nation System—In the world community of nations, different nations have different political systems and ideologies. Thus, there are countries which refuse to make the height of their tariffs dependent on agreements with other countries while some other countries are firmly wedded to the ideology of free trade. For such countries, unconditional most-favoured-nation agreement offers the best trade policy. In fact, there is much to be said in favour of this principle which neither seeks nor grants any special advantage. The principle of equality of treatment in matters of trade and tariff policy *per se* is beneficial in maintaining peace and harmony between countries, regardless of their political system or ideology because in the absence of such equality of treatment numerous potential causes of conflict would emerge. For small countries it is of special and vital importance that the world should adhere faithfully to this policy. An added substantial advantage of the unconditional most-favoured-nation system, if universally applied, is that a country will have a single uniform duty on any given import, regardless of its source of supply. Consequently, there is no need to furnish the customs authorities with proofs of origin resulting in the simplification and cheapening of the collection of import duties.

Most Favoured Nation (MFN) Clause and the World Trade Organisation (WTO)

For almost fifty years, key provisions of the GATT—predecessor of WTO—outlawed discrimination among members and between imported and domestically-produced merchandise. According to Article I, the famous "most-favoured-nation" (MFN) clause, members are bound to grant to the products of other members treatment no less favourable than that accorded to the products of any other country. Thus, no country is to give special trading advantages to another or to discriminate against it; all member countries are on a equal basis and all share the benefits of any moves towards lower trade barriers. There are, however, certain exceptions to Article I, particularly those covering customs unions and free trade areas. However, the most-favoured-nation treatment generally ensures that the developing countries and others with little economic leverage are able to benefit free from the best trading conditions wherever and whenever they are negotiated.

A second form of non-discrimination, known as "national treatment", requires that once goods have entered a market they must be treated no less favourably than the equivalent domestically-produced goods. This has been provided in Article III of the GATT. Apart from these two fundamental features of non-discrimination, several other WTO agreements contain important provisions relating to MFN and national treatment. The agreement on Trade-Related Aspects of Intellectual Property Rights (TRIPS) contains, with some exceptions, MFN and national treatment requirements relating to the provision of intellectual property protection by the WTO members. Similarly, the General Agreement on Trade in Services (GATS) requires the WTO members to offer MFN treatment to services and service suppliers of other members. The other WTO agreements with non-discriminatory provisions include those on rules of origin, pre-shipment inspection, trade-related investment measures, and the application of sanitary and photosanitary measures. In short, the WTO—successor to the GATT—with its more than 150 countries' membership and accounting for over 90 per cent of the total merchandise exports ensures the development of the non-discriminatory multilateral trading system in the world by subscribing to the provisions of the MFN clause in the sphere of trade.

Suggested Readings

Gottfied Von Haberler, *The Theory of International Trade,* English Translation, Third Impression, 1950, Chapters XX and XXI.
WTO, *Trading into the Future* and *Focus.*

Questions

1. Discuss the content and form of the most-favoured-nation treaty. How far does it ensure the promotion of a non-discriminatory multilateral trading system in the world? Discuss fully.
2. Explain to what extent the World Trade Organisation (WTO) supports the system of most favoured-nation treaty. Has it been successful in promoting the multilateral trading system? Discuss.

CHAPTER 26

World Trade Organisation

The World Trade Organisation (WTO) was established on 1 January 1995 as successor to the General Agreement on Tariffs and Trade (GATT) which was established in 1947. The member governments of GATT had concluded the Uruguay Round Negotiations on 15 December 1993 and the Ministers had given their political support to the results by signing the Final Act at a meeting held in Marrakesh, Morocco in April 1994. The Uruguay Round, concluded at the end of 1993, had reduced the tariff and non-tariff barriers to trade in merchandise goods, liberalised trade in major areas such as agriculture, textiles and clothing and extended multilateral rules to new areas, notably services and intellectual property. The Marrakesh Declaration of 15 April 1994, affirmed that the results of the Uruguay Round would "strengthen the world economy and lead to more trade, investment, employment and income growth throughout the world". The establishment of the WTO is the embodiment of the Uruguay Round results.

As the legal and institutional foundation of the multilateral trading system, the World Trade Organisation provides the principal contractual obligations determining the manner in which the member governments shall frame and implement the domestic trade legislation and regulations. It is the platform on which the trade relations among the member countries evolve through collective debate, negotiations and adjudication. At present the membership of the WTO is over 153 and its secretariat is based in Geneva, Switzerland. The WTO is headed by a director-general (currently Mr. Pascal Lamy) who is assisted by four deputies from different member states. The director-general is appointed by the General Council for a four-year term after consultations among member countries

Functions

The essential functions of the WTO are:
1. To administer and implement the multilateral and plurilateral trade agreements which together make up the WTO;
2. To act as a forum for multilateral trade negotiations;
3. To resolve trade disputes among members;
4. To oversee national trade policies, and
5. To cooperate with other international institutions involved in global policy-making.

Brief History of the GATT

Since the World Trade Organisation is successor to the General Agreement on Tariffs and Trade, it is only proper to discuss the GATT briefly. As the WTO's proud predecessor, the GATT was established on a provisional basis after the Second World War in the wake of other new multilateral institutions dedicated to international economic cooperation—notably the "Bretton Woods" twins now known as the World Bank and the International Monetary Fund. The GATT's original 23 signatory countries were among over 53 countries which agreed to a draft charter for establishing an International Trade Organisation (ITO) under the aegis of the United Nations. The proposed ITO did not come into existence due to non-ratification of its Charter by the US administration in 1950.

In an effort to give an early boost to trade liberalisation after the Second World War—and to begin to correct the large overhang of the protectionist measures which had been prominently in existence from the early 1930s—tariff negotiations were started among the 23 founding GATT "contracting parties" in 1946. This first round of negotiations resulted in 45,000 tariff concessions affecting $10 billion worth, or about 20 per cent, of world trade. It was also agreed that the value of these concessions should be protected by early—and largely "provisional"—acceptance of some of the trade rules in the draft Charter of the ITO. The tariff concessions and rules together came to be known as the General Agreement and Tariffs and Trade and entered into force in January 1948.

Trade Rounds

The GATT trade rounds, or multilateral trade negotiations, have served as an important vehicle of boosting international trade liberalisation, the Uruguay Round being the latest and most extensive. Most of the GATT's early rounds were devoted to reduce the tariffs. The Kennedy Round (1964–1967) included, however, a new anti-dumping agreement. The Tokyo Round (1973–1979) with 102 participants was a continuation of the GATT's efforts to progressively reduce tariffs. The Uruguay Round (1986–1993) in which 123 members participated was the most comprehensive and important of all the eight GATT rounds as it covered beside tariffs, non-tariff measures, rules, services, intellectual property rights, dispute settlement, textiles and clothing, agriculture, establishment of the WTO etc. The following chart casts a bird's eye-view on the eight GATT rounds beginning from 1947 to 1993.

Table 26.1 GATT Trade Rounds

Year	Round	Subjects Covered	Participating Countries
1947	Geneva	Tariffs	23
1949	Annecy	Tariffs	13
1951	Torquay	Tariffs	38
1956	Geneva	Tariffs	26
1960–1961	Geneva (Dhillon Round)	Tariffs	26
1964–1967	Geneva (Kennedy Round)	Tariffs and Anti-dumping Measures	62
1973–1979	Geneva (Tokyo Round)	Tariffs, non-tariff measures and "framework" agreements	102
1986–1993	Geneva (Uruguay Round)	Tariffs, non-tariff measures, rules, services, intellectual property rights, dispute settlement, textiles and clothing, agriculture, establishment of the WTO etc.	123

Source: WTO

Achievements of GATT

Given its provisional nature and limited field of action, the success of the GATT in promoting and securing the liberalisation of world trade during its 47 years' life cannot be disputed. The continual reductions in tariffs alone spurred very high rates of growth in world trade of around 8 per cent a year on average during the 1950s and 1960s. Throughout the GATT era, the momentum of trade liberalisation ensured that trade growth consistently outpaced output growth promoting globalisation. The rush of new members during the Uruguay Round proved beyond dispute that the multilateral trading system represented by the GATT was perceived as an anchor for the development and an instrument of economic and trade reforms.

In the Uruguay Round, which was the last round within the framework of the GATT, the developed countries agreed to slash down their average tariffs on industrial goods by 40 per cent over five years. The share of the duty-free imports in the total imports also rose. The developing countries agreed to raise from 13 per cent to 61 per cent the share of their imports covered by the bound tariff. In agriculture, most of the non-tariff barriers were to be replaced by tariffs while the share of imports covered by the bound tariffs rose. Further, domestic price supports or "floor prices" were to be reduced by 20 per cent (relative to a 1986–88 base) over six years in the case of the developed countries and by 13 per cent over 10 years by the developing countries. Commitments were also made to reduce the export subsidies.

The Uruguay Round established that trade in textiles and clothing would be gradually integrated into the World Trade Organisation. Earlier, such trade was governed by a separate Multifibre Arrangement which fixed quotas for the developing countries' exports to the developed countries. It was agreed that these quotas will be phased out by 2005. However, the phase-out was to be back-loaded with about two-thirds of the developed countries' textiles and clothing imports to remain subject to quotas upto 2002. In brief, during the 47 years of the GATT era world trade was substantially liberalised through a continuous process of tariff reductions and these tariff barriers at the border were reduced from around 40 per cent in 1947 at the time of GATT's establishment to a mere around 3 per cent at the end of 1994 when its successor the World Trade Organisation took over. The chart in Figure 26.1 shows the single achievement of the GATT in liberalising the world

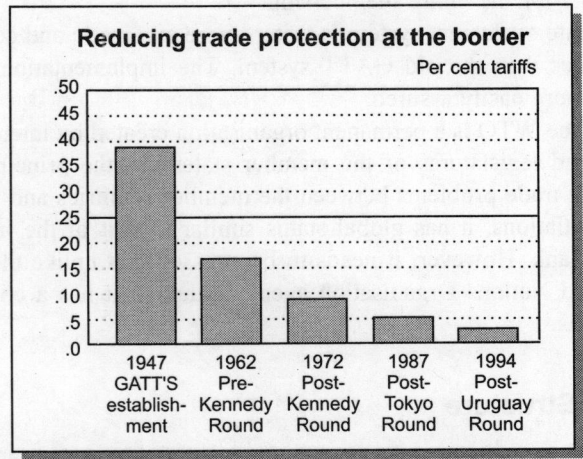

Source : WTO

Figure 26.1

trade by substantially curbing the tariff restrictions. It shows that following the establishment of the GATT in 1948, the average tariff levels fell progressively and dramatically through a series of eight trade rounds. In the Uruguay Round over 123 countries helped increase the overall market access through the liberal and massive scale tariff reductions. Over 30 per cent of agricultural produce had been subject to quotas or import restrictions. The "tariffication" of all non-tariff import restrictions for agricultural products has substantially increased market accessibility for agricultural products.

Difference between and Superiority of the WTO over the GATT

It may be legitimately asked as to what was the compelling need to replace the GATT which was fairly successfully functioning for over the past 47 years by the WTO. In other words, in what respects has the replacement of GATT by WTO improved the future world trading system? In fact, it may be said that the WTO is different from and superior to the erstwhile GATT in several ways. The WTO is not a simple extension of the GATT. While completely replacing the GATT, it has a very different character. The following are the principal differences between the two institutions.

1. While the GATT was a set of rules, a multilateral agreement without institutional foundation with only a small associated secretariat which had its origins in the attempt to establish an International Trade Organisation (ITO) in the 1940s which, however, did not see the light of the day, the WTO is a permanent institution with its own full-fledged secretariat located in Geneva headed by a director-general, four deputy directors-general and over 600 regular staff drawn from 64 members.

2. Even after more than four decades, the GATT was applied on a "provisional basis". On the contrary, the WTO commitments are full, enduring and permanent.

3. The GATT rules applied only to trade in merchandise goods. The scope of the WTO is much wider because in addition to goods it covers trade in services and trade-related aspects of intellectual property.

4. While initially the GATT was a multilateral instrument, by the 1980s many new agreements of plurilateral and, therefore, of selective nature were added rendering its functioning complex and often difficult. The agreements which constitute the WTO are almost all multilateral involving commitments for the entire membership.

5. The WTO dispute settlement system is faster, more automatic and consequently much less susceptible to blockages than the old GATT system. The implementation of the WTO dispute findings will also be more easily assured.

6. Unlike GATT, the WTO is a permanent organisation created by international treaty ratified by the governments and legislatures of the member states. As the principal international body concerned with solving trade problems between the member countries and providing a forum for multilateral trade negotiations, it has global status similar to that of the International Monetary Fund and the World Bank. However, it needs to be stressed that unlike the IMF and the World Bank it is not a United Nations Organisation agency although it has a cooperative relationship with the United Nations.

Membership and Structure

At present the WTO has a membership of 153 countries ranging from the 'Quad Group' of top four world trade powers—the United States of America, the European Union, Japan and Canada—to the increasingly influential emerging economies of Asia to some of the world's poorest countries, like

Bangladesh, Guinea, Gambia, Solomon Islands, Togo, Nepal and Cambodia. The accession of Cambodia and Nepal, who represented the first least-developed countries, as members of the WTO on September 11, 2003 at the Fifth Session of the Ministerial Conference held at Cancun, Mexico, represented an important further step towards the goal of universal membership and a significant achievement in fulfilling the commitment undertaken by Ministers at the Doha Ministerial Conference to facilitate and accelerate the accession of the least-developed countries to the WTO.

The structure of the WTO is dominated by its highest authority the Ministerial Conference (MC) composed of representatives of all the WTO members. The Ministerial Conference must meet at least every two years. It can take decisions on all matters under any of the multilateral trade agreements. The first regular bi-annual meeting of the WTO at the Ministerial level was held in Singapore from 9 to 13 December 1996. The second Ministerial Conference was held in Geneva in 1998. The third Ministerial Conference was held in Seattle, USA in December 1999 (which broke up without achieving much because of wide protests); the fourth Ministerial Conference was held in Doha, Qatar on 9-13 November 2001 while the fifth Ministerial Conference was held in Cancun, Mexico from 11 September to 14 September 2003.

The day-to-day work of the WTO is principally handled by the General Council which is also composed of all the WTO members and which reports to the Ministerial Conference. Apart from conducting its regular work on behalf of the Ministerial Conference, the General Council convenes in the two particular forms—as the Dispute Settlement Body (DSB) and as the Trade Policy Review Body (TPRB). The DSB, on which all members can sit, usually meets twice a month to hear complaints of violations of the WTO rules and agreements. It sets up expert panels to study disputes and decide if the rules are being broken. The DSB's final decisions cannot be blocked. The TPRB is a forum for the entire membership to review the trade policies of all WTO member countries. Major trading powers' trade policies are reviewed every two years while the trade policies of the other individual members are reviewed every four years.

The three other bodies established by the Ministerial Conference and which report to the General Council are the Committee on Trade and Development, the Committee on Balance of Payments and the Committee on Budget. While the Committee on Trade and Development is concerned with issues relating to the developing countries and, especially, to the "least developed" among them, the Committee on Balance of Payments is responsible for consultations between the WTO members and countries which take trade-restrictive measures under Articles XII and XVIII of GATT in order to cope with their balance-of-payments difficulties. The Committee on Budget, Finance and Administration deals with issues relating to the WTO's financing and budget. Beside these bodies there is the Council for Trade in Goods, which is assisted by 12 committees with each being concerned with separate subject. The Council for Trade in Services is assisted by six separate groups. Finally, there is the Council for Trade-related aspects of the Intellectual Property Rights.

In the eight years of WTO's functioning, the Dispute Settlement Body (DSB) has already handled over 400 disputes brought before it. The WTO members see the DSB as producing fair and more or less enforceable rulings. Developing countries regard it as a strong line of defence against more powerful economies and its first ruling was against a US gasoline tax and Washington agreed to amend its law as a result. The United States has taken Japan to the DSB in a dispute over the Japanese Photographic-film market rather than impose unilateral sanctions against Japan which it might have done in the past.

WTO, the Developing and the Least-Developed Countries

At present the developing countries account for the four-fifths of WTO's total membership. The Committee on Trade and Development (CTD), assisted by the Sub-Committee on Least-Developed Countries (LLDCs), monitors all aspects of the participation of these countries in the multilateral system. The developing countries, especially the least-developed among them are helped with trade and tariff data relating to their own export interests and to their participation in the WTO bodies. The WTO secretariat has also continued GATT's programme of courses for these countries. These courses are conducted twice a year for the officials of developing countries. Since their inception in 1955 and up to the end of 2003, nearly 3,000 trade officials have been benefited by these courses.

Although the share of developing countries in international trade has almost doubled in past 40 years, and these countries have diversified their exports, at the same time participation in world trade by WTO's 29 least-developed members (LLDCs) had fallen from about 1.4 per cent in 1960 to below 0.4 per cent in 1995. Apart from trade, this "marginalisation" is also evident in foreign investment for the least-developed countries receive less than two per cent whereas developing countries as a whole attract 37 per cent. The LLDCs have miserably failed in diversifying their exports depending as they do on exports of a few commodities, minerals or tropical products and they trade mainly under regional preferential arrangements or generalised systems of preferences granted by the developed countries.

Realising that special urgent measures were needed to help the LLDCs to develop and diversify their trade, to draw benefit from the WTO trade system and to join the world economy on the eve of twenty-first century, the Singapore Ministerial Conference adopted the Plan of Action for the Least-Developed Countries which mandates the WTO, in coordination with other international organisations, to take an active role in assisting the least-developed countries to remove bottlenecks in their production capacity and diversify exports. It also called on the developed countries, and the developing countries on an autonomous basis, to explore the possibilities of granting duty-free access to the least-developed countries' exports.

WTO, IMF and the World Bank

A far-reaching globalisation of economic activity has led to an ever-increasing interdependence between different areas of economic policy. Trade, financial policies and development are increasingly interlinked within countries and in the way they affect other countries. Consequently, given their responsibilities in these areas, the International Monetary Fund (IMF) and the WTO on 9 December 1996 moved in Singapore to strengthen their relations when WTO Director-General Renato Ruggiero and IMF Managing Director Michel Camdessus signed an Agreement for future cooperation and collaboration between the two international institutions.

The Agreement provides the basis for carrying forward the WTO Ministerial mandate to achieve greater coherence in global economic policy by cooperating with the IMF as well as with the World Bank. A cooperation agreement between the WTO and the World Bank was also signed. Reflecting the synergies in the work and responsibilities of the IMF and the WTO, the Agreement provides channels of communication to ensure that the rights and obligations of members are integral to the thinking of each organisation. Finally, in keeping with enhanced cooperation, the Agreement accords observer status to the IMF and WTO in certain of each other's decision-making bodies.

The benefits of the Agreement include better access for both organisations to each other's information and data. As the institutional footing has now been put in place, better coherence will be achieved in global economic policy-making with the WTO, the IMF and the World Bank each playing their distinctive roles.

Ministerial Conference

The Ministerial Conference (MC) is the highest authority of WTO and is composed of representatives of all the WTO members. It is required to meet at least every two years and is competent to take decisions on all matters under any of the multilateral trade agreemen's. The first regular bi-annual WTO Ministerial Conference was held in Singapore from 9 to 13 December 1996. The Conference was opened by Singapore's Prime Minister Goh Chok Tong who eulogising the WTO observed in his opening remarks that "countries in the region are active supporters of the WTO process because they have experienced first-hand the benefits of maintaining a conducive global trading environment which is free, transparent, inclusive and stable".

At the close of the first Ministerial Conference, the Ministers:
* Adopted the Singapore Ministerial Declaration reaffirming WTO members' support of the multilateral trading system and setting out the WTO work programme for the next few years;
* Adopted a Comprehensive and Integrated WTO Plan of Action for the Least-Developed Countries, and
* Took note of the report of the General Council and endorsed the WTO bodies' recommendations contained in the report.

The Conference also provided an opportunity for 28 member countries representing 85 per cent[1] of global trade to negotiate the epoch-making zero-tariff Information Technology Agreement (ITA) aiming at eliminating tariffs in the $500-billion trade in computer products by the year 2000. The ITA is regarded as a key achievement of the biennial ministerial conference. While reaffirming its commitment to internationally recognised labour standards, the conference rejected their use for protectionist purposes. Further, in deference to the wishes of WTO's developing member countries the eight-page document comprising the Singapore Ministerial Declaration took note of the concerns of the Third World on linking labour issues with trade and recognised that the International Labour Organisation (ILO) was the competent forum for setting and dealing with labour standards. In their Declaration, the Ministers stated: "We reject the use of labour standards for protectionist purposes and agree that the comparative advantage of countries, particularly low-wage developing countries, must in no way be put into question".

The Ministerial Conference is the cornerstone of the global trading system. The Singapore Ministerial Conference had been an outstanding event in all respects being attended by all the WTO members. Unlike most of its GATT predecessors (Rounds), it was not about the start or completion of a major round of trade negotiations. Indeed, the Singapore Ministerial Conference represented an important point on a continuum in the growth and evolution of the multilateral trading system.

As stated earlier the second, third, fourth and fifth Ministerial Conferences were held in Geneva, Seattle, Doha (Qatar) and Cancun (Mexico) in 1998, December 1999, November 2001 and September 2003 respectively. While the third Ministerial Conference held in Seattle (U.S.A) ended in a fiasco without any achievement, at the fourth Ministerial Conference held in Doha

---

1. The total number of member countries has since risen to 42 accounting for 93.3 per cent of the world market in information technology products such as computers and software.

from 9 to 14 November 2001, the Ministers adopted a Ministerial Declaration setting out a broad work programme for the WTO for the coming years. This work programme, called the Doha Development Agenda, incorporated both the expanded negotiations–going beyond the mandated negotiations in agriculture and services which started in 2003–as well as other activities and decisions designed to address the challenges facing the trading system and interest of the diverse membership of the WTO.

The fifth Ministerial Conference was held in Cancun, Mexico from 10 to 14 September 2003. Preparations for the conference started in Geneva in May 2003 under the supervision of the General Council which is the executive body of the WTO between the Ministerial Conferences. The Cancun Conference was to take decisions and provide guidance in several areas of work under the Doha Development Agenda (DDA) including decisions by explicit consensus on modalities on the four "Singapore" Issues—relationship between trade and investment; interaction between trade and competition; transparency in government procurement and trade facilitation.

India participated in Cancun Ministerial Conference proactively forging very useful and effective coalitions, bringing the concerns of the developing member countries to the forefront. India played an important role in forging coalitions of G-20 on agriculture and G-16 on Singapore issues. She with G-20 countries focussed on agriculture as the key issue and sought elimination of distortions in world agriculture created through high level of subsidies in the developed countries. She pointed out urgent need to bring down the high tariffs and non-tariff barriers on products of export interest to developing countries to enable these countries to reap sufficient gains from globalisation. India supported the tariff reduction formula devised by the chairman of the negotiating Group on market access negotiations on non-agricultural products.

Results of the Cancun Ministerial Conference

While the substantive results of the Conference did not result in providing necessary political guidance and the adoption of decisions as necessary regarding the on-going WTO work programme under the Doha Development Agenda, the progress achieved through the deliberations enabled the members to come much closer to a true Development Round. The debate at Cancun was focussed on negotiations for liberalising agriculture and developing new multilateral disciplines on the four Singapore issues. The developing countries, including the four central African countries of Benin, Burkina Faso, Chad and Mali whose economies are heavily dependent on cotton, demanded that all export subsidies and domestic support provided by the developed countries, particularly by USA and the EU countries must be eliminated in three years while in the interim they must be given financial compensation for the loss of export earnings. The developing countries also severely criticised the functioning of the WTO itself which was heavily tilted against the poor member countries. The sixth Ministerial Conference to be held at Hong Kong (China) during Dec. 2005 is expected to take care of the developing countries interest.

Sixth Ministerial Conference of W.T.O. Hongkong Conference

The Sixth Ministerial Conference of W.T.O. was held at Hongkong from 13-18 December 2005. It was opened by Donald Tsang, Chief Executive Hongkong Administrative Region (HKSAR), in the gracious presence of Pascal Lamy, Director General of W.T.O. The conference was attended by about 5800 delegates from W.T.O.'s 149 members (Tango became 150th member) and 2100 representatives from non-governmental organisations. Despite strong demonstration outside the meeting hall from thousands of antiglobalisation protestors coming from Korea, Vietnam, Phillipines, Indonesia and India and treating multilateral trading system as 'anti-poor' and 'anti-development'. However, the conference came to an end with happy notes and a draft text covering 44 pages was finally approved.

Contentious Issues

The contentious issues between developed and the developing countries were mainly as follows :

1. The developed countries are giving heavy export subsidies to their farmers. It is almost 300 times of developing countries. Consequently, the developing countries which are depending heavily on agriculture for their livelihood, are not only unable to face competition in international market, but also their food security is in danger as the developed countries have complete control over agriculture from seed to market. The developing countries, therefore, demanded complete elimination of farm export subsidies without any reciprocity.

2. The developing countries wanted easy market access in developed countries for their industrial products, for which they demanded removal of tariff and non-tariff barriers and end of abuse of antidumping laws as most of the developing countries are busy in economic reforms.

3. The developed countries also asked the developing countries to open the gates of their market for industrial and consumer goods from the developed countries.

4. The developed countries laid much emphasis on the free trade in services. We should keep in mind here that the developed countries depend on services about for 60 p.c. of their G.N.P.

From begining to end hot discussions took place in the conference hall. India formed NAMA (Non-Agriculture Market Access) along with other seven countries and joined hands with A.C.P. (Africa Caribbean and Pacific) a group of 33 countries. Ultimately 110 developing countries formed a strong group to put united pressure for their demands on developed countries. The collective leadership of the grand alliance of 110 developing countries salvaged global trading system from embarrassing failure like Seatle and Cancun. A group of 3 countries including India, Brazil and U.S.A. was formed to work on elimination of farm export subsidies. The conference came to a happy end and fixed April 30, 2006 to compute the full modalities of the agreements.

Major Agreements

The major agreements in the conference are as follows:

1. The developed countries have agreed to phase out their farm export subsidies by 2013. The draft has also prevented developed countries from shifting taxes from one box to another to continue farm export subsidies in different farms.

2. On Non-Agriculture Market Access, the draft makes it clear that tariff cuts by developed countries on industrial goods from developing countries will be based on Swiss formula. According to this formula the higher rates of tariff will take more time than lower rates for their elimination.

3. The developed countries have agreed for granting duty free and quota free market access to Least Developed Countries (L.D.C. Package) by 2008.

4. The draft text has also agreed to eliminate cotton subsidies in 2006 giving some much needed relief to poor Western African cotton growing countries.

5. Developing countries including India will also be able to designate on 'self selection' basis an appropriate number of farm items as (Special Product — S.P.). The items under S.P. could be put under 'Special Safeguard Mechanism'. Accordingly government authorities can impose price and volume restrictions on import surge to protect domestic concerns.

6. Agriculture related projects in developing countries are put out of the preview of W.T.O. It means that expenditure cut on these projects is not necessary.

Evaluation

Views for and against the Hongkong Conference have been expressed from Several corners. The supporters are of the view that the conference has succeeded in salvaging global trading system from an other embarrassing failure (Seatle and Cancun Conferences were less successful). According to the supporters though the draft is much lower ambitious agreement than the trade negotiators had originally expected to achieve at the Sixth Ministerial Conference here, it paves the way for completion of the Doha Development Round negotiations by the end of 2006. India along with the G-20 and G-33 welcomed the text. Meanwhile, civil society groups denounced the draft text saying it was full of peanuts and full of time tables. Leading international NGO Green Peace said the text was a long way from equity and sustainability. Action Aid said that negotiators were on wrong track with poor countries concerns shunted to the side lines. The developing countries failed to protect their traditional knowledge of medicinal plants, Jari and Buti. If the terms and conditions of service sectors are accepted by the developing countries they will have to allow foreign companies in sectors like education, health, sewage, watersupply, insurance and communication etc. which will prove to be disastrous. The only praiseworthy achievement for the developing countries is their unity reflected in the grand alliance of 110 countries, for the first time on any international forum. The Indian Trade and Commerce minister Mr. Kamal Nath is happy that in the case of surge in agricultural imports the country will be able to protect domestic concerns by selecting an appropriate number of farm items on 'Self Section' basis as special product and putting, them under 'Special Safeguard Mechanism'. But he should note that only 30 types of crops grown in America whereas 260 types of crops are grown in India. In this situation the selection of special Product will not be an easy task in a democratic country like India.

Seventh WTO Ministrial Conference : Geneva Conference

The Seventh Ministerial Conference was held in Geneva from 30 November to 02 December, 2009. It was attended by 3000 representatives from all 153 WTO Members, 56 Observers, numerous civil society organizations, Secretary General of UNCTAD attended the conference on behalf of the U.N. Secretary General Ban Ki-Moon. About 100 Trade Ministers from different member countries took part in the conference.

The conference was inaugurated by Chile's finance minister, Andres Velasco followed by WTO Director General Pascal Lamy who reminded of a Chinese proverb which means that 'unity is strength'. This became in a way the moto of the whole meeting which fastidiously tried to avoid any possible conflict. ;Thus, the conference was given the broad title "The WTO, the Multilateral Trading System and the Current Global Economic Environment." Consequently no major decisionswere taken, no major surprises occurred and no major demonstration disturbed the meeting.

After inauguration of the conference two sessions took place in the conference. The first session on 1st December was entitled "Review of WTO activities", including the Doha work programme, and dealt with the issue of the Doha Round. The second session on 2nd December was entitled "The WTO's contribution to recovery, growth and development" and in which members mainly reaffirmed the value of WTO Secretariat's Report on trade and trade related developments in 2009. Various proposals on strengthening and reforming the WTO were submitted before the Ministerial Conference but they all were faint hearted attempts. Thus, the only final document was the Chairman's Summary, a short two page document comprising the essence of the statement delivered during the conference.

Evaluation

Considering the major outcome the conference was unsatisfactory. Considering the two political parts towards the conclusion of the Doha Round one could assess it insufficient. Considering the half hearted attempt to reform the WTO it was disappointing. However, there was satisfactory progress on one of the WTO's longest running dispute on the European bananas tariff system which has been challenged by Latin American countries. At the Ministerial Conference the EU achieved a break through with Latin American bananas supplying countries, the African, Caribbean and pacific (ACP) Group and the U.S. The resulting Geneva Agreement on trade in Banana was concluded on 5 December and is expected to end the banana wars'.

Another strong signal was Brazil's announcement to grant LDCs Duty Free and Quota Free (DFQF) access for 80 percent of all tariff lines and after 4 years 100 percent. Brazil expressed the hope that developed countries would follow suit.

From the above it is clear that the WTO is a valuable asset. The next conference was decided to take place from Dec. 15 to 17 in 2011.

WTO and the MFN Clause

For 47 years, the key provisions of GATT outlawed discrimination among members and between imported and domestic goods. According to Article I, the famous "most-favoured-nation" (MFN) Clause, members are bound to grant to the products of other members treatment no less favourable than the treatment accorded to the products of any other country. Thus, no country will give special trading advantages to another country or discriminate against it; all WTO members are on an equal basis and all share the benefits of any moves towards lower tariff barriers. Thus, Article I of the GATT which has been incorporated as Article I of the WTO guarantees implementation of the MFN clause by all the WTO members.

There are, however, a number of exceptions to Article I, particularly those covering customs unions and free trade areas. However, the most-favoured-nation treatment generally ensures that developing countries and others with little economic leverage benefit freely from the best trading conditions wherever and whenever they are negotiated. A second form of non-discrimination known as "national treatment" requires that once goods have entered a market, they must be treated no less favourably than the equivalent domestically-produced goods. This provision of national treatment to goods is contained in Article III of the GATT.

Apart from the MFN Clause and national treatment clauses contained in Articles I and III of the revised GATT (known as GATT 1994), several other WTO agreements contain important provisions relating to the MFN and national treatment. The agreement on Trade-Related Aspects of Intellectual Property Rights (TRIPS) contains, with some exceptions, MFN and national treatment requirements relating to the provision of intellectual property protection by the WTO members. Similarly, the General Agreement on Trade in Services (GATS) requires the WTO members to offer, with certain listed exemptions, the MFN treatment to services and service suppliers of other members. In those cases in which listed exemptions are taken, they should be reviewed after five years and should not be maintained for more than two years. The other WTO agreements with non-discrimination provisions include those on rules of origin; pre-shipment inspection; trade-related investments measures and the application of sanitary and photo-sanitary measures.

While quota restrictions are generally outlawed, tariffs are legal in the WTO and are commonly used by the governments to protect domestic industries and to raise revenues. These are, however, subject to disciplines—for instance, these are non-discriminatory among imports and are largely 'bound'. Binding means that a tariff level for a particular product becomes a commitment by a WTO member and cannot be raised without compensation negotiations with its main trading partners. Thus, in the event of extension of a customs union leading to higher tariffs in some areas compensation negotiations are necessary.

The WTO is not, however, the free trade institution in the usually understood sense because it permits tariffs and, in limited circumstances, other forms of protection. It is more accurate to say that the WTO is a system of rules dedicated to open, fair and undistorted competition. The economic case for an open system of trading based upon multilaterally agreed rules is simple enough and rest largely on commercial common sense.

All countries of the world, including the poorest, have assets—human, industrial, natural and financial—which can be used to produce goods and services for their domestic markets or to compete overseas. Comparative advantage means that countries can prosper by taking advantage of their assets in order to concentrate on what they can produce best. If the world trading system is allowed to operate without the constraints of protectionism under conditions of open, fair and undistorted competition firms are encouraged to adopt in an orderly and relatively painless way to focus on new products, finding either a new "niche" in their current area of economic activity or expanding into new product areas. One of the objectives of the WTO is to prevent a self-defeating and destructive drift into protectionism occasioned by import protection and perpetual government subsidies that lead to bloated, inefficient firms supplying consumers with outdated unattractive products.

India and the WTO

The establishment of the World Trade Organisation is a forward step towards establishing a globalised set-up dedicated to promote free exchange of goods and services. Within the framework of WTO, efforts have been made to remove prohibitive restrictions of various kinds in the arena of international trade. India, being a signatory to the WTO declaration, has to pay heed to its obligations in shaping its future macroeconomic policies. While the withdrawal of tariffs on imports will provide India access to cheaper inputs of production, producers of domestically-produced identical goods will face tougher competition. Such industries which do not have appropriate technology to match the product quality of their foreign counterparts will suffer from adverse fallout of foreign competition.

This apart, the phasing out of subsidies on exports will render India's exports internationally more expensive. Since subsidies are levied mostly on those exports which do not have an inbuilt cost advantage, withdrawal of export subsidies will amount to taking away the existing cost advantage. With quotas on their way out, exports can no longer find their way unhindered into any country. In a couple of years, the low-wage advantage which India has at present will also be nullified. India's interests at present rest in adopting a go-slow policy and before pledging itself to a decision, India should take stock of the situation in the immediate short and medium terms because if industrial growth, export performance and other indigenous macroeconomic variables perform well in the coming few years, consequences of a contemplated phase-out of restrictions would become more transparent.

Phasing-out of Quantitative Restrictions on Imports

India has been asked to phase-out all quantitative restrictions on imports in the coming six years. India had imposed quantitative restrictions on several items mainly on grounds of its balance of payments constraints and dipping forex reserves. It has been argued by the G-7 countries that India has far too many quantitative restrictions on her imports and now that India's balance of payments and forex position is comfortable she should phase-out quantitative restrictions in not more than five years.

The United States of America has contended that quantitative restrictions maintained on more than 2,700 agricultural and industrial products, are inconsistent with Article 11.1 of general elimination of QRs, Article 18.11 of BoP provisions of GATT 1994, Article 4.2 of the WTO agreement on agriculture, and Article 309 of the import licensing agreement. The United States has moved the dispute settlement body of the WTO for consultations with India on the phase-out of QRs on imports. Canada and Australia have also endorsed the US move. The US has argued that both the World Bank and the IMF had pointed out that Indian reforms require elimination of QRs to succeed and that India's continuance of QRs using its erstwhile BoP deficit problems was an excuse that was no longer credible. Consequent upon the elimination of physical quota on the import of textiles by the USA and EU, India's exports of cotton textiles, multifibres, woollens and garments will get a big boost.

Evaluation

The establishment of the WTO as first major institutional institution in the post-cold war era is an epoch-making event. Its creation has resulted in several important advantages for all. The biggest gain of the post-Marrakesh era is the existence and expansion of a trading system based on internationally agreed and enforceable rules and disciplines to both oversee and guarantee progress in the sphere of international trade. The establishment of the WTO is deservedly seen as the single most outstanding international achievement of the decade to which all member countries have made substantial contributions.

The implementation of the epoch-making zero-tariff Information Technology Agreement (ITA) aiming at eliminating tariffs in the $ 500-billion trade in computer products by the year 2000 by 42 members who account for more than 95 per cent of the world market in information technology products such as computers and software and the liberalisation of trade in financial services when the landmark agreement on financial services was implemented on 12 December 1997 are sterling achievements of the WTO. A total of 102 members representing 95% of trade in banking, insurance, securities and financial information have brought financial services into the realm of international rules. With this agreement, the WTO has completed a golden year.

It should, however, be stressed that the credibility and effectiveness of the new multilateral trading system based on open, just and undistorted competition rests on the WTO member governments' full compliance with the rules, disciplines and commitments resulting from the historic Marrakesh Agreement. The Ministerial Conference which is the highest authority of the WTO should pledge itself to achieve the target of global free trade by 2025 because fixing of a target date will facilitate the liberalisation process. It is equally necessary in the quest of a perfectly free trade regime to take effective measures to limit the discriminatory effects of the preferential trade arrangements as a fallout of the regional trade bodies like the customs unions and free trade areas. There is much to reform in both the process of negotiations and in the functioning of the WTO. The current feeling of the developing countries that they have been forced by the developed countries into trade liberalisation patterns which have had de-industrialising effects and created agrarian crises, even as the much-promised benefits of increased market access in agricultural commodities and textiles have been denied to these countries must soon end. The Dispute Settlement procedures have also become another hurdle, especially for the small developing countries who find them extremely expensive, cumbersome and unduly prolonged giving an edge to the developed countries.

The G-20 coalition of developing countries has warned the giant economies like the United States of America and the European Union that unless the farm subsidies are eliminated by the developed countries and the issues related to agriculture are resolved the WTO talks will not succeed. The US and the EU have to eliminate export subsidies so that market access for farm goods of the developing countries becomes a reality. According to developing countries the rich nations provide export subsidies to the tune of US dollar 279 billion every year thereby distorting international trade.

Suggested Readings

GATT, The Role of GATT in Relation to Trade and Development.
GATT, International Trade.
WTO, The World Trade Organisation.
WTO, Focus, Newsletter.
WTO, Annual Reports.

Questions

1. Discuss the origin, purposes and the working of the GATT.
2. Discuss the purposes, membership and organisation of the World Trade Organisation. In what respects is it different from and superior to the GATT? Discuss fully.

3. To what extent the WTO ensures implementation of the MFN clause? How does it safeguard the interests of the developing and the least-developed member countries of the WTO?

4. Discuss the purposes and functioning of the WTO. How far has it been successful in achieving its objectives? Discuss fully.

CHAPTER 27

State Trading

Introduction

Writing more than two centuries and a quarter ago in his classic *An Inquiry into the Nature and Causes of the Wealth of Nations* published in 1776, Adam Smith had strongly pleaded for free trade. His was an unqualified and clear philosophy of *laissez-faire, laissez-passer* in economic domain. Free trade resulted in opulence, asserted Smith. According to Smith, state should not conduct commercial activities which were best conducted by individual spurred by self-interest. Smith assigned government only three functions of (1) defence from attack by other nations; (2) the maintenance of Justice and order; and (3) the duty of erecting and maintaining certain public works and institutions which either an individual or small number of individuals would not be interested to erect or maintain because the profit was inadequate. In Smith's opinion, only post office was the best commercial project which was successfully managed by government. Barring such a project, no commercial activity should be undertaken by state or government because no two characters were more inconsistent or dissimilar than those of trader and sovereign. In Smith's opinion, the East India Company's operations exemplified this principle. In short, trading, domestic and foreign, by state was a forbidden activity.

In recent years, however, there has been witnessed a growing tendency toward the conducting of international trade by governments or their agencies in place of private importers and exporters. State trading refers to the socialisation of economic policy pertaining to foreign trade. Under state trading, government arranges all the details regarding the bulk purchases and sales of goods, fixing their prices, making payments, negotiating credit, etc. State trading, therefore, implies either a partial or complete monopoly of state in the sphere of international trade. When the private trade is completely taken over by the state this is known as pure state trading. The extent of state trading varies considerably among world's different countries. On the one hand, there are the socialist countries like the former Soviet Union and its satellites where the foreign trade is an exclusive state monopoly while at the other extreme end are the free economies like the United States of America where only a small fraction of foreign trade is directly handled by the government. In between are countries like the United Kingdom which, through bulk purchase agreements, imports all or most of certain types of products through a state agency with the largest part of foreign trade being conducted by private traders.

State trading is not amenable to any precise definition as it can take many forms. It can take the form either of a government agency or monopoly which works according to the same principles as a private firm or it can be a government department or organisation which completely controls the country's foreign trade. In a sense, state trading is an old-fashioned subject of international trade. Especially in central Europe and before World War I there used to be state monopolies of such commodities as tobacco and alcohol. For instance, in Sweden the state liquor authority is the sole importer of wine and liquor. It is the largest single buyer of French wines in the world. But state participation was not then very significant. The significant participation of state in the sphere of foreign trade was seen at the time of World War I when states regulated and controlled the foreign trade for defence and security purposes. After World War I, the restoration period of free trade started. The process of restoration had, however, hardly been completed when the Great Depression of the 1930s gave a severe blow to the free trade policy. The final blow to the free trade movement was given by the Second World War.

In the postwar period the need for speedy economic reconstruction of the war-shattered economies and the emergence of welfare states brought the entire field of economic activities within the orbit of government intervention. Today state trading is a world-wide phenomenon. In the centrally planned economies of communist countries like the former U.S.S.R., China and other east and central European countries both exports and imports are complete monopoly of the state. In England, the birthplace of free trade, 50 per cent of country's imports are in the hands of government. In the U.S.A., Canada, Australia and Sweden also some part of the foreign trade is conducted through the agency of state trading. State trading has also become quite important in many developing countries where in order to control the country's foreign trade and improve the terms of trade, the governments have established control over the imports and exports. According to the United Nations Commission for Asia and the Far East, in the early 1960s about 80 per cent of exports and 40 per cent of Burmese imports were controlled by state agencies. In India too, state trading through its organ the State Trading Corporation of India, until recently had been playing a significant role and before the era of economic reforms about 50 per cent of imports had been conducted by the State Trading Corporation of India.

Purposes of State Trading

There may be different purposes behind the state participation in the field of foreign trade. A few important purposes behind the state trading are mentioned below.

1. State trading agencies are considered to be rational tools and potentially powerful weapons for preserving the foreign exchange equilibrium and safeguarding the external balance of payments position of a country.

2. By monopolising foreign trade, state trading strengthens the bargaining power of the state.

3. During the process of development economic planning, state trading in commodities can be resorted to for increasing the government revenues.

4. State may participate in international trade for stabilising the prices of essential raw materials and for diversifying the country's foreign trade.

5. State may also establish monopoly over foreign trade for providing protection to the domestic industries against unfair foreign competition.

6. State trading explores greater and new export outlets and earns profit for financing the development projects in the country.

7. State trading is a powerful tool for the elimination of malpractices adopted by the private traders in the sphere of foreign trade.

Advantages of State Trading

The rationale of state trading in foreign trade lies in the benefits which it confers on the country. The principal advantages of state trading have been discussed below.

1. The development of a proper vision of international trade and a continuous watch for safeguarding the national interests are possible only through a state-owned trading agency.

2. State trading can bridge the gap between the demand price and the cost price because the state is usually aware of the demand and supply situations and can suitably adjust the prices of products in which it trades so as to minimise the gap between the demand for and supply of the goods handled by the state trading agency. Thus, state trading in strategic commodities can serve as an effective concomitant of comprehensive development planning programme in a backward economy. In an inflationary situation, the technique of state trading can be utilised for adjusting the supply price to the demand price and the profit so obtained may be spent for financing the various development projects in the economy.

3. State trading coordinates the entire country's trading machinery and assumes added responsibility for rationalisation and institutionalisation of trade policies.

4. It eliminates several evils like tax evasion, unauthorised dealings in foreign exchange, speculation, black marketing and other such malpractices indulged in by private traders.

5. State trading encourages export promotion by supplying essential raw materials at reasonable prices and at assured time, and selling the country's export goods at better prices through enhanced monopolistic bargaining power.

6. State trading in the capitalist and mixed economies is intended to overcome the difficulties encountered while the country is trading with the socialist countries where foreign trade is a monopoly of state.

Among the benefits of state trading the inculcation of a sense of financial discipline is more important. By making its management more efficient and careful, state trading minimises the cost of inputs and maximises output.

Disadvantages of State Trading

In spite of the above advantages state trading has the following disadvantages.

1. State trading establishes monopoly in international trade and monopoly, whether it is public or private, is always injurious for public interests. A monopolist is always interested in maximising his net monopoly gain which is not possible without the exploitation of consumers. In fact, experience shows that a public monopoly is worse than its private counterpart because it is far more easier for the omnipotent state to bypass public criticism than it is for a private monopolist who is always afraid of the powerful grip of the state on his malpractices.

2. By eradicating competitors, state trading invites inefficiency and idleness. The talented private entrepreneurs are not given opportunity to enter the foreign trade sector.

3. State trading functions against the principle of comparative cost advantage. Relative price differences are no longer the guiding principles for the allocation of productive resources.

4. State trading, due to its centralised bureaucratic administration, lacks in quick business decisions which are essential requirements in a dynamic international trade.

5. State trading hurts private traders and generates unemployment.

6. Non-economic considerations become more important than economic considerations in the conduct of foreign trade when trade is in the hands of state than when it is in the private hands.

This is more so when the state has a monopoly of import and export trade and the state may arbitrarily shift its sources of supply of imports or its export markets. The foreign trade of Eastern Europe in the postwar period has been deliberately directed toward the Soviet Union for political reasons.

7. Political considerations apart, state trading on a large scale introduces public monopoly elements in the sphere of international trade. Although from a national point of view, a country may succeed in obtaining a larger share of gains from trade by pursuing monopolistic policies in her international economic relations through state trading reflected in the charging of higher prices for her exports and paying lower prices for her imports but when other countries also engage in identical state trading activities very soon bargaining replaces price comparisons as the guiding principle of trade. In an atmosphere surcharged by bargaining spree, the ability of anyone country to obtain the lion's share of the gains from trade becomes uncertain and dependent upon its relative bargaining strength.

8. The effects of state trading on the volume and efficiency of international trade apart from depending upon the extent of monopoly or monopsony in exporting or importing also depend upon the internal marketing policies. If the goods imported through state trading are sold in the country at prices that are either above or below the international market prices, the relative cost comparisons lose their important function of guiding international specialisation.

State Trading in India

After independence, the Government of India has attached great importance to the issue of state trading in commodities. The subject was examined by two Committees. The first Committee under the chairmanship of Dr. P.S. Deshmukh was appointed in 1949. The Committee recommended that the government might entrust its trading activities in foodgrains and fertilisers to a statutory organisation which should be called the State Trading Corporation. The Deshmukh Committee also suggested that the State Trading Corporation might take over all the import and export operations of a commercial nature which were then handled by different departments of the Central Government. The Committee, however, pointed out that in the case of certain luxury articles whose prices generally remain high, state trading could be resorted to for the purpose of raising additional revenue.

The second Committee, with Shri S.V. Krishnamurthy Rao as its Chairman, examined the question of state trading in the country during the second phase of the First Five Year Plan. The Rao Committee, however, was of the opinion that in view of the changed world market conditions, it was not necessary to undertake imports of foodgrains, cotton and fertilisers through state-sponsored trading agency although it favoured the setting up of such an agency for dealing with the exports of handloom cloth and products of certain selected small-scale and cottage industries. Thus, both the Committees were in favour of prescribing a limited sphere for state trading in the country.

The Taxation Enquiry Commission (1953–54) also probed into the question of state trading and did not oppose state trading but it pointed out that there was dearth of "personnel with specialised experience of business".

With the launching of ambitious development planning in India under the auspices of the Second Five Year Plan, the government was compelled to make a frantic effort to find additional sources of revenue for the public exchequer. For this purpose, positive steps were taken by the government in the direction of state trading. Almost simultaneously with the commencement of the

Second Five Year Plan, the State Trading Corporation of India Ltd., was established with the blessings of the government.

The State Trading Corporation of India Ltd., otherwise known as the STC, was registered in May 1956 under the Indian Companies Act. Not being a statutory corporation it has, however, to comply with all the obligations and requirements of the Companies Act like any other limited company.

The Corporation has its main objective the broadening and enlarging of the scope of country's exports and the arranging of essential imports. The activities of the Corporation are directed toward diversification of country's exports, expansion of existing markets for country's exports, development and promotion of exports of certain bulk commodities on a long-term basis and handling canalised imports of bulk commodities. The Corporation often undertakes price support and buffer stock operations in certain commodities on the directions from Government of India. The Corporation works in close association with the trade and industry in the country. The STC group now comprises the STC, the Cashew Corporation of India (CCI), the Handicrafts and Handloom Export Corporation (HHEC), the Project and Equipment Corporation of India (PEC), the State Chemicals and Pharmaceuticals Corporation (CPC) and the Central Cottage Industries Export Corporation.

The Corporation has made a rapid growth since its inception in 1956 as is well reflected in the increase of its turnover from Rs. 9.2 crores during 1956–57 to over Rs. 3,500 crores during 1992-93. Particularly, on account of the unique export performance of the Corporation, the STC has emerged as an important exporter accounting for more than 30 per cent of the total exports from the country.

The export programme of the Corporation is divided into five main groups comprising the consumer goods, leatherware, industrial products, development exports and producers' exports. The Engineering and Railway Equipment Divisions of the holding company the STC were taken over by its subsidiary PEC in April 1971. The exports of the STC comprise canalised and non-canalised items. To develop exports of the products of small and medium-scale industries, the STC has organised assistance to small-scale industry through a separate State Marketing Division which liaises with the Small Industries Development Corporation both at the state and apex levels. The Corporation has been arranging imports of some capital goods and industrial raw materials and also of certain scarce commodities required for the country's economic and industrial development. Large quantities of soyabean oil, hops, chemicals, raw wool, art silk yarn, newsprint, alkalies, tractors, printing and textile machinery, copra, palm oil and many other essential goods have been imported by the Corporation. In pursuance of the government's policy to ensure fair prices to growers of agricultural commodities and to maintain internal production at reasonable level, the Corporation has from time to time undertaken price support and buffer stock operations in respect of jute, rubber and tobacco. To keep in constant touch with the changing trends of trade in international markets, the Corporation has a network of offices located at Sydney, Hongkong, Bangkok, Singapore, Colombo, Beirut, Tehran, Nairobi, Lagos, Moscow, Belgrade, Budapest, Prague, East Berlin, London, Paris, Frankfurt, New York, Dar-es-Salaam, Buenos Aires and Dacca. The share of the STC in India's exports has increased from 4.6 per cent in 1970–71 to more than 35 per cent in 1992–93.

The paid-up capital of the Corporation which was Rs. 1 crore in 1956-57 at the time of the establishment was doubled to Rs. 2 crores in 1958–59 and was again raised to Rs. 10 crores in 1973-74 through the issue of bonus shares. During more than four decades up to 2001–02, the Corporation has contributed more than Rs. 500 crores to the public exchequer. During 2003–04, the annual turnover of the STC was Rs. 8,349 crores while the net profit for the year was Rs. 30 crores. The net worth of the Corporation as on 31-3-2004 was Rs. 296 crores.

The Corporation achieved a turn cover of the order of ₹ 20,000 crore during 2010-11 with a trading profit of ₹ 178 crore. The net worth of the corporation in 2010-11 was ₹ 679 crore. The value of exports by the corporation is dwindling year by year. In the year 2009-10 the value of export was the order of ₹ 1504 crore which came down to ₹ 492 crore in 2010 Similarly domestic trade was reduced from ₹ 956 crore in 2009-10 to ₹ 555 crore in 2010-11.

Evaluation of Corporation's Working

The State Trading Corporation of India devoted the first year of its existence largely to the exploration of markets, formulation of policies, experimentation with business methods and techniques and development of contacts with buyers and sellers both in India and abroad. The Corporation has been trying to negotiate link arrangements with foreign firms so that essential items may be imported against the exports of Indian manufactures. Several such arrangements were concluded with the western countries like Germany, Vietnam and Hungary.

Within a short period of four decades of its establishment, the Corporation has succeeded in concluding large and valuable business contracts. On the exports side, vigorous efforts have been made to consolidate and organise the trade in mineral ores which have so far constituted the bulk of the export business of the Corporation. The Corporation has been successfully developing new markets for Indian shoes, woollen fabrics and handicrafts. The Corporation takes part in international fairs to promote India's exports. The Corporation has also established offices in many foreign countries. Economical arrangements have been made for the import of cement, raw silk, skimmed milk powder, caustic soda, soda ash and fertilisers, etc.

Apart from promoting India's export trade, the STC has exercised a healthy restraint on the sharp fluctuations in prices especially in the field of fats and oils. The Corporation has also built up business ties with important trading bodies in the foreign countries like Japan and the U.S.A.

In short, the State Trading Corporation of India has distinctly made its mark in the field of foreign trade. Faced with the grave foreign exchange crisis, India requires an effective integration of her export promotion and import substitution policies. The State Trading Corporation of India can play an important role in this sphere. Apart from reshaping the country's foreign trade policies, the Corporation can also help in achieving the country's planned economic growth with social justice. However, in the changing context of economic reforms aiming at liberalisation of country's trade, the role of the Corporation since 1992 has been marginalised. Consequently, in the context of economic liberalisation, the future of the State Trading Corporation of India is far from being bright and the possibility of its being eventually wound-up cannot be ruled out.

The State Trading Corporation of India should, however, venture into virgin territories where private traders are reluctant to enter either because these areas of activity are not sufficiently lucrative for private enterprise to enter or because huge capital outlay well beyond the individual's resources is required. For the successful functioning of state trading, it is necessary that it should function on purely commercial basis and there should be no undue interference of the government in the functioning of the Corporation. The Corporation has been criticised for slow moving, red tapism, inflexibility and insensitivity to consumers' needs.

Suggested Readings

Charles P. Kindleberger, *International Economics,* Fifth Edition, 1973, Chapter 8.

Delbert A. Snider, *Introduction to International Economics,* Fourth Edition, 1977, Chapter 12, pp. 243–244.

Bo Sodersten, *International Economics,* Second Edition, Reprinted 1981, Chapter 14, pp. 191–192.

State Trading Corporation of India, *Annual Reports.*

Questions

1. What is state trading and on what grounds is it justified?
2. What are the objectives of the State Trading Corporation of India? How far has it been successful in achieving these objectives?
3. Explain the working of the State Trading Corporation of India and state how it is helping the export trade of the country.

CHAPTER 28

United Nations Conference on Trade and Development (Unctad)

In the present era of mutual understanding and international cooperation, the staggering poverty of the teeming millions of people inhabiting world's developing countries has aroused the conscience of the whole world. The proof of this consciousness is the birth and growth of many international finance organisations like the World Bank, International Development Association, International Finance Corporation, and the regional financial institutions like the Asian Development Bank, the African Development Bank, and the Inter-American Development Bank in the postwar period. Apart from these international financial institutions whose function is to provide institutional financial assistance for economic development of the poor areas of the globe, other international bodies including the UNESCO, FAO, WHO and UNCTAD had been created under the banner of the United Nations Organisation. The United Nations Conference on Trade and Development, otherwise popularly known as UNCTAD, is the result of the U.N. resolution on 'Development Decade' of 1961. This organisation is a forum of nations for finding and resolving the various knotty international problems of trade and development.

The United Nations Economic and Social Council (ECOSOC) had passed a resolution in 1946, calling for the convening of an international conference on trade and development. An effort was made for adoption of the Charter for international trade at the Havana Conference on Trade and Employment. However, the attempt proved abortive and certain provisions of the Charter were incorporated in the General Agreement on Tariffs and Trade (GATT) which came to be accepted as the nucleus for dealing with the trade problems. The International Trade Organisation (ITO) envisaged in the Havana Charter never came into being. In the meantime, the membership of the GATT which was originally subscribed by 23 members in October 1947, increased from time to time and at the end of 1994 when it was replaced by the World Trade Organisation (WTO) stood at 123 countries as full members of the GATT as the "Contracting Parties". The GATT had made several efforts to remove the obstacles in the path of the expansion of trade and economic growth of the less developed countries of the world. Since the cardinal principle of the GATT, the "Most Favoured Nation Principle" treated the developed and the developing countries on par and put ban on any discrimination and special facilities in favour of the developing nations, the GATT failed in facilitating the economic development of the poor countries.

As a consequence of this failure, the critics had criticised the working of the GATT by nick-naming it as a "rich nations' club". Moreover, the GATT was actually a forum for discussing only the tariff problems of the member countries. Consequently, numerous pressing economic problems of commodities' prices, non-tariff barriers imposed on the exports of the developing countries by the developed nations, shipping transport, economic aid, economic development and several other issues relating to the development of the less developed countries and harmonisation were beyond the limited scope and purview of the GATT. All these vast pressing economic problems of the underdeveloped countries which could not be effectively discussed and resolved through the existing machinery of any existing international economic organisation resulted in the realisation of the need for setting up an independent international body under whose aegis all the relevant problems relating to trade and development of the developing countries could be effectively discussed and resolved as a consequence of mutual understanding and free frank discussions.

Birth of UNCTAD

The Cairo Conference of the developing countries held in July 1962 on the problems of economic development passed the 'Cairo Declaration on Developing Countries' calling for the convening of the United Nations Conference on Trade and Development and constituting an "International Trade Organisation" (ITO) which would consider vital questions relating to the international trade of the poor nations. The United Nations Economic and Social Council agreed to convene such a conference—the first UNCTAD—and passed Resolution on August 3, 1962 which was endorsed by the United Nations General Assembly in its Resolution of December 8, 1962. The historic decision of the United Nations General Assembly to name 1960–69 as a 'Development Decade' was a further recognition of the deep world-wide concern with the urgent necessity of raising the living standards of the people of the developing countries. All these developments led to the convening of the United Nations Conference on Trade and Development in Geneva from March to June 1964. In July 1963, the United Nations Economic and Social Council passed a resolution for UNCTAD to be convened at an interval of not more than three years. The United Nations General Assembly accepted the recommendation and the UNCTAD was established as a permanent organ of the U.N. General Assembly. The U.N. General Assembly also defined the functions, activities, and membership of the UNCTAD.

The UNCTAD has set up a Trade and Development Board as a policy-making body to take policy-making decisions when the conference is not in session. The Board is composed of 55 members elected on the basis of equitable geographical distribution. The Board is also helped by subsidiary committees which deal with the problems of primary products, manufactured and semi-manufactured goods, development finance and questions relating to invisible services, including shipping and insurance etc.

Functions of UNCTAD

The main purpose of creating the UNCTAD was to promote speedy development of the underdeveloped countries by solving the problems of the sluggish expansion of their export trade, deficits in their external balance of payments and excessive burden of foreign debt, etc. The principal functions of the UNCTAD are as stated below.

1. To promote international trade, especially with a view to accelerating the economic development of underdeveloped countries, particularly trade between countries with different systems of economic and social organisation taking into account the functions performed by the existing international organisations.

2. To formulate the principles and policies of international trade and related problems of economic development.

3. To make proposals for putting the said principles and policies into effect and to take such other steps within its competence as may be relevant to this end.

4. Generally, to review and facilitate the co-ordination of activities of other institutions within the United Nations system in the field of international trade and related problems of economic development and in this regard to cooperate with the General Assembly and the Economic and Social Council in respect of the performance of their charted responsibilities.

5. To be available as a centre for harmonising the trade-related development policies of governments and regional economic groupings in pursuance of Article 7 of the United Nations Charter.

UNCTAD I and Directive Principles

The first UNCTAD was held in Geneva from March 23 to June 16, 1964. The Conference was attended by delegates from 120 countries, 13 specialised agencies, and 32 non-government bodies. The aim of the Conference was to provide means of international cooperation and to find appropriate solution to the problems of world trade in the interest of the whole world and particularly recognising the urgent need of the developing countries. For this purpose, the Conference laid down a number of principles, policies and recommendations to bring about basic changes in the working and set-up of trade relations between the advanced and poor nations. Following were the important directive principles laid down and accepted by the Conference.

1. Economic development and social progress should be the common concern of the whole international community for which peaceful relations and co-operation should be sought.

2. National and international economic policies should be directed toward the attainment of the division of labour consistent with the needs and interests of the developing countries in particular, and the world as a whole in general.

3. Developed countries should reduce restrictions on trade that hinder the trade of the under developed countries and should increase the market for the products of developing nations.

4. Developed countries should extend new preferential concessions, both tariff and non-tariff, to the developing countries also. These should not be limited to the developed countries only.

5. Assistance and aid from the developed countries should not be subject to political or military considerations.

6. Economic relations between nations should be based on the principles of equality and non-interference in their internal affairs. No distinction should be made on the basis of economic systems.

It was decided to hold a periodic meeting of UNCTAD after every three years. The Trade and Development Board was authorised to take policy-making decisions when the Conference was not in session.

Achievements of UNCTAD I

Those who were optimistic about the role of UNCTAD thought that UNCTAD would be the third powerful international organisation after the International Monetary Fund (IMF) and the World

Bank (IBRD) and that it would bring peace, prosperity and high standard of living for the underdeveloped countries. At the other end of the thinking were those who felt that UNCTAD was merely flying in the aeroplane of ideas and the aeroplane of action was far away from it. In their opinion, UNCTAD was a forum of ideas rather than a stage of action. We should, however, keep in mind that for growing a better crop, better ploughing of field is inevitable. UNCTAD is a better way of mutual understanding. Manubhai Shah, India's Minister for Commerce, who led India's delegation to the first UNCTAD had correctly remarked: "The first UNCTAD had succeeded in increasing the interest in developed countries regarding the problems of the developing countries.... For the first time the countries with free market economies and the centrally planned economies came together and cooperated in examination of the problems of the developing countries. The 77 developing countries had shown a remarkable sense of unity in forming the common platform for working in concert and for confronting the developed countries".[1]

However, when we come to action and look to the recommendations and actual performance of the "Development Decade" the results were not very encouraging and left much to be desired. Even the minimum annual growth rate of five per cent set for the Development Decade had been achieved by very few developing counties. The economic disparity between the developing and the affluent countries has been widening. Every year the developed countries had added up approximately $ 60 to the per capita income of their people. The share of the developing countries in total world exports, which was 34 per cent (including exports from all countries) in 1950 had fallen to 20 per cent in 1988 while that of the developed countries had increased from 60 to 68 per cent and that of the centrally planned economies had risen from 8 to 11 per cent during the same period. The reason for this was that no easy access was given to the exports of the developing countries to the markets of developed countries. Import and export restrictions were not removed by the developed countries for the developing countries. These were removed only for the developed countries. Nearly two-fifths of the developed markets got preferential tariffs. The Kennedy Round had resulted in reducing the tariffs on those commodities which are produced by and traded between the developed countries. Raul Prebisch, the well-known advocate of the cause of developing countries, when he was asked: what was wrong with UNCTAD? had said: "The financial aid from developed countries was set at one per cent of their GNP but in 1961 it was 0.87 per cent and in 1964 it was only 0.66 per cent". It shows clearly that the main obstacles in the economic growth of the developing countries are the developed countries which are reluctant to give them financial assistance which is urgently needed for the development of the backward areas of the world.

Before the start of the meeting of UNCTAD II, when Raul Prebisch was asked to comment on the achievements of UNCTAD I, he commented: "Nothing in the field of action but a considerable advance in the field of ideas". Raul Prebisch was stating the hard truth because UNCTAD I miserably failed in narrowing the gap between the world's poor and the rich.

UNCTAD II

It was decided by the U.N. General Assembly to hold a periodic meeting of UNCTAD after every three years. However, the second meeting of UNCTAD took place in 1968 four years after the first meeting was held in 1964 in Geneva. The United Nations General Assembly gave its final approval to hold the second meeting from February 1, 1968 to March 30, 1968 in New Delhi. After the final

1. Manubhai Shah, *The Underdeveloped Countries and UNCTAD*, 1968, p. 5.

approval of the General Assembly, the fifth meeting of the Board of Trade and Development took place in Geneva to discuss the broad objectives of this proposed historic conference. After thoughtful consideration, the following objectives were unanimously resolved.

1. To re-evaluate the economic situation and its implications for the implementation of the recommendations of UNCTAD I;
2. To achieve specific results by initiating appropriate form of negotiations which would ensure real progress in international co-operation for development; and
3. To explore and investigate matters requiring thorough study before fruitful agreements could be envisaged.

Just after one month of the meeting of the Board of Trade and Development, members of the underdeveloped countries assembled in Algiers to discuss UNCTAD II's programme.

Algiers Meet

In September 1967, the countries in the ECAFE region met at Bangkok to discuss the Delhi Conference and the problems related to trade and development. A similar meeting of the LAFTA countries was also held in Bogota. The African countries also met at Algiers. Ultimately, the ministerial meeting of the group of 77 underdeveloped countries was held in Algiers from October 10, 1968 to October 24, 1968 to go through regional calculations and prepare a charter of general programmes. The "Group of 77"[2] agreed on a series of unanimous recommendations seeking to expand trade and accelerate economic development of the poor countries of the world. These recommendations are contained in a document which is known as the "Charter of Algiers" of the '77'. This Charter is treated as the *Magna Carta* of trade and development for the developing countries.

UNCTAD II was held at New Delhi from February 1 to March 30, 1968 to evolve a concrete 'Action Programme' which was to be implemented in 1968 and 1970 well before the UNCTAD III meet in 1971. The Conference which was attended by delegates from 121 countries was presided by Shri Dinesh Singh and was inaugurated by India's Prime Minister late Shrimati Indira Gandhi. The Conference which began on February 1, 1968 had the following items on agenda.

1. Trends and problems in world trade and development.
2. Commodity problems and policies of different nations.
3. Growth, development finance, and aid to developing nations, including synchronisation of national and international policies in this regard.
4. Expansion and diversification of exports of manufactured and semi-manufactured goods of the developing countries.
5. Problems of the developing countries in regard to invisible services including shipping etc.
6. Problems and measures for economic integration and trade development among the developing nations.
7. Special measures to be taken in favour of the least developed among the developing countries engaged in improving their economic and social development.
8. General review of the work and functions of the UNCTAD.

The Conference discussed every item on the agenda with clear vision and cool mind and efforts were made for concrete materialisation. The results of the Conference were, however, very disappointing for the developing countries because no concrete agreements could emerge on most of

---

2. The actual membership of the Group totalled 112 countries at the time of UNCTAD IV in May 1976. India is a member of the Group of 77.

the controversial issues. The important reason for the poor achievements of the Conference was the prevailing unhealthy and unfavourable political and economic climate in which the Conference was held, i.e., the Vietnam War, gold crisis, external balance of payments problem faced by the U.S.A. and the U.K., fluid situation in West Asia etc. It will, however, be wrong to say that the Conference wasted time in fruitless discussions achieving nothing. The Conference had certainly some achievements to its credit. The achievements of the UNCTAD II have been briefly discussed below.

It was for the first time at UNCTAD II that the developed countries realised that the reduced export earnings of the developing countries would reduce their external purchasing power and consequently the importing capacity of these countries would be reduced. As a result, the export earnings of the developed countries would fall. Consequently, the aggregate world trade would shrink. This realisation on the part of developed countries was very important from the standpoint of the developing countries. To avoid this misfortune, the developed countries may adopt specific measures like the removal of import tariffs and non-tariff barriers to their imports from developing countries, providing technological and financial assistance to developing countries etc. in order to expand the export earnings of the developing countries.

The final resolution, therefore, stressed the need for an early establishment of a mutually acceptable generalised, non-reciprocal and non-discriminatory system of preferences beneficial for the developing countries. It is popularly known as the Generalised System of Preferences (GSP)[3]. The objectives of such a system are to accelerate the rate of economic growth, to promote industrialisation and to increase the export earnings of the underdeveloped countries.

Furthermore, the Conference recommended that each developed country should contribute one per cent of her GNP at market price in the form of actual disbursement to the developing countries which are net importers of capital. It was further agreed that if there was any gap between one per cent of GNP and actual assistance that should be recovered by means of commodity transfers, etc. Several developed countries, however, expressed their inability to reach this target by 1972. France, the Netherlands and Sweden agreed to contribute one per cent of their GNP by 1972 while others only agreed to make efforts towards this goal.

As regards the commodity agreements, it was decided that the Conference should be reconvened to evolve international agreement on cocoa not later than June 1968 and the Sugar Agreement should come into operation before January 1969. As regards other commodities, the Conference recommended that further studies should be undertaken.

The Conference asked the socialist countries to expand and diversify their trade with the developing countries. On the other hand, the developing countries urged the developed countries to provide them with conditions for trade not inferior to those granted to the developed countries. The permanent machinery of UNCTAD has been entrusted with the responsibility of promoting trade relations between the socialist and the developing nations.

The Conference stressed the need for trade expansion and economic integration among the developing countries. The Declaration of Intent by the developing countries was matched by the declaration of support by the developed countries. The developed nations declared their support

---

3. The Generalised System of Preferences (GSP) envisages a system of preferential tariff rates by the developed countries favouring the manufactured and semi-manufactured imports from the less developed countries. It represents one of a package of global trade measures instituted to promote the development process in the less developed countries. It calls upon the developed countries to eliminate or reduce tariffs on all the manufactured imports from the less developed countries. All the developed countries that had agreed to provide preferential treatment to the less developed countries under the GSP scheme have by now introduced their individual GSP scheme, thereby assuring either free entry or entry at reduced tariff rates to manufactures, semi-manufactures, and a few primary and agricultural products which are exported from the less developed countries.

through financial and technical assistance to the integrated endeavours of development. The poor countries expressed their determination for the mutual economic integration. The Conference, however, entrusted the dealing of unsettled issues to a subsidiary body of the Trade and Development Board.

It is, therefore, obvious that the achievements of UNCTAD II were quite modest. The Conference, however, succeeded in keeping alive the "torch of international economic co-operation".

UNCTAD III

Before discussing the UNCTAD III, let us have a bird's-eye view of what happened to trade and development during the interval of four years between UNCTAD II and UNCTAD III and how far the resolutions of UNCTAD II were implemented. After UNCTAD II, several developed countries accepted the Generalised System of Preferences (GSP), especially to the imports of manufactured and semi-manufactured products from the less developed countries. The U.S.A., however, did nothing in this matter. On the contrary, the U.S. Government imposed a 10 per cent surcharge in December 1971 on her imports to face the dollar crisis. The story did not end here.

The U.S.A. also fixed the import quotas for cotton textiles coming from Asian countries. This was harsh for the developing nations. United Kingdom also imposed import tariff of 15 per cent on her imports of cotton textiles from the Commonwealth countries. Under these circumstances, the Conference failed to secure any special privileges for the underdeveloped countries. This is why it is generally believed that the preferences secured by the underdeveloped countries would have been secured even if there had not been any UNCTAD because liberalisation in trade policy had progressed much in the seventh decade of the twentieth century. The failure of UNCTAD II becomes all the more obvious when we find that the decade of sixties witnessed a boom in the world trade but recorded a continuous fall in the share of developing countries in the expanding world exports which fell from 21.3 per cent in 1960 to 18.1 per cent in 1969, and further to 17.6 per cent in 1970. Had the developing countries maintained their share in world's exports from 1960 to 1969, they would have cumulatively earned additional export earnings of $ 40 billion. The so-called first Development Decade had, therefore, been a decade of development for the developed countries and one of frustration for the developing economies.

As regards the resolution relating to the contribution of one per cent GNP from the developed countries to the developing countries, it had not been implemented successfully. The contribution of GNP which was 0.96 per cent in 1961 fell to 0.74 per cent in 1971. The United States of America, world's most affluent nation, contributed a bare 0.24 per cent of her GNP to developing countries. All these and other related problems were before UNCTAD III.

The third UNCTAD was held at Santiago in Chile (South America) from April 13, 1972 to May 17, 1972 and 99 developing countries forming the Group of 77 attended the Conference with new hopes and fresh enthusiasm. Individual personalities from or speaking for developed countries had gone to Santiago to assist imposing on its discussions grand design of ends and means. International statesmen of the calibre of the President of the World Bank, Mr. George McNamara, had come to attend the Conference to safeguard the common interests of both the developed and the developing countries and to avoid sterile confrontation between the two blocks of nations. But the six-week discussions at Santiago ended in an atmosphere of great distrust between the 'haves' and 'have-nots'. The poor countries attacked vehemently the rich countries for their unsympathetic and callous attitude towards them. The delegate from Venezuela in particular gave strong expression to the resentment of the poor countries. He said that "when the rich nations put their minds to solve their own problems, they were swift and diligent" but when they discussed problems of interest of poor nations, they were "miserly and selfish."[4]

At the Santiago meet, 40 proposals relating to different issues were tabled. The key issues on which the success of UNCTAD III depended were, however, only four: (i) link between the Special Drawing Rights and development finance; (ii) access to the markets of the developed countries and price policy for primary commodities; (iii) special measures for the development of the least developed countries; and (iv) international code of conduct for liner conferences. The 'Group of 77' tried to establish an institutional link between UNCTAD on the one hand and the IMF and the GATT on the other. The industrialised countries treated it as an indirect takeover of the Fund and the GATT and refused to endorse the link and eventually agreed to "draw attention of the IMF to the many statements made at the third session of the UNCTAD on the relationship between SDRs and the development finance, many of the statements indicating both the desirability of such a link and the possibilities that might provide for channeling new resources for development finance."[5]

This was a considerably diluted version of the draft resolution originally submitted by the "Group of 77" and even this was unacceptable to the United States. Although the compromise resolution was carried by 67 votes to 0, the U.S. abstained along with six other countries thus reducing the prospects for any early action. This may be regarded as small victory for the developing countries.

The draft resolution of the developing countries regarding the access to the markets of developed countries, pricing policy, international price stabilisation measures and mechanisms and inter-governmental consultations on primary commodities was very comprehensive. But none of this was acceptable to the developed countries. According to the rich countries, negotiations regarding the primary products could be held only under the auspices of the GATT. In this way, the developing countries were dozed off by the developed countries.

The only achievement of UNCTAD III was the unanimous decision to adopt special measures for the development of 25 least developed countries designated by the U.N. as the "Hard Core" countries. The hard core countries included sixteen African countries, eight Asian countries and one Latin American country.

The resolution for assistance to "land-locked" countries might be important for the countries concerned but this was a relatively small gain for the Third World.

Of the four issues, the lone one which ended in something like a victory for the developing countries was the subject of shipping at Santiago. The Conference agreed on an international conduct for liner conferences. The disagreement had been over the procedure to be followed in elaborating and implementing such a code. In short, the urgent demand of the developing countries had been denied. The developing countries had gone to the Santiago meet with sanguine hopes and burning desires but returned back with frustration, despair and disappointment.

The reason of this failure was on the one hand the attitude of negativism and obstruction of the U.S. government and on the other hand the inadequate leadership in the ranks of the developing countries. Evaluating UNCTAD II and UNCTAD III, the *Eastern Economist* in its editorial dated June 2, 1972 stated that "without exaggerating the impact of the delegation (UNCTAD II), on the course of the Conference, it would be claimed that its clear-eyed vision and cool-headed purposiveness

4. *Eastern Economist*, May 26, 1992, p. 1024.

5. Malcolm Subhan, "UNCTAD III: A Balance Sheet", *Commerce*, June 19, 1972, p. 1514.

contributed a great deal towards steering the course of consultations and negotiations and even the controversies and debates into ultimately productive channels. It is not suggested here that UNCTAD II was a famous victory but its success in carrying through the concept of Generalised System of Preferences (GSP) to the point of acceptance in principle and subsequent implementation in detail did result in saving UNCTAD from the real danger of infant morality. In Santiago, there was apparently no such thrust of practical leadership to achieve major results."

UNCTAD IV

The fourth quadrennial session of the United Nations Conference on Trade and Development (UNCTAD) was held at Nairobi, Kenya from May 3 to May 31, 1976 attended by more than 170 representatives from 153 countries, specialised agencies, and other inter-governmental and private organisations. The Conference provided an opportunity for the developing countries led by the Group of 77 (which in fact now has 112 members) to take stock of progress in various forums, such as the Paris Conference on International Economic Cooperation (CIEC), the General Agreement on Tariffs and Trade (GATT), and the Development Committee. The demands of the developing countries as voiced by the Group of 77 were enunciated for the industrialised world in the Manila Declaration.

The demands in the Declaration included the integrated programme for commodities and the common fund for buffer stock financing; improved market access for the developing countries' manufactured and semi-manufactured products through improvement of the existing generalised system of preferences schemes and preferential treatment to the developing countries; targets of official development assistance (ODA); aid flows and debt relief[6]; a binding code of conduct on the transfer of technology; increased cooperation based on collective self-reliance among the developing countries; and the strengthening of UNCTAD as a forum for international negotiations on economic issues. The Manila document stated the maximum demands on most issues, partly to take account of the diversity of large membership in the Group and partly to leave room for compromise.

The major decisions taken by UNCTAD IV reflected the desire of its members to influence the restructuring of the world economy as well as the economies of the developing countries. Below are given, in brief, some of the major results that emerged from the Nairobi meeting.

1. It was agreed to convene a negotiating conference under UNCTAD not later than March 1977 on a common fund for buffer stock financing; it was also agreed that negotiating conference on specific commodities, preceded by preparatory meetings on each commodity, should finalise commodity agreements by the end of 1978.

2. It was agreed that the integrated commodity programme should ensure, among other things, stable conditions for commodity trade and development of export products from the developing countries.

3. A draft resolution on international resources bank submitted by the industrialised countries (Group B) was not adopted.

4. Existing international forums were asked to study the debt problems of the developing countries with a view to providing guidelines for future action on individual cases and the Trade and Development Board of UNCTAD was asked to review progress on this subject in 1977.

---

6. The Third World debt to the industrialised countries, to agencies like the International Monetary Fund (IMF) and to private banks is approximately $ 350 billion. Interest on this debt burden almost nullifies the impact of fresh aid flow. At Nairobi, the Third World countries demanded "generalised debt relief" —a moratorium on debt to the poorest countries and stretching out of the repayment schedules for all the developing countries.

5. No resolution was passed on the issue of finance and transfer of real resources. Furthermore, no agreement was reached on the period during which 0.7 per cent official development assistance target would be achieved.

6. It was agreed that the Generalised System of Preferences should be improved and its terms extended beyond the originally envisaged ten years.

7. It was recommended to participants at the Multilateral Trade Negotiations in Geneva that barriers to the improvement of export trade from the developing countries should be lifted or reduced.

8. The World Bank and regional development institutions were invited to consider facilities which would provide export credit refinancing for the developing countries.

9. On the question of transfer of technology, it was resolved that a draft code should be completed by the middle of 1977 and an advisory service should be set up in UNCTAD for the purpose of helping the development of technology in the less developed countries.

10. It was also agreed that the least developed, land-locked, and island developing countries should receive high priority for and a greater share of ODA and that aid agencies should adopt suitable criteria for such assistance to these countries.

11. It was also agreed that the strengthening of the UNCTAD was essential to the creation of improved conditions for international trade and related issues of international economic cooperation.

Assessments of UNCTAD IV vary. According to Gamani Corea, UNCTAD Secretary General, the Nairobi Conference "took a very real step toward meaningful reform and strengthened UNCTAD's role as an effective instrument of the U.N. system." The *Financial Express* described it as a "success of sorts", but believed that historians might regard it as "90 per cent semantics, 10 per cent dramatics." Perhaps India's Commerce Minister Professor D.P. Chattopadhyaya considered extension of GPS as a welcome feature. India's *Samachar Views* agency summed up the Nairobi Conference best in these words: "Neither a glowing success nor an exercise in futility."

UNCTAD V

UNCTAD V was convened on May 7, 1979 and ended on June 3, 1979 in Manila, the Philippines. The Conference was attended by about 150 representatives from member countries and international organisations. The discussions which continued for four weeks were carried on among the world's developing industrial market and socialist economies. In intense negotiations at the end, consensus agreements were hammered out on the transfer of resources to developing economies, commodities, protectionism, and a number of other issues. The Conference, however, failed to produce agreements among the three groups of the countries on such central issues as monetary reform and a proposed complementary financing facility for the stabilisation of commodity export earnings.

The UNCTAD agenda contained 12 substantive items—interdependence of trade, money, finance and development, and structural changes in the world economy; developments in international trade; commodities; manufactures and semi-manufactures; monetary and financial issues; technology, shipping; least developed countries; land-locked and island developing countries; trade relations between countries having different economic and social systems; economic cooperation among developing countries and institutional issues. From the beginning of the Conference, areas of agreement were mostly on secondary issues such as reverse transfer of technology, the brain drain, and the industrial property system; but on virtually all core issues, profound cleavages developed which were reflected in the submission of proposed resolutions on every agenda item. The Conference negotiations were deadlocked on nearly all the core issues.

On *Interdependence,* the proposal of the Group of 77 for the establishment within the framework of the UNCTAD of an inter-government group of global consultation on policies of trade, money, finance and development was opposed by the industrial countries which rejected any encroachment by UNCTAD in areas of competence of other international organisations. On *Monetary issues,* the developed market countries rejected the proposals of the Group of 77 recommending to the IMF to modify its conditionality and to make further improvements in its compensatory financing facility; the preparation by the UNCTAD of a report on the establishment of a medium-term financing facility to assist countries facing structural balance of payments deficits arising from external factors; the establishment in UNCTAD of a group of experts to examine the reform of the international monetary system, including a link between SDR creation and development finance; and the agreement in principle to convene an international monetary conference. The developed market countries rejected all these proposals on the plea that they would duplicate studies being made by the World Bank and IMF for Development Committee (Joint World Bank-Fund Committee on the Transfer of Real Resources to Developing Countries) and would unduly interfere with the current work of other organisations. The industrial nations, instead, endorsed the concept of an evolutionary reform of the monetary system, welcomed the adoption by the IMF of new guidelines for conditionality and invited the IMF to review ways of improving its extended facility.

The resolution on monetary issues was ultimately adopted by 69 developing countries (opposed by 17 developed countries, with 13 socialist countries and a few developed market economy countries abstaining). The resolution (1) invited the IMF to examine the overall size of quotas in relation, inter alia, to the current levels of international trade, the magnitude of the balance of payments deficits of its members, and the need to finance these deficits in the context of the adjustment process with due regard to increasing the quota share of developing countries; (2) stressed the necessity to apply conditionality in a "flexible and appropriate manner taking into account the domestic, social and political objectives, the economic priorities and circumstances of the members of the IMF, including the causes of balance of payments problems, using especially those of developing countries, so as to encourage timely recourse to the Fund's facilities, and the much higher rate of utilisation of its resources"; (3) emphasised that the approach of the IMF to adjustment programmes should be such as "to seek reconciliation between a country's short-run and long-run objectives and not be disruptive of development, and should take into account factors attributable to external elements beyond the control of developing countries"; (4) invited the IMF to study ways to improve the terms and use of the extended facility; (5) invited the IMF to study the need for the establishment of a longer-term facility for the balance of payments support for the programming of adjustment over longer periods; (6) agreed that the existing compensatory facility should be improved and liberalised and invited the IMF to take into account the Arusha programme in its ongoing review of this facility; (7) invited the IMF to consider an interest subsidy account for developing countries making use of the supplementary financing facility; and (8) established an ad hoc inter-governmental high level group of experts within UNCTAD to examine the evolution of the international monetary system.

On the *debt problems* of the developing countries, the industrial countries opposed the proposal of the Group of 77 for the establishment of an Internal Debt Commission since it was argued by these countries that the existing machinery of creditor groups could be improved. The industrial countries proposed that any interested developing country could seek impartial assessment of its debt situation from either the IMF or the World Bank. On *commodities,* the developing countries (the Group of 77), the developed market economics, the socialist countries of Eastern Europe and the People's Republic of China welcomed the agreement on the fundamental elements of a Common Fund and agreed on the desirability of accelerating negotiations on individual commodities and on the need to emphasise the development aspects of commodity problems. There was, however, sharp disagreement over the Group of 77's proposal for an UNCTAD study of a complementary compensatory facility for export shortfalls, measured in real terms, in each commodity. The developed countries took the position that the IMF and World Bank were already involved in a study of export earnings stabilisation.

On *protectionism,* the Group of 77 proposed a mechanism within UNCTAD for periodic review of the patterns of the production and trade of industrial countries and the identification of sectors in these countries needing structural adjustments. The industrial countries rejected such a role for UNCTAD and emphasised the continuous nature of structural adjustments in both developed and developing countries. A consensus resolution called for continued resistance to protectionist pressures and urged the industrial countries to reduce or eliminate quantitative restrictions on imports from developing countries and to improve market access for exports of manufactures and semi-manufactures from developing countries. The resolution also requested the Trade and Development Board to continue reviewing trade restrictions in order to make recommendations concerning protectionism. The resolution acknowledged that structural adjustments were a constant process which should be facilitated by conscious efforts of the international community, and it authorised the Trade and Development Board to arrange for an annual review of the evolution of world production and trade, to be used by national governments in their adjustment assistance measures.

A consensus resolution on respective business practices reaffirmed a previous decision covering decision by countries to eliminate or deal effectively with practices, including those of transnational corporations, which adversely affected international trade and economic development. A majority vote resolution on protectionism in the service sector called upon industrial countries to eliminate all discriminatory and unfair practices in the service sector, particularly in transport, banking, and insurance. The resolution also requested the UNCTAD to analyse urgently the effects of the discriminatory and unfair civil aviation practices used by the industrial countries on the growth of air transport, including air cargo and tourism, in the developing countries. Another consensus resolution invited the UN General Assembly to strengthen UNCTAD, giving "clear recognition to UNCTAD as a principal instrument of the General Assembly for negotiations on relevant areas of international trade and related issues of international economic cooperation, particularly in the context of negotiations on the establishment of the new international economic order." Other consensus resolutions related to the transfer of technology to developing countries, strengthening of technical capacity in these countries, reverse transfer of technology and brain drain; the industrial property system and its impact on the development process and a convention on a code of conduct for shipping conferences.

Three consensus resolutions dealt with special categories of countries. The resolution dealing with the least-developed countries called for the launching of "a comprehensive new programme of action" in two phases: an immediate action programme for 1979/81 and a substantial new programme of action for the 1980s covering structural change, social needs, transformational investments and emergency support. The resolution on land-locked developing countries sought to reduce the costs of access to and from the sea and to world markets, the improvement in transport services, and the restructuring of land-locked countries' economies. For the island developing countries, the resolution listed economic diversification, export promotion, and attraction of foreign investment as areas where specific action was required. The resolution requested UNCTAD to undertake studies of common problems of island economies and invited the international civil aviation organisation to study the policy issues involved in the development of air transport services in the island countries.

Other resolutions adopted by majority vote included greater participation of developing countries in the world shipping financing facilities and technical assistance for the acquisition of ships by developing countries; and recommendation requiring all countries to refrain from exploiting the seabed resources until an international regime was adopted by the U.N. Conference on the Law of the Sea.

UNCTAD VI

UNCTAD VI was held in Belgrade on June 6, 1983. It was preceded by a preparatory meeting of the Group of 77 at Buenos Aires in April 1983. The demands of the developing countries incorporated in the 20-odd resolutions fell into three categories of trade, finance and commodities which were mutually interlinked and presented as a package. The adverse balance of trade of the developing countries stood at about $ 185 billion and the demand was made to lower the walls of protectionism and to expand the Generalised System of Preferences on the part of the developed countries. How far the rich countries would concede the point was problematical particularly in the context of mounting unemployment in the developed countries.

The second issue raised by the developing countries related to their worsening terms of trade which was reflected in the fact that compared to the mid-seventies, the developing countries had to give 20 per cent more of their exports in order to get the same amount of goods of the developed countries. To stabilise the prices of their exports, the developing countries pressed for the speedy creation of a multi-billion dollar Common Fund for financing the international buffer stocks of different commodities covered under the United Nation's Integrated Programme for Commodities. As many as 18 commodities from bananas to bauxite were to be covered by the Common Fund. India's interest was mainly in tea, coffee and jute.

In the sphere of finance, the major issue related to an accumulated debt of $ 700 billion of the developing countries which was well beyond the repaying capacity of many of these countries. The demands in this sphere related to debt rescheduling, conversion of the loans into grants in the case of 'core' least developed countries and raising up of the official development assistance (ODA) of the developed countries from the present 0.3 per cent to 0.7 per cent of their gross domestic product in terms of the UN resolution.

Apart from these demands, the 'Group of 77' stressed the imperative need for a new international economic order which could not be achieved without structural reforms. The Group of 77 also pressed for structural reforms in the World Bank and IMF in order to give more voice to the developing countries. It was also pointed out that the least developed countries needed the greatest attention because the per capita income in at least one-third of these countries had been falling over the last two decades. There was also made the demand of diverting more funds for development from the international financial institutions.

UNCTAD VII

UNCTAD VII held in July–August 1987 in Geneva, was attended by delegates from over 150 countries and international agencies. It was concluded on 3 August 1987 after adopting a package of proposals to solve the difficult problem of debt of the developing countries, arrest the growing trend of protectionism in international trade, and improve the economic and social conditions in the least-developed countries. Most of the agreements on the four main agenda items of international trade, Third World debt, commodities and assistance to least developed countries were a far cry from the 'Havana Declaration' of developing nations in May 1987. However, one of the most tangible achievements of the Conference was the agreement among the rich and poor nations on the integrated commodities programme (ICP) which had never taken off after its launching in 1976. The ICP could now start operating as the rich nations had at least recognised its existence. Considering the difficulties in achieving an international cooperation agreement in a gloomy situation, the accord on ICP was a great achievement and progress on ICP paved way for new agreements on commodities which the developing countries had been long striving to achieve.

UNCTAD VIII

UNCTAD VIII which was held in Cartagena de Indias, Columbia during February 1992 was attended by 170 member countries and international organisations. It took place against a backdrop of divergent economic and financial development of the 1980s. Established in 1964, the UNCTAD was established to serve as a forum in which trade-related development issues could be discussed and analysed to lead to the negotiations of international understandings on those issues which were in dispute. UNCTAD VIII, unlike its earlier predecessors, was free from polemics and tensions between the North and the South and the East and the West. A major issue at UNCTAD VIII was about the role of UNCTAD itself. Members attending the conference agreed on the broad features to revitalise the UNCTAD by making it more effective in dealing with the issues related to development. UNCTAD VIII reached a consensus to provide a new structure to the UNCTAD. The new structure included (1) the Conference which meets every four years; (2) the Trade and Development Board (TDB) which meets twice a year in regular session, and in special session as required; (3) an Executive Committee of the Board which comprises the permanent representatives in Geneva to UNCTAD to meet periodically; (4) new Standing Committees on commodities, poverty alleviation, economic cooperation among developing countries and services; (5) continued special committee and groups; and (6) ad hoc working groups to deal with investment and financial flows, non-debt creating finance for development, new mechanisms for increasing investment and financial flows, trade efficiency, comparative experiences with privatisation, expansion of trading opportunities for the developing countries, and inter-relationship between investment and technology transfers. The spirit of Cartagena was imbided in the 'partnership for development'.

UNCTAD IX

The ninth session of the United Nations Conference on Trade and Development—UNCTAD IX — was held at the Gallagher Estate Conference Centre in Midrand, Gauteng Province, Republic of South Africa from 27 April to 11 May 1996. The representatives of 138 member states of UNCTAD participated in the Conference. The Trade and Development Board served as the preparatory committee for the ninth session of the Conference. The Ministerial Round Tables at the Conference were centered on the following four themes:

1. Globalisation: development, instability and marginalisation.
2. International trade as an instrument for development in the post-Uruguay Round World.
3. Enterprise development: national strategies and international support.
4. Future work of UNCTAD in accordance with its mandate; institutional implications.

UNCTAD IX was characterised by frank assessment of UNCTAD's functioning for development made during the round tables of Heads of State, Multilateral Agencies and Ministers. This inspired member States to build a more effective organisation capable of implementing its mandate in a changing world. UNCTAD VIII held in 1992 heralded *The Spirit of Cartagena,* a partnership for development recognising the need for a new approach to assisting development. After four years, at the UNCTAD IX it was realised that further vigorous initiatives were called for in order to translate that spirit into reality. Since UNCTAD VIII held at Cartagena, the United Nations had held several important global conferences on major economic and social themes. In addition, the creation of the World Trade Organisation (WTO) has strengthened the rules-based multilateral trading system promoting further the process of liberalisation, opening new opportunities and sustainable development and growth in its wake. The UNCTAD IX had responded to these changes by initiating important reforms aimed at giving new and real meaning to the partnership for development.

On globalisation, it was recognised that the rules-based trading system of the WTO will facilitate positive integration of countries into the global trading system provided the commitment to this objective was strengthened. On the theme of partnership for development need was expressed to base it upon a clear definition of the roles, the establishment of common objectives and development of joint action. On the issue of institutional reform of the UNCTAD, it was recognised that in a rapidly changing environment it was essential to improve the accountability of UNCTAD based on assessment, review and transparency of operation. It was also agreed to hold UNCTAD X in the year 2000 in Thailand.

UNCTAD X

The tenth session of the United Nations Conference on Trade and Development (UNCTAD) started on February 12,2000 in Bangkok with the election of Dr. Supachai Panitchapakdi, Thailand's Deputy Prime Minister as the President of UNCTAD X. The aim of UNCTAD X was to move the trade and development issues forward and to define the complex relationship between globalisation and development, trade, investment and finance. The conference, among other things, discussed the problem of depressed commodity prices of the low-income poor countries over the past two decades leading to huge trade losses and resulting in the rising debt for the producers of agricultural goods. The tenth session of Conference was attended by 146 countries.

According to the Bangkok Declaration of UNCTAD X, globalisation, if properly managed, could become a powerful and dynamic force for growth and development by laying the foundations for enduring and equitable growth. The Bangkok Declaration, while stating that the conference had brought the development partners together to propose practical and meaningful solutions, hoped for a fairer and better world economic system, poverty alleviation and offering all people security and growing opportunities to raise their living standards. The UNCTAD meeting emphasised the merits of a multilateral approach, with pledges from institutions such as the World Bank, IMF and ILO to work for a better world.

According to the UNCTAD X, the two categories of imbalances needed to be redressed if globalisation was to deliver equitable and sustainable growth for all peoples. The first related to the structure of international systems governing development, trade and finance. The second stemmed from the severe poverty and structural constraints which accompany under-development in most developing countries in the world.

UNCTAD XI

The eleventh session of the United Nations Conference on Trade and Development (UNCTAD) was held in Sao Paulo, Brazil from June 14 to June 17, 2004. Its total membership has now grown to 188 which includes all the UN member states. The members agreed on a declaration which they called the "Spirit of Sao Paulo". The decisions adopted at UNCTAD XI, in addition to the Bangkok

Plan of Action, form the solid basis to build upon and are essential instruments in the member states continued commitment to support UNCTAD in fulfilling its mandate as the focal point within the United Nations for the integrated treatment of trade and development on the road to UNCTAD XII in 2008.

UNCTAD XII

The 12th ministiral conference of the United Nations Conference on Trade and Devolpment (UNCTAD XII) was held in Accro (Ghana) on 20-25 April 2008. More Than 4000 participants including representatives from UNCTAD'S Member States, international organization, civil society, private secretor, academia and the media attended the conference. The welcome address was delivered by the President of Ghana, John Agyekum Kufuor. The conference was inaugurated by the U.N. Secretary General Ban Ki-moon. Mr Ban insisted on the need for 'fresh thinking' and tresh approaches' to meet the global disenfranchisement.

After an intense negotiations, Member States adopted by consensus on 25 April, two major outcome documents: a short ministerial declaration called the 'Accro Declaration' and a longer text called 'Accro Accord'. In the Accro Declaration, Member States pledged, to take immediate steps to boister the world food security. The Member States asserted in the same text that they would take all necessary steps to meet urgent humanitarian needs, specially in africa, least developed countries and net food importing developing countries. The 'Accro Accord' reaffirms UNCTAD's unique role in multilateral system as "the focal point of the United Nations for the integrated treatment of trade and development and interrelated issues in the areas of finance, technology, investment and sustainable development".

Speaking at the end of the conference, Dr. Supachai secretary General of UNCTAD, vowed to strengthen the orgainzation work on commodity-including by promoting steps that boister agricultural sector in developing countries. The 13th UNCTAD ministerial conference (UNCTAD XIII) will be held in Doha(Qatar) from 20-26 April 2012.

Suggested Readings

Indian Institute of Foreign Trade, *International Trade and Development,* 1972, pp. 1–18.
Raul Prebisch, *Towards A New Trade Policy for Development,* United Nations, Geneva, 1964.
Manubhai Shah, *The Developing Countries and UNCTAD,* Second Edition, 1968.
Bo Sodersten, *International Economics,* Second Edition, Reprinted 1981, Chapter 18.
UNCTAD, *Annual Reports.*
UNCTAD, *Report of Ninth Session.*

Questions

1. What are the main purposes of the UNCTAD? Discuss its role in promoting international economic cooperation.
2. Write a short note on trade barriers to the exports of the less developed countries. Discuss the role of the UNCTAD in this connection.
3. What are the essential functions of the UNCTAD? How far has it been successful in protecting the economic interests of the less developed countries in the sphere of trade? Discuss fully.

CHAPTER 29

Foreign Aid

Foreign aid has become an important international economic issue in the context of economic development of world's less developed countries. Naturally, the issues centring around the problem of foreign aid relate to the purposes to be served by such aid, the terms and conditions on which such aid should be made available and the amount in which it should be made available in order to serve a meaningful purpose for those poor countries to which it is given. As against the flow of private capital which is mainly induced by considerations of profit, foreign aid largely depends on the policy decisions of governments of the aid-giving countries. Consequently, apart from the economic considerations, political, defence and other considerations weigh substantially in the aid-giving decisions of the rich countries from whom such aid mainly flows towards the poor countries. Sometimes the conditions attached to foreign aid may be so stringent that the poor borrower countries may find it difficult to accept these conditions. The poor countries have in general raised their voice against the stringent conditions which have frequently been attached to foreign aid with the result that the slogan "trade and not aid" has been frequently heard from the side of these countries in international conferences and debates.

Foreign aid has been given to world's less developed countries both by the multinational financial institutions like the World Bank, International Development Association, International Finance Corporation, International Monetary Fund etc., the regional financial institutions such as the Asian Development Bank, the African Development Bank, the Inter-American Development Bank etc., the multinational agencies such as India Development Forum (IDF), Help Pakistan Club etc.—consortia of the developed aid-giving countries—and governments of the friendly countries. Foreign aid has been given for a variety of purposes, such as gaining the support of allies in the cold war, gaining the support of the aid-receiving countries on political and other strategic issues, obtaining strategic materials for promoting a country's economic development. In fact, in today's world foreign aid to the developing countries has become an important plank of developed nations' foreign policies and no developed nation can afford to ignore the foreign aid aspect of her foreign policy.

Economic development has multi-dimensional aspects which include vastly increased investment, use of improved technology, acquisition of various skills, removal of diseases, squalor and slums in world's poor countries. Consequently, foreign aid for economic development in the poor countries could be given for any of these and other allied purposes. Whether any amount of foreign aid given to a country has succeeded or failed in achieving its purpose is generally judged from its effectiveness in raising the per capita income in the country although such a general criterion is not in all situations an infallible measure of the effectiveness or otherwise of the foreign aid given to the country.

Foreign Aid and Economic Development

A programme for planned economic development and an aid programme go together. Since any planned economic development programme extends over several years encompassing various aspects, its fulfilment requires coordinated concerted efforts combined with a continuous financial aid flow. Without a well-thought-out development programme or planning, individual projects will not have much meaning and will lead to haphazard and unrelated development that may create more problems than it may solve. On the other hand, an individual development project undertaken as a part of planning will become an intelligible part of a whole which has a purpose and a direction. However, under planned economic development foreign aid will prove wholly effective only if it is utilised for a scheduled programme. This requires that the aid-giving agencies should keep this factor in mind and should not make aid ineffective by tying its use to some particular project that does not fit into the framework of planning. For example, if a development planning programme requires at a particular time the establishment of a major power project whose output will serve as an input for the expansion of fertiliser industry on which will depend the expansion of agricultural production and following it of industrial production in the country, the insistence of the aid-giving country in such a situation to supply only telecommunications equipment does not have much significance for the poor country and reflects an uncooperative attitude of the lender country. The massive loan assistance of more than US $ 650 billion given by the World Bank and the IDA to the developing countries has contributed significantly in the economic development of these countries.

Apart from the issue of the purpose of foreign aid and its relationship with the planned economic development of the country, the other equally important issues in the context of foreign aid relate to the form of foreign aid—should foreign aid take the form of grants or of loans or even of the equity investment capital or should loans be given at normal market rate of interest, concessional terms of low interest rate and easy repayment schedule in instalments spread over longer period? The other equally important issues that arise in the context of foreign aid are: the relative merits and practicability of depending on multilateral as against bilateral foreign aid, the administration of the aid—should or should not the aid-giving country or agency have any say or control in the manner in which foreign aid is to be used—soft currency and hard currency loans, burden of servicing the foreign loans or debt burden for the poor debtor countries and the impact—political and economic—of foreign aid on the economy of the borrower recipient country.

The developing countries which are mostly recipients of foreign aid given by the developed countries have vehemently opposed any administrative control of the aid-giving developed creditor countries in the matter of utilisation of aid given by these countries. They have also shown anxiety at their growing external debt burden which has eroded a considerable part of their total meagre foreign exchange earnings. The nominal value of total external public debt of world's developing countries at the end of 1991 aggregated US $ 1,280 billion. Of this total, outstanding long-term debt amounted to US $ 1,060 billion. The debt-to-export ratio in 1991 stood at 176 per cent; the debt-to-gross national product ratio was 38 per cent; and the debt service-to-exports ratio stood at 21 per cent. These aggregate ratios, however, hide significant regional differences. The following table shows the growth of external debt of world's developing countries from 1980 to 1991.

Although the entire external public debt shows the amount of foreign aid which the developing countries have received from the developed countries for development purposes, it seems almost certain that foreign aid in the form of the foreign loans, be it bilateral or multilateral, does not offer an easy solution of the dilemma of poverty for the poor developing nations. In its efforts to help the poor countries in their economic development the United Nations Organisation celebrated 1960s as the "Development Decade" by setting the target of 0.7 per cent on the gross domestic product of the developed countries as official development assistance to be made available as development assistance to the poor countries. However, even this low ratio was never reached and at the end 2004 it was only 0.25 per cent.

Official Development Assistance (ODA)

Development Assistance Committee of the 22 DAC countries had provided 66 billion U.S. dollars in the year 2000 as ODA to developing countries which rose to only 78 billion dollars in 2004. This is on account of the fact that economically big and most developed countries belonging to G-8 are very miser in giving ODA to developing countries. For instance the U.S.A., are of the richest country in the world, had, contributors only 0.16 percent of its GNI in 2004 as against 0.7 percent international target set in the "Development Decade". The following table giving the disbursement of net official development assistance from DAC countries to developing countries and multilateral organisations as percentage of donors country's. GNI for 3 years reveals the generosity (sarcastic) of the developed countries in giving ODA to developing countries.

Table 29.1 ODA as Percentage of GNI

Country \ Year	2000	2001	2002
1. U.S.A	0.10	0.11	0.12
2. U.K.	0.32	0.32	0.30
3. Japan	0.28	0.23	0.22
4. Norway	0.76	0.80	0.31
5. Sweden	0.80	0.77	0.74
6. Denmark	1.06	1.03	0.96
7. Luxembourg	0.71	0.82	0.82
8. Netherlands	0.84	0.82	0.82

Source: (UNO) Statistical Year Book, forty-eight issue, 15 Dec. 2003

It is evident from the table that only 5 small countries have met the set international target, while the 3 big countries belonging to G-8 are lagging much behind. Not only the amount of ODA is low but also its distribution is biased. For example, in 1990-91 Japan gave 60 percent of its ODA to Asia, France gave 50 percent to mainly to its territories, U.S.A. gave 50 percent to middle east and North Africa and the European Union gave 58 percent to Sub-Sahara Africa. It seems as if geographical and political affinity is more powerful than economic backwardness in the allocation of official development assistance.

The economics of Brazil, India and China are very big and growing rapidly but per capita foreign aid is very low, however, Pakistan is comparatively, better placed. In absolute terms Pakistan received 1923.6 U.S. million dollar aid in 2001 while India received only 1710 U.S. million dollars.

The socio-economic development assistance provided through UN system organisations like UNDP, UNICEF, UNFPA (Population fund) and WFP (World Food Programme) etc. is considerable, The amount allocated under this system for the world as a whole was about 7132902 thousand

U.S. dollars in 2001 which rose to 7338331 thousand U.S. dollars in 2002. Out of this fund India received 208181 thousand U.S. dollars in 2001 which came down to 103085 thousand U.S. dollars in 2002.

From other international institutions including UNDP, UNICEF, UNFPA etc India recived a grant of 76.2 US$ million in 2010–11 which rose to 179.2 US$ million in 2011–12. During the same period over all external assistance came down from $ 8218.4 million to $8019.2 million. This is on account of a fall in grants.

In the era of glabalisation and liberalisation foreign aid will help change in production efficiency, upgradation of technology and innovation in developing countries. The combined, gross national income of the developed countries is about 32 thousand billian dollars at present (2004). If they provide 135 billian dollers as aid to developing country the poverty in those countries will be reduced to half by the end 2015. It is not a difficult task as they have provided about 279 billion dollars to their farmers as subsidy in 2004. The developed countries should always remember that poverty any where is dangerous to prosperity every where. The developing countries have made this view concerns repeatedly at the WTO, UNCTAD, IMF and IBRD meetings and several other international conferences. The developed countries should not forget that in the 1950's Italy and Japan were among poor countries. But at present on account of inflow of liberal foreign aid they are members of G-8. If liberal foreign aid flows rapidly to developing countries like China and India they will not only become the members of the club of rich nations but also be in a position to provide foreign aid to other developing countries.

Suggested Readings

R.E. Asher, *Development Assistance in the Seventies: Alternative for the United States,* 1971.
—*Grants, Loans and Foreign Currencies,* 1961, pp. 83–84.
R.F. Mikesell, *The Economics of Foreign Aid,* 1968.
J.A. Pincus, *Trade, Aid and Development: The Rich and Poor Nations,* 1967.
World Bank, *Global Development Finances* 2004.

Questions

1. Explain the contribution of foreign aid in the economic development of the developing countries. In this context, discuss the problem of foreign debt servicing burden of these countries.
2. Discuss the importance of increased flow of foreign aid in the economic development of world's developing countries. In what form should foreign aid be given to the developing countries?

CHAPTER 30

International Debt

Introduction

The phenomenon of international debt crisis, particularly the low-income third world countries' difficult external debt situation, which is a constant danger to lasting international peace and stability was not created overnight. The external debt crisis of the developing countries has its origin in the unusually large capital inflows which took place in the postwar period, more particularly during the two decades of the *seventies and eighties* from the developed countries to the developing countries. The massive borrowing of capital from the developed countries by the low-income developing countries was motivated by their desire to raise the level of their gross domestic product beyond the level that could be achieved with domestic resources alone.

An inflow of foreign capital, particularly if it is in the form of loans, creates the difficult problem of debt servicing for the borrower country which can be solved only through the creation of required export surplus on current account. Since the borrowing developing countries did not enjoy the happy position of having current account surpluses, their external debt mounted with the passage of time. At the early stage of their development, the developing countries experienced the inflow of concessional capital in terms of the official grants and loans. When the official capital, both bilateral and multilateral, shrank they were forced to borrow in the international capital markets at high interest rates and also at variable interest rates which carried high risk. Thus, the scenario for many developing countries changed from low borrowing and low debt-servicing burden to high borrowing and high debt-servicing burden.

The total external debt of the capital importing developing countries grew at a rapid rate throughout 1973–82 exceeding the compound rate of 20 per cent. The proportion of total external debt of the less developed countries (LDCs) at floating interest rates rose steadily from 21 per cent annual average in 1973–74 to more than 53 per cent in 1984–85. The position of the 15 highly indebted countries was still worse as these percentages in their case were 42 and 73.6 respectively. In due course of time most of the low-income borrowing developing countries landed themselves in the so-called 'foreign debt trap'.

The international debt crisis of 1982 was built on the continuous reckless borrowing on the part of governments of the LDCs most of which were weak and soft in handling efficiently the difficult problem of productive utilisation of foreign loans that had to be repaid with interest on the expiry of the loan maturity term. The governments of these countries forgot the hard fact that any loan has to be repaid and a nation can be rendered bankrupt by an externally-held debt if it is misutilised or wasted.

Size, Growth and Composition

The total external debt of the developing countries went from 1996 to 2001 is given in the following table.

Table 30.1 Total Long Term External Debt of Developing Countries

(in US Billion dollars)

Source/Year	1995	1996	1997	1998	1999	2000	2001
Official Creditors	858.20	825.20	783.20	829.20	849.40	821.80	791.40
Multilateral	287.20	283.60	282.80	101.50	108.50	111.30	112.10
IBRD	111.67	105.30	101.50	108.40	111.30	112.10	112.50
IDA	71.50	75.10	77.40	84.00	86.60	86.80	89.20
Bilateral	571.00	541.50	500.30	511.70	515.60	489.10	452.60
Private Creditors	550.00	567.20	590.70	623.30	620.10	612.00	603.00
Bonds	246.70	279.30	388.70	323.00	341.00	372.20	373.30
Commercial Banking	167.70	164.60	200.20	211.40	132.20	167.60	164.00
Other Sources	135.80	123.30	101.60	88.70	86.80	72.00	65.50
Private Non-Guaranteed	219.4	275.30	348.70	493.30	523.40	534.50	513.40
Total	3218.80	3240.40	3375.10	3375.50	3440.90	3379.40	3877.00

Note: Figures are rounded, source: Statistical Year Book (UNO) Dec, 2003 p. 837.

It is evident from the table that the contribution of the official creditors comprising national governments, public subdivisions and public antonomous bodies is either constant or declining year to year, where as the contribution of private creditors comprising manufacturers, exporters, and the suppliers of goods is increasing. The amount under private non-guaranteed debts has more that doubled during a period of 7 years. The debt provided by IBRD and IDA is very small and constant year to year. The bilateral debt is also declining whereas the bond market debt is expanding. The total external debt rose from 3218.8 billion U.S. dollars in 1995 to 3877.00 billion U.S. dollars in 2001. It disapponting that the contribution of IDA, which is known as 'soft loan window' is very small. Similarly the contribution of IBRD is very small and constant year to year.

Debt and Debt-Service Ratios

The following table (30.2) shows the total debt, debt ratio and debt-service ratio of the ten most indebted developing countries for the year 2011.

Among the top ten indebted developing countries China is on no. one, Russian Federation is on no. two and India is placed on fifth position. However, the debt to GNI in China is less than Inida. The short term debt to total external debt is approximately three and a half times higher in China than Inida. The short term debt is supposed to be burdensome than long term debt. So far as concessional debt is concerned India was in a better position than China in 2008. The share of concessional debt in China was only 10.8 p.c. whereas in India it was 20.5 p.c. Kazakhstan is severly indebted (94.3 per cent of GNI) and it is followed by Turkey and Argentina.

Solution of External Debt Problem

From the foregoing description of the size, composition and growth of the external debt of the developing countries, it is obvious that any solution of the problem requires the serious concerted

Table 30.2 International Comparison of Top Ten Developing Debtor Countries, 2011.

No.	Countries	Total External Debt Stock (US$ million)	Total Debt to GNI (per cent)	Short-term to Total External Debt (per cent)	Foreign Exchange Reserves to Total Debt (per cent)
	2	3	4	5	6
1.	China	548551	9.3	63.4	531.2
2.	Russian Federation	384740	26.9	10.1	124.6
3.	Brazil	346978	16.9	18.9	83.2
4.	Turkey	346978	40.4	26.6	29.3
5.	India	293872	16.9	19.4	103.5
6.	Mexico	290282	19.5	19.5	60.3
7.	Indonesia	200081	26.1	17.5	53.7
8.	Argentina	179064	36.1	27.4	40.8
9.	Romania	127849	76.4	20.6	39.5
10.	Kazakhstan	121505	94.3	7.6	23.8

Source: World Bank, Global Development Finance, 2012

efforts on the part of the debtor developing countries, creditor countries, international financial institutions, and the commercial bank creditors. Most of the developing countries are net exporters of primary commodities whose prices in the international market have been continuously declining. There are 84 developing countries which derive 50 per cent of their export earnings from only one or two primary commodities' exports. The Third World countries exporting commodities had suffered an average annual loss of $ 8 billion during 1981–86 on account of declining prices. The loss on account of unfavourable terms of trade was even greater. According to UNCTAD estimates the net barter terms of trade of developing countries showed an annual average deterioration of 3.9 per cent. The cumulative loss suffered due to this by all developing countries amounted to $ 93 billion and for the least developed countries it was $ 5 billion.

Broadly, three approaches have been advocated to solve the developing countries' external debt crisis. According to the first approach, it is the responsibility of individual debtor countries to find solution to their debt problem. Several techniques have been suggested and implemented. Rescheduling of debt, asking for forgiveness, suspending payment of interest etc. have been some of the techniques used. For example, in 1986, 18 countries negotiated for rescheduling of their debt through the Paris Club. Some Latin American debtor countries converted their debt into equity investment at discount rates. According to the second approach, it is the duty of both the debtor and the creditor countries to cooperate in finding an effective solution to the problem of debt. The commercial banks must also cooperate as "no debt strategy can succeed without the active support and broad participation of commercial banks." According to the third approach, the debt strategy calls for case-by-case study.

Baker Debt Strategy

At the World Bank-Fund annual meetings held in October 1985 at Seoul, the US Treasury Secretary James A. Baker III announced the U.S. debt initiative which was characterised by its author Mr. Baker as a programme for sustained growth. The Baker debt strategy aimed at strengthening the strategy which had been followed since 1982. The essential features of the Baker strategy included the following:

1. Adoption by the debtor countries of comprehensive macro-economic and structural policies to promote growth, balance of payments adjustments, and lower inflation.
2. Increased lending by the World Bank and other multilateral development bank—amounting to an additional $ 9 billion in net lending during 1986–88—as well as more effective structural adjustment lending by these institutions, in conjunction with a continued central role for the Fund; and
3. An increase in net new lending by private banks over the next three years in support of comprehensive economic adjustment programmes.

While Baker identified 15 indebted countries that required net new lending from commercial banks of $ 20 billion over the next three years, the U.S. Treasury indicated that the list was not exclusive. In addition, Baker stressed that countries now receiving adequate financing from banks on voluntary basis should continue to do so provided that these countries maintained sound policies. The Baker debt strategy received the backing of the international banking community and banks in the major financial centres showed their willingness to play their part in implementing the strategy on a case-by-case basis and in collaborating with all other relevant parties—including debtor and creditor governments and the international institutions.

On January 15, 1986 in Washington Mr. H. Onno Ruding, Chairman of the Interim Committee of the Fund's Board of Governors on the International Monetary System asked the commercial banks, international institutions, and governments of industrial and developing countries to cooperate in resolving the world debt problem along the lines suggested by the US Treasury Secretary James A. Baker III. Mr. Ruding identified five "essential conditions" for a solution to the world's external debt problem. According to Mr. Ruding:

1. The Fund must promote adjustment "with an eye for growth" while the World Bank must promote growth "with an eye for adjustment." Cooperation between the Fund and Bank was necessary and Fund conditionality remained essential. While the Bank should expand its role as a development institution, it should not become a balance of payments institution which the IMF already was. Apart from Baker strategy, a case-by-case approach to the problem of debt solution was as essential as was the "central, catalytic role" of the IMF, as an advisor and an honest broker.
2. The commercial banks, in their own interests, should help the debtor countries to improve their growth prospects. The banks should coordinate their efforts so that growth-oriented policies and Fund guidance are not undermined by lack of confidence on the part of individual banks.
3. In order to solve the problems of capital flight and attracting new money faced by the debtor countries, these countries must carry out the necessary adjustment in their domestic economies.
4. Apart from providing money, industrial countries must also pursue adequate policies themselves. They should reduce interest rates and budget deficits, increase savings, and reverse protectionism.
5. The U.S. development aid was very low by international standards and should be exempted from budget cuts. According to Mr. Ruding, the Baker strategy could not succeed without the US support for a general capital increase for the World Bank.

Progress of External Debt Strategy

Several steps have been taken within the overall framework of the debt strategy to solve the problem of external debt faced by the developing countries. During 1987, 19 countries renegotiated a total of $ 103 billion of debt. This amount included $ 48 billion in previously rescheduled debt for which terms were modified to the great relief of debtor countries. In the Paris Club, official creditors provided longer-term debt rescheduling for selected low-income debtor countries that were making adjustment efforts. The commercial bank consortiums also restructured debt with longer terms of repayment and lower spreads than before. Some new money packages were also arranged in connection with commercial bank restructurings. Several multiyear rescheduling arrangements which were previously concluded with official and commercial bank creditors were renegotiated during 1987.

Additionally, the positive developments included the increase in official lending following the Baker strategy, the agreement in principle of creditors to support a general capital increase for the World Bank, the financial involvement of the Fund, and the expansion of financing options for commercial banks. The establishment of the Structural Adjustment Facility (SAF) in March 1986 initially with SDR 3.7 billion (later raised to SDR 9 billion) under which loans have been given to low-income member countries which are facing protracted balance of payments problems in support of medium-term macro-economic and structural adjustment programmes, $ 8 billion debt retired through debt-equity swaps over the past three years, increased trade and project loans, lending of funds to specific end-users, new money bonds, notes or bonds convertible into local equity, exit bonds, external debt conversions, interest capitalisation and balance of payments loans are those positive developments which give rise to the cherished hope that external debt problem will continue to be managed in a way that will maintain and foster the essential path of cooperation between the debtor and creditor countries in the interest of promoting sustained economic growth.

In 1991, the Paris Club creditor countries agreed to reduce 50 per cent debt burden of Egypt and Poland. The total face value of eligible debt covered under the agreements was $ 30 billion for Poland and $ 28 billion for Egypt. At the end of 1991, nonconcessional but special extended rescheduling terms had been extended by Paris Club creditor countries to Congo, Cote d' Ivoire, the Dominican Republic, El Salvador, Honduras, Jamaica, Morocco, Nigeria, Peru and the Philippines. In December 1991, the Paris Club agreed to implement a new menu of enhanced concessions for the low-income rescheduling countries. Under this arrangement, Nicaragua, Benin, Tanzania and Bolivia benefited to the extent of 50 per cent reduction in their debt service burden. The agreements with these countries increased the degree of concessionality in comparison with Toronto terms.

Following the modification of the debt strategy to allow for official support of debt-reduction operations, a significant number of heavily indebted developing countries, including Costa Rica, Mexico, Morocco, the Philippines, Uruguay and Venezuela reached agreements with commercial creditors in 1989 and 1990. In 1991, however, only Niger, Mozambique and Nigeria reached agreement with their commercial creditors on a debt reduction programme. These agreements involved buybacks of the bulk of their commercial bank debt at a deep discount and a menu of options, including a buyback and a par exchange for reduced interest bonds. In February 1992, the Philippines reached agreement with her bank creditors on a term which provided for a financing package that provided for a comprehensive restructuring of its foreign commercial debt that remained after the 1990 buyback. Debt-equity swaps, particularly in Argentina, to the tune of $ 7 billion and cash payment of interest arrears, particularly by Brazil and Nigeria, also played a limited role in reducing the external debt. The agreement with the Philippines provided for a menu of options: additional cash buybacks, new money combined with debt-conversion bonds and different types of debt exchanges at par for bonds with either permanent or temporarily reduced interest rates. In April 1992, Argentina reached an agreement in principle on a debt and debt-service reduction package with its creditor banks. The agreement which covered $ 31 billion of commercial bank debt, including $ 8 billion in arrears, allowed banks to choose between floating-rate discount bonds and fixed-rate par bonds. Several of other countries, including Brazil, have also negotiated debt-restricting agreements which included debt and debt-service reduction with their commercial bank creditors. A substantial fall in the average annual inflation rate in the developing countries in Western Hemisphere and South Asia is a hopeful sign toward increasing the export potential of these countries which will help these countries to repay their foreign loans through the creation of substantial export surplus in their balance of trade.

External Debt Initiative

Under the joint IMF-World Bank debt initiative for the heavily indebted poor countries (HIPCs), which was launched in 1996, bilateral and multilateral creditors of indebted poor countries pursuing sound policies provide debt relief to help these countries put their external debt burdens on a sustainable basis over the medium term. The IMF is participating in the initiative through its enhanced structural adjustment facility (ESAF)—a concessional financial facility that supports comprehensive macro-economic and structural reform in low-income countries. The IMF Executive Board had assisted Uganda under the HIPC initiative plan by sanctioning a total of $ 340 million in debt relief. Under the initiative $ 340 million had been provided as debt relief from Uganda's all creditors. The IMF's share was $ 70 million. All multilateral creditors reduced their claims on Uganda in net present value terms by about 20 per cent. In nominal terms, the resulting total reduction in Uganda's debt service was likely to be about $ 700 million.

Preliminary agreement on the eligibility of debt relief for Bolivia, Burkina Faso, Cote d'Ivoire had also been reached owing to the commitment of these countries to sound policies. The Group of 24 developing countries has urged quicker application of, and more flexible terms under the RIPC initiative.

The debt relief programme has significantly reduced the debt stock in heavily indebted poor countries from $ 69 billion in 1999 before the HIPC Initiative to $ 28 billion in 2003 while the debt service as percentage of exports was reduced from 14.5% to only 9.8%. At the end of April 2004, about $ 52 billion had been committed in debt-service relief to the 27 countries of which 13 countries had already reached the completion points while the remaining 14 countries had reached their decision points. Debt relief, especially through the enhanced HIPC Initiative, is essential to enable low-income countries to free resources for the social and infrastructure spending that they will need to achieve the Millennium Development Goals (MGDs)[1].

It should, however, be remembered that while the active cooperation of the creditors is necessary in any successful resolving of the external debt problem, the primary responsibility in the matter is that of the debtor. For the developing countries to accelerate their pace of economic development and to grow out of their debt burden, these countries will have to implement effectively the appropriate macroeconomic policies supplemented by an adequate flow of foreign capital in a supportive international economic environment.

---

1. The eight Millennium Development Goals sought by 2015, are to (1) halve the extreme poverty and hunger relative to 1990; (2) achieve univeral primary education; (3) promote gender equality; (4) reduce child mortality; (5) improve maternal health; (6) combat HIV/AIDs, malaria, and other diseases; (7) ensure environmental sustainability; and (8) establish a global partnership for development. The IMF and World Bank have a key role to play in helping the member countries in achieving the MDGs through the various channels.

Table 30.4 India's Key External Debt Indicators (per cent)

Year	External Debt (USS billion)	Total External Debt to GDP	Debt-Service Ratio	Foreign Exchange Reserves to Total Debt	Concessional Debt to Total External	Short-term External Debt to Foreign Reserves	Short-term External Debt to Total Debt
1	2	3	4	5	6	7	8
1990-91	83.8	28.7	35.3	7.0	45.9	146.5	10.2
1995-96	93.7	26.9	26.2	23.1	44.7	23.2	8.4
2000-01	101.3	22.5	16.6	41.7	35.4	8.6	14.1
2005-06	139.1	16.8	10.1#	109.0	28.4	12.9	16.4
2006-07	172.4	17.5	4.7	115.6	23.0	14.1	20.4
2007-08	224.4	18.0	4.8	138.0	19.7	14.8	19.2
2008-09	224.5	20.3	4.4	112.1	18.7	17.2	19.2
2009-10	261.0	18.3	5.5	106.9	16.8	18.8	20.0
2010-11PR	306.4	17.8	4.2	99.5	15.5	21.3	21.2
End-June 2011 PR	317.5	–	4.6	99.6	15.1	21.7	21.6
End-Sept. 2011 QE	326.6	–	–	95.4	14.7	22.9	21.9

Source : Ministry of Finance and RBI. PR: Partially Revied; QE : Quick Estimates.

India's External Debt

India's external debt increased from $83.8 billion at the end of March 1991 to $326.6 billion at the end of Sept. 2011. The rise in external debt is mainly because of NRI deposits, Resurgent India Bond and World related credit selecting larger import demand. However, external debt to GDP declined sharply from 28.7 per cent in 1991 to 17.8 per cent in 2011 Similarly debt-service ratio also declined from 35.3 per cent in 1991 to 4.6 per cent in 2011 The Foreign exchange reserves to total external debt increased from 7.0 in 1991 to 95.4 in 2011. Thus Foreign exchange reserve position at present is confortable. However, concessional debt to total external debt declined from 45.9 in 1991 to 14.7 in 2011. The debt share of short-term debt to total external debt has increased from 10.2 in 1991 to 21.9 in 2011. The short-term debt is supposed to be burdensome.

Composition of India's External Debt

Table 30.5 Composition of External Debt

Sl.	Component	March 2010	September 2011-QE
1	2	3	4
1	Multilateral	16.4	15.0
2	Bilateral	8.7	8.4
3	IMF	2.3	2.0
4	Export credit	6.5	6.0
5	Commercial Borrowings	27.1	30.3
6	NRI Deposits	18.3	16.0
7	Rupee Debt	0.6	0.4
8	Long-term debt (1 to 7)	79.9	78.1
9	Short-term debt	20.1	21.9
10	Total External Debt (8 + 9)	100	100

Source : Ministry of Finance and RBI.

Over the years India's external debt composition has witnessed sturctural change. The external commercial borrowings in India's total external debt has increased substantisally. The increase in ECBs in the recent period caused some concern given the depreciation of the rupee This would mean a heigher debt service burden. Similarly NRI deposits are aslo increasing. The IMF loan is hardly 2 per cent of the total external debt. This is on account of the allocation of SOD. Short term debt which is supposed to be burdensome is also increasing. The rupee debt is nominal. It is only about helf a per cent. The present Finance Minister of India, Mr. Prnab Mukherjee is of the opinion that main reason of India's debt burden is external commercial borrowing and short-term credit.

The share of ECB was 30.3 p.c. in 2011 and the share of short-term credit was about 22 percent. Thus, these two accounted for more than 50 p.c. of total external debt of India.

A prudent external sector policy, particularly in relation to external debt, pursued since 1991 placed India's external debt position at a comfortable level. The policy focus has been on concessional and relatively less expensive source of funds, preference for long maturity loans, monitoring of short-term debt and emphasis on non-debt creating capital flows. Recent initiatives towards external debt moderation include, *inter alia*, prepayment of costly Government and non-Government loans, rationalisation of interest rates as well as structure of NRI deposits, end-use stipulations for ECB and restriction on trade credits. As regards external debt statistics, continuous efforts are being made to bring in refinement in coverage, classification, presentation and technological upgradation in the computation of external debt data.

Suggested Readings

IMF, *Annual Reports.*
World Bank, *Annual Reports.*
World Bank, *World Debt Tables, 1987–1992.*
World Development Report, 1986.
Indian Economic Association, *Conference Volume,* 1987.
World Bank, Global Development Finance, 2004.

Question

1. Discuss the origin and growth of external debt of world's developing countries. What measures have been suggested and taken to solve the problem of external debt? Discuss.
2. Discuss, the trend and composition of India's external debt.

PART FIVE

INTERNATIONAL MONETARY INSTITUTIONS

CHAPTER 31

International Monetary Fund (IMF)

Introduction

The International Monetary Fund (IMF) was established for promoting international economic stability by promoting the balanced growth of free international trade and the multiconvertibility of national currencies. The Fund is a pool of the central bank reserves and national currencies which are made available to the Fund members under certain conditions. In a way, the pool may be regarded as an extension of member countries' central bank reserves.

The pre-1914 gold coin standard was abandoned during World War I. After the cessation of hostilities, desire was manifest among the leading world nations to return to the old gold standard which had for long fostered the growth of stable international trade and economic relations. After the war, the United States of America was the first among the gold standard nations to return to the gold standard in 1919. The fear was, however, prevalent that other countries might resort to competitive currency depreciation in order to attract the gold inflows.

Deflation decreases the money supply increasing money's scarcity and thereby raising the purchasing power of country's money unit. Dear money policy pursued consequent upon deflation pushes interest rates upward in the country. The rise in interest rates accompanied by the fall in prices due to deflation attracts foreign capital. Consequently, it reduces the deficit of a country's external balance of payments. As a method of correcting the adverseness of country's external balance of payments, deflation is not, however, adopted by a country if easier methods are available because it causes unemployment, fall in production and incomes in the country. The Economic Conference convened in Geneva in 1922 had recommended that the member countries should adopt gold standard suggesting that world's total gold reserves should be held at two or three leading financial centres like London, New York and Paris, with the other countries meeting their foreign exchange requirements by holding bank deposits and other liquid assets at these leading international financial centres. The Conference also recommended the regulation of credit in the interest of international peace and economic prosperity.

Since the postwar international gold standard was adopted at the 'hit and miss' parities, at the time of its restoration the French franc was undervalued in relation to the British currency pound-sterling. Consequently, France experienced surplus in her external balance of payments and accumulated large gold and foreign exchange reserves. This huge surplus in her external balance of payments should have been invested abroad if financial equilibrium was to be maintained. However, due to the reluctance of French investors to make long-term investments of their foreign balances abroad, only a part of these foreign balances was invested abroad in the form of short-term and call deposits creating the difficult problem of 'hot money' for the debtor countries whose external balance of payments situation was already fragile. A substantial part of foreign balances held by the foreign creditors in those debtor countries whose balance of payments position was weak was converted into gold. The basic fact that eventually international trade is a barter trade whereby ultimately the payment for a country's imports is made by her exports and *vice versa* was ignored by these creditor countries, particularly by France and the United States of America. Even the richest country has only limited reserves of gold to use in emergency to tide over the deficit in her external balance of payments. However, the deficit in the external balance of payments experienced by the debtor countries was serious arising from the existence of fundamental disequilibrium in their external balance of payments. To correct this deficit, massive gold outflows exhausting the entire official gold holdings of the debtor countries took place without reaching the equilibrium in the balance of payments position of these countries.

France and the USA did not honour their obligations as the world's leading creditor countries. These obligations were fulfilled with sincerity by England before the World War I as the world's leading creditor nation. Gold was imported by France and the USA in large quantity only to be locked in the central bank cellars. Consequently, the money supply did not expand due to these massive gold inflows. This in effect meant thwarting the working of the price-specie-flow adjustment mechanism of the international gold standard. Gold imports did not exert any pressure on the domestic price level in the USA and France. In the postwar world, America had emerged as the world's leading creditor nation. A creditor country must be ready to receive the payment of her loans by creating the necessary import surplus in her balance of trade requiring her to follow an open door policy allowing free imports in the country. But unfortunately the USA became highly protectionist shutting her sea shores for imports from the debtor countries by erecting the high tariff walls. An international monetary standard cannot function in a mad nationalist world. An export surplus creditor country and protectionist policy ill-go together. Consequently, the international gold standard broke down causing great panic and confusion in the world.

After the breakdown of the postwar international gold standard during the *thirties,* a make-shift arrangement was evolved between England, the USA and France to achieve the much-needed foreign exchange rate stability by establishing the exchange stabilisation funds by the three countries. The arrangement visualised by the Tripartite Agreement of 1936 worked smoothly until 1939. During the war no international monetary arrangement existed. After the breakdown of the international gold standard, the world lost the most efficient automatic monetary standard on which world nations had for long relied for restoring equilibrium in their external balance of payments. No alternative arrangement comparable to the old gold standard, however, emerged to replace it. Instead, each country dealt with its external balance of payments deficit in her own way in a manner that resulted in the shrinkage of the volume of world trade in a world characterised by trade and payments restrictions.

In the period of growing trade restrictions, world countries resorted to exchange clearing arrangements, blocked accounts, multiple exchange rate practices and other restrictions on international trade and payments. These restrictions imposed on multilateral trade and payments increased in severity during the war. The enlightened public opinion and world statesmen feared that these restrictive trade and payments practices would continue after the war unless concerted efforts were made to create some effective international machinery whereby the foreign exchange rate stability which was essential for smooth flow of free world trade could be guaranteed. It was the outcome of such conviction shared by the experts during the war that they prepared comprehensive plans of international monetary cooperation for implementation after the war. The British Plan, prepared by the well-known British economist John Maynard Keynes, took the shape of 'Keynes Plan' while the plan prepared by the American expert Harry Dexter White was known as the 'White Plan'. The basic features of these two plans were fused into a common plan evolved at the United Nations Monetary and Financial Conference of 44 nations held at Bretton Woods, New Hampshire in the USA in July 1944. The Conference gave birth to the International Monetary Fund (IMF) and International Bank for Reconstruction and Development (IBRD) now more popularly known as the World Bank.

According to the Conference, three main economic problems dominated the postwar period. Firstly, in order to ensure world economic order and peace it was necessary to restore stability in the monetary systems of those countries which had been forced by the exigencies to abandon all conventional rules of monetary discipline observed under the gold standard. Secondly, it was necessary to find effective means to reconstruct the war-ravaged economies of the European countries. Thirdly, it was realised that stable peace could never prevail in a world in which the developed nations were unconcerned about the untold miseries of vast sea of humanity living in the undeveloped and underdeveloped poor countries of Asia, Africa and Latin America. The world was to be made a better place to live for the masses of the poor Afro-Asian nations. This could be achieved only by diverting a part of the world resources to the development of the economies of the Afro-Asian countries. The effective solution to the problem of economic reconstruction and development was necessary for ensuring an expanding world economy and for dismantling the complex trade and exchange restrictions that had grown during the previous decade. The IMF was established in order to abolish effectively all exchange and trade restrictions and to promote the multilateral trading system while the International Bank for Reconstruction and Development was established to solve the problems of economic reconstruction of the war-ravaged economies of Europe and the long-term development of world's backward countries.

The establishment of the International Monetary Fund is a great landmark in the sphere of international monetary cooperation. Although the creation of the Fund does not represent the first of the international monetary cooperation since there had been instances of international monetary cooperation even before World War I, e.g., the cooperation between the central banks in 1910s, nevertheless it is a most comprehensive and firmly determined effort ever made to organise orderly conducting of international monetary affairs. Notwithstanding that international monetary cooperation was not unknown before World War I (for example, the prewar gold standard was based on international cooperation), such cooperation as was there, relied only on the working of the impersonal market forces and not on the establishment of man-made institution like the International Monetary Fund (IMF) reflecting honesty of purpose on the part of organisers with directors, staff and powers of action. While the prewar international monetary cooperation was automatic, the international monetary cooperation represented by the IMF is deliberate and conscious, created to achieve certain mutually agreed international economic goals. The motivating purpose for creating the IMF was to promote postwar economic growth by establishing an international institution which would prevent the world from relapsing into autarky and protectionism and not just only to avoid a recurrence of the Great Depression of the thirties.

The year 2004 marked the sixtieth anniversary of the Bretton Woods Conference held during 1–22 July 1944 in a small town in New Hampshire ushering the birth of the Bretton Woods twins—the IMF and the IBRD. The IMF came into existence on December 29, 1945 when 29 countries signed the Fund's Articles of Agreement—IMF's Charter. During these 60 years the international monetary system has undergone a fundamental change from a system of adjustable fixed foreign exchange rates to a diversity of arrangements ranging from floating to fixed exchange rates. This shift has been accompanied by another important structural change in the international economy involving the rapid expansion of private international financial markets. In both these facets of the international monetary system, the IMF has played an important role through the exercise of its surveillance function.

Purposes

According to Article I of the Fund's Articles of Agreement which entered into force in December 1945, the purposes of the International Monetary Fund are:

1. To promote international monetary cooperation through a permanent institution which provides the machinery for consultation and collaboration on international monetary problems.

2. To facilitate the expansion and balanced growth of international trade, and to contribute thereby to the promotion and maintenance of high levels of employment and real income and to the development of the productive resources of all members as primary objectives of economic policy.

3. To promote exchange stability, to maintain orderly exchange arrangements among members, and to avoid competitive exchange depreciation.

4. To assist in the establishment of a multilateral system of payments in respect of current transactions between members and in the elimination of foreign exchange restrictions which hamper the growth of world trade.

5. To give confidence to members by making the general resources of the Fund temporarily available to them under adequate safeguards, thus providing them with opportunity to correct maladjustments in their balance of payments without resorting to measures destructive of national or international prosperity.

6. In accordance with the above, to shorten the duration and lessen the degree of disequilibrium in the international balance of payments of members.

The Fund is guided in all its policies and decisions by the purposes set forth in this Article.

Structure

The Fund's organisational structure is mentioned in its Articles of Agreement. Article XII states that "the Fund shall have a Board of Governors, an Executive Board, a Managing Director, and a staff." The highest authority of the Fund is the Board of Governors, in which each of the 187 member countries is represented by a Governor and an Alternate Governor. The Fund's Governors, whose members are usually ministers of finance or heads of central banks in their countries, normally meet once a year and vote on major institutional decisions such as whether to increase the Fund's financial resources or admit new members.

The International Monetary and Financial Committee (IMFC) comprising 24 members is primarily responsible for the Fund's political oversight. The IMFC meets twice a year and advises the Fund on the broad direction of policies. A Development Committee which also has 24 members of the ministerial rank advises the Board of Governors of the IMF about the issues facing the developing member countries. It meets twice a year.

The head of the IMF is the Managing Director who is selected by the Executive Board of Directors which consists of 24 members of whom five members are appointed by the five largest quota holder countries—the United States of America, Japan, Germany, France and the United Kingdom. Apart from these five countries, three other countries—China, Russia and Saudi Arabia—have large enough quotas to elect their own Executive Directors. The rest 176 countries are organised into 16 constituencies, each of which elects an Executive Director. In 2005, the IMF had over 2,714 staff 45.9 p.c. drawn from the developed countries and 54.1 p.c. from the developing contries most of whom work at the IMF headquarters in Washington, D.C. The IMF staff is organised into departments with each department headed by a director who reports to the Managing Director. At present Christine Lagyarde is the Managing Director. She is the first lady Managing Director in the history of the IMF.

Membership and Quotas

The IMF has successfully completed sixty years and during this six-decade period its membership has become more than six-fold from mere 29 on the eve of its establishment in 1945 to 1984 (as on July 31, 2005). All those countries which agree to subscribe to Fund's Articles of Agreement are eligible for Fund's membership. As on April 30, 2012, the Fund had 187 members with total paid in quotas of SDRs 813.5 billion. The membership of the Fund ranges from the largest quota holder member in the United States with 17.5 per cent of the total Fund quota to Palau being the smallest quota holder with mere 0.001 per cent of the total Fund quota. Members' quotas in the Fund determine (1) their subscriptions of the Fund; (2) their drawing rights on the Fund under both regular and special facilities; (3) their voting power; and (4) their share of any allocation of the SDRs. A country's quota is broadly determined by its economic position relative to other members and takes into account the size of members' GDP, current account transactions, and official reserves. Quotas determine the members' capital subscriptions to the IMF and the limits on how much they can borrow. Quotas also help determine the members' voting power.

Every Fund member must subscribe to the Fund an amount equal to its quota. While before the Second Amendment of Fund's Articles of Agreement which became effective from April 1, 1978 usually 25 percent of the quota of a member had to be contributed in gold or foreign exchange, under the present arrangement that became effective from April 1, 1978 a member's quota that must be contributed in reserve assets may be less than 25 per cent. The remainder of the quota may be contributed in member's own currency. The proportion of a member's quota that must be contributed in reserve assets is related to the 'norm', a formula which has the effect, over time and with successive quota increases, of reducing to zero the part of a member's quota that must be subscribed in reserve assets.

The voting power of a member is determined by 250 'basic votes', plus one vote for each SDR 100,000 of quota. The U.S.A. with 13.08 p.c. voting rights is on the first place followed by Japan with 6.13 p.c. voting right. When the special drawing rights facility was established in 1969, it was decided that the amount of SDRs to be allocated over time would be determined after consideration of the demand and supply of international liquidity. Allocations of SDRs are in proportion to quota and are made only to participants in the Special Drawing Rights Department. In determining the quotas of members, several factors such as national income, reserves, export variability and the ratio of exports to national income are taken into consideration. Reviews of the quotas of Fund members are made at intervals of not more than five years to determine whether quotas should be increased looking to the growth of world economy and different rates of development among the members.

In the First and Second Quinquennial General Reviews of Quotas held in 1951 and 1956 the general increases in quotas were recommended. However, as a result of special review held in 1958–59 quotas were increased by 50 per cent. The introduction in 1963 of the compensatory financing scheme was followed by a 25 per cent general increase in quotas as a result of the Fourth Quinquennial Review in 1965. The Fifth Quinquennial General Review of Quotas in 1970 made a further 25 per cent general increase in the quotas of all members. In addition, there were special increases in the quotas of some members looking to their rapid development. The Sixth General Review was considerably affected by the far-reaching developments in the international monetary system, including the increase in oil prices in 1973. Consequently, it was decided to increase the total of quotas by 32.5 per cent as well as to double the quotas of the major exporting countries. The Sixth General Review came into effect in 1978 and by the end of 1979 the quotas of 140 Fund members stood at SDR 39.0165 billion. The Seventh General Review of Quotas under which the members' quotas increased by 50 per cent became effective in the latter part of 1980. It provided for a 50 per cent general quota increase for most members and additional increases for 11 members. As result of the Seventh General Review, total quotas of 157 Fund members had risen to SDR 91.2215 billion. An agreement was reached to increase Fund's quota from SDR 91.06 billion to SDR 97.4 billion under the Eighth General Review of Quotas and the increase in quotas had come into effect at the end of 1983.

During the financial year 1991–92, the IMF had made major progress towards universal membership with more than twenty countries—including all the fifteen states of the former USSR applying for Fund's membership. Consequently, Fund's membership has now swelled to 184 as a consequence of all the states of former USSR becoming its members. Moreover, as result of the Eleventh General Review of Quotas, which took effect on January 22, 1999, members' total quotas have risen to SDR 212.8 billion from SDR 145 billion. The Eleventh General Review of Quotas was the last review to result in an increase in IMF quotas. The Board of Governors concluded the Tenth Review without an increase in quotas. The distribution of increase in members' quotas under the Eleventh General Review of Quotas was equi-proportional and corrected the most important anomalies in the current quota distribution. The Twelfth General Review of Quota in January 2011 left the maximum size of quota unchanged at SDR 213.7 billion.

Use of Fund Resources

Fund members may draw on Fund's financial resources to meet their balance of payments needs. Under tranche policies, members may use the reserve tranche and the four credit tranches. In addition, members have access to three permanent facilities for specific purposes—the facility for compensatory financing of export fluctuations (established in 1963 and liberalised in 1975 and 1979); the buffer stock financing facility (established in 1969); the Extended Fund Facility (established in 1974) and the Structural Adjustment Facility (SAF) established in March 1986. The SAF provides additional concessional balance of payments assistance at 1/2 per cent interest rate repayable in 10 equal semi-annual instalments over 5-1/2–10 years to 62 low-income countries which are eligible for assistance under this scheme. Apart from these permanent borrowing facilities, members may also avail of the temporary facilities established by the Fund and borrowed resources. For example, following the sharp rise in oil prices, the Fund provided in 1974 and 1975 assistance under a temporary oil facility designed to help members meet the increased cost of the imports of petroleum and petroleum products. Since then, a supplementary financing facility has been established with borrowed resources amounting to SDRs 7.784 billion from 13 member countries and the Swiss National Bank. For use by low-income developing countries and as a supplement to Fund resources, there was the Trust Fund administered by the Fund and Trustee. The Trust Fund was designed to provide balance of payments assistance on very concessionary terms to 61 eligible developing member countries that qualify for assistance. Trust Fund loan carried only 1/2 per cent annual interest and were repayable in ten semi-annual instalments which started after five years from the beginning of the sixth year of the loan. Since July 1976 up to the end of April 1980, the Fund had disbursed a cumulative total of loans amounting SDRs 1.64 billion. The Trust Fund was terminated on April 30, 1981.

The Fund also administers a Subsidy Account to alleviate the high cost of interest payments on purchases under the 1975 oil facility by the most seriously affected developing countries. The most seriously affected (MSA) members, numbering 25, received total assistance from the Subsidy Account of SDR 59.7 million during the fiscal year ended April 1987 bringing the cumulative total of the assistance given to the members to SDR 366.8 million since the inception of the Supplementary Financing Facility Subsidy Account in December 1980 to reduce the cost of using the supplementary financing facility for the low-income developing member countries. The Subsidy Account was funded by contributions from 24 members and Switzerland. Its object is to reduce the effective rate of annual charge payable on drawings under the 1975 oil facility by about 5 percentage points per annum. India has received total assistance under the Subsidy Account Scheme of SDR 107.4 million. There is also the supplementary financing facility for members' use in support of programmes under the stand-by arrangements reaching into the upper credit tranche or beyond, or under the extended arrangements, subject to relevant policies on conditionality and phasing performance criteria.

The compensatory financing facility is available to members facing the balance of payments difficulties due to temporary export shortfall for reasons beyond the members' control. The buffer stock facility is available for building international buffer stock in order to prevent fluctuations in members' export earnings. The Extended Fund Facility (EFF) is designed to provide medium-term assistance of up to four years to overcome structural balance of payments maladjustments. Resources under the scheme are provided in the form of extended arrangements which include performance criteria and drawings in instalments. The scheme, as a result of liberalisation made in 1979, permits the repurchases to be made by a member within a maximum period of ten years. The lengthening of the maximum period of repayments from 8 to 10 years under this facility has been beneficial for the developing member countries seeking to correct the structural imbalances in their production, trade and prices.

When a member draws on the Fund, it uses its own currency to purchase the currencies of other member countries or SDRs held by the General Resources Account. Consequently, as a result of drawing, the Fund's holdings of the purchasing member's currency increase while there is a corresponding decrease in the Fund's holdings of the currencies or SDRs that are sold. A member must reverse the transaction by buying back its own currency with SDRs or currencies specified by the Fund within a prescribed time or earlier when its balance of payments and reserves position improves. Usually, repurchases should be made within 3 to 5 years after the date of purchase; but under the extended fund facility the period for repurchases has been extended to take place within 4 to 10 years; under the oil facility within 3 to 7 years; and under the supplementary financing facility within 3-1/2 to 7 years.

Under the Structural Adjustment Facility (SAF) and the Enhanced Structural Adjustment Facility (ESAF), the Fund provides financial support to low-income members on concessional terms. The Fund's special facilities consist of the Compensatory and Contingency Financing Facility (CCFF) and the Buffer Stock Financing Facility (BSFF). Under CCFF, the Fund provides resources to members to cover shortfalls in export earnings and services receipts and excesses in cereal imports that are temporary and arise from events beyond their control. Under BSFF, the Fund provides resources to help finance members' contributions to approved buffer stocks. The SAF and ESAF enable the Fund to provide resources to low-income countries on concessional terms to support their medium-term macroeconomic adjustment and structural reforms. If the Fund's holdings of a member's currency are less than its quota, the difference is called the *reserve tranche*.[1] Purchases in the reserve tranche are not subject to prior challenge and the facilities are almost automatic. Further purchases are made in the four *credit tranches* each of 25 per cent of the member's quota. Normally, the total purchases made under credit tranche policy are confined to an amount that raises the Fund's holdings of the member's currency to 200 per cent of its quota. The Fund may, however, in exceptional circumstances allow a member to purchase larger amounts. A member's indebtedness to the Fund, i.e., its obligation to repurchase begins when Fund holdings of its currency exceed 100 per cent of the member's quota. For making any drawing, a member must represent to the Fund that the desired purchase is needed because of its balance of payments or reserve position or developments in its reserves. All requests for the use of the Fund's resources, other than use of the reserve tranche, are examined by the Fund to determine whether the proposed use would be consistent with Fund's Articles and policies.

Under the credit tranche policy, use must be in support of the economic measures designed to overcome a member's balance of payments difficulties. The Fund applies charges for the use of its resources, except for reserve tranche purchases. A service charge of 1/2 per cent is payable on all purchases other than the reserve tranche purchases. Over the 4-year period (1977–80), the Fund had sold to 126 members 24.5 million ounces of gold at a price of SDR 35 per ounce in accordance with August 1975 agreement.

Special Drawing Rights (SDR)

The Special Drawing Rights or SDRs are an international reserve asset created by the IMF in 1969 to supplement the existing official reserves of member countries. The SDRs are allocated to member countries in proportion to their quotas. The SDR also serves as the unit of account of the IMF on April 30, 2005 one SDR was equal to 1.561678 & U.S. dollars or one U.S. dollar was equal to 0.659291 SDR. The composition of SDR was represented by 0.4260 Euro, 21 year, 0.0984 Pound-starling and 0.5770 U.S. dollar. Its value is based on a basket of key international currencies. The Fund allocated SDR 4,033.3 million on January 1, 1980 to 139 members who were participants in the Fund's SDR Department on December 31, 1979. A total of SDR 4,032.7 million was allocated to the 137 members on January 1, 1979. The third and last allocation of SDRs 4,052.5 million was made on January 1, 1981 to 141 members. Since the first allocation made in 1970, the total amount of SDRs allocated and issued has remained at SDR 21.4 billion as there have been made no fresh allocations since January 1, 1981.

---

1. Prior to the entry into force of the Second Amendment of the Articles of Agreement of the Fund on April 1, 1978, it was known as the *gold tranche*. Purchases made under this tranche by a member have been legally automatic and the Fund can challenge only *ex post* the member's representation of the balance of payments or reserves need.

The IMF members may use SDRs in transactions and operations among themselves, with 15 "prescribed institutions holders", and with the IMF itself. The SDR is IMF's unit of account. It is used as the unit of account or as a basis for the unit of account by several other institutions, regional organisations and international conventions. The value of the SDR is determined daily on the basis of a basket of five currencies—the US dollar, the German deutsche mark, the Euro, the Japanese yen, and the British pound-sterling. The value of the SDR tends to be more stable than that of any single currency in the basket since movement in the exchange rate of anyone component currency will tend to be partly or fully offset by movements in the exchange rates of the other currencies in the basket. The SDR valuation basket is revised every five years.

Since the last allocation of SDRs was made on January 1, 1981 when the total strength of IMF membership was 141, at present more than one-fifth (43) of IMF member countries have never received an SDR allocation because these countries joined the IMF after the last SDR allocation. Understandably, the new members who have been left out have pleaded that the IMF should make a special one-time allocation of SDRs to correct this unbalanced distribution.

Consequently, in September 1997, the IMF Board of Governors proposed an amendment to the Articles of Agreement to allow a special one-time allocation of SDRs to correct for the fact that more than one-fifth of the IMF members who joined after the last general allocation have never received an SDR allocation. The special allocation of SDRs would enable all members of the IMF to participate in the SDR system on an equitable basis and would double the cumulative SDR allocations to SDR 42.9 billion. The proposal will become effective when three-fifths of the IMF membership (111 members) representing 85 per cent of the total voting power have accepted the proposal. As of April 30, 2005, 131 members having 77.33 per cent of the total voting power had agreed, and only acceptance by the United States of America was required to implement the proposal.

Exchange Rate Policy

In accordance with the purposes of the Fund as expressed in Article I of the Articles of Agreement, members are under general obligation to collaborate with the Fund and with other members to assure orderly exchange arrangements and to promote a stable system of exchange rates. The New Article IV of the Second Amendment of the Articles requires each member to follow the exchange rate policies compatible with its commitment to

(*i*) "endeavour to direct its economic and financial policies toward the objective of fostering orderly economic growth with reasonable price stability, with due regard to its circumstances;

(*ii*) seek to promote stability by fostering orderly underlying economic and financial conditions and monetary system that does not tend to produce erratic disruptions; and

(*iii*) avoid manipulating exchange rates or the international monetary system in order to prevent effective balance of payments adjustment or to gain an unfair competitive advantage over other members."

To oversee the compliance by each member of these obligations and to assure the effective operation of the international monetary system, the Fund, under the Second Amendment, is required to exercise firm surveillance over the exchange rate policies of members and to adopt specific principles for the guidance of all members with respect to these policies. The three principles for the guidance of exchange rate policies are enunciated here as below.

"A. A member shall avoid manipulating exchange rates or the international monetary system in order to prevent effective balance of payments adjustment or to gain an unfair competitive advantage over other members.

B. A member should intervene in the exchange market if necessary to counter disorderly conditions which may be characterised *inter alia* by disruptive short-term movements in the exchange value of its currency.

C. Members should take into account in their intervention policies the interests of other members, including those of the countries in whose currencies they intervene."

Members are free to adopt the exchange arrangements of their choice, except for a prohibition against the maintenance of a currency in terms of gold. On the entry of the Second Amendment into force on April 1, 1978, members notified the Fund of the exchange arrangements they intended to apply in fulfilment of their obligations under Article IV. For the future, Article IV provides that a high majority of 85 per cent of the total voting power of the Fund will be required to take decisions recommending general exchange arrangements that accord with the development of the international system, including a system based upon stable but adjustable par values. Even then, however, the members would continue to have the right to choose their own exchange arrangements, and will at all times be subject to the general obligations and specified undertakings of Article IV.

Under the Second Amendment of the Fund's Articles of Agreement—which became effective on April 1, 1978—the SDR has become the principal reserve asset in the international monetary system. Under the new arrangement, the SDR is the unit of account for all purposes of the Fund and members of Fund have pegged their currencies to the SDR. When a member pegs its currency to the SDR, the value of its currency is fixed in terms of the SDR and is determined in terms of other members' currencies by reference to the SDR value of these currencies as calculated and published by the Fund. Since the first allocation made in 1970, the SDRs have been used in a wide variety of transactions with total transfers exceeding SDRs 50 billion up to the end of April 2004. Now SDRs can be used for a wide variety of transactions and operations among official holders, in accordance with rules established by the Fund. In order to make the SDR principal reserve asset in the international monetary system, during the three years 1979, 1980 and 1981 an amount of SDRs 12.1 billion had been created through fresh second lot of allocation to members raising the total allocations of SDRs to 21.4 billion.

Technical Assistance

The technical assistance provided by the IMF to its members constitutes an integral part of its activities. It takes many forms, operates at all levels of authority within member countries and covers a wide array of topics, ranging from broad policy issues to narrow technical problems. Much of the technical assistance provided by the Fund is in the nature of regular annual consultations with the members. These consultations provide an occasion for review and appraisal of the country's economic and financial situation and also an opportunity for the member to draw upon the expertise of the Fund's staff for advice and assistance. Similarly, when the Fund and a member country develop a financial stabilisation programme, the Fund, apart from providing financial assistance for the programme drawn up by a member, assists in its execution and in monitoring its effectiveness. The Fund also receives requests for technical assistance on members' specific economic and financial problems encompassing aspects of general economic policy, problems arising from inflation, exchange and trade systems, balance of payments programmes, debt management systems, accounting and valuation problems, macroeconomic modelling, computer programming for economic analysis and data processing.

The IMF provides technical assistance and training to its members in four broad areas of (*i*) designing and implementing fiscal and monetary policies; (*ii*) institution building (like the development of central banks, treasuries, tax and customs administration); (*iii*) collecting and refining the statistical data; and (*iv*) drafting and reviewing the financial legislation. This technical assistance is provided by the IMF through its Monetary and Exchange Affairs Department (MEA), the Fiscal Affairs Department, the Statistics Department, the IMF Institute, the Legal Department and the Treasurer's Department. A key feature of IMF's technical assistance and training over the past few years has been its application in countries emerging from crises of different types. Such technical assistance programmes have been implemented in the post-crises countries such as Albania, Angola, Cambodia, Haiti, Lebanon, Malawi, Namibia, Rwanda and Yemen.

In 1964, the Fund had organised its technical assistance activities in the training field by establishing the IMF Institute which conducts courses of officials from member countries in Washington. These courses cover a broad spectrum of subjects in financial analysis and policy making, as well as such technical subjects as public finance, government finance, statistics and balance of payments methodology. Since its establishment in 1964, the Institute has trained more than 25,000 officials in Washington from more than 160 member countries. The IMF Institute trains officials from member countries through courses and seminars in the core areas of macroeconomic policy management and financial sector, fiscal and external sector policies. During 2004, the Institute, with the assistance of other departments, offered 120 courses attended by 3,846 participants. Much of the training was provided through IMF's four regional centres located in Brazil, Austria, Singapore and Tunisia. Function-wise during 2004, 26% of the technical assistance was related to fiscal affairs, 33% to monetary and financial systems, 16% to statistics, 15% to IMF Institute, 6% to legal affairs and 4% to others. Since 1999 upto 2004, the IMF Institute had conducted 673 courses and seminars in which the total number of 20,589 officials participated. During FY 2005 the IMF conducted 124 courses and seminars in which 3904 officials participated.

In order to provide the members technical assistance in the field of central banking, the Central Banking Service was started in 1963 and since then more than 150 members or their dependent territories and 20 multinational organisations have received technical assistance under the Fund's central banking programme. The Fund also assigns experts under the central banking programme to serve in the executive and advisory capacities. Since 1963, over 1,500 panel experts have been made available to the members under the central banking technical assistance programme.

The Fiscal Affairs Department was established in 1964 to provide the members technical assistance for providing policy advice on tax and customs administration, public expenditure management and budgeting, tax policy issues, pension reform and social safety net design, and public expenditure reviews. Upto 2004 more than 160 member countries and 15 inter-governmental organisations were provided technical assistance in the field of public finance. Advice has been given on the reform of countries' tax structures, specific forms of taxation, tax rate structures, investment incentive codes and organisation of fiscal research units. Advice has also been given on the reorganisation of tax departments, on improving assessment and collection procedures and on implementation of new taxes. Apart from this, short-term assignments are carried out by Fund staff members and experts.

Fund's Legal Department has been rendering technical assistance in banking, central banking, currency, exchange and negotiable instruments in close association with the Central Banking Service. The department has also been rendering technical assistance in fiscal affairs in close association with the Fiscal Affairs Department, involving primarily the drafting of legislation on individual and corporation income taxes and on direct taxes, capital gains taxes and land taxes. The Central Bank Bulletin Project, started in 1969, has assisted more than 110 countries and 10 inter-governmental organisations up to 2004 leading to the establishment of 40 new central bank bulletins.

The regular programme of Fund publications includes the publication of *Annual Report of the Executive Directors* with shortened versions in French, German and Spanish; *Balance of Payments Year Book, Annual Report on Exchange Restrictions; International Financial News Survey* (weekly); *International Financial Statistics* (monthly); *Schedule of Par Values; Proceedings of Annual Meetings of Governors* and *Staff Papers.* Besides these regular publications, in 1964 was published a collection of speeches of Per Jacobson under the title *International Monetary Problems, 1957–63.* During the fiscal year ended April 30, 1969 were published the first two volumes in new series of books of African economies and three new pamphlets. The Fund also publishes a quarterly periodical titled *Finance and Development* in conjunction with the World Bank in English, Spanish and French. In September 1966, was published the second report on the compensatory financing of fluctuations in exports of primary producing countries titled *Compensatory Financing of Export Fluctuations, Development in Fund's Facility* in English, French and Spanish. The first report on this subject was published in 1963. In April 1967, the Fund published *Central Banking Legislation; A Collection of Central Bank, Monetary and Banking Laws: Volume II, Europe,* selected and annotated by Mr. Hans Aufrichst of Fund's Legal Department. The first volume was published in 1961. In April 1977, the Fund published *The International Monetary Fund, 1966–1971: The System Under Stress* in two volumes authored by Margaret G. de Vries. The Fund's Bureau of Computing Services received and trained officials from 40 central banks, regional and international organisations during 1981–2004.

Beginning in 1989, the IMF formally coordinated its technical assistance policies and cooperated with other multilateral and bilateral agencies to minimise the conflicting advice and redundant activities. The IMF also started exploring ways of complementing its own resources through various financing arrangements with other technical assistance supplies. This cooperation has led to the implementation of comprehensive technical assistance programmes with the United Nations Development Programme (UNDP), the World Bank and the European Union.

In recent years the IMF has concluded general technical assistance agreements with the United Nations Development Programme (UNDP) and the Japanese government and several individual country agreements with the World Bank and the European Union. Several bilateral contributors have also supported the IMF-administered technical assistance by making cash contributions to UNDP-IMF projects. To accommodate the growing interest of other potential contributors in supporting the IMF technical assistance, the Executive Board of Directors has established a technical assistance Framework Account in April 1995 under which separate sub-accounts are created for individual contributors to support the IMF technical assistance activities. These sub-accounts can be established easily and quickly and can be used to finance a wide variety of short or long-term technical assistance and training activities which a contributor wishes to support through the IMF.

Liquidity

The liquid resources of the IMF consist of the usable resources and SDRs held in its General Resources Account. Usable currencies which constitute the largest component of Fund's liquid resources, are the currencies of those members whose balance of payments and reserve positions are sufficiently strong to warrant the use of their currencies in financing the IMF operations and transactions. At the end of April 2003, the IMF's liquidity remained adequate to meet the needs of its members. The one year forward commitment capacity (FCC) amounted to SDR 61 billion at the end of financial year 2003. The FCC—a new measure of liquidity introduced in FY 2003—indicates the amount of quota-based resources available for lending over the next 12 months. It is the primary measure of IMF liquidity. It is calculated by adding loan repayments scheduled in the coming year to available resources and then subtracting a prudential balance (set at SDR 32.8 billion for 2004). At the end of FY 2004, the IMF liquidity (one year forward commitment capacity) was SDR 62.1 billion ($ 91 billion) up by SDR 1.1 billion over FY 2003. To meet the requirement of additional liquidity, the IMF has access to the two facilities of General Arrangements to Borrow (GAB) and the New Arrangements to Borrow (NAB) under which it can borrow upto SDR 34 billion. The NAB which became effective in November 1998 was renewed for a further period of five years from November 2003. Like the NAB are a set of credit arrangements under which 26 participants have agreed to provide resources to the IMF to forestall or cope with an impairment of the international monetary system or to deal with an exceptional situation which threatens the stability of that system.

The GAB which have been in existence since 1962, are a set of credit arrangements under which 11 participants (industrial countries or their central banks) have agreed to provide resources to the IMF to forestall or cope with an impairment of the international monetary system. The potential amount of credit available to the IMF under the GAB totals SDR 18.5 billion (including additional SDR 1.5 billion available under an associated agreement with Saudi Arabia). The GAB have been activated ten times, the latest being in July 1998 for SDR 6.3 billion. The GAB decision has been renewed nine times, most recently in November 2002 when the IMF Executive Board approved its renewal for a further period of five years from December 2003.

Activities of Fund

In the early years of the working of the Fund, it was feared that the resources of the Fund would be inadequate to meet the heavy postwar deficits in the balance of payments of the members. Most members faced acute dollar shortage and the mounting deficits in their external balance of payments. With its limited financial resources, the Fund could not assist the members to secure dollars to remove the deficits in their balance of payments. Consequently, these countries could either stop their imports from the USA in which case a major political upheaval could be expected in many of these countries, or make separate settlement with the USA outside the Fund framework so that their balance of payments deficits would be covered under some aid or loan from the USA. It was the second alternative which was followed after 1948 when Marshall Aid helped many West European countries. The Fund can help members to secure foreign currencies only in limited amount. If many countries simultaneously demand the currency of one country and if this demand is very large the Fund cannot do anything in the matter.

External Debt Relief Initiative to Assist Low-income Countries

The IMF and the World Bank launched a joint initiative in providing external debt relief for the heavily indebted poor countries (HIPCs). The joint IMF-World Bank initiative for heavily indebted poor countries enables countries that have established good policy track record to achieve substantiality in their overall external debt burdens with the participation of bilateral and multilateral creditors along with the IMF and World Bank. Thus, the debt reduction assistance under the initiative permits eligible countries to exit from the debt-rescheduling process and to focus their energies on striving for sustainable growth and development.

Under the debt initiative, in April 1997 Uganda was the first country to receive a total of $ 340 million in debt relief, in net present value terms (a reduction of 20 per cent) assuming the participation of Uganda's all creditors and its implementation of sound reforms. The IMF's share of this assistance was of the order of $ 70 million. Bolivia had also been considered eligible for exceptional assistance to reduce its debt burden to a sustainable level. As a result of the assistance provided by Bolivia's all creditors (bilateral and multilateral) and the IMF's share of $ 29 million Bolivia's external debt has been reduced by $ 448 million in net present value. The IMF's exceptional assistance of $ 29 million approved on September 10, 1997 was provided at the completion point (after the second stage of adjustment and reform) in the form of a grant into an escrow account to be used exclusive to pay debt service to the IMF. Under the HIPCs external debt relief joint initiative, Burkina Faso and Cote d'Ivoire had also been considered eligible owing to their commitment to sound policies. In April 1998, Uganda reached the completion point under the HIPC Initiative and SDR 51.5 million was disbursed in the form of a grant. During 1998–99, it was decided to commit assistance to seven countries that had reached their "decision points" under the HIP Initiative. Two of these countries—Uganda and Bolivia—had reached their 'completion points' under the Initiative by April 30, 1999 and had received assistance from the IMF in the form of grants.

In 1999, the IMF and the World Bank announced the (*i*) introduction of Poverty Reduction Strategy Papers (PRSPs) prepared by each borrowing country to provide the basis for concessional lending; and (*ii*) enhancement of the debt reduction programme under the HIPC Initiative— introduced in 1996 to boost their support for the low-income countries. The IMF encourages low-income countries to use the Poverty Reduction Strategy Paper process to set out realistic plans to achieve the Millennium Development Goals (MDGs) by strengthening domestic policies and securing additional external financing. The MDGs seek, by 2015, to halve the extreme poverty relative to 1990; promote gender equality; reduce child mortality; achieve universal primary education; combat HIV/AIDs, malaria and other diseases; ensure environmental sustainability; and establish a global partnership for development.

Following a comprehensive review undertaken in September 1999, several enhancements were approved to provide faster, deeper and broader debt relief and to strengthen the links between debt relief, poverty reduction and social policies. Under recent programmes, the 27 countries receiving HIPC relief have increased their outlays on health care, education and other social services to almost four times the amount of debt service payments. At the end of April 2004, about $52 billion had been committed in debt service relief to 27 countries which had met the criteria to start receiving aid. The debt stocks of these countries are likely to decline by about two-thirds as a result. The cost to the IMF of this debt relief initiative will be $5 billion.

In Fy 2011 the fund committed loan amounting to SDR 1.1 billion to its low-income member countries under the Poverty Reduction and Growth Trust (PRGT). The total outstanding concessional loans amounted to SDR 4.9 million at April 2011. The fund provides debt relief to poor countries under the Heavily Indebted Poor Countries (HIPC) initiative and the Multilateral Debt Relief Instiative (MDRI) In total the IMF has provided debt relief of SDR 1.5 billion under the HIPC Instiative and SDR 2.3 billion under the MDCS in FY 2011.

Financial Operations

The IMF uses its financial resources to help members redress their balance of payments problems and to help cushion the impact of adjustment. The IMF's financing is provided through both its general resources and its concessional financing facilities which are administered separately. The IMF financing is subject to Executive Board's approval and, in most cases, to the member's commitment to address the causes of its balance of payments imbalances — members using IMF's general resources "purchase" or draw other members' currencies or SDRs with an equivalent amount of their own currencies. The IMF levies charges on these drawings and requires that members "repurchase" their own currency from the IMF (or repay) with other members' currencies or SDRs within a specified time.

Regular Facilities

Reserve Tranche: The regular borrowing facilities to members are provided by the IMF under Reserve Tranche, Stand-by Arrangements and the Extended Fund Facility. A member has a reserve tranche position if the IMF's holdings of its currency in the General Resources Account—excluding those holdings that reflect the member's use of IMF resources—are less than its quota. A member may draw upto the full amount of its reserve tranche position at any time, subject only to the member's representation of a balance of payments need. A drawing from the reserve tranche does not constitute a use of IMF credit and is not subject to charges or to an expectation or obligation to repay.

Credit Tranches: Apart from the reserve tranche drawing which is almost automatic and is not subject to any condition, a drawing made by a member from credit tranches is subject to certain conditionality requirements depending on whether it is made available in the first credit "tranche" (segment) of 25 per cent of a member's quota or in the upper credit tranches (any segment above 25 per cent of quota). For drawings in the first credit tranche, members must demonstrate reasonable efforts to overcome their balance of payments problems. Drawings in the upper credit tranches are made in instalments, or phased manner, and are released when performance targets are met. Such drawings are normally associated with Stand-by or Extended Arrangements which are intended to resolve the balance of payments difficulties and to support structural policy reforms where appropriate. Performance criterion and periodic reviews are used to assess the policy implementation.

Stand-by Arrangements: Under a stand-by arrangement, introduced in 1952, a member has a right to draw upto a specified amount of IMF financing during a prescribed period. This facility is intended to meet members' balance of payments difficulties that are of short-term nature. Drawings are normally made in a phased manner on a quarterly basis, with their release being conditional or meeting performance criteria and the completion of periodic reviews. Performance criteria generally cover bank credit, government or public sector borrowing, trade and payments restrictions, foreign borrowing and international reserve levels. These criteria allow both the member and the IMF to assess progress and may signal the need for further corrective policies. Stand-by Arrangement typically cover a 12–18 months period, although these can be extended upto three years. Repayments are to be made within 3-1/4 to 5 years of each drawing.

In FY 2003, a stand-by arrangement amounting to SDR 22.8 billion ($31.5 billion), being the largest arrangement in IMF history, dominated new IMF lending commitments to its members. This arrangement along with other large arrangements for Colombia and Argentina and the augmentation of an existing arrangement for Uruguay kept the commitments in FY 2003 relatively high. New commitments totalled SDR 29.4 billion. In FY 2004, the IMF's commitments stood at SDR 14.5 billion for five new stand-by arrangements to seven countries which included Argentina, Brazil, Dominion Republic, Guatemala, Paraguay and Ukraine. At the end of FY 2004, thirteen stand-by and extended arrangements were in effect. At the end of April 2004, undrawn balances under the arrangements still in effect amounted SDR 19.8 billion. During FY 2005 six standly aggreements were approved for 1.8 billion SDR. In FY-2011 six stand by agreements were approved for SDR 36.6 billion.

Extended Fund Facility (EFF): Under the EFF, the IMF provides assistance to its members for adjustment programmes over longer periods and generally for larger amounts of financing than is provided under the Stand-by Arrangements facility. The Extended Arrangements which normally are for three years, extendable for a fourth year, are meant to rectify members' balance of payments difficulties which stem largely from structural problems and require a longer period of adjustment. Member countries using EFF resources must repay the currencies drawn by them, within 4-1/2 to 10 years of the drawing. The normal access is 100 per cent of member's quota annually and can go upto 300 per cent of quota on a cumulative basis. In 2003, two EFF agreements were approved for 794 million SDR but no approval was made during 2004 and 2005.

Under the Extended fund facility two agreement mere approved for SDR 19.5 billion in the year 2010.

Flexible Credit Line: This facility was created in 2009. During FY 2011four agreements were approved for SDR 2.4 billion

Precanutianary Credit Line: This facility was created in the year 2010. In the year 2011 Macedonia-forner Yugoslav Republic was approved a credit of SDR 413.4 million.

Special Facilities

The special facilities of the IMF include the Complementary and Contingency Finance Facility (CCFF) and the Buffer Stock Financing Facility which has not been in use since 1983.

Compensatory and Contingency Financing Facility (CCFF): The export compensatory element of the CCFF provides financing to those members which experience a balance of payments need related to temporary shortfalls in their export earnings. This element of the facility has been used particularly by commodity exporters. The cereal element compensates for the temporary excesses in cereal import costs attributable to factors largely beyond the members' control. The contingency element helps members with IMF arrangements keep their adjustment programmes on track when faced with unforeseen adverse external shocks largely beyond their control. During 1996–97, Algeria and Bulgaria made use of the CCFF with drawings amounting to SDR 0.3 billion.

Buffer Stocks Financing Facility: Under this facility, the IMF helps in financing members' contributions to approved international buffer stocks if the member can demonstrate a balance of payments need. For the past 13 years, however, no drawings have been made under this facility.

During the financial year ended April 2003, the IMF disbursed SDR 21.8 billion in loans to members from its General Resources Account. As this total exceeded loan repayments of SDR 7.8 billion, IMF's credit outstanding amounted to a record high of SDR 66 billion.

Concessional Facilities

IMF's concessional assistance is provided under the PRGF and the HIPC Initiative. During the FY 2003, ten new PRGF arrangements totalling SDR 1.2 billion were approved and the same amount was also disbursed. The adjustment and reform efforts of 36 low-income countries involving SDR 4.5 billion were supported by PRGF arrangements. By the end of FY 2003, eight countries had reached their completion points under the enhanced HIPC Initiative and another 18 had passed their decision points and had started to receive interim relief. In FY 2004, 27 HIPC members had reached their decision points while 13 countries reached their completion points. During FY 2005, 31 countries received disbursements under the RPGF while 27 countries had received financial commitments under the enhanced HIPC Initiative.

The IMF also provides emergency loan assistance to countries emerging from conflict. In FY 2003, seven donor countries had pledged SDR 11.5 million in subsidies for such loans and disbursements to seven affected countries totalled SDR 1.4 million. In September 2004, the IMF approved SDR 300 million (S436 million) post-conflict assistance to Iraq in support for her reconstruction efforts. During 2005 six countries made purchases under emergency assistance. Three purchases were made under ENDA (Emergency Natural Disaster Assistancy) and another three were made under EPCA (Emergency Past-conflict Assistance) by PRGF eligible countries.

In the year 2010 three new concessional credit facilities–

(i) *Extended Credit Facility*

(ii) *Standly Credit Facility* and

(iii) Rapid credit facility, were created for low-income countries under the Poverty Reduction and Growth Trust. Under these facilities a total amount of SDR 113.3 million was approved during FY 2011.

Independent Evaluation Office (IEO)

The Independent Evaluation Office (IEO) was set up in July 2001 to conduct an objective and independent assessment of issues related to the IMF mandate. The IEO has produced a series of detailed reports on aspects of the IMF's work. These reports are used to evaluate how the IMF does its job and to help formulate desirable changes in policies and practices. Studies completed during 2002–04 evaluated IMF's prolonged use of its resources; evaluated IMF's role in recent capital account crises in Korea, Indonesia and Brazil; the role of IMF in Argentina during the crisis of 2000–02; and the effectiveness of the Poverty Reduction Strategy Paper (PRSP) process and the Poverty Reduction and Growth Facility (PRGF). During 2004–05, the IEO's evaluation efforts will include Financial Sector Assessment programme and Financial Sector stability assessments.

Evaluation

Notwithstanding that the IMF has not fully succeeded in achieving its objectives due to various limitations from which all international institutions based on international cooperation suffer, the IMF can be proud of solving many problems of its motley group of members. It has strengthened the monetary discipline among its members by timely assisting them to tide over their balance of payments deficits. It has helped members in technical matters relating to fiscal and monetary policies, debt servicing, balance of payments, currency devaluation, etc. The importance of the Fund in enforcing the code of monetary discipline among its members had been frankly expressed by Mr. Per Jacobson in the following words:

"If we then examine how the Fund has actually operated, I think it is fair to say that the Fund has increasingly been a factor not for weakening, but rather for strengthening monetary discipline, been able to grant assistance only when countries make reasonable efforts to solve their problems and present programmes, holding out the hope of enduring stability at realistic rates of exchange. Indeed, the mere possibility of having access to the Fund's resources, even if no drawing is actually made, may make countries more confident in their attempts to restore balance, and may indeed induce them to take stricter measures than if they had to rely on their own resources alone".

However, for making a detached evaluation of Fund's working it is necessary to mention the concrete cases of its shortcomings. Fund's working during the past 57 years since 1947 shows that all has not been well with it. It has met only with limited success and members have violated Fund's rules regarding the alteration in the par values of their currencies. Nor had the countries whose currencies were in short supply taken any material steps to rectify it. Consequently, in January 1948, France devalued her currency by 44 per cent in the face of opposition from the Fund. In addition, a free market in gold, US dollars and Portuguese currency was created in Paris giving emergence to perfectly legal free rates of exchange alongside the official exchange rates fixed by the Fund. Similarly, the dollar was not declared 'scarce' currency in 1949 despite the fact that dollar was scarce throughout the world. Had the Fund declared the US dollar a scarce currency, the USA would have been forced to revalue the dollar mitigating the dollar shortage. Rather than declare the dollar as a scarce currency, the Fund advised England to devalue her currency.

Although the lessening of exchange restrictions, simplification of exchange control, the current account currency convertibility by an increasing number of member countries and fall in the number of bilateral payments are not encouraging developments but it cannot be denied that progress toward the establishment of multilateral system of trading and payments in future will depend upon the extent to which the problems associated with the growth of the developing countries are solved. Even today the developing countries face the problem of instability of export prices giving rise to fluctuations in their export earnings on the one side, and excess imports far beyond the available foreign exchange resources occasioned by their heavy capital outlays and development on the other. The success of the Fund toward the establishment of multilateral system of trading will depend upon the degree to which the underdeveloped countries can implement their development plans within the framework of financial stability free from the distorting effects of inflation.

The performance of the International Monetary Fund has on the whole been satisfactory. It has been able to solve effectively many problems. The Fund has successfully exercised surveillance over the exchange rate policies of its members. Following the Mexican financial crisis, the Fund has strengthened its surveillance operations in the areas of banking soundness, transparency and good governance. The Fund has rendered useful service to the members whose number today stands at 184 as against only 29 in 1945 at the time of its establishment. This shows that the Fund has now become a real international financial institution. This, however, presents a challenge to the Fund since the growing membership of the countries will subject the limited resources of the Fund to growing pressure of demand rendering the problem of international liquidity more acute. The establishment and expansion of the African Department and the institution of the two new services in the areas of central banking and fiscal affairs are matters for which the developing countries will remain ever grateful to the Fund. The Fund has diversified its training programmes and conducts courses in the French language as a special service to many of its new members in Africa. The creation of the new facilities for compensatory financing of export fluctuations, buffer stocks, the extended fund facility, and oil purchase financing have proved a much-sought-after real boon for many developing countries which export primary products and import petroleum and petroleum products. The Special Drawing Rights scheme is an important step taken to solve the problem of international liquidity for its members. The contribution of the Fund in solving the problem of international liquidity has been no less significant. The Fund has offered increased unconditional and conditional liquidity to its members to help them tide over the temporary deficit in their international balance of payments.

India and IMF

India is a founder member of the International Monetary Fund in 2005 India's quota amounted to 4.158 Billion SDR much less than Saudi Arabia (6.98b.SDR) and China (6.369b.SDR). Consequently the voting power of India (1.33 p.c.) is also more less than Saudi Arabic (3.22 p.c.) and China (2.94 p.c.) and has played an important role in the formulation of Fund policies. India has stood for liberalisation of Fund's lending policies and has criticised Fund's scale of charges in respect of drawings.

The initial par value of the Indian rupee, established with the Fund on December 18, 1949, was 0.2086 gramme of fine gold or 30.2250 US cents and became effective on that data. On September 18,1949 the par value of the rupee was changed to 0.186621 gramme of fine gold or 21.0000 US cents and became effective on September 22, 1949. With the devaluation of the rupee on June 6, 1966 the par value of the rupee had been reduced to 0.118489 gramme of fine gold or 13.33 US cents. After the devaluation of the US dollar this rate had fallen and the exchange rate had been kept floating.

To meet the balance of payments deficit, India purchased from the Fund $ 100 million during 1948 and 1949. By 1956–57, the entire purchase obligations were liquidated and in February 1957 India entered into another stand-by agreement with the Fund for drawing $200 million to tide over her temporary balance of payments difficulties. This amount had been drawn in three instalments between February and June 1957. Fund's attitude towards India has been sympathetic to her problems. In the case of India, the Fund waived the usual condition that the purchases by a member of other currencies from the Fund during any one year should not exceed 25 per cent of that country's quota on the sound plea of food scarcity and natural calamities that India had faced. In July 1961, the Fund permitted India to draw $250 million in six currencies comprising $110 million in US dollars, $65 million in pound-sterlings, $45 million in German marks, $15 million in French francs, $10 million in Italian liras and $5 million in Japanese yens. India was the first country to draw Japanese yens from the IMF.

The financial assistance provided to India by the Fund has been in the form of a continuing flow. On July 9, 1962 India entered into one year stand-by agreement with the Fund for $100 million. On the expiry of this agreement, it was replaced by a fresh stand-by agreement of the same amount. This agreement expired in July 1964. On March 19, 1974 the Fund extended the 'timely assistance' to India by granting a stand-by credit of $200 million to tide over her foreign exchange crisis. India drew $100 million worth of foreign currencies under this stand-by credit during 1964–65. The balance of $100 million India could draw up to March 19, 1966. During 1965–66, India purchased from the Fund $75 million in foreign currencies under the stand-by arrangement and $187.50 million by direct purchases. In December 1967 and February 1974, India purchased from the Fund $90 million and SDR 62 million under the Decision on Compensatory Financing of Export Fluctuations to meet her balance of payments difficulties caused by temporary shortfall in her export earnings. India had been initially allocated $126 million from the Fund under Special Drawing Rights for development purposes. The Fund asked India in July 1970 to take on some obligation under the SDR scheme by accepting up to $14 million from other participants of the SDR scheme and provide convertible currency in return. During 1973–74, India used SDR 310,000 from the General Resources Account. In the third allocation made on January 1, 1970 India had received SDR 99.60 million. India received allocations of SDR 119.08 million on January 1, 1980 and SDR 116.79 million on January 1, 1981 under the scheme which allocated SDR 12.1 billion to Fund members in 1979, 1980 and 1981.

Under oil facility, India purchased a total of SDR 401.34 million from the Fund in 1974–75 and 1975–76. Apart from this, she also received up to April 30, 1982 a sum of SDR 26.95 million from the Subsidy Account established to assist the Fund's most seriously affected 31 members to meet the cost of using resources made available under the oil facility of 1975. Moreover, India is one of the 61 developing member countries of the Fund which are eligible to make drawings from the Trust Fund established on May 5, 1976 to finance their balance of payments deficit at a very nominal cost of 0.5 per cent rate of interest. India had borrowed SDR 529 million from the Trust Fund. The Fund also distributed in sales at price equivalent to SDR 35 per ounce 804,429 ounces of gold in accordance with the Interim Committee's agreement in August 1975 to restitute 25 million ounces of gold to 126 countries that were members of the Fund on August 31, 1975. India also received from the Fund SDR 149 million as her share of profits accrued from the gold sales made to 104 developing member countries that were eligible to participate in the direct distribution of profits. India had also received from IMF a massive credit assistance of SDR 5 billion under extended arrangement in November 1981 to enable her to make structural adjustments in her economy to achieve balance of payments viability in the medium term. The amount of SDR 5 billion was 291 per cent of India's quota of SDR 1,717.5 million.

IMF has come to India's rescue in tiding over her balance of payments deficit. India made substantial drawals from the IMF from 1990–91 onwards under different financing facilities. Besides the Reserve tranche drawings of SDR 487 million, India drew SDR 1,269 million in 1990–91. In 1991–92 the drawal was SDR 905 million. India also made a 20-month stand-by arrangement with the Fund in October 1991 for a total loan of SDR 1,656 million; two instalments worth SDR 270 million were drawn in 1991–92. During April 1992–May 1993, SDR 1,386 million had been drawn. The share of IMF in total external debt of India was 2.3 percent at the end of March 2010 which came down to 2.1 percent in 2011. However, the contribution of the IMF in the form of SDR allocation rose from SOR 6041 million in 2010 to $6308 million in 2011.

IMF on India's Economic Reforms

The Fund has guided India in many ways. It has helped India in sustaining her economic growth through undertaking economic reforms. However, according to the IMF, India needs more comprehensive reforms to sustain her economic performance. It has been stated by the IMF that "while significant progress has been achieved in reforms since 1991-92—such as in trade and exchange liberalisation, reduction of tax rates and industrial delicensing—a wide range of reforms remain to be implemented." Recent tax cuts should be followed by measures to expand the tax base, to cut subsidies and increase efficiency in the Central and State government-run enterprises, including an increase in privatisation efforts. Deeper trade reforms, including lowering average trade tariff rates to global levels, are required for a strong medium-term export performance.

The IMF's reckoning India can achieve the tiger status and reduce poverty substantially if she continues her process of reforms. As a result of initiating economic reforms process in the early 1990s, the Indian economy has been undergoing a profound transformation and the results are evident in stronger economic performance. During the past three years economic growth has averaged around robust six per cent with inflation contained to a single digit and a sound external balance of payments position. The foreign exchange reserves stand at hefty figure of over $120 billion and there is no problem of liquidity in the economy. This should not, however, make India complacent and she should strive to maintain and further augment the economic growth. A second further round of economic reforms in India is necessary in order to achieve Asian Tiger status and to make major inroads on poverty. Fiscal adjustment should be broadened and extended to include states and public enterprises. Additionally, bold structural reforms encompassing further trade reforms, addressing infrastructure bottlenecks and accelerating efforts to strengthen the financial system are needed. In March 2005 Managing Director of IMF Mr. Rodrigo de Rato made a goodwill visit to India.

It is obvious that but for the timely adequate financial aid provided by the IMF India would have found herself in jeopardy and financial bankruptcy. The country is sincerely grateful to the Fund for coming to her timely rescue in her period of crisis.

Suggested Readings

1. IMF, Annual Reports and Staff Papers.
2. IMF and World Bank, *Finance and Development,* Volumes 1–42.
3. Bo Sodersten, *International Economics,* Second Edition, 1980, Chapters 30–31

Questions

1. Explain the purposes of the International Monetary Fund. How far has the International Monetary Fund been successful in achieving these purposes? Discuss fully.
2. Discuss the role of the International Monetary Fund in solving the problem of international liquidity.
3. Critically appraise the assistance provided by the IMF to India.

CHAPTER 32

International Liquidity

During the past six decades, world's leading countries and international financial institutions have been increasingly concerned about the worsening situation of international liquidity which has assumed a serious form with the passage of time. It is so because international liquidity, which serves as a means of international payments, has failed to keep pace with the expansion in the volume of world trade. This failure of the international monetary reserves to expand *pari passu* with the expansion of world trade is reflected in the fall of total reserves as a percentage of total imports. The ratio of official monetary reserves to imports which for the world as a whole was 67 per cent in 1951 had fallen to 33 per cent in 1968. While the world official monetary reserves had increased from 1952 to 1968, albeit at a slower rate in the later years, they showed a continuous and substantial decline throughout the greater part of this period in relation to the magnitude of the world trade in value terms. The inadequacy of official monetary reserves necessitated their augmentation by establishing the Special Drawing Rights Scheme in early 1970 due to which the official global monetary reserves initially increased by SDR 9.3 billion. However, in the face of the growing external balance of payments imbalances and continuing expansion in the volume of trade, the global official monetary reserves proved inadequate. This inadequacy was evident by the unduly heavy pressures to which the international monetary system was subjected. For a sample of 60 countries for which data have been continuously available since 1954, the ratio of aggregate reserves to aggregate imports, which had interrupted its decline of many years by rising from 29 per cent in 1970 to 38 per cent in 1972, fell to 34 per cent in 1973 and 24 per cent in 1974 thereby reaching the lowest level ever recorded. For the world as a whole, however, this ratio which stood at 37 per cent in 1966 fell to 25 per cent in 1976 and further to 24 per cent in 1977. In fact, the problem of shortage of international liquidity became so serious that in the seventies the international monetary system came to the brink of collapse. To prevent it from collapsing, concerted efforts were made in the form of short-term measures to ease the problem of international liquidity while the Committee of Twenty was appointed to suggest the long-term measures for the reform of the present international monetary system.

There is, however, nothing new in the world community's serious concern about the shortage of international liquidity; in fact, it existed as far back as the end of World War I. One of the basic reasons for the breakdown of the prewar international gold standard was the inability of the leading gold standard countries to solve the problem of shortage of international liquidity—inadequacy of official monetary gold reserves with the debtor and those countries which experienced substantial deficits in their external balance of payments combined with the reluctance to lend on the part of the countries having surplus in their external balance of payments. The nature and magnitude of this problem have, however, substantially changed since then. The problem of international liquidity at present is of a larger magnitude and of a more permanent nature arising from large chronic deficits in the external balance of payments of many countries, including many developing countries, particularly the non-oil developing countries. The problem of international liquidity is partly of the absolute shortage—its total amount available to support the means of international payments being inadequate for financing the present volume of world imports—and partly of the uneven distribution with the bulk of world official monetary reserves being concentrated in the industrially developed countries. Of the total official holdings of reserve assets, including gold amounting SDRs 2,553 billion in 2004, the industrial countries alone possessed SDRs 1,126.3 billion. According to underdeveloped countries, the existing imbalance in the distribution of international liquidity has been accentuated by an agreement providing for gold revaluation and domination of reserve currencies in the use of international liquidity. The Special Drawing Rights which have been accepted as the principal reserve asset have not been strengthened in the past twenty three years after 1981. The SDR is a reserve asset created by the IMF in 1969 to supplement the existing reserve assets and allocated to members in proportion to their IMF quotas. As a result of special one time allocation proposal becoming effective, the total SDR allocations will be doubled to SDR 42.9 billion.

Meaning of International Liquidity

In the discussion on international liquidity, a great deal of the financing of international trade is excluded, i.e., excluded from the concept of international liquidity are the private holdings of foreign exchange banks and trade credits that are used to finance the routine international trade transactions. The concept of international liquidity also excludes the government credit supplied for export purposes by institutions such as the Import-Export Bank of Washington and the long-term international financing routed through the private international capital markets and/or through the international institutions such as the World Bank, the International Finance Corporation and the International Development Association. International liquidity refers to all those financial resources and facilities that are available to the monetary authorities of individual countries for financing the deficit in their international balance of payments—resources that are used to make the residual payments in foreign currencies when all the other sources of supply of foreign funds prove inadequate to bring the international payments into balance.

The assets which constitute part of such liquidity range from those financial assets which are readily available to a country to those financial resources which become available to a country only after protracted negotiations with the creditors. Consequently, international liquidity includes the well-known components of official reserves, the gold held by the central banks (gold held by individuals and businesses does not constitute a part of official reserves and is, therefore, excluded), foreign currencies held by central banks or treasuries, borrowing facilities available from international monetary institutions, drawing facilities available from the IMF under different schemes, credit facilities available under the swap and related credit arrangements of central banks and treasuries and other financial assets which are mobilisable when the need arises. Consequently, international liquidity also includes those elements which are not readily amenable to statistical measurement, such as a country's total borrowing capacity in the international money markets or in the event of the country happening to be a reserve currency centre, the willingness of world's other countries to accumulate their further official monetary holdings of its (reserve currency country's) currency.

In recent years, the role of international capital markets as a source of liquidity supply has substantially increased as these markets, particularly the international banking system, have played a very active and effective part in recycling funds from countries with substantial current account surplus (such as the OPEC countries) to countries having current account deficit. The intermediation of the international banking system, particularly from 1974 onward, has been both timely and substantial in scale and this fact is amply reflected in the amazingly rapid growth of international bank lending in response to both the supply and the demand factors. On the supply side, the availability of funds increased partly on account of the weak demand for loans in the industrial countries and partly on account of the increased fund placements by some oil exporting countries. On the demand side, international credit increased due to the increased demand for funds by many developing countries, including several non-oil primary producing countries in order to finance their current account deficit and to increase their holdings of international reserves.

"International liquidity thus covers a wide spectrum of availabilities, and any classification into broad categories is to some extent arbitrary. Two types of classifications have come into use in recent years. The first distinguishes between 'owned' reserves and borrowing facilities; the second, between liquidity that is available automatically or without prior conditions that significantly restrict the user's right to access, and liquidity that is available only on prescribed or negotiated conditions as to use or as to policies to be pursued by the country using it".

"These two classifications do not quite coincide. For example, in some countries the use of gold reserves is restricted by domestic legislation. On the other hand, borrowing facilities, once granted, may be usable without further question. Again, the drawing facilities in the International Monetary Fund cannot properly be classified as either owned liquidity or borrowing facilities".[1]

Unconditional liquidity comprises countries' gold holdings and foreign exchange reserves in freely convertible currencies, SDR holdings, super-reserve and reserve tranche positions in the Fund and, in many instances, bilateral mutual credit or swap arrangements. Most other forms of international liquidity belong to the conditional liquidity although in some instances the degree of conditionality involved may be small. The recent increasing importance of the availability of international liquidity arises partly from its effects on the confidence in the stability of exchange rates which, by influencing the size and direction of short-term international capital flows, influence the magnitude of deficit in the external balance of payments of the countries. However, the chief impact of such liquidity is felt through its effects on individual countries' policies relating to their external balance of payments reflected in their domestic demand, foreign exchange rate, trade, capital flows and aid policies.

Inadequacy of International Liquidity

The inadequacy of international liquidity is evident from the fact of increasing balance of payments difficulties of the individual countries. The ever mounting pressure on the external balance of payments of most countries of the world, particularly the developing countries, clearly shows that the means of international payments—gold, SDR and the acceptable foreign exchange reserves—have not grown at a fast enough rate required to meet the growing demand for them for payment purposes to finance the world trade. While the world trade in merchandise and commercial services

1. IMF, *Annual Report*, 1964, p. 26.

has been growing at an annual average rate of about four per cent, the global official holdings of the reserve assets have not recorded any commensurate growth resulting in the overall shortage of international liquidity.

It should, however, be remembered that international liquidity is not an end in itself. The world community is keenly interested in the adequacy of liquidity because an adequate amount of international liquidity—means of international payments—evenly distributed between the different groups of world countries—industrial countries, developed primary producing countries, and the less developed debtor countries—contributes to the expansion and the balanced growth of international trade, the promotion and maintenance of high levels of employment and real income, the development of productive resources of all members, foreign exchange rate stability, orderly exchange arrangements among members and avoidance of competitive exchange depreciation, promotion of multilateral system of payments in respect of current transactions and elimination of foreign exchange restrictions which hamper the orderly growth of world trade. Looking at the substantial expansion of international trade over a period of five decades from 1955 to 2004, the pace of growth of official holdings of reserve assets has been rather very slow resulting in the continuous decline in the ratio of official reserves to imports pointing out to the growing shortage of international liquidity. The table on next page shows the trend in the various sources of supply of international liquidity represented by the official monetary reserves during 1955-end to March 2004.

The table shows that during the past 60 years the total world holdings of official reserve assets representing the international liquidity have increased by more than thirty times. However, in spite of the large increase in the official monetary reserves, there has been witnessed a growing shortage of international liquidity as is amply evident by the constantly falling ratio of the total official monetary reserves to total imports for the world as a whole over the past five decades, except during 1961–73 when it rose marginally. Moreover, the inadequate official monetary reserves constituting the international liquidity have been unevenly distributed between different groups of countries of the world as is clearly reflected in the unequal ratios of reserves to imports for different groups of countries of the world. Of the total official holdings of reserve assets (including gold) of SDR 2,862.3 billion at the end of March 2005, industrial countries' share of these official holdings was SDR 1,121.3 billion representing 39.5 per cent of the total official reserve assets while all the developing countries' share was SDR 1,740.9 billion representing 60.5 per cent of the total official holdings of reserve assets including gold. It clearly shows that the problem of shortage of international liquidity has basically two aspects—firstly, an absolute shortage of the total international means of payments in relation to the volume of world trade and secondly, a relative shortage resulting from uneven distribution of the total official reserves (including gold) between the developed and the developing world countries. The table on page 426 shows the trend of international liquidity for the world as a whole and for the different groups of countries during 1966–1981.

It is brought out from the table that while for the world as a whole, the situation of official monetary reserves continued to be a cause for worry in as much as the growth of official reserve assets lagged behind the expansion of world trade, during the *seventies* (1972–78) the monetary reserves had been growingly concentrated with the oil exporting primary producing countries due to the frequent hike in the price of petroleum. The OPEC countries had repeatedly raised the price of petroleum and it is reflected in the further growing concentration of inadequate international liquidity with the exporting countries. From the standpoint of world community, this trend of

Tabel 32.1   Official Holdings of Reserve Assets (End of Year 1955–End March 2010)

(In billions of SDRs)

End-year	Gold	SDRs	Reserve positions in the IMF	Foreign exchange	Total
1955	35.0	—	1.9	18.1	55.0
1965	41.5	—	5.4	25.4	72.3
1975	121.9	8.8	12.6	137.8	281.1
1985	282.6	18.2	38.7	348.2	687.8
1995	236.1	19.8	36.7	890.6	1,183.1
2005	253.0	20.2	50.7	2,538.4	2,862.3
2010	254	21.9	60.4	3266.3	3502.6

Table 32.2   Ratios of Reserves to Imports, 1966-1981

(In per cent)

Year	World	Industrial countries	Primary producing countries		
			More developed	Oil exporters	Other less developed
1966	37	40	31	43	27
1967	36	38	29	46	28
1968	33	34	30	45	28
1969	30	30	30	43	28
1970	29	28	28	43	29
1971	32	33	33	52	28
1972	33	37	48	63	32
1973	34	31	47	59	34
1974	26	21	29	78	25
1975	28	22	26	93	23
1976	25	21	25	95	29
1977	24	20	24	79	33
1978	25	21	25	80	32
1979	24	20	25	76	32
1980	24	20	24	75	31
1981	24	20	24	75	31

concentration of reserves with world's few major oil producing and exporting countries is as bad as was the similar trend of concentration of reserves with the industrially developed countries witnessed in the *fifties* and *sixties,* in as much as the distribution of official reserves among countries affects the degree of ease afforded by a given volume of reserves. In this connection, it is relevant that while the global reserves grew by one-third from the end of 1973 to end of April 1982, the reserves of non-oil primary producing countries did not show any net increase at all during this period. During 1974, the entire global increase accrued to the major oil producing countries; in 1975 and 1976 it was shared approximately equally by the oil-exporting and industrial countries. The substantial recent growth in official reserves being most unevenly distributed added, therefore, less to the degree of global reserves ease than it would have done if the relative growth in official reserves had been more evenly distributed among the countries. The picture of 1979 and 1980 was more gloomy in as much as the distribution of the international reserves which were already inadequate to support the expanding world trade had been increasingly skewed in favour of the 13 OPEC countries. The ratio of reserves to imports had shown no signs of improving during the decade of eighties. In 2004, as against the total official reserve assets (including gold) of SDR 2,553.3 billion, the total value of world merchandise imports and commercial services was SDR 7,800 billion giving the ratio of reserves to imports of only about 32.7 showing that the problem of international liquidity had not eased; if anything it had become more grim.

A factor continuing to exercise a positive influence on the degree of global reserves ease is the adaptability of the supply of reserves to demand. The reduction in the need or demand for owned reserves also resulted from two features of the present international monetary arrangements; widespread floating of exchange rates and the possibility of financing the balance of payments deficits through official borrowing. However, these features do not merely reduce the demand for reserves but also facilitate the acquisition of owned reserves when needed, or the reduction of holdings of excess reserves. In some circumstances, a country with a floating exchange rate which finds its reserves holdings insufficient may raise them by intervening more strongly in the exchange market as a buyer of foreign exchange as and when its foreign exchange rate appreciates, and less strongly as a seller when it depreciates A country may also borrow in the international capital markets, apart from financing its balance of payments deficit, to strengthen its reserves position. Since these reserves can in turn themselves be placed in the capital markets, the cost of adding to its reserves for a country by resorting to borrowing in the international capital market may be relatively small. However, due to the increased use of short-term borrowing to finance deficit, the net reserves for many countries have developed much more unfavourably than the gross reserves.

The two main components of international liquidity are the official reserve assets and credit facilities that are available to a country for the temporary balance of payments financing. Although an absolute demarcation between the two forms of liquidity is difficult to make, distinction is, nevertheless, clear in principle and practice. Reserve assets or 'reserves' are at the disposal of the country owning them without any need for negotiations, without any conditionality as to the country's policies, and without any significant limitation as to the circumstances in which these can be used. Access to credit is always subject to negotiations which at times may be protracted; it may also be subject to other restrictions mentioned. Consequently, prior to successfully negotiating credit line, the availability of credit provides less assurance to a country compared to the ownership of an equal amount of reserves that it will be able to meet the possible balance of payments deficit. In as much as the two forms of liquidity perform somewhat different functions, the composition of liquidity is a matter of concern both for the individual countries and for the international community as a whole.

IMF and International Liquidity

The IMF is an international monetary institution dealing in the global liquidity and is the principal source of supply of temporary balance payments credit for most member countries. The Fund has been established to provide a pooling of SDRs and members' currencies which could be used when needed by the members to supplement their other reserves. With its total quotas of SDR 212.8 billion and the borrowings made from the members under different schemes, the Fund is an important source of supply of international liquidity for its members. The Fund provides its members two kinds of liquidity which member countries use by exercising their drawing rights on the Fund. The drawing rights of the members are related to their quotas. The first kind of liquidity provided by the Fund consists of the members' drawing rights in the reserve tranche—

unconditional liquidity—which members can regard like other types of reserves. A member has a reserve tranche position if the IMF's holdings of its currency in the General Resources Account—excluding those holdings which reflect the member's use of the IMF resources—are less than its quota. Borrowing facilities from the Fund in the reserve tranche are available to the member unconditionally without any performance test being insisted upon by the Fund on the part of the borrowing member. A member may draw upto the full amount of its reserve tranche position at any time, subject only to the member's representation of a balance of payments need. In other words, liquidity within the reserve tranche position is available to the members automatically.

The second kind of liquidity provided by the Fund to its members consists of the drawing rights of the members in the credit tranches—conditional liquidity—and the exercise of these drawing rights requires good performance or the promise of improved performance on the part of the borrowing members. For drawings in the first credit tranche of 25 per cent of a member's quota, a member must demonstrate reasonable efforts to overcome her balance of payments difficulties. Drawings made in the upper credit tranches (above 25 per cent of quota) are made in instalments and are released when performance targets are met.

The size of the unconditional liquidity which the Fund can provide is, however, small relatively to the conditional liquidity. It should not, however, be thought that conditional liquidity is not useful because it is conditional. Members know those circumstances in which they can use their conditional drawing rights and their ability to do so is a source of strength to them. At present, the Fund's unconditional liquidity supply capacity comprised of the usable currencies and SDRs held in the General Resources Account is SDR 83.7 billion while the conditional liquidity exceeds SDR 80 billion. The unconditional and conditional liquidity with the Fund has been growing over time. At the end of April 2004, as a result of general increase in members' quotas, the total quotas of the members stood at SDR 212.8 billion, increasing the capacity of the Fund to supply liquidity to its members. The reserve tranche positions in the Fund stood at SDR 63.3 billion at the end of April 2004. With the increase in Fund members' quotas to the extent of SDR 70.0 billion, Fund's liquidity has considerably increased.

The IMF's liquid resources consist of usable currencies and SDRs held in the General Resources Account (GRA). Usable currencies—the largest component of liquid resources—are holdings of currencies of members whose balance of payments and reserve positions are considered sufficiently strong to warrant the inclusion of their currencies in the quarterly operational budget for use in financing IMF operation and transactions. The IMF's usable resources increased sharply toward the end of financial year 2003–04 as a result of quota payments amounting to SDR 46.0 billion in usable currencies and SDRs. Consequently, the IMF's usable resources increased to SDR 83.7 billion at the end of April 2004 from SDR 47.3 billion at the end of April 2003. The stock of uncommitted usable resources more than doubled to SDR 70.6 billion at the end of April 2004 from SDR 32.0 billion at the end of April 2003. The IMF's net uncommitted usuable resources amounted to SDR 56.7 billion while the liquidity ratio rose to 89.2 per cent at the end of April 2004.

Fund members may draw on Fund's financial resources to meet their balance of payments needs under tranche policies—members may use the reserve tranche and the four credit tranches. In addition, members have access to three permanent facilities for specific purposes—the facility for compensatory financing of export fluctuations (established in 1963 and liberalised in 1975 and 1979); the buffer stock financing facility (established in 1969), the extended Fund facility (EFF); the structural adjustment facility (SAF) established in March 1983 and the enhanced structural adjustment facility (ESAF) established in 1987. Apart from these regular borrowing facilities, members may also avail of the temporary facilities established by the Fund with borrowed resources. For example, following the sharp rise in oil prices, the Fund provided for 1974 and 1975 assistance under a temporary oil facility designed to help members meet the increased cost of imports of petroleum and petroleum products. Since then, a supplementary financing facility has been established with borrowed resources amounting to SDR 7.784 billion from 13 member countries and the Swiss National Bank. For use by low-income developing countries and as a supplement to Fund resources, there is also the Trust Fund administered by the Fund as Trustee. The Trust Fund is designed to provide balance of payments assistance on very concessionary terms to 61 eligible developing member countries that qualify for assistance. Trust Fund loans bear only 1/2 per cent annual interest and are payable in ten semi-annual instalments which start after five years from the beginning of the sixth year of the loan. Since July 1976 upto the end of April 1982, the Fund had disbursed a cumulative total of the loans amounting SDR 1.64 billion. The Fund also administers a Subsidy Account to alleviate the high cost of interest payment on purchases made under the 1975 oil facility by the most seriously affected developing countries. The most seriously affected (MSA) 25 members received assistance from the Subsidy Account of SDR 9.32 million during the fiscal year ended April 1982 bringing the cumulative amount to SDR 175.52 million under the Subsidy Scheme. The Subsidy Account is funded by contributions from 24 members and Switzerland and its object is to reduce the effective rate of annual charge repayable on drawings made under the 1975 oil facility by five percentage points per annum. Apart from this, upto April 1984 the IMF had made a total amount of SDR 366.8 million available to 23 eligible low-income developing countries under its supplementary financing facility subsidy account that was established in December 1980 to reduce the cost of using the Fund's reserves for very poor members.

There is also the supplementary financing facility for members' use in support of programmes under the stand-by arrangements reaching into the upper credit tranche or beyond or under the extended arrangements, subject to relevant policies on conditionality, phasing and performance criteria. The Compensatory and Contingency Financing Facility (CCFF) is available to members facing balance of payments difficulties due to temporary export shortfalls for reasons beyond the members' control. During 1998–99, Azerbaijan, Jordan, Pakistan and Russia made use of this facility with drawings totalling SDR 2.6 billion. The Buffer Stock Financing Facility is available for building international buffer stocks in order to prevent fluctuations in members' export earnings. No drawings have, however, been made under this facility for the past 13 years. The Extended Fund Facility (EFF) is designed to provide medium-term assistance of upto four years to overcome structural balance of payments maladjustments. Resources under the scheme are provided in the form of extended arrangements which include performance criteria and drawings in instalments. The scheme, as a result of liberalisation made in 1979, permits the purchases to be made by a member within a maximum period of ten years. The lengthening of the maximum period of repayments from 8 to 10 years under this facility is very beneficial for developing member countries seeking to correct the structural imbalances in production, trade and prices. During 2003–04, the IMF had approved five new Extended Arrangements totalling SDR 14.7 billion for Indonesia, Bulgaria, Jordan and Ukraine. As of April 30, 2004, four countries had Extended Arrangements with commitments totalling SDR 14.1 billion.

When a member draws on the Fund, it uses its own currency to purchase the currencies of other member countries or SDRs held by the General Resources Account. Consequently, as a result of drawings, the Fund's holdings of the purchasing member's currency increase while there is a corresponding decrease in the Fund's holdings of the currencies or SDRs that are sold. A member must reverse the transaction by buying back its own currency with SDRs or currencies specified by the Fund within a prescribed time or earlier when its balance of payments and reserves position improves. Usually, repurchases should be made within 3 to 5 years after the date of purchase; but under the Extended Fund Facility, the period for repurchases has been extended to take place within 4 to 10 years; under the oil facility within 3 to 7 years, and under the supplementary financing facility within 3-1/2 to 7 years.

Under its Enhanced Structural Adjustment Facility (ESAF), established in 1987 and extended and enlarged in February 1994, the IMF provides financial support in the form of highly concessional loans to low-income poor member countries facing protracted balance of payments problems. As of April 30, 2004, 83 ESAF arrangements were in effect. In 2003–04, the IMF's 10 new ESAF arrangements totalled SDR 0.9 billion.

If the IMF's holdings of a member's currency are less than her quota, the difference is called the *reserve tranche*.[2] Purchases in the reserve tranche are not subject to prior challenge and the facility is almost automatic. Further purchases are made in the four *credit tranches* each of 25 per cent of the member's quota. Normally, the total purchases under credit tranche policies are confined to an amount that raises the Fund's holdings of the members' currency to 200 per cent of its quota. The IMF may, however, in exceptional circumstances allow a member to purchase larger amounts. A member's indebtedness to the Fund, i.e., its obligation to repurchase begins when Fund holdings of its currency exceed 100 per cent of the member's quota. For making any drawing, a member must represent to the Fund that the desired purchase is needed to correct its adverse balance of payments or reserves position or developments in its reserves. All requests for the use of the Fund's resources, other than use of the reserve tranche, are examined by the Fund to determine whether the purported use would be consistent with Fund's Articles and policies. Under credit tranche policies, use must be in support of economic measures designed to overcome a member's balance of payments difficulties. The IMF applies charges for the use of its resources, except for the reserve tranche purchases. A service charge of 0.5 per cent is payable on all purchases other than the reserve tranche purchases. Over the 4-year period (1977–80), the IMF sold to 126 members 25 million ounces of gold at a price of SDR 35 per ounce in accordance with August 1975 agreement. During the four years (1976–80), the Fund had distributed US $1.29 billion to 104 developing member countries as their share of profit from the gold sales.

The IMF also provides financial assistance to members in the form of stand-by arrangements. This facility was created in 1953 in response to a general feeling that a technique was needed whereby members, not needing immediate use of Fund's resources but feeling that they might need Fund's resources in the near future, could be assured of prompt financial assistance if and when the need arose. The quantum of assistance provided by the Fund under the stand-by arrangements has been quite substantial. During the period 1995–2004, the IMF granted SDR 79.4 billion in the form of 141 stand-by arrangements to members. The Fund is also a source of supply of liquidity under the General Arrangements to Borrow Scheme which was started in October 1962 and was initially activated in December 1964 for a four-year period authorising the Fund to borrow upto SDR 6.4 billion in the currencies of ten main industrial member countries

---

2. Prior to the entry into force of the Second Amendment of the Articles of Agreement of the Fund on April 2, 1978, it was known as the *gold tranche*. Purchases made under this tranche by a member have been legally automatic and the Fund can challenge only *ex post* the member's representation of the balance of payments or reserve need.

when need arose, to avert a major foreign exchange crisis. The GAB scheme had been activated seven times. The largest borrowing under the facility amounting SDR 897.5 million was made by Germany with the United States of America occupying second position with borrowing of SDR 777 million made in November 1978 in connection with a reserve tranche purchase.

In order to augment its lending capacity, the IMF's Executive Board approved the New Arrangements to Borrow (NAB) scheme under which the IMF may borrow upto SDR 34 billion as and when it needs additional resources to forestall or cope with an impairment of the international monetary system or to deal with an exceptional situation which poses a threat to the stability of the system. There are 26 participants in the NAB including for the first time a number of developing countries. The NAB has not replaced the GAB. The NAB credit lines may be drawn on for the benefit of all NAB participant countries or for the non-participants under circumstances similar to but more flexible than those under the GAB. With the NAB having become effective, the amount that has become potentially available for supplementary resources with the IMF has doubled. In November 2002, both the GAB and NAB were renewed for five years effective from November and December 2003 respectively. As of April 30, 2004, the IMF had no outstanding debt.

Under the oil facility, the Fund had provided SDR 6.9 billion to 55 members during the three years 1974 to 1976. Apart from these various schemes under which the Fund provides credit facilities to its members, the Fund has restituted in four annual instalments 15 million ounces of gold to members at the official price of SDR 35 per ounce. Further, the Fund acts as the Trustee for the Trust Fund established with the proceeds of profit earned on the market sales of 25 million fine ounces of gold. As on May 7, 1980, the total amount available for the benefit of 61 developing member countries in the Trust Fund net of payment by the Trust for the gold (at a price equivalent of SDR 35 per ounce) was US $64 billion. Loans equivalent of SDR 1.9 billion had been made to the 50 developing members, and an additional amount of approximately SDR 1.9 billion had become available for the same purpose. Further, of the total profit of US $4.64 billion, US $1.29 billion were transferred directly to 104 developing countries in proportion to their quotas on August 31, 1975.

Under the Extended Fund Facility during 1995–2004, the IMF had committed a total of SDR 232 million spread over 80 Arrangements; under the Poverty Reduction and Growth Facility (PRGF), IMF had committed SDR 39.15 billion covering 433 Arrangements during 1995-2004. In FY 2004, IMF had approved SDR 967 million under the PRGF. Under the Structural Adjustment Facility (SAF), IMF had committed SDR 3.46 billion in 38 Arrangements during 1987-96; no amount was committed thereafter. The IMF related assets comprising reserve positions in the IMF and SDRs amounted SDR 83.8 at the end of March 2004. As of April 30, 2005, total disbursements of the HIPC (highly indebted poor countries). Initiative assistance made by the IMF amounted SDR 1.5 billion to 34 members.

Special Drawing Right

The Special Drawing Right (SDR) scheme was first initiated by the IMF in 1969. It is reserve asset created to meet a long-term global need to supplement the existing reserve assets and allocated to the members in proportion to their quotas in the IMF. A member may use SDR to obtain the foreign exchange reserves from other members and to make payments to the IMF. Such use does not constitute a loan. Members are allocated SDRs unconditionally and may use these to meet a balance of payments financing need without undertaking economic policy measures or repayment obligations. A total of SDR 9.3 billion was allocated to the members in 1970–72 and an allocation of SDR 12.1 billion was made in 1978–81 bringing the total to SDR 21.4 billion.

Under the scheme, the pool of SDRs 21.4 billion is a kind of *paper gold* which is world's international money backed by cooperation. According to Fund's amended Articles, the Fund conducts its activities under two separate accounts. All the present operations of the Fund, including the sales of currencies to members, are conducted through the General Resources Account (GRA). Transactions involving special drawing rights are conducted through the Special Drawing Rights Department. Under the SDR scheme, each member is entitled to participate in the Special Drawing Rights Department if it so chooses. The member entitled to participate in the SDRs is able to use its holdings of Special Drawing Rights. Accordingly, it is provided that a participant in the SDR scheme who is in balance of payments or in reserves difficulties is entitled to transfer its Special Drawing Rights to another participant who has been designated by the Fund and obtain convertible currency from her. A member participating in the scheme can transfer its Special Drawing Rights to a designated participant but is not expected to do so unless it is faced with the balance of payments or reserves difficulties. Nevertheless, if a participant uses its Special Drawing Rights under this entitlement, the transfer will not be challenged in any circumstances. Thus, the use of SDRs is unchallengeable; in short, for the participants the SDRs are a form of unconditional liquidity. Further, a participant is able to use its SDRs until none remains and this could not be objected to under any of the provisions relating to use. However, use beyond a certain average proportion over time may require the participant to restore its holdings to a certain average level.

When a participant transfers its SDRs to a designated participant (transferee) in return for a currency, the transfer e must provide on demand 'currency convertible in fact'. The choice of currency is left to the transferee provided that it meets the definition of 'currency convertible in fact'. The 'currency convertible in fact' may be either transferee's own currency or any other member's currency. It has been done in order to provide the transferer promptly and with minimum inconvenience the currency usable for the support of its own currency. A member participating in the SDR scheme may transfer its SDRs to another participant, even though not designated by the Fund, if the latter so agrees.

Each participant's allocation in the new reserves assets has been based on its size and economic strength. The first allocation of SDR 3,414 million was made on January 1, 1970 to 105 members who were participants in the Special Drawing Rights. The second allocation of SDR 2,949 million was made on January 1, 1971 to 110 members who were participants in the scheme. As a consequence of these two allocations made on January 1, 1970 and on January 1, 1971, the official world reserves increased by 5 per cent in 1970 and by 3.6 per cent in 1971. The third and final allocation of SDR 2,952 million was made on January 1, 1972. As a consequence of these three allocations of SDR 9,315 million, the problem of international liquidity eased considerably. Under the second round of the scheme, in accordance with the Board of Governors resolution which became effective on December 11,1978, a total of SDR 4,032.7 million was allocated on January 1, 1981 to 141 members who were participants in the Fund's SDR Department at the time of making the earlier two allocations.

Decision on general allocations of SDRs are made in the context of five-year basic periods and require a finding that an allocation would meet a long-term global need to supplement the existing reserve assets. A decision to allocate the SDRs requires an 85 per cent majority of the total voting power.

One of the principal goals of the IMF is to facilitate the expansion and balanced growth of international trade which requires adequate levels of international reserves. The IMF has the authority to create unconditional liquidity by allocating SDRs to all member countries in proportion to their quotas. At present, more than one-fifth (43) of IMF members have never received the SDR allocation because these countries joined the IMF after the last SDR allocation made on January 1, 1981. In addition, other members have not participated in every allocation. Understandably, members who have been without SDR allocation have argued that the IMF should make a special one-time allocation of SDRs to correct this unbalanced distribution. The Executive Board has broadly agreed to it. With the implementation of the one-time allocation of SDRs, the cumulative SDR allocations would increase to SDR 42.9 billion from SDR 21.4 billion at present and the liquidity problem of the members will be eased.

A recent development in the supply of international liquidity has been the phenomenal growth of private international capital markets as a result of which there has been a rapid expansion of private institutions—commercial banks—which lend to countries facing the serious balance of payments difficulties. This new source of supply of liquidity is, however, important only for those countries which can borrow in the international money and capital markets. A large number of the low income developing countries do not have access to this source of supply of international liquidity as they cannot borrow on easy terms in the private international capital markets as their creditworthiness is rated low in the international financial markets.

Suggested Reading

1. IMF, Annual Reports.
2. Bo Sodersten, *International Economics,* Second Edition, 1980, Chapter 30 and 31.

Questions

1. Explain the meaning of international liquidity. Discuss the recent developments in international liquidity.
2. Discuss the recent trends in international liquidity. Explain the role of International Monetary Fund as an important source of the supply of international liquidity.

CHAPTER 33

Eurocurrency (Eurodollar) Market

INTRODUCTION

The term 'Eurocurrency market', also synonymous with the Eurodollar market, is a broad catch-all for a number of specific markets separated by the type of transactions, institutional arrangements, by the use of various instruments for financial dealings, sometimes geographical location and even by the lack of standardisation on the part of European central banks in their use of terminology. The market, which has recorded an extremely rapid growth during the past four decades, has been an enigma to most observers, including the eminent economists. For many, the origin and development of the Eurocurrency market as an international phenomenon is shrouded in mystery. For instance, to Fritz Machlup the development of the Eurocurrency market is a mystery story and Eurodollars are magicians' rabbits. Milton Friedman attributes the development of the Eurodollar market to the magic of a mere book-keeper's pen—Eurodollar banks create money, primarily dollars, like the ordinary commercial banks by making loans which are redeposited in the Eurodollar market. Charles P. Kindleberger makes a mention of the two puzzles of the Eurodollar market. Whatever may be the riddle that still remains unresolved behind the development of the Eurocurrency market, the fact remains that the development of the Eurocurrency market has been astounding. This is amply borne out by the fact that during the decade 1965-75, the total net deposit liabilities of banks in the Eurocurrency market increased by about US $140 billion. And this covers only that part of the market which is accounted for by the banks or countries reporting to the Bank for International Settlements (BIS). Consequently, the Eurocurrency market has today become the focus of wide attention as it has become an important component of the international economic scene. In mid-1996, the total identified holdings of Eurocurrencies in the official holdings of foreign exchange aggregated over US $500 billion being about one-third of the total official holdings of foreign exchange reserves of over US $2,000 billion.

Ever since its beginning in the late 1950s, the Eurodollar market has expanded to financial centres outside Europe and to currencies other than the American dollar. Consequently, the term "offshore currency market" is more appropriate to describe the market's location and its range of activity. Along with the coverage of the currencies transacted in the market at different locations, the array of financial products offered in the market has also expanded. At present, the Eurocurrency deposits extend to maturities of five years and beyond while the lending activity ranges from direct bank to customer credits to multi-billion dollar Eurocurrency syndications involving complex financing terms.

Meaning, Origin and Growth

The Eurocurrency market is an international banking market whose major location is in London[1] and which specialises in the borrowing and lending of currencies outside their countries of issue. Besides the US dollar, which dominates the market in transactions, transactions take place in other Eurocurrencies whose share in the total assets and liabilities of the Eurobanks has tended to increase over the years. Those who participate in the Eurocurrency market include the commercial banks, monetary authorities, business firms (particularly large multinational corporations), government agencies and semi-governmental entities, including central and international organisations. Although in the absence of any reliable comprehensive official statistics, it is hazardous to say anything definite about the size of the market, it is, however, estimated that the gross volume of Eurocurrency assets of banks in major international banking centres range between US $350 and $800 billion. On the basis of available statistics, it can be said that inter-bank transactions dominate the market constituting the largest and the fastest growing part of the Eurocurrency market. The non-bank sources and uses of funds channelled through the market account for nearly one-third of the total Eurocurrency transactions in the market. Since the major part of the market is located in Europe and since transactions are mostly in dollars, the market is commonly called the 'Eurodollar market'. However, with the increasing use of non-dollar currencies, for example, the deutsche mark and the Swiss franc, the market is more appropriately labelled as the 'Eurocurrency market'. However, even so the use of the prefix 'Euro' is misleading because a significant part of the transactions is conducted at centres outside Europe, such as the Bahamas, Panama and Singapore.

Although international banking centres had conducted some operations in the offshore currency assets and liabilities for a long time, two developments toward the end of 1950s contributed significantly in inducing the banks to engage in large-scale transactions in Eurodollars. *Firstly,* in 1959, following the sterling crisis, the British government imposed tight controls on the non-resident sterling borrowing and lending by the British banks. In order to retain their position in the financing of international trade, the British banks turned to dollars as a substitute for sterling. *Secondly,* toward the end of 1958 the currencies of the major west European countries were made convertible for non-residents offering scope for the banks in these countries to buy and sell dollars freely and to use dollars in the financing of international trade.

Several factors contributed to the early development of the Eurodollar market. In the late 1950s, following the Suez Canal crisis, England started to curtail the use of pound-sterling to finance international transactions between the third parties. To overcome this restriction, the British commercial banks started using the dollar to conduct transactions from accounts based in London. Since the Bank of England regulations did not cover the US dollar, banks were free to set competitive interest rates to attract deposits and offer dollar denominated external loans. Furthermore, the continuation of the cold war and East-West tensions encouraged the Soviet Union and other Eastern Block countries to shift their dollar deposits to London and Paris often with the affiliates of the state-owned Soviet banks. A more important stimulus to the growth of the Eurocurrency market was the setting up of credit restrictions and capital controls by the USA in response to an undesired building up of dollars overseas.

---

1. This is not to suggest by any means that London is the exclusive location of the Eurocurrency market. Besides London mention may be made of Luxembourg and other offshore Eurocentres such as the Bahamas, Singapore, Panama, Bahrain and Beirut as the major banking centres.

The tremendous growth of the Eurocurrency market has raised interest in the offshore banking. The market grew from essentially zero in the late 1950s to $16.4 trillions on a gross basis and $7.8 trillions on a net basis at the end of 1995. By far the most important currency in the market is the US dollar although its market share has now declined from about 80 per cent in the 1970s. Although the geographical reach of the market is now worldwide but Europe (with 65 per cent market share) and London (with 23 per cent market share) remain the dominant offshore banking centres. Japan's share has substantially increased and it now stands at about 20 per cent.

While deposit maturities extend upto five years, about 70 per cent of these deposits are short-dated with maturities of less than three months and 90 per cent of less than one year. Most of these deposits are issued on fixed-rate terms although longer-term maturities are issued as floating rate notes or floating rate certificate of deposits. Loan maturities range from short-term trade financing to beyond five-year commitments. The pricing of Euro loan of longer than six months' duration is typically structured on a roll-over basis under which at the start of each 3 or 6-month period, the interest rate is reset at a fixed amount above the ruling London Interbank Offered Rate (LIBOR).

In the early 1960s, there were two major international dollar markets centering at New York and at London. Although in the matter of volume of international transactions the New York market was larger compared with the London market, the London banks enjoyed a competitive advantage. In a way, this reflected 'natural' locational advantages, for example, time differences and closer proximity to some important customers. A major factor giving a competitive advantage to London over New York international dollar market was, however, the absence of regulations of banking operations in currencies other than pound-sterling in the London market. Unlike the banks operating in the United States, banks operating in England were not subject to legal reserves requirements or official interest rate ceilings in the matter of their dollar transactions enabling them to pay higher interest rates on dollar deposits as well as to operate with narrower reserves requirements compared with their American counterparts which were subject to such regulations. The return to external convertibility of currencies and substantial relaxation of exchange controls in 1958 contributed substantially to the development of Eurocurrency banking operations. The removal of these exchange and capital controls combined with the competitive edge enjoyed by the Eurocurrency banks over their US counterparts in being able to operate without any restrictions in the form of interest rate ceilings and legal reserves requirements played an important part in the subsequently rapid growth of the Eurocurrency market.

It is ironical that the imposition of exchange and capital controls in some developed countries in the late 1960s and early 1970s played the role of key factor in promoting the rapid growth of the Eurocurrency market during this period. Since, by and large, the controls that were instituted during this period did not aim at completely choking off international transactions, these could be, and were in fact, circumvented in one or the other way. In the process of circumventing these controls, the volume of Eurocurrency transactions expanded. For example, the imposition of a host of capital controls in the United States in 1968 together with the tight monetary policy pursued by the monetary authorities during this period, encouraged the US banks to circumvent these restrictions by expanding the activities of their overseas branches tremendously in 1968 and 1969. Consequently, the assets of the US banks' overseas branches more than doubled during these two years from $15.7 billion 1967 to $41.1 billion in 1969 while the number of overseas branches rose from 295 to 459 during the same two-year period. Similarly, the activities of foreign branches of German banks, particularly in Luxembourg, recorded rapid expansion in the early 1970s following the imposition of the minimum reserves requirements on the German banks' external liabilities. At the end of 1971, the consolidated balance sheet of the subsidiaries of large German banks in

Luxembourg alone was around DM 6 billion compared with DM 2 billion at the end of 1970. The bulk of the DM 4 billion increase in the deposit liabilities of German banks' subsidiaries was the result of successful efforts to circumvent the exchange and capital controls imposed by the authorities in Germany, France and other European countries.

It is, therefore, natural to conclude that but for these exchange and capital controls, New York and other smaller money markets such as Frankfurt and Paris would in all probability not have been able to keep pace with the growth of London, Luxembourg and other Eurocurrency centres as the volume of international trade and capital transactions expanded. The imposition of controls created a favourable climate for concentration of these free transactions in short-term capital in London and to a lesser degree in Luxembourg and other 'offshore' Eurocentres.

The growth of the Eurocurrency market continued to be very rapid until mid-1974, with an estimated growth rate exceeding 49 per cent in 1973 and early 1974. Thus, the growth of the market continued undiminished even after the abolition in January 1974 of restrictions on capital exports in the United States of America. After mid-1974, the growth of the market received some temporary setback causing concern about its future until in 1978 it was resumed at a significantly reduced rate. In 1975 the annual growth rate was only 15.8 per cent as against 43.5 per cent in 1973 and 34.1 per cent in 1974. However, the annual rate of increase revived in 1976 when it was 20.5 per cent and rose to 25 per cent in 1978 and further to 26 per cent in 1979. Between the end of 1975 and mid-1979 nearly a third of the $140 billion increase in the global exchange reserves was deposited in the Eurocurrency market.

Changing Market Structure

Notwithstanding that throughout the history of the Eurocurrency market the US dollar has accounted for by far the largest share of total transactions, the share of the non-dollar currency assets and liabilities has continued to increase over the years. The share of the non-dollar transactions in the total market transactions increased from 17 per cent in 1969 to around 33 per cent at the end of 1978. In other words, the trend over the years has been toward a decline in the domination of the dollar transactions relative to the non-dollar transactions. This trend was mainly due to an increase in the demand for the German mark and the Swiss franc denominated financial assets as these currencies were expected to appreciate. Since investments made by the non-residents in the Federal Republic of Germany and Switzerland were increasingly restricted by capital controls, the demand for German deutsche marks and Swiss francs was diverted to the offshore banking centres. Consequently, Eurobanks became increasingly active in the deutsche mark and Swiss franc segment of the market leading to the change in the structure of the market. Apart from the change in the composition of currencies transacted in the market, the regional or geographical composition of the market has also undergone a substantial change over time. Although in the absence of comprehensive statistics this regional change cannot be detailed, estimates made by the Bank for International Settlements (BIS) indicate an increasing relative importance of countries outside North America and Western Europe as lenders and borrowers in the Eurocurrency market since 1964. Japan's presence in the market has risen to 20 per cent.

Banks of the Eurocurrency market transact in widely dispersed international banking centres. Notwithstanding the still dominating position commanded by London as the main centre of the offshore banking business, its share in the total Eurocurrency banking business has declined since 1970 to the advantage of the Bahamas, Singapore, Panama and Beirut. Apart from the change in the currency composition and regional composition of the market, a significant change has also been witnessed in the distribution of Eurocurrency business among the banks according to their national origin. From 1963 to 1969, the share of the US banks in the London Eurocurrency market had substantially increased from about 25 per cent to 54 per cent of the total Eurocurrency business. The rest of the business was mainly undertaken by the UK banks. In other words, till 1969 only the US and UK banks operated in the Eurocurrency market. However, since 1973 the share of the US banks in the London market has declined to less than 40 per cent, with banks having their headquarters located outside the United States of America and the United Kingdom gaining in the share of the total market transactions. At the end of 1978, the banks of non-British and non-American origin had accounted for about one-third of the total Eurocurrency business transacted in the London centre of the Eurocurrency market. Since the mid-1960s, there has also been a growing number of mostly multinational consortium banks operating in the Eurocurrency market. In 1980, these multinational consortium banks handled about 10 per cent of the total Eurocurrency business centred in London.

General Features of the Market

In the Eurocurrency market, the Eurobanks have developed certain characteristic lending and borrowing practices which have in a way become the general features of Eurocurrency market. These practices or techniques were largely the result of the particular structure of the market and the change in this structure over time. The Eurocurrency market has the following general features.

1. The Eurocurrency market is a wholesale market in so far as most final borrowers are large companies or official entities and the average unit size of transaction is large. Consequently, the overhead expenses of banks are relatively low.

2. The largest segment of the Eurocurrency market is the interbank market which is essentially short-term in nature. Consequently, more than three-fourths of the foreign currency liabilities and assets of the reporting Eurobanks are against other banks. In the last few years, increase in the gross size of the market represents to a very large extent increase in the interbank deposits.

3. The Eurocurrency market is a highly competitive market with unrestricted entry for the newcomers in the market. Consequently, the margin between interest rates on deposits and loans has tended to decline and the Eurobanks have to remain contended with lower rates of return on the Eurocurrency assets than those rates of return which can be had on the domestic currency assets in national markets. During periods of increasing competition among banks operating in the Eurocurrency market, some Eurobanks have also taken higher risks, both in the choice of borrowers and in the acceptance of longer maturities of loans generally extended on an unsecured basis. More recently, however, the market has adopted a more cautious attitude both with regard to exposure to individual borrowers and to longer maturities of loans. Although resembling in many ways the national money markets, the Eurocurrency market is basically different from them since it is a market without any central monetary authority and free from controls.

Loans and Deposits

Euromarket loans are almost similar to the loans that are made in the domestic or onshore market. One important difference, however, is that borrowers in the Eurodollar market are typically large, well-known firms with high credit standing. Consequently, credit evaluation and documentation are less rigorous. Loan maturities range from short-term trade financing to 10-year commitments.

With a view to spreading the default risk generally inherent in large-scale loans over a large number of banks, the Eurobanks have increasingly resorted to the technique of syndicating the medium-term loans extended outside the interbank market. There are different methods for syndicating a loan. The common characteristic of all the methods is to involve a large number of participating banks (as many as 95 banks in one case) with only one bank—the lead bank—managing the loan. By this method many small and medium-sized banks that would have found it difficult or even impossible to extend the loan singly and consequently to operate in the market are enabled to operate in the Eurocurrency market. The same factors which have promoted the growth of syndicated loans have also in recent years encouraged many banks to join in consortia. In order to limit the default risk resulting from indirect loan relationship often involving many banks, Eurobanks typically place limits on the amount of outstanding advances which they will extend to any single borrower and to borrowers in any single country. The syndicated credit market developed during the late 1960s and was a major factor in facilitating the access of developing countries to Eurocurrency funds.

Yet another feature of Eurocurrency dealings outside the interbank market is the use of 'floating rate' medium-term credit arrangements which represent a type of roll-over credit with interest rates determined periodically, usually every three or six-monthly, on the basis of interest rates prevailing in the interbank part of the Eurocurrency market. In a way, by resorting to the use of 'floating rate' medium-term credit arrangements, Eurobanks try to reduce the interest rate risk which is inherent in unmatched maturities of assets and liabilities. Most loan agreements specify lending rates as a margin over the deposit rates of 'reference banks', usually taken to represent the London Interbank Offered Rate (LIBOR). In this way, the lending rates of Eurobanks are pegged to the deposit rates plus a specified margin and the banks virtually run no risk that would result from changes in the deposit rates while their lending rates remained unchanged in the case of medium-term loans.

In many loan agreements, a multi-currency clause is also incorporated in order to avoid the risk of loss resulting from exchange rate fluctuations. Prior to the recent upheavals in the exchange markets, it was usual to denominate loans in the US dollars. However, as the uncertainty concerning exchange rates grew, the risk for the lenders and borrowers facing the choice of currency also grew. The multi-currency clause provides an important element of flexibility in this respect by offering a choice of currencies in which the whole or parts of a loan may be drawn upon.

Eurobanks do not provide facilities to withdraw deposits through cheques. This does not, however, matter much because the relatively short maturities of most Eurocurrency deposits make these deposits very close substitutes for liquid assets. An important new facility was introduced in 1966. London-located branches of the US banks started issuing negotiable dollar certificates of deposits (CDs). These are negotiable receipts for the US dollar deposits made with a London bank. The advantage of this instrument is that it combines the yield opportunities of time deposits with a high degree of liquidity. The liquidity of CDs, of course, depends upon a well-functioning secondary market which exists and happens to function very efficiently.

Recent Development and Future Prospects

In 1974 the Eurocurrency market experienced a difficult period of adjustment as the world economy strived to adjust itself to a new environment following the quadrupling of oil prices in late-1973. By mid-1974, confidence in the market was shaken by the announcement of foreign exchange losses and a number of bank failures although these losses were not directly connected with the Euromarket business. It also began to be felt that Eurobanks would not be able to continue to play a major role in recycling oil surpluses on account of the capital inadequacy of commercial banks and the difficulty of maturity transformation needed for oil financing. It was also felt that in order to overcome the difficulties arising from the problem of maturity transformation and the deteriorating creditworthiness of some borrowers in the market it was essential to have some clearly well-defined lender-of-last-resort in the Eurocurrency market to keep the market going in times of panic and thin confidence. In the absence of any real lender-of-last-resort, it was argued by some experts that the future of the Eurocurrency market was gloomy. It was also argued that the lifting of capital and exchange controls in the United States in January 1974 would deprive the Eurocurrency market of its *raison d'etre* as a substitute for the New York market. In fact, the gross side of Eurocurrency market suffered a setback in the third quarter of 1974 for the first time in its history. The external foreign currency liabilities of the banks in the eight reporting countries—Belgium, France, the Federal Republic of Germany, Italy, the Netherlands, Sweden, Switzerland, and the United Kingdom—fell by US $12 billion (more than five per cent) from the level at which these stood in June 1974. The fears of the critics and sceptics about the shaky future of the Euromarket, however, soon subsided following the announcement of support to the market from the central banks after the Bank for International Settlements (BIS) meeting in September 1974 and the growth of the market, although moderate, was resumed in the fourth quarter of 1974 and the first-half of 1975.

A very significant development in the Eurocurrency market is the emergence of SDR-linked certificates of deposits and bonds. In terms of volume, the most important SDR market is the deposit market in which banks index the value of deposits to the SDR. The SDR deposit market is a wholesale market with a typical deposit of approximately SDR 12 million. The principal depositors are the central banks, Middle East countries' governments or agencies and institutions in the international oil business.

Major International Role

The developments during 1974 were, in a way, suggestive of the true nature and role of the Eurocurrency market in the world economy. The Eurocurrency market, which originated as a substitute for US markets and which would have been regarded as an anomaly or aberration thirty years ago, is now firmly established as the major cornerstone of international capital and the monetary scene. In the absence of enormous expansion of the Eurocurrency market in the first-half of 1974, it would have been very difficult to recycle the oil funds, at least in the short run. Consequently, world economy would have been subjected to additional strains and stresses. The sag in the market in the third quarter of 1974 simply reflected the fact that the Eurocurrency market was a short-term and a competitive market which cannot bear the burden of medium-term and long-term functions of recycling of funds and that even a competitive market needed overall framework supported by international official authorities. In any case, the re-emergence of New York as international capital and money centre cannot deprive the Eurocurrency market of the major role which it has played during the past three decades. The international bond issues and placements in Eurobonds in 1977 stood at US $19,335 million as against US $15,368 million in 1976 recording an increase of more than 25 per cent during the year. The Eurocurrency market will continue to function as a major financial centre of the world and will be increasingly integrated with the US markets now that the capital restrictions on the US markets have been removed.

However, in the interest of an orderly growth of the world economy and international monetary system, the Eurobanking operations, which hitherto had been free of any regulation and control, should be governed by systematic regulation achieved through international agreement. However, there is little chance that progress can be made in designing specific regulatory measures in the absence of any agreement among the principals involved about the ultimate objective of Eurocurrency market regulation. Unfortunately, this has so far proved elusive.

Eurocurrency Market and Developing Countries

During the past two decades, a growing number of developing countries have borrowed increasing amount of funds from the Eurocurrency market. This is, however, far from suggesting that either the access for the developing countries to the market funds has been easy or that the funds have been obtained from the market at cheap cost or have always been suitable for purposes of economic development. While some developing countries have had access to international and foreign bond markets as well as bank loans, borrowings have been largely made through medium-term syndicated loans with maturities ranging between 3 and 10 years with floating interest rates. According to the conservative estimate made by the Bank for International Settlements, the medium and long-term lending from banks in the Group of Ten countries and Switzerland to developing countries rose from about US $2 billion in 1971 to US $10 billion in 1976 and to over US $40 billion in 1995.

The substantial increase in the borrowed funds through the Eurocurrency market recorded in recent years has, however, not been widely dispersed among the developing countries. The borrowed funds have been largely concentrated among relatively few high-income countries like Argentina, Brazil, Mexico and Peru. There was, however, a marked shift in favour of lending to middle-income developing countries in 1973 and 1974 although the capacity of these countries to raise large amounts of Eurocurrency loans diminished somewhat toward the end of 1975. Borrowings made by the other developing countries have been erratic and insubstantial. On the whole, the net lending position of banks in relation to non-oil exporting developing countries in 1977 showed a fall of US $5.6 billion from US $35 billion in 1976 to US $29.5 billion in 1977. In 1977, the total Eurobond issues and placements made by the developing countries totalled more than US $200 million.

The importance of the Eurocurrency market in international economy has increased so overwhelmingly that its future is linked with the future of the world economy as a whole. Indeed, so much amazingly rapid has been the development of the market that the Eurocurrency market has now become the focus of considerable and at times of diffused controversy. Its size, rapid growth and freedom from national regulations have attracted interest in the macroeconomic implications of this market and its impact on the policies of individual countries. So long as world's international transactions or the integration of the world economy proceed as smoothly as they have in the past, the Eurocurrency market will continue to play a major role in the world economic scene. The important role of the Eurocurrency market in the international lending is well brought out by the fact that of the total international bond issues and placements aggregating US $34,904 million more than one-half of this total amount (US $19,335 million) was in the Eurodollar issues and placements. It is, therefore, obvious that the sound development of the economy cannot be maintained without the smooth functioning of the Eurocurrency market.

The extent of development of Eurocurrency market is also an indicator of the extent of monetary interdependence of the economies of the world. The degree of the development of the Eurocurrency market is such that for the proper management and monitoring for the world economy effective coordination between the monetary and other policies of, at least, the major countries of the world is essential. To bring this about, agreement on some form of coordinated surveillance and official intervention in the Eurocurrency market may well prove the beginning of laying the foundation for further cooperation in economic policies of the countries of the world. There is, however, little chance of progressing in the direction of formulating specific regulatory measures for the market until there is some consensus agreement among the concerned countries regarding the overall role of the Eurocurrency market as an international capital market.

As the Eurocurrency market has developed rapidly and with relatively little regulation from national authorities, the activities of the market have at times been a source of concern. In the 1960s and 1970s, the primary concern was whether the relatively uncontrolled lending activity in the Eurocurrency market led to worldwide inflationary bias and to greater exchange rate instability. However, since the collapse of Herstatt Bank in 1974 and until now, policy concern has focused on matters of supervision and prudential control—particularly the lender-of-last-resort for offshore banks, the capital adequacy of offshore banking operations and the adequate supervision of offshore loan portfolio.

A committee was set up to address these supervisionary and prudential concerns under the auspices of the Bank for International Settlements (BIS)—the Basic Concordant (1975). The first statement from this committee proclaimed that supervision of offshore banks should be the joint responsibility of the host and parent authorities. While the supervision of liquidity would be responsibility of the former, the supervision of solvency would be latter's responsibility. In 1980, the BIS announced an agreement that required the commercial banks to consolidate their worldwide accounts. This agreement enables the bank examiners to regulate offshore and onshore operations on a consistent basis. The BIS capital adequacy guidelines applied on a worldwide consolidated basis are an important illustration of this concept. The notion about the Eurocurrency banks that they operate beyond the reach of any government was absolutely incorrect. The challenge before the regulators is to bring the Eurocurrency market within the international safety net without endangering the innovation and dynamism that have been its salient features ever since their inception.

Suggested Readings

1. A. Crockett, "The Eurocurrency Market: An Attempt to Clarify Some Basic Issues", IMF *Staff Papers,* Volume 23, July 1976, pp. 375–386.
2. Edward J. Frydi, "The Debate over Regulating the Eurocurrency Market", *Quarterly Review,* 1979, Federal Reserve Bank of New York.
3. Paul de Grauwe, "The Development of the Eurocurrency Market", *Finance and Development,* Volume 12, No. 3, September 1975.
4. Ishan Kapur, "The Supply of Eurocurrency Finance to Developing Countries", *Finance and Development,* Volume 14, No. 3, September 1977.
5. Charles P. Kindleberger, *International Economics,* Fifth Edition, 1973, pp. 288-290, and p. 409.
6. Eisuke Sakakibara, "The Eurocurrency Market in Perspective", *Finance and Development,* Volume 12, No. 3, September 1975.

Questions

1. Describe the origin, development and the general features of the Eurocurrency market. What, in your opinion, is the future of this market?
2. In what respects is the Eurocurrency market a highly competitive market? What major role has this market played in the smooth functioning of the world economy?

CHAPTER 34

International Bank for Reconstruction and Development (IBRD)

Introduction

The Internatinal Bank for Reconstruction and Development, popularly known as the World Bank, owes its birth to the deliberations of the United Nations Monetary and Financial Conference which met at Bretton Woods, New Hampshire, to prepare the final text of the Articles of Agreement of the International Monetary Fund (IMF) and the International Bank for Reconstruction and Development from July 1 to July 22, 1944. The Bank was established on December 25, 1944 when the Articles of Agreement of the Bank were ratified by the requisite number of member governments. The global war had completely dislocated the multilateral trade and had caused massive destruction of life and property. The economies of England and other countries in Europe had been completely shattered. While the need for promptly reconstructing the war ravaged economies of European countries was recognised, it was also recognised that the stable world peace was threatened by the presence of great disparities in incomes and wealth manifested in the wide differences in the standards of living between the developed and the underdeveloped countries. Consequently, the problem of raising the standard of living of the vast masses of people of the underdeveloped countries brought to fore the need to develop the economies of these countries. Thus, the Bretton Woods Conference was also responsible for establishing the International Bank for Reconstruction and Development.

The term 'World Bank' and 'Bank' refers to IBRD and IDA and 'World Bank Group' refers collectively to IBRD, IDA, IFC, MIGA and ICSID. As a true global community the World Bank comprises more than 10,000 staff from 168 countries. More than 38 per cent staff works in the Bank's 124 country offices with their Head quarter at washingtons D.C. U.S.A.

Purposes

The World Bank, which comprises the International Bank for Reconstruction and Development (IBRD) and the International Development Association (IDA) has one central purpose of the promoting economic and social progress in the developing member nations by helping raise productivity in order to enable their people to live a better and fuller life. The Bank is an international corporate institution whose capital stock is owned by its members. The principal purposes or functions of the IBRD are:

(*i*) to assist in the reconstruction and development of the territories of its member governments by facilitating investment of capital for productive purposes;

(*ii*) to promote foreign private investment by guarantees of or through participation in loans and other investments of capital for productive purposes;

(*iii*) where private capital is not available on reasonable terms, to make loans for productive purposes out of its own resources or out of the funds borrowed by it; and

(*iv*) to promote the long-range growth of international trade and the maintenance of equilibrium in the balance of payments of members by encouraging international investment for the development of the productive resources of members.

(*v*) to achieve equitable and sustainable economic growth in the national economies

(*vi*) to achieve global environmental sustainability

(*vii*) to over come global poverty by providing loan and risk management products.

The IBRD's loans are directed to help the members to build the foundation of sound economic growth. Loans made or guaranteed by the Bank are, except in special circumstances, for the purpose of specific projects of reconstruction and/or development. The Bank ensures that the proceeds of any loan are used only for the purpose for which the loan is granted. The Bank lends to only creditworthy borrowers. Loans are given for only those projects which promise high real rates of economic return to the country. About 14.5 per cent of IBRD's total cumulative loans have been made for the development of electric power and other energy; 15.0 per cent for the development of transportation; 15.0 per cent for agriculture; 6.0 per cent for industry; 10.0 per cent for finance; 5.0 per cent for education; 4.4 per cent for urban development; 4.0 per cent for water supply and sanitation; 11.6 per cent for multisector; 3.0 per cent for oil and gas; 2.2 per cent for population, health and nutrition; 3.0 per cent for public sector management and the balance 6.2 per cent for social sector, telecommunications, environment and mining, other extractive and urban development projects.

Membership and Organisation

Any country is eligible for membership of the IBRD if it subscribes to its Charter under the Bank's Articles of Agreement. Only those countries which are members of the International Monetary Fund (IMF) can be considered for membership of the IBRD. Subscriptions by member countries to the capital stock of the IBRD are related to each member's quota in the IMF, which reflects the country's relative economic strength. As of June 30, 2011 the total membership of the Bank comprised 187 countries. A member can withdraw at any time its membership of the Bank. Its withdrawal is, however, effective upon receipt by the Bank of a written notice from the member to that effect. Failure to fulfil its obligations toward the Bank may lead to suspension of a member. Even when a government ceases to be a member, it is obliged to repay on demand its portion of the losses, if any, sustained by the Bank on its operations till the date when that government ceases to be a member.

The Bank has a Board of Governors, Executive Directors, a President and other staff. All powers of the Bank are vested in the Board of Governors consisting of one governor and one alternate appointed for five years by each member. No alternate can vote except in the absence of his principal. Each governor has the voting power which is related to the financial contribution of the government which it represents. Although even the smaller member gets a minimum number of votes, however, the voting power of the smaller shareholder in the Bank is far outweighed by the voting power which the big shareholders enjoy. The United States of America with her subscription of US $31,964.5 million has 16.39 per cent of the total votes; Japan with her subscription of US $15,320.6 million has 7.86 per cent of total votes; Germany with her subscription of US $8,733.9 million has 4.49 per cent of total votes; while France and the United Kingdom each with her subscription of US $8,371.7 million have 4.30 per cent of the total votes. The Board of Governors meets once every year. Although mainly dealing with matters requiring only formal action, the annual meeting of the Board of Governors of the Bank is an important occasion for the informal exchange of views at high level on the major international financial and monetary problems.

Of the total of 24 Executive Directors who direct IBRD's general operations, five are appointed by the five biggest shareholders—the United States of America, United Kingdom, Germany, Japan and France—and the remaining nineteen are elected by the remaining members. Each Director holds voting power[1] in proportion to the shares held by his government. With certain exceptions,[2] the Board of Governors has delegated all its powers to the Board of Executive Directors which is responsible for the conduct of the general operations of the Bank. The Executive Directors function in continuous session and meet regularly once a month. A majority of the Executive Directors exercising 50 per cent or more of the total voting power constitutes a quorum. The President of the Bank acts as Chairman of the Board of Directors. At present (2012) Rebert B. Zoclliate 5 the present. He has no vote except a deciding vote in case of an equal division. He is chief of the operating staff of the Bank and is responsible to the Board of Governors of the Bank for the conduct of the ordinary business of the Bank and its organisation. He is assisted by a number of departments with each department under the charge of a separate department head.

The Executive Directors decide on International Bank for Reconstruction and Development (IBRD's) loan and guarantee proposals and International Development Association (IDA's) credit, grant and guarantee proposals made by the President, and they decide on policies that guide the Bank's general operations. They are also responsible for presenting to the Board of Governors, at the Annual Meetings, an audit of accounts, an administrative budget and an annual report on the Bank's operations and policies as well as other matters.

Capital

At the time of establishment in 1945, the authorised capital of Bank was US $10,000 million which was divided into 1,00,000 shares of US $1,00,000 each. Of this US $9,400 million was subscribed. A member's total subscription in the capital of the Bank is divided into following three parts.

I. Two per cent of the subscription is payable in gold or US dollars and is freely available for lending;

II. 18 per cent of the subscription is payable in member's own currency and is available for lending with the consent of the member whose currency is involved; and

III. The remaining 80 per cent of the subscription is not available for lending and is subject to call as and when required to meet the Bank's obligations.

1. Each member has 250 votes plus one additional vote for each 1,00,000 shares of the capital stock subscribed by it.

2. The exceptions include the admission of new members, the increase or decrease of the capital stock, the suspension of a member, decisions of appeals from interpretations of the Articles of Agreement made by the Executive Directors, approval of the formal agreements with other international organisations, decisions on distribution of the net income of the Bank and its liquidation. (The World Bank, *Principles and Policies*, p. 14.)

The capital of the Bank provides it with substantial lending resources from its paid-up capital (20 per cent) and much more sizeable *guarantee resources* to enable it to mobilise private capital for international investment 'either through the sale of Bank's obligations to private investors or through the Bank guarantees of private international credit.'

On June 30, 2005, the authorised capital of IBRD was $190,811 million, of which $189,718 million had been subscribed capital, $11,483 million had been paid-in and $178,235 million was callable. Of the paid-in capital, $9,032 million was available for lending and $2,451 million was not available for lending. Initially, $2,796 million of IBRD's capital was paid in gold or US dollars or was converted from the currency of the subscribing members into US dollars. This amount can be freely used by IBRD in its operations. Of the remaining amount of $8,682 million of the paid-in capital has been paid in the national currencies of the subscribing members and can be used for funding loans with the consent of the member whose currency is involved. Presently the IBRD obtains most of its funds by issuing bond in international capital market. In fiscal 2011 it raised USDeq. 29 billion by issuing bond in 26 currencies. As a corporate institution IBRD seeks not to maximise profit but to earn income to give loan to member countries The IBRD's allocatable income rose to $996 million in fiscal 2011 up from $706 million in fiscal 2010. To enhance IBRD's financial capacity the Development Committee endorsed a package of measures in closing an $86.2 billion general and selective capital increase with $5.1 billion in paid in capital and the board of Governers approved the capital increase resolution in March, 2010.

The principal purpose of increasing the Bank's capital resources is to increase its ability to lend for financing projects for economic development. Under the Bank's Articles of Agreement, the amount of Bank's disbursed and outstanding loans cannot exceed the total of its subscribed capital and reserves. Now that the Bank's capital resources have been augmented in a large measure, the Bank is in a stronger position to extend its lending activities. All members, especially the less developed nations, feel gratified at this step to augment the capital resources of the Bank.

Activities

The fundamental aims underlying IBRD's activities are:[3]
1. The Bank is not intended "to provide the external financing required for all meritorious projects of reconstruction and development (but) to provide a catalyst by which production may be generally stimulated and private investment encouraged..."
2. "The Bank should encourage necessary action by the member governments to ensure that the Bank's loans will actually prove productive. The promotion of sound financial programmes, the removal of unnecessary barriers, and the regional integration of production loans, where appropriate, are some of the fields in which the Bank may be able to exert a helpful influence", and
3. "The Bank must play an active rather than a passive role (and take advantage of its international cooperative charter) to initiate and develop plans to the end that the Bank's resources are used not only prudently from the standpoint of its investors but wisely from the standpoint of the world."

3. World Bank, *Op. cit.*

The Bank is more than the usual type of lending institution. Its concern is primarily to ensure that its loans make the greatest possible contribution to increasing the production, raising the living standards of people in the borrowing member country and opening opportunities for further investment in the borrowing member country. As a matter of policy, the IBRD does not reschedule payments. Under its Articles of Agreement, the Bank cannot allow itself to be influenced by the political character of a member country; only economic considerations are relevant. It also seeks to ensure that the developing country gets full value for the money it borrows from the Bank. Consequently, Bank's assistance is untied in that it can be used to purchase goods and services from any member country.

Recently, the scope of Bank's assistance to its members in their economic development has been considerably widened. From the monolithic, engineering-oriented projects of the late 1940s and the early 1950s the concept of 'bank projects' has developed to the multifaceted oriented-to-policy and demonstration projects of today and the limitations supposed to be inherent in the concept of lending for 'projects' have been greatly 'modified'. In its early years, most of the Bank's lending was for 'hardware' projects which simply added to physical works or goods in the economy, like dams, roads and power plants. Today, however, Bank's lending is frequently for 'software' like education, health and other social services, law and justice, population and nutrition, multisector, social sector, urban development, water supply, flood control and sanitation etc.

Bank's Lending Operations

The IBRD makes loans to members in anyone or more of the following ways:

1. by granting or participating in the direct loans out of its own funds;
2. by granting loans out of the funds raised in the market of a member or otherwise borrowed by the Bank; and
3. by guaranteeing in whole or part loans made by private investors through the investment channels.

The total outstanding amount of the loans made or guaranteed by the Bank is not to exceed 100 per cent of its total unimpaired subscribed capital, resources and surpluses. Before a loan is made or guaranteed to a member, the Bank ensures that the

(*i*) project for which the loan is asked has been carefully examined by a competent committee as regards the merits of the proposal;
(*ii*) borrower has reasonable prospects for the repayment of loan;
(*iii*) loan is meant for productive purposes; and
(*iv*) except in special circumstances the loan is meant to finance the foreign exchange requirements[4] of specific projects of reconstruction and development.

The Bank normally makes the medium and long-term loans, the term being related to the estimated useful life of the equipment or plant being financed. The Bank keeps itself informed on the projects which it finances by means of periodic reports received from the borrower and also

---

4. The Executive Directors have now confirmed the principle that the Bank should be prepared to provide for local expenditure on high-priority projects in cases where only the funds needed for direct imports would not provide adequate support.

through on-the-spot inspections made by its representatives. The interest rate charged by the Bank on its loans is the estimated cost to the Bank of borrowing money for a comparable term in the money market and is uniform without any distinction being made among the borrowers. In addition to the rate of interest, the Bank charges on all loans a commission of one per cent for the purposes of creating a special reserve against losses and 0.5 per cent charges for meeting the administrative expenses.

The Bank has made loans for specific development projects in the fields of agriculture, electric power, energy and mining, industry and trade, oil and gas, environment, finance, transportation, health and other social services population and nutrition, water, sanitation and flood protection, social sector, urban development, education etc. In view of the fact that underdeveloped countries need basic transportation facilities to develop their domestic economies and to provide new incentives for production, the Bank has also made loans for the development of transport. Such lending includes the financing of highway construction, railway rehabilitation and development, power and port development. During the past 60 years up to June 30, 2004, the IBRD had approved a total of 4,810 loans, net of cancellations, totalling $344,546 million to 129 countries representing more than 70 per cent of its members[5] for more than 3,500 projects in addition to granting six loans of $750 million to the International Finance Corporation.

The IBRD is now 67 years old. During this period of 67 years, it has lent a cumulative total amount of $407.4 billion, net of cancellations (upto June 30, 2005), in the form of 4,810 loans mostly to the developing countries for development projects. The total amount of outstanding loans was $109,610 million while the undisbursed balances amounted $24771 million.

IBRD lending generally falls into one of two categories: investment or adjustment lending. Investment lending is generally used to finance goods, works and services in support of economic and social development projects in a broad range of sectors. In contrast, adjustment lending generally supports social, structural and institutional reforms. In the past, majority of IBRD loans were for investment projects or programmes. However, the percentage of IBRD loans approved for structural adjustment lending over the past seven years occasionally exceeded 50 per cent. In FY 2004, new IBRD commitments for structural adjustment lending accounted 40.3% of total commitments. Its loans are long-term loans granted at conventional rate of interest for projects of high economic priority. The Bank has grown stronger by way of its lending activities year after year. During the fiscal year ended June 30, 2005, IBRD had given loans amounting $13.6 billion for a total of 118 projects, in 37 countries. During fiscal year 2006 to fiscal year 2011 the IBRD lent $144.2 billion. Thus, the total loan upto fiscal year 2011 amounts approximately to $621.6 billion.

In fiscal year 2005-08 the average lending by the IBRD was 13.5billion per year. This follows the record $44.2 billion in fiscal year 2010 when global financial carisis was at its peak. In fiscal year 2011 new lending commitments reached $26.7 billion including 132 operations in different countries The region wise distribution has been given in the following table.

From the table it is clear that Latin America and the Caribbean received the largest amount of loan followed by East Asia and Pacific.

---

5. The borrower may be a member government, a political subdivision or business, industrial or agricultural enterprise. Where the borrower is other than a government, the loan must be guaranteed by the member government in whose territory the project is located or by its central bank or some other comparable agency.

Table 34.1 Regionwise Destribution of Lending in 2011

Region	Percent	Amount *(US $ billions)*
Africa	<1	56
East Asia and Pacific	23.8	6370
South Asia	13.9	3790
Europe and Central Asia	20.5	5470
Latin America and the Caribbean	34.3	1196
Middle East and North Africa	7.3	1542
Total	100	26737

The following table shows the percentage share of IBRD's total lending of $26.7 billion during FY 2011 by sector.

Table 34.2 Sectorwise Distribution of lending

Sector	Share in Total Lending *(Percentage)*
Law and Justice and Public Administration	22
Health and Other Social Services	16
Transportation	20
Education	4
Finance	2
Water, Sanitation and Flood Control	11
Agriculture, Fishing and Forestry	5
Energy and Mining	14
Industry and Trade	5
Information and Communication	1

It is evident from the table that the IBRD is laying more emphasis on law and justice, transportation and health and social services in recent years.

In addition to its financial assistance operations, IBRD provides technical assistance to its member countries, both in connection with, and independently of, the lending operations. To assist its developing member countries, IBRD—through the World Bank Institute and its partners—provides courses and other training activities related to economic policy development and administration for governments and organisations which work closely with the IBRD. The IBRD, alone or jointly with the IDA, administers on behalf of donors, funds for specific uses and held in trust.

Technical Assistance and Other Activities

Apart from giving massive loan assistance to its members for various economic development projects, the IBRD has also been giving technical assistance to members on matters relating to loan operations, particularly in regard to[6]

1. defining priorities among different projects;

6. The World Bank, *Op. cit.*, p. 70.

2. modifications in the technical plans for project designed reduce its cost or to make it more efficient; and

3. administrative or organisational arrangements for a project or as to plans for its financing, including the raising of local capital.

The Bank has also provided technical assistance in development programming through various Survey Missions which make intensive studies of national resources of developing member countries and make recommendations to serve as the basis of long-term development programmes. The Bank has made surveys in the member countries providing valuable unbiased expert advice to governments of the surveyed countries on economic development and matters relating to policy formulation. The provision of technical assistance continues to be an integral and important element in Bank's work. By far the largest element is the advice and assistance given by the Bank in the normal course of economic and sector work, and identification, preparation, appraisal and supervision of projects supported by the Bank. For this reason alone, some of the Bank's member countries which no longer require a net transfer of resources from the Bank nevertheless want the Bank to continue project lending for key sectors of the economy in need of technical assistance, including assistance in establishing strong national institutions to promote development.

During 1998–99, the IBRD subjected its technical assistance to considerable scrutiny as its performance had not been satisfactory. The Bank-supported technical assistance in the early years and through the 1970s and much of the 1980s had focused on engineering assistance in designing bridges, dams, highways, telecommunication systems. In recent years, however, technical assistance has increasingly been directed at capacity building entailing a more complex process of creating and disseminating knowledge for development purposes at all levels of society. It is largely culture bound and process oriented. The Institutional Development Fund (IDF) established in 1993, has provided grants for capacity building activities not directly linked to Bank operations. Since its inception upto 1999, the IDF had made over 380 grants to 108 member countries. The Bank also maintains close working relationship with the United Nations Development Programme (UNDP). During FY 2004, the Economic Development Institute (EDI) held more than governance-related activities in more than a dozen countries across the five regions. In FY 2004, the World Bank Institute (WBI) delivered programmes in the Key Corporate priority areas of human development, poverty reduction and economic management, environmentally sustainable development and finance and private sector development.

During 1998–99, IBRD jointly with the IMF took the initiative in providing relief to the heavily indebted poor countries (HIPCs) by reducing their external debt burden by establishing the HIPC Debt Initiative Trust Fund. Preliminary country documents have been prepared for a number of countries. These will form the basis for consideration of eligibility for assistance under this initiative. So far Uganda, Bolivia, Guyana and Mozambique have benefited from this external debt initiative for heavily indebted poor countries.

In FY 2004, efforts to provide relief to the world's poorest and most heavily indebted countries continued to make good progress. As one part of comprehensive development strategy, the HIPC Initiative is well on its way to achieving its fundamental goal of giving a fresh start to HIPC by cutting their external debt to a manageable level. Twenty-six countries—two-thirds of the eligible HIPCs are now receiving relief which will amount to more than $40 billion from all creditors over time. Fourteen of these countries have completed the programme.

On March 29, 1965, the IBRD had established a machinery to settle disputes between member nations and foreign investors. This machinery known as the Convention on the

Settlement of Disputes between member States and nationals of other States started its operations on October 14, 1966. Under the Convention, an International Centre for Settlement of Investment Disputes (ICSID) had been established in 1966 for providing facilities for the settlement by voluntary recourse to conciliation or arbitration of investment disputes between contracting States and foreign investors who are nationals of other contracting States. By June 30, 2004, 140 member countries had ratified the Convention on the Settlement of Investment Disputes between member States and nationals of other States and were members of ICSID. The ICSID publishes a semi-annual law journal *"ICSID Review—Foreign Investment Law Journal"* and multi-volume collections of *"Investment Laws of the World'* and *"Investment Treaties".* Upto June 30, 2004, the total cases registered stood at 159 while during the FY 2004, 30 fresh dispute cases had been registered with the ICSID.

The Bank has also promoted international peace by successfully resolving difficult international disputes. It settled the dispute between the United Kingdom and the United Arab Republic on the nationalisation of Suez Canal. The Bank added another important feather to its cap when it successfully liquidated in September 1960 one of the toughest and most frustrating disputes between India and Pakistan over the sharing of the waters of the Indus System of rivers. But for the sincere endeavours which culminated in the creation of Indus Basin Development Fund, the dispute would have never been resolved. In bringing about this settlement of a most knotty problem the World Bank had been motivated by a sincere desire to remove sources of threat to lasting world peace.

In order to promote foreign direct investment in developing countries by providing guarantees to investors against non-commercial risks, such as expropriation, currency inconvertibility and transfer restrictions, war and civil disturbance and breach of contract the Multilateral Investment Guarantee Agency (MIGA) was established in 1988 through the efforts of IBRD. Its total membership at the end of 2005 stood at 165 and the amount of cumulative guarantees issued stood at $14.7 billion. In FY 2005, the MIGA has issued guarantees of $1.2 billion.

Bank's activities have been further diversified by taking urban development, population planning, and tourism within its activities and a Population Projects Development and Population Studies Division have been set up in the Economic Department of the Bank. Bank's first population mission, consisting of three staff members and three outside experts visited Jamaica in 1976-77 to assist that country's government in preparing a long-range family planning programme. To assist the developing countries in increasing their foreign exchange income through the development of tourism, the Bank has established a new Tourism Projects Department which provides technical assistance to the International Finance Corporation on possible new IFC tourist investment. Bank's tourism missions have visited many member countries.

The Millennium Development Goals (MDGS)

In September 2000 the global community through U.N.O. decided to accelerate economic growth by removing poverty, illiteracy and declared Millennium Development Goals which has the following objectives :
    1. to reduce present global poverty by half by 2015
    2. primary education to all children by 2015
    3. to remove gender disparity in education.

4. to reduce infant motality rate by two-third by 2015
5. to reduce mother mortality rate by three-forth by 2015
6. complete check of HIV/AIDS and other epidemics.
7. to achieve environmental sustainability and provide drinking water to all by 2015.
8. to develop non-discriminatory trade policy.

Since 2000 World Bank has been making sincere efforts to achieve the above MDGs. With a view to removing poverty WB has been giving sustainable financial support to member countries. However, according to World Development Indicators at the country level 49 out of 87 countries are on track to achieve poverty removal goals. In South Asia two-thirds of its 1.5 billions people still live on less than $2 a day which is international poverty line. However poverty rate is dropping in South Asia and in East Asia and Pacific.

Bank's support for education is commendable. The bank has committed nearly $25.3 billion for education since 2000. During fiscal year 2011 about 4 percent of total lending is allocated to education. The Bank is a key player in global effort to reach the education MDGs.- universal primary education and gender parity and to achieve quality learning for all, Thank in part to Bank's support, three quarters of the countries in East Asia and Pacific, Europe and Central Asia and Latin America and Caribbean have met or are on track to meet the education MDGs. During the past decade the total number out of school children worldwide declined from 106 million to 69 million and net primary enrolment in Africa rose from 58 percent to 76 percent.

If we look at a country level we find that in a populous country as India, over 98 percent Indian children now have access to primary school with in one Km. of their homes, 5 million children remained out of school in 2011 compared with 25 million in 2004, transition rate from primary to upper primary rose from 75% in 2002 to 84% in 2007. However poor children and children in rural areas have benefited less than children elsewhere.

The deaths of children under age 5 have declined in developing countries, falling from 101 per 10000 live births in 1990 to 73 in 2008. Similarly mother mortality rate has also gone down. The Bank is also actively engaged to reverse the spread of HIV/AIDs by 2015. A total of one million of new lending has been designated for operation that supports HIV/AIDs related prevention and treatment in Niger, Kenya, Swaziland, Argentina, Lesotho etc. which are prone to HIV/AIDs. However 26 million people are still leaving with HIV/AIDs in South Asia alone. From the above discussion it is clear that the Bank has adopted multipronged approach to achieve MDGs by 2015.

IBRD and the Developing Countries

In addition to the conventional loans which it has made available for development projects, IBRD has made sincere efforts to secure outside assistance from developed countries for underdeveloped countries. Due to the sincere efforts of the Bank, a consortium of 12 lending western nations known as the 'Aid India Club' comprising the UK, the USA, West Germany, Japan France, Canada, Italy, Sweden, Austria, Belgium, the Netherlands and Holland was formed to help India out of her foreign exchange difficulties. The 'Aid India Club' had provided $5,472 million financial assistance to India during the Third Plan period to aid India's economic development. The India Development Forum (IDF) which is a consortium of international donors had pledged further liberal total assistance of $66.9 billion to India during the fiscal years beginning from 1988–89 to 1997–98. The Bank also established a 'Help Pakistan Club' with the USA, the UK, Japan and other West European countries as members, with a view to inducing these countries to raise finances for Pakistan's planned economic development. At the request of both the donor and recipient governments, the Bank has taken the lead for many years in organising aid coordination mechanisms for a number of developing countries which received assistance from several bilateral and multilateral sources. The establishment of International Development Association, nicknamed as the 'soft loan window', from which underdeveloped countries have borrowed in hard currencies without being worried to repay in the same currencies and the amendment to the Charter of International Finance Corporation to enable it to provide equity capital to private industrial undertakings in the underdeveloped countries are important developments to which the World Bank has made substantial contribution.

The Bank's lending to poor countries among the developing countries with annual per capita income of below $296 has risen more than ten times from the low annual average of $800.0 million in 1981–82 to more than $10,000 million in 2001–02. Further, the sectoral shift in Bank's lending has been increasingly in favour of agriculture and rural development, tansportation, education, multi-sector, urban development, social sector and industry—sectors whose development is vital for the growth of borrower developing countries' economies. The poverty alleviation programme initiated by the Bank and the joint initiative by the Bank and IMF to reduce the external debt burden of heavily indebted poor countries are indeed praiseworthy.

Judging from all angles, the role of the Bank has been laudable. That the popularity of the Bank has increased as years have rolled by is proved by the rapid increase in Bank's membership. Taking an optimistic view of the future of the Bank, Mr. Illiff, Vice-President of the Bank had once remarked: "The time is not far when new countries entering the community of independent nations will have brought the total number of sovereign states to more than 100. Most of these new nations will be ill-supplied with development resources and determined to do something about their poverty. Most of them will, therefore, probably become members of the World Bank". The anticipation made by Mr. Illiff has become true and the Bank's membership has far exceeded the 100 mark, being now 187.

Criticisms

i. The *modus operandi* of the Bank has been criticised on various grounds. It is alleged that the bank charges a very high rate of interest on its loans even when its loans are guaranteed by governments of the borrowing member countries and there is no risk of loss of capital. There is truth in it because the latest loans that India has received from the Bank bear interest of over seven per cent including the commission of one per cent which is credited to the Bank's special reserve fund. While there is no harm in the Bank being run on sound business principles, it should not be devoid of missionary helping spirit vis-a-vis the borrowing countries because it was created to be an active instrument in the establishment of lasting international economic peace by making it possible for the economically weaker nations to come up by mobilising their resources. This purpose of the Bank can be fulfilled only when the rate of interest charged by the Bank is low enough for all countries to enable them to take loans from the Bank more frequently. Besides, the commission of one per cent which the Bank charges on its loans does not have much sense in it. Recently, however, the Bank has adopted a new more rational formula for determining the interest rate which it charges on new loans after July 1, 1976. Under the new formula, the Bank's lending rate is reviewed quarterly and will be 0.5 per cent above the weighted average cost of funds borrowed by the Bank in the preceding 12 months. As a consequence of this new approach to charge the rate of interest on the funds lent to members, the Bank's lending rate has become flexible and varies from time to time according to conditions prevailing in the international money and capital markets. For example, the Bank has lowered its lending rate from 7.9 per cent to 7.45 per cent.

2. The Bank's insistence of the presence of transfer or repaying capacity in the borrower country before granting the loan is faulty. The Bank should not apply orthodox standards of judging the transfer capacity of a borrower country. The transfer capacity follows rather than precedes the utilisation of the loan. It is created as the projects financed by loans materialise. In an underdeveloped country with its vast untapped resources awaiting exploitation, search for the transfer capacity before granting the loan is a misnomer that falls considerably short of wisdom on the part of the Bank.

3. The Bank is a non-political and non-partisan institution. It is expected to treat all members equally being enjoyed not to discriminate against some and in favour of others in the granting of loans. It should give loans purely on merit basis. However, in practice loans have not been given purely on merit and economic considerations. It is only recently that the economically backward countries of Asia and Africa have caught the eye of the Bank. And even now the position is alarming. While the Bank claims to have given loans in an increasing amount to the Asian and African countries during the past 60 years (from 1944 to 2004) with the result that the total percentage of loans to these two areas has risen from 23 per cent to more than 80 per cent, there is another aspect of the picture also. Asia and Africa taken together have the largest population area and unexploited economic resources in the world. Furthermore, the people are poor notwithstanding the richness of the resources. Consequently, the phenomenon of "actual poverty amidst potential plenty" is in full swing in these continents. On the contrary, Europe and Western Hemisphere are smaller both from population and area considerations and even then they have received huge amounts of loans. All this cannot be defended on economic consideration alone.

4. While it will be ungenerous on our part to belittle the importance of the Bank in reshaping and moulding economic structures of the countries to which the Bank's loans have flowed, it is still far from playing an effective role in developing properly the economies of the member countries. The financial help given by the Bank does not amount to more than a drop in the vast ocean of financial requirements so essential for various development projects.

India and IBRD

India is a founder member of IBRD. She had the privilege of presiding at the annual general meeting of the Bank held in Paris in September 1950. The Bank has sent several missions to India to assess the country's development programmes and also for field surveys of various projects. In November 1951, a World Bank mission came to India to review the progress made by the country in economic development and to assess what further assistance could be recommended to the Bank for consideration. In 1952, president of the Bank visited India and his visit was followed by that of other officials of the Bank for investigations into specific projects considered suitable for loan assistance. In February 1952, a Bank mission visited India to explore the possibilities of establishing a privately owned and operated corporation to finance the expansion and modernisation of private industry. At the invitation of the Government of India, a Bank mission visited India in April-June 1956 to review the economic situation with particular reference to India's progress made under the First Five Year Plan and to study the Second Plan. The Bank has appointed a Resident Representative in New Delhi who remains in close contact with the

Government of India in regard to the country's development plans and projects. A three-member Bank mission consisting of Sir Olive Franks, Mr. Allan Proual and Dr. Herman Abs came to India in February 1960 to survey the development programmes of country with particular reference to India's requirements of external resources. In January 1961, a Bank mission under the leadership of Joseph Rucinsky visited the country to prepare report for the May 1961 meeting of countries friendly to India to discuss the amount of foreign assistance to be given to India for the implementation of the Third Plan and to examine the projects for which the Bank and the International Development Association could extend aid. In November 1967, a Bank team headed by Mr. William Gilmartin visited India to make a survey of India's economy. In June 1970, a Bank mission visited Madhya Pradesh to examine State Government's request for a Rs. 50 crore project to level up and reclaim 2.25 lakh acres of the relatively shallow ravine land along the Chambal river and its tributaries. A seven-member Bank mission came to India in August 1970 to make a study in depth of the family planning programme in Uttar Pradesh to assess the feasibility of launching a special project for intensive family planning work.

India is the biggest single borrower of IBRD. Beginning since August 18, 1949 upto the end of June 2004, the Bank had lent India a hefty amount of $30,726.4 million in 190 loans. Of this loan assistance, more than 50 per cent was given for the improvement of transportation—railways, ports, roads and aircraft. Electric power development claimed about 20 per cent of the total loan assistance. Industry's share came to about 28 per cent while agriculture got about eight per cent of the total loan assistance given by the Bank to India.

What is noteworthy about the Bank's assistance to India is reflected in the fact that each year is indicator of the Bank's increased interest in India's economic development. India received $5,472 million financial help during the Third Plan period from the 12-nation Aid India Consortium formed through the sincere efforts of the Bank. She also received assistance of $66.9 billion from the India Development Forum (IDF) during eight years from 1988–89 to 1996–97. The signing of the Indus Water Treaty in September 1960 which ended the 13-year old dispute between India and Pakistan was a great triumph for the Bank's honest intentions to help the member countries. Judging from its past performance we can hope that the Bank will play a more progressive role in the future economic development of India.

For India, IBRD means many things. The Bank has not been merely a lending institution for India but has also served as a worthy counsel whom India has often approached for advice in difficulties. She has also successfully played a mediator's role in the Indus water system dispute with Pakistan.

India owes a deep debt of gratitude to World Bank. Not only does the Aid India Club, now renamed India Development Forum (IDF), a consortium of aid-giving nations owe its existence to the promptings of the Bank, but what is much more important is the fact that but for the keen interest taken by the World Bank in India's economic development, the massive assistance of $5,472 million which she received from the members of the Consortium for the development programmes during the Third Five Year Plan and the non-project aid of $3,700 million would never have been made available. The help provided by the Bank and friendly nations has been in the nature of a continuing flow. In the massive assistance of $5,472 million provided by the members of the Consortium during the Third Plan period, the United States was by far the largest contributor. Subsequently, the India Development Forum (IDF) had provided liberal total assistance of $66.9 billion to India during 1988–89, 1989–90, 1990–91, 1991–92, 1992–93, 1993–94, 1994–95, 1995–96 and 1996–97 to meet her financial needs. The massive financial assistance pledged by the consortium members is the largest foreign aid commitment since the Marshal Aid and is a landmark in the history of cooperative western aid to underdeveloped countries. Apart from providing the massive loan assistance to India, the Bank also paid the foreign exchange cost amounting to $862,000 of a study aimed at improving the transport of coal in India. The Bank also paid part of the cost of a study of the feasibility of constructing a new crossing for the road traffic over River Hoogly. A new crossing was urgently needed to relieve congestion on the famous Howrah Bridge which forms at present the chief link between Calcutta and the highly industrialised area to the West. The study was completed in 1965 at an estimated cost to the Bank of $87,500. In 1965, the Bank joined the Government of India in financing a survey of all modes of transport in the eastern region of the country in order to enable the Government to formulate transport investment programme for the Fourth and Fifth Five Year Plans (1966–76). The Bank paid the foreign exchange cost of $285,000. By June 1999, more than 1,400 Indians had participated in the various courses conducted by the EDI, the Bank staff college for senior officials from the developing countries whose work involves decisions on economic policy and the formulation of development programmes. In 1964, the Institute joined the Indian Institute of Management, Calcutta, in organising a project evaluation course in Jaipur for participants from the four Asian countries.

The Bank also gave a loan of $39 million in 1971 and 1972 to the Punjab Agricultural Project to boost farm production. Of this loan, $5.3 million had been earmarked for the purchase of 8,000 tractors during the two years. The machinery components cost was $9.4 million; special harvestors cost was $1.1 million and the cost of spare parts was worth $3.2 million. Projects financed with the help of Bank have already brought new productive sinews to strengthen India's economy. In short, the World Bank's role in India's economic development has indeed been more than substantial and but for the timely loans given to India by the Bank, the success of the country's economic development plans would have been considerably delayed.

During 1996-97, the Bank had lent India $626.5 million in four loans for agriculture and transportation raising the total cumulative loan assistance to $24,360 million in 164 loans which was 8.2 per cent of the IBRD's total lending. The International Development Association had also lent a total of $26,161.3 million in 219 soft credits for various development projects upto June 30, 1999. In FY 2003, IBRD had granted two loans of $836.0 million while IDA had granted five development credits of $686.6 million. In FY 2004, the World Bank (IBRD and IDA) had lent India $1,220 million in the form of two loans and three credits. This shows that upto June 30, 2004, the World Bank (IBRD and IDA) had granted massive combined loan and credit assistance of $61,796.6 million in the form of 91 loans and 255 credits to India for the planned economic development of the country's economy. In fact, India occupies the pride of place of being World Bank Group's largest borrower. The World Bank Group (IBRD and IDA combined) had upto June 30, 2004 lent India US $61,796.6 million in the form of 446 loans and credits. Due to the sincere efforts and interest shown by the World Bank in India's economic development, India had been promised $3 billion aid by the IBRD and IDA during 1997-98. During 1998-99, India received three loans amounting US $40.0 million from IBRD and four credits of US $654.8 million from the IDA.

During 2002–2003, IBRD lent India $836.0 million in two loans while IDA granted $686.6 million in the form of five development credits taking World Bank group's cumulative financial assistance to $61,796.6 million. In FY 2004, IBRD len India $460 million for two projects, one in Uttar Pradesh and the other in Andhra Pradesh. For the three-year period 2005–2007, the World

Bank has promised to len $9.0 billion mostly for the development of railways, power, roads and water resources. On June 30, 2005 the total loans to India amounted to $9.3 billion.

In fiscal year 2011 India was the Bank's largest borrower with a total commitment of $5.5 billion. India is facing the problem of adequate infrastructural facility. To overcome this problem the government has taken serval large scale programmes. This includes the Pradhan Mantri Gram Sarak Yojana, the Prime Minister Rural Road Programme for which the Bank approved $1.5 billion loan in fiscal year 2011, the project plans increase road comectivity in several states (Himachal Pradesh, Jharkhand, Meghalaya, Punjab, Rajasthan, Uttra Khand and Uttar Pradesh) over five years, constructing 24200 km all weather roads which will benefit 6 million people In addition to it $350 million have been approved for the Second State High Way Improvement Project. In May 2011 the Bank approved $1 billion credit and loan to support the government 'Mission Clean Ganga.' The Bank also sanctioned $975 million loan to Indian Railways Thus, the Bank is keenly interested is the development of this country.

Evaluation

Although IBRD may have belied the expectations of some nations, in appraising the Bank's role we should not forget the limitations within which the Bank works. The Bank has been largely instrumental in accelerating the pace of economic development in different countries of the world. Although the IBRD has failed to finance all the development projects, it has nevertheless financed a large number of them which have proved a remarkable success. The Bank has also played a significant role outside the sphere of finance by serving as a mediator between different countries on major knotty issues, e.g., between the United Kingdom and the United Arab Republic on the nationalisation of Suez Canal and between India and Pakistan in resolving the Indus Basin water dispute. It can be hoped that in future IBRD will be in a stronger position to render financial assistance to the member countries with its increased capital resources and with the active co-operation of its affiliates—International Development Association and the International Finance Corporation.

The Bank is actively engaged with global partners on Climate change which threatens to erode development gains around the world In fiscal year 2011 in partnership with other multilateral development banks (MDBs) the Bank ramped up operations and imprementation of project under Cilmate Investment Fund (CIF), Clean Technology Fund (CTF) and Strategic Climate Fund (SLF). At Ca. and N. climate change conference in Dec, 2010 in cancun the Bank announced to bring developed under developing countries together on International Forum.

The Operation Evaluation Department (OED) is an independent unit in the World Bank that reports directly to the Bank's Board of Directors. It provided an objective basis for assessing the Bank work and allow the Bank staff to learn from past experiences.

Suggested Readings

1. World Bank, *Annual Reports.*
2. World Bank, *Principles and Policies.*
3. IMF and World Bank, *Finance and Development,* Volumes 1–39.

Questions

1. State the functions of the International Bank for Reconstruction and Development. How far has the World Bank been successful in performing these functions?

2. Explain the role which the World Bank has played in the economic development of its less developed countries. How far has been India benefited from her membership of the World Bank?

3. Explain the Millenium Development Goalo and discuss the performance of the IBRD in this respect.

CHAPTER 35

International Development Association (IDA)

Introduction

The IBRD unanimously adopted a proposal on October 1, 1959 for setting up in principle the International Development Association. It was "resolved that with respect to the question of creating an international development association as an affiliate to the Bank, the executive directors, having regard to views expressed by the governors and considering the broad principles on which such an association should be established and all other aspects of the matter, are requested to formulate articles of agreement of such an association for submission to the member Governments of the Bank".

The idea to establish an international institution to give loans on liberal terms to underdeveloped countries was first mooted by Senator Monroney which was later approved by President Eisenhower. In August 1959, President Dwight Eisenhower publicly supported the idea of establishing the IDA. He included it as one of the three proposals for raising international liquidity and directed the Administration to study the feasibility of implementing it in cooperation with the International Monetary Fund and IBRD. Thus, the credit for taking initiative for the establishment of the IDA goes to the United States of America and IBRD which felt that the burden of development assistance should be more widely shared by the more industrialised countries of the world.

The IDA, nicknamed as the 'soft loan window' at which the underdeveloped countries can borrow in hard currencies without worry to repay in the same currencies, formally commenced its operations on November 8, 1960. The IDA gives development credits more generously to the poorer developing countries and its loans are more flexible than the IBRD loans, being for 15 years at least. The IDA finances a certain percentage of the cost of a project that is meant not only for meeting the foreign exchange component of the project but also a part of the cost of local currency. Many countries which cannot borrow from the IBRD for projects because these are not regarded creditworthy by the Bank are able to secure credit from the IDA. The credits granted by the IDA are free of interest and only an administrative charge is levied on the IDA credit. The IDA has financed wider range of projects than the World Bank has been able to finance. The only criterion for the IDA to grant credit for a project is that the project to be financed should be of a 'high development priority'. The other salient feature of the IDA credits is that these credits can also be repaid in local currencies of the borrowing countries. Consequently, borrowers have not to worry about finding the scarce foreign exchange at the time of repayment of credits. In short, the IDA provides loans on terms which do not bear heavily on the external balance of payments position of the borrower members.

IDA is the world's largest single source of concessional financial assistance for the poorest countries, and it invests in basic economic and human development projects. Eligibility process to IDA resources is governed by two basic criteria: a country's relative poverty (as measured by per capita income) and its lack of creditworthiness for IBRD resources. The IDA provides development finance to less developed members on easy terms that bear less heavily on the balance of payments position of the debtor countries than do IBRD loans which are granted on conventional terms. All the credits so far extended by IDA have been interest-free repayable in foreign exchange over 50 years, the repayment to begin in easy instalments after a grace period of ten years spread over the remaining forty years. To meet IDA's administrative cost, a small service charge 3/4 of one per cent per annum is payable on amounts withdrawn and outstanding.

All those countries which are members of IBRD are eligible for IDA's membership. At the end of June 2004, 165 countries were members of IDA and the total subscriptions and contributions of IDA aggregated $123,029.7 million. Subscriptions are payable in five annual instalments. For purposes of subscriptions, distinction has been made between the more industrialised (Part I) and less developed (Part II) countries. The 27 developed countries have been designated as Part I countries and the rest 138 counties have been named as less developed countries and included in Part II. Members in Part I list consist of those countries whose subscriptions may be freely used or exchanged by IDA and who have participated in the replenishment of the IDA's resources. Part II countries are required to pay only 10 per cent of their subscriptions in gold or freely convertible currencies and the rest in their own currencies. The subscription in national currencies from the less developed countries is not to be converted into other currencies or used to finance exports from these countries without the consent of the country concerned. The USA, UK, Germany, Japan and France are the five biggest subscribers and providers of supplementary resources to IDA. These five biggest subscribers have each provided US $28,691.8 million; US $9,750.3 million; US $14,196.9 million; US $26,515.2 million, and US $8,667.3 million respectively in the form of subscriptions and supplementary resources. India's subscription to the IDA's capital is US $56.6 million; of this 10 per cent has been paid in gold or convertible currencies and the balance has been in rupees. The total subscriptions and contributions of 28 Part I members aggregated US $119,114.0 million while those of the 137 Part II members aggregated US $3,915.7 million on June 30, 2005.

Lending Operations

Upto the end of June 2004, the total cumulative amount of development credits extended by IDA aggregated US $151,390.6 million in the form of 3,745 development credits for various development projects in 112 member countries. In FY 2005, IDA's total lending of US $8.7 billion was committed for 160 projects to 66 countries including 23 adjustment lending projects worth $1,698 million. During fiscal years 2006-2011 the total lending of IDA amounted to $77.4 billion. In the fiscal year 2011 total lending was $16.269 billion which was much higher than $14.550 billion in fiscal year 2010. Regionwise distribution of $16.269 billion in fiscal year 2011 is given belwo.

Table   35.1   Regionwise   Distribution   of   Lending.

Region	Amount ($ billion)	Per cent
South Asia	6.4	39.3
Middle East and North Africa	0.123	0.75
Latin America and the Caribbean	0.46	2.8
Europe and Central Asia	0.65	4.0
Africa	7.0	43.0
East Asia and Pacific	1.62	10.2
Total	16.26	100.0

The following table presents total lending by the World Bank in fiscal year 2011 in term of percentage.

Table   35.2   Sectorwise   Distribution   of   Lending

Purpose	Per cent
Water, Sanitation and Flood Protection	11
Transportation	20
Energy and Mining	14
Industry and Trade	5
Health and Other Social Services	16
Agriculture, Fishing and Forestry	5
Law, Justice and Public Administration	22
Information and Communication	1
Education	4
Finance	2

It is evident from the tables that the largest of Bank's (IBRDIDA) resources went to Africa followed by South Asia. The smallest share is that of Middle East and North Africa. So far as sectoral distribution is concerned, Law, Justice and Public Administration were leading sectors followed by Transportation and Health.

IDA's Resources

FY 2004 marked the second year of IDA-13, which will fund commitments for fiscal years 2003 to 2005. IDA-13 will provide a total of SDRs 18 billion (about $23 billion) in concessional resources to IDA-eligible borrowers over the three-year period. Its amount includes SDR 10 billion (approximately $13 billion) in new donor contributions; IDA's internal resources including repayments of principal from past credits and service charges of the order of SDR 7.3 billion (about $9 billion); IBRD net income transfers of SDR 0.7 billion (about $0.9 billion) and a small carryover of donor resources from the previous replenishment. Negotiations for the 14th Replenishment of IDA (IDA-14) was concluded in Feb. 2005. IDA-14 provides 24.2 billion SDR (about $ 35.3 billion) which will be available to eligible borrowers over the fiseal years 2006-8.

IDA is financed Largely by contributions from donor government. Additional financing comes from tranfer from IBRD's net income, grants from IFC and borrowers repayment of carlier IDA credits. Every three years governments and representatives of borrower countries meet to discuss IDA's policies and priorities and agree on volume of new resource required to fund its lending per year over the subsequent three years. During the 15th Replishment (IDA-15) which covers fiscal year 2009-11 total resources were $43.7 billion. Discussion for IDA-16 replishment concluded in Dec. 2010, resulting a record level of finance of $32.8 billion for fiscal year 2013-14 from 52 countries including 7 new donors.

India and IDA

India has immensely benefited from the credits granted to her by the IDA. Up to June 30, 1999, India had received massive credit assistance of US $26,161.3 million in the form of 229 credits to finance her various development projects. This is by far the largest credit assistance that the IDA has given to any single member country so far. The first credit of $60 million was granted in June 1961 to assist a programme of road construction and improvement. The second credit of $6 million was given in September 1961 for the expansion of tubewell irrigation in Uttar Pradesh. The economic benefits of this project financed through the IDA's credit had been considerable. The credit had provided finance for the drilling and equipping of 800 tubewells which have irrigated 320,000 acres of land in Uttar Pradesh. This has increased the value of farm production in the area to the tune of Rs. 10 crores. The IDA granted another credit of $10 million for financing the flood control and drainage scheme in Punjab. As a result of the credit, the surface drainage problem has been solved over an area of about eight million acres in the state. In addition, two more credits each of the amount of $8 million and $4.5 million had been given by the IDA to assist the Salandi Project in Orissa for the irrigation of 2,25,000 acres of land and to finance the completion of the Shetrunji Project in Gujarat respectively. In June 1962, the IDA gave a $15 million credit for improving the Sone River irrigation system in Bihar. The scheme financed by this credit has provided irrigation facilities to one lakh acres of land in western Bihar. In June 1962, the IDA extended $13 million credit for Puma River valley irrigation project in Maharashtra State. The project has provided irrigation facilities to 152,000 acres of dry land and generates 15,000 KW electric power in the State of Maharashtra. The 50-year credit is interest free. In August 1962, IDA gave another credit of $17.5 million for Koyna Power Project in Maharashtra. The credit has helped to increase the production of electric power to 580,000 KW.

In September 1962, India received two more credits of $42 million and of $18 million for the development of telecommunications and port of Mumbai respectively. In March 1963, the IDA gave a $67.5 million credit to Indian railways to complete the electrification and purchase of other essential equipment. In May 1963, a credit of $20 million was granted for the Kothagudam Power Project. In June 1964, the IDA surpassed all previous records by granting $90 million credit to enable companies in three major industries—commercial vehicles, industrial machinery, and construction equipment—to import components and materials on a large scale and to produce more capital goods. As a result of this credit, output in the concerned industries has increased by about 30 per cent. During third quarter of 1964, the IDA granted $33 million credit for the development of telecommunications and $62 million credit in the fourth quarter of 1964 for the development of Indian railways system. During 1966-67, the IDA extended a total credit of $215 million to India as the IDA's contribution to the $900 million non-project assistance agreed to by the members of the Aid India Club during the Indian fiscal year ended on March 31, 1967. The assistance was designed to support economic policy changes, including the easing of import and other controls, initiated by the Government of India. In June 1964, the IDA extended four credits to help the selected Indian industries manufacturing industrial and electrical machinery, heavy construction equipment, commercial vehicles, machine tools and cutting tools, agricultural tractors, fertilisers and pesticides. A credit of $68 million was made for the railways in June 1966.

During the fiscal year 1968–69, IDA extended two credits to India—one credit of $27.5 million for the development of telecommunication facilities and another credit of $125.0 million for the industrial imports. During 1969-70, the IDA had granted four credits totalling $200 million to India consisting of one credit of $55 million for the expansion programme of Indian railways, one credit of $7 5 million for the industrial imports and two credits of $70 million (of $35 million each) for the development of irrigation facilities. During 1971-72, IDA provided $412.2 million in the form of 11 credits for the diversified economic development of country's economy. These included a $75 million credit for industrial imports to enable the selected Indian industries to expand their production of fertilisers and capital goods essential for the development of key sectors of the economy—agriculture, electric power and transport—a $83 million credit for the development of shipping transport; a $12 million credit for agricultural education for developing two new agricultural universities in Assam and Bihar; a $25 million credit to meet the development needs of the small-scale industry sector over the next two years; a $58 million credit to increase the production of fertilisers; a $85 million credit for the expansion of power transmission facilities in Assam, Delhi, Haryana, Kerala, Madhya Pradesh, Maharashtra, Mysore, Orissa, Punjab, Rajasthan and Tamil Nadu; and a $8 million credit for the development of agricultural wholesale markets in the State of Karnataka. During 1972–73, among others, IDA gave India a credit of $100 million to help 700 medium and large-scale enterprises to maintain and expand production by enabling them to make industrial imports. Another credit of $80 million was given for the expansion of telecommunications and manufacture of telephone equipment. India also received a credit of $85 million from IDA for financing the imports of equipment from abroad and for technical know-how. During 1973–74, the IDA's credit assistance to India totalled $390 million spread over seven credits. Of these seven credits, three credits involving a total of $75 million were given for the development of a dairy project in Karnataka, for investments in minor irrigation, tubewells and pump-sets to farmers in Bihar and for improvement of apple processing and integrated marketing facilities in Himachal Pradesh. One credit of $150 million was given for non-project lending to medium and large-scale business enterprises in selected priority industries to maintain and expand production. A $80 million credit was also given for the expansion of railways during the Fifth Five Year Plan period. The IDA also gave a credit of $50 million for expanding the capacity of a large fertiliser plant near Mumbai. A credit of $235 million was given for 44 schemes covering water supply, sewerage and drainage, roads and traffic improvement, environmental hygiene, garbage disposal and housing and area development. During 1975-76, India received the total credit assistance of $684 million for the development of railways, power, rural electrification, for cotton development and for industrial imports.

During 1976-77, India received nine credits of $481 million from IDA. Of these nine credits, eight credits of $314 million were given for the agricultural and rural development projects while one credit was given for power development in the country. In 1977–78, India received from the IDA total assistance of $951.5 million in 13 credits out of which 11 credits for $694.5 million were given for flood control, irrigation, fisheries, research and extension projects having direct bearing on agriculture and rural development. One credit of $200 million was given for power development while another credit of $87 million was given for rural development. During 1978–79, India received $1,920.0 million in 11 credits for agriculture and rural development, power, transportation and water supply and sewerage projects. The provision of irrigation and flood control, development of forestry, railways and power generation in the country have attracted the IDA's attention. During 1980-81 and 1981–82, India received 11 credits amounting to $1,281.0 million and five credits amounting to $900.0 million for the development of irrigation and drainage, energy, fertiliser and other chemicals, research and extension projects. During 1986–87, India received four credits amounting to $677.7 million. India received two credits amounting to $182 million for irrigation and drainage, one credit of $85 million for research and extension and one credit of $119.6 million for the development of highways.

During 1991–92, India had received five credits of $1,023.5 million from the IDA for forestry, population, health and nutrition and non-project. During 1996-97, India received from the IDA six development credits of the total amount of $903.0 million for agriculture, environment, health, population and nutrition programmes. During 1998-99, the IDA gave India four credits of total US $654.8 million for agriculture, education, population, health and nutrition programmes. During 2002–03, IDA provided India five development credits of US $686.6 million for environmentally and socially sustainable development ($370.6 million) and for human development ($316.0 million). In FY 2004, IDA granted four credits of $1,000 million to the states of Uttar Pradesh, Andhra Pradesh, Maharashtra and Rajasthan for reducing transport congestion, drinking water and sanitation, supporting a health system project and to strengthen economic reforms. Upto June 30, 2007, cumulative development credits stood at 260 with cumulative amount of US $30,531.2 million. The IDA along with IBRD gives financial support to India to increase road connectivity and helps 'Mission Clean Ganga' with financial supports.

This is an impressive account of IDA's generous help in country's economic development. Without the massive soft credit assistance made available by the IDA to India the pace of her economic development would have been considerably slow. Judging from past and present, the future holds many bright promises for India from IDA.

The IDA supplements the financial efforts of the IBRD by giving 'soft loans' to the poor countries. There are many sound projects which IBRD as a matter of policy does not finance. The IDA finances these projects. Many economically sound projects like the steel plants, fertiliser factories and machine tools factories besides several other projects for creating 'social capital' like the construction of roads, bridges, slum clearance, education and urban development have been financed by the IDA. The developing countries of Asia, Africa and Latin America look to the IDA with great hopes for their economic development.

Suggested Readings

1. IDA, *Charter of the Articles of Association of the IDA* and *Annual Reports.*

Questions

1. Explain the purposes of the IDA. How far has the IDA helped the less developed countries in their economic development? Discuss with particular reference to India.
2. Explain the role of the IDA in promoting the economic development of world's less developed countries. How has India been benefited by her membership of the IDA? Discuss fully.

CHAPTER 36

Asian Development Bank (ADB)

Introduction

The Asian Development Bank, popularly known as the ADB, is a multinational development financial institution established for the purpose of lending funds, promoting investment and providing technical assistance for promoting the economic and social progress of its developing member countries (DMCs) in the Asia and Pacific region. The Charter of the Asian Development Bank was ratified by the majority of the signatory countries (31) in early 1966 and it started its operations in Manila in December 1966. The Bank operates under the aegis of ECAFE. Since the idea of establishing the ADB was first proposed by the United Nations Economic Commission for Asia and the Far East (ECAFE) in 1963, the speed with which it has been established represents great achievement by its architects and sponsors. The ADB has two significant features. Firstly, it is an Asian Bank conceived by the ECAFE. Its headquarters are located in the ECAFE region in Manila, the Philippines. The ADB's capital is subscribed by its 48 regional members and 19 countries from outside the region. The President and seven of the ten directors also come from the region. Secondly, unlike certain regional financial institutions, the membership of the Bank extends beyond the region. Consequently, many countries outside Asia have contributed to the capital of the Bank and are represented on the Board of Directors and professional staff of the Bank. Such members numbering are designated as non-regional members 19. The Bank is engaged in promoting the economic and social progress of its developing member countries in the Asia-Pacific region. As of 31 Dec. 2010 the share of voting power of regional numbers was 71.42 p.c. and those of non-regional member was 28.58 p.c. Because of the highest share of Japan in subscribed capital (17.72 p. c.) its share of voting power (17.47 p.c.) is also largest followed by China Republic (6.15 p.c.) and India (5.24 p.c.). Among the non-regional members the share of voting power of Canada (5.02 p.c.) in the highest followed by U.S.A (5.03 p.c.) and Germany (4.22).

At the end of 2010 total staff of ADB was 2833 from 59 out of 67 member countries. Women accounted for 29.1 p.c. of the international staff.

Rationale of the Bank

The justification for the establishment of the ADB is firmly based on the hypothesis that accelerated economic growth in poor countries is functionally related to the rate of capital formation in these countries. Since a high rate of capital formation in the low per capita income countries of the Asian region is not possible through the low levels of domestic saving, financial resources from outside sources must be made available to these poor countries. The idea to establish a regional development bank for Asia emerged from the discussions at ECAFE's First Ministerial Conference on Asian Economic Cooperation held in Manila in December 1963. This initiative was taken up by a working group of experts whose findings were submitted to the ECAFE session in Wellington in March 1965 which, in turn, led to the formation of a Consultative Committee. The Committee drafted the Agreement establishing the Asian Development Bank which was subsequently adopted at the Second Ministerial Conference on Asian Economic Cooperation held in Manila in November-December 1965.

The Charter of the Bank was formally signed on December 4, 1965. It was, however, left open for additional signatures until January 31, 1966 when 31 countries had signed the document. The Charter became effective on August 22, 1966 with the ratification by 15 of the 31 signatories. By September 30, 1966, 30 nations had satisfied conditions for the ADB's membership and had remitted the first instalment of their paid-in capital subscription. Preliminary arrangements for the Bank's establishment were made by a Preparatory Committee formed in December 1965 and the Inaugural Meeting of the Bank's Board of Governors was held in Tokyo on November 24-26, 1966. The Bank formally commenced business on December 19, 1966. During the past 38 years, the Bank has maintained its role as a catalyst in promoting the development of the most populous and fastest growing region in the world today.

Functions

The ADB's principal functions are:

(*i*) to make loans and equity investments for the economic and social advancement of the developing member countries (DMCs);

(*ii*) to provide technical assistance for the preparation and execution of development projects and programmes and advisory services;

(*iii*) to promote investment of public and private capital for development purposes; and

(*iv*) to respond to requests for assistance in coordinating development policies and plans of DMCs.

In its operations, the Bank gives special attention to the needs of the smaller or less-developed countries and priority to regional, sub-regional, and national projects and programmes which will contribute to the harmonious economic growth of the region as a whole and promote regional cooperation.

Sources of Funds

Initially, the Asian Development Bank had an authorised capital of US $2,985.71 million of which $1,091.75 million had been subscribed. Of the subscribed capital, one-half was in the form of 'paid-in' capital and the other half remains as 'callable' shares to serve as security for the obligations of the Bank. The 'callable' capital constitutes, in effect, a guarantee for the Bank's securities and thus facilitates the Bank's borrowing of funds in the capital markets of the world. The 'paid-in' portion is to be paid in five equal annual instalments, one-half of each instalment must be paid in local or convertible currency and the other half may be paid in local currency. The Bank may also accept non-interest bearing demand notes in lieu of the amount payable in local currency provided such currency is not needed by the Bank for the conduct of its operations. The Bank's authorised capital was for the first time increased in November 1972 by 50 per cent to

$3,707 million. The second general increase in Bank's authorised capital was approved by Bank's Board of Governors on November 8, 1976. As a result of the second general increase, the Bank's authorised capital increased from $3,707 million to $7,965.1 million. At the end of 2003, 10 per cent of the increase had been paid-in while the balance 90 per cent was callable. At the end of 2003, the ADB's authorised capital stood at equivalent of $51.996 billion while the subscribed capital amounted $51,997 million. The Bank's callable capital at the end of 2003, accounted for about 90 per cent of its subscribed capital. Of the total subscribed capital amounting $51,997 million, $3,6.571 million was paid-in while $48,339 billion was callable capital. At the end of 2003, 57 member countries had subscribed to the Fourth General Capital Increase (GCI-IV) amounting $24,675.4 million. The financial resources of the Bank consist of the ordinary capital resources (OCR), comprising the subscribed capital, reserves, and funds raised through borrowings; and special funds comprising contributions made by member countries, accumulated net income and amounts previously set aside from the paid-in capital. Loans from OCR on non-concessional terms account for 70 per cent of the cumulative Bank lending. Loans from the Asian Development Fund are made on highly concessional terms and almost exclusively to the poorest borrowing member countries.

The Bank has borrowed funds for its ordinary operations from the capital markets of Asia, Europe, Middle East and North America as well as international capital markets and from certain member countries' central banks. Its borrowings in the capital markets are backed by the Bank's callable capital which at the end of 2003 accounted for about 90 per cent of its subscribed capital. Of the total subscribed capital of $51,997 million at the end of 2003, the 45 regional members' share stood at $32,883 million comprising 63.241 per cent while the 18 non-regional members contributed the balance $19,114 million comprising 36.759 per cent of subscribed capital.

At the end of 2010 authorised and subscribed capital amounted to $163.8 billion and $143.9 billion respectively. Other resources in the OCR in the form of revenue and net realised gains amounted to $1.2 billion. Resources in the form of contribution and revenue during 2010 in ADB's special fund totaled about $ 0.6 billion This included the Asian Development Fund, Technical Assistance Special Fund, Japan Special Fund, Asian Tsunami Fund, Pakistan Earth Quake Fund, Regional Cooperation and Integration Fund, Climate Change Fund etc. In addition to it ADB also raised $ 14.9 billion in medium and long term funds through public bond issues.

The Bank may increase its funds by increasing its capital, issuing bonds or accepting contributions to what are known as 'Special Funds'. The Board of Governors, subject to a two-thirds vote, may decide to increase the authorised capital. A member may elect to increase its subscription provided that such an increase does not have the effect of reducing the percentage of shares of regional members below 60 per cent of the total subscribed shares. The Bank may also sell its bonds in the international money markets with the approval of the countries in which the bonds are to be sold.

Membership

The membership of the Asian Development Bank is open to:

1. Members of ECAFE,
2. Associate members of ECAFE, and
3. Other countries in the ECAFE region which are members of the United Nations or any of its specialised agencies.

Admission to the membership of the Bank may be obtained upon the affirmative vote of two-thirds of the total number of Governors representing not less than three-fourths of the total voting power of the members. At the end of 2012, 67 countries were members of the Bank comprising 48 regional members and 19 non-regional members.

Loans and Technical Assistance

The lending operations of the ADB are divided into ordinary and specialised operations. The ordinary operations relate to those lending activities of the Bank which are financed out of the ordinary capital resources of the Bank. Ordinary loan operations of the Bank ordinarily consist of the financing of the foreign exchange or local currency component of the cost structure of specific projects in so far as the Bank can make such currencies available to the borrower countries from its resources. In addition to giving direct loans for particular development projects, the Bank also lends to national development banks or other suitable entities to enable them to relend for specific projects which in themselves are not large enough to warrant direct supervision by the Bank.

Special lending operations are financed from the various Special Funds such as the Technical Assistance Special Fund, Asian Development Fund, Agricultural Special Fund, and Multipurpose Special Fund which are managed by the Bank. The Charter of the Bank contemplates assistance for projects of high development priority and requiring loans for longer deferred commencement of repayment and lower interest rates than those established for ordinary operations. The Bank may earmark up to 10 per cent of its 'paid-in' capital as Special Funds which it may use for purpose of 'soft' lending on the lines mentioned above. The allocation of such funds, however, is conditional upon an authorisation by two-thirds of the total number of Governors. The Bank is authorised to accept Special Funds which the Bank may administer on terms agreed upon with the donors so long as the purposes are consistent with the Bank's development objectives and authorised functions. The Bank has received contributions to its Special Funds from the United States of America, the United Kingdom, Japan, Germany, Austria, Denmark, India, Pakistan, the Netherlands, Finland, Switzerland and Canada. The Charter of the ADB stipulates that the Special Funds of the Bank must be held, used, invested or otherwise disposed of entirely separate from the ordinary capital resources of the Bank.

In addition to lending for specific projects to the members in the ECAFE region, the ADB also provides technical assistance to its member governments, their agencies or subdivisions, private firms in their territories and regional institutions. The technical assistance provided by the Bank is designed to help in (i) the preparation, financing and execution of development plans and projects, including the consideration of priorities and the formulation of loan proposals for specific national or regional development projects; and (ii) the functioning of existing institutions or the creation of new institutions on a national or regional basis in such fields as agriculture, industry, and public administration among others.

In addition to lending and technical assistance activities, the Asian Development Bank also organises technical assistance missions with the approval of the potential recipient member countries. These missions may also be sponsored in collaboration with international organisations if they suit the convenience of the countries concerned. The technical assistance provided by the Bank may be either on a grant or on a loan basis. The Charter of the Bank limits the funds available for technical assistance of a non-reimbursable nature to the Bank's income to any special Funds resources received by the Bank for this purpose.

Organisation

The Bank is governed by a Board of Governors, a Board of Directors, a President, a Vice-President and other officers and staff. The Board of Governors is the highest policy making body of the Bank. At the end of 2003, it consisted of 63 members representing 45 regional and 18 non-regional member countries. Each member nominates one Governor and one Alternate Governor. Under Article 28 of the Bank's Charter all the powers of the Bank are vested in the Board of Governors which delegates its powers to the Board of Directors except on certain matters such as admission of new members, change in the authorised capital stock of the Bank, election of Directors and the President and amendment of the Charter. The total voting power of each member consists of the sum of its basic votes and its proportional votes. The basic votes are made up by the equal distribution among all members of 30 per cent of the aggregate sum of the total voting power of all members. The number of proportional votes is equal to the number of shares that members hold of the Bank's capital stock. In the Board of Governors, each Governor is entitled to cast the number of votes of the country he represents.

The Board of Governors meets in a formal session once a year for the Bank's Annual Meeting. Under Article 28 of the Bank's Charter, the Board of Governors elects a 12-member Board of Directors, eight elected by the regional members and four by the non-regional members. Each Director appoints an Alternate. The President of the Bank is the Chairman of the Board of Directors. The responsibility for the general direction of the operations of the Bank rests with the Board of Directors. The Board of Directors exercises all powers delegated to it by the Board of Governors. It takes, in conformity with Bank's Charter, decisions concerning loans, guarantees and other investments by the Bank, borrowing programmes, technical assistance and other operations of the Bank; it also approves the administrative budget of the Bank and submits accounts pertaining to each financial year for approval by the Board of Governors. The Board of Governors conducts an election for the Board of Directors every two years.

The Directors hold office for a two-year term subject to re-election. Each Director appoints an Alternate Director and is entitled to cast the number of votes originally determined for the purpose of his election. All matters are to be decided by majority vote except where expressly provided otherwise in the Charter. The President, under the direction of the Board of Directors of which he is the chairman, is responsible for the organisation and operations of the Bank. He serves for a five-year term but may be reflected. The Vice-President acts as the deputy of the President in the management and operations of the Bank. In the absence of the President, he exercises the authority and performs the functions of the President. The Bank has been functioning with the simplest possible framework for future expansion. The Bank has 21 departments. The operational functions of the Bank are carried out by the Operations Department and the Project Department. The service functions are performed by the Administration Department and by the Office of the Secretary. The Treasury Department, the Office of the General Counsel, the Economic Office, the Financial Advisor's Office, the Internal Auditor's Office and the Information Office assist in both the operational and service functions. ADB's staff as of 31 Dec, 2010, totaled 2033, from 59 out of 67 countries 6 members of management, and 1024 international and 1803 national and admintstrative staff. At percent H. kuroda is the president of the Bank.

Operations

The Asian Development Bank started functioning from January 1, 1967 and has completed its 36 years. The birth of the ADB is an important milestone of the first Development Decade. The ADB is a living example of the mutual cooperation, independence and complementary efforts based on multilateral co-ordination necessary for the development of the Asian member countries. Starting in 1968 during the past 35 years upto 2003, the Asian Development Bank (ADB) has made a cumulative lending of $105,069.60 million to its developing members.

Starting from 1966 during the part 45 yrs. upto 2010 the ADB has made a cummulative lending of $ 201.392 billion. In the year 2010 alone a lending of $ 17.51 billion was made. The sectorwise distribution of loan and grants in 2010 in percentage had been given below :

Table 36.1   Sectorwise Distribution of Lending

Sector	Loan (%)	Grants (%)
Agriculture and Natural Resources	5.4	11.7
Education	0.6	2.3
Energy	21.4	21.5
Finance	11.0	1.7
Health and Social Protection	1.5	4.0
Industry and Trade	–	–
Public Sector Mangement	7.8	3.0
Transport and ICT	33.4	47.5
Water Supply and Other Services	5.3	2.2
Multisector	13.5	6.1

ICT = Information and Communication Technology

From the above table it is clear that top priority has been given to transport and ICT followed by energy. The largest share of both loans and grants has been received by transport and energy.

ADB's Approach to Millenium Development Goals (MDGS)

At a U.N. World Summit on the Millennium Development Goals (2010) the ADB's President Kuroda held strategic discussion with development partners and governments to achieve MDGs by 2015. Since then ADB is on a fast track to achieve Eight MDGS.

The scale of the task of achieving the Millennium Development Goals (MDGs) is daunting. ;The Asia and Pacific region is home to more than 900 million poor comprising more than two-thirds of the World's population in extreme poverty. It contains 470 million people who have no clean water which is more than half of the world's population without clean water. It also has 70 percent of the developing world people who have no basic sanitation, two-thirds of the world undernourished children and more than 60 percent people infected with tuberculosis.

Overcoming such huge challenges will require stronger basic infrastructure particularly road transport, water supply, sanitation, electricity, information technology, telecommunication and urban low housing. ADB is on right track to face these sechallenges.

If we look at the regional scenario in Asia and the Pacific in reference to Millennium Development Goals we find that South Asia has made good progress on eight MDGs namely indicators of gender equality, HIV prevalence, tuberculosis prevalence, forest cover, protected areas, ozone depleting substance consumption and safe drinking water.

Performance in primary school enrolment has increased and India, the Maldives and Srilanka are on track or likely to exceed MDGS. However, much remains to be done in other South Asian countries. In Bangladesh ADB is helping revitalize education through the Second Primary Education Development Program (PEDP II). In Nepal with the help of ADB's support for primary education net enrolment rate in primary education increased from 72.4 percent in 2004 to 93.7 percent in 2009.

Providing sustainable access to safe drinking water and basic sanitation continues to be important for ADB's operation in South Asia. In Bangladesh, the Chittagong Hill Tracks Rural Development Project provided 180 villages with tap water. In Nepal, the Small Town Water Supply and Sanitation Sector Project brought a clean 24 hour supply of drinking water to nearly 2000 households. In Srilanka the Third Water Supply and Sanitation Sector Project gave 1.4 million people better services.

Thus, ADB is keenly interested in achieving Millennium Development Goals and its operations are on right track.

Asian Development Bank and India

The ADB's operational strategy in India is to assist the Government in achieving its objective of increased economic efficiency and high levels of sustainable economic growth. These objectives, according to the Bank, are to be achieved through (*i*) support for structural reforms; (*ii*) promotion of competition and increased private sector participation; (*iii*) development of regulatory mechanisms and institutional capacities; (*iv*) corporating and restructuring of public sector undertakings, and (*v*) enhanced resource mobilisation through financial sector and capital market reforms. Consistent with the Bank's overall strategy, the Bank's operational programme is guided by the principle of selectivity in both sectoral and geographical aspects. At the sectoral level, the Bank has facilitated development in the energy, transport, communications and urban infrastructure sectors. Geographically, state-level operations are underway in the states of Gujarat and Madhya Pradesh and the state-level operations account for about 50 per cent of annual lending. To mobilise resources, Bank's support for financial and capital market reforms is being made continuously available. The Bank's strategy also advocates support for those selected states which show commitment to policy reforms and have adequate implementation capacity. This geographic focus enables the Bank to maximise the development impact of its assistance. Emphasis has also been given to improve the incentive environment for enhanced private sector participation.

To alleviate the country's serious infrastructure bottlenecks, the Asian Development Bank has actively pursued sectoral reforms in the power, transport and communication sectors through projects and dialogues. The power sector has remained one of the focal points in the Bank's operations with emphasis on the structural reforms, cost recovery and an enhanced role for the private sector.

During 1996, the Asian Development Bank approved five project loans—including one private sector loan without government guarantee—amounting to $788 million in the energy, financial and telecommunication sectors. The Bank also approved a programme loan for $2.50 million in the form of 46 project loans. During 1997, the Bank approved six loans amounting to $563 million in the housing, energy, ports and financial sectors. In addition to these six loans, the Bank also provided ten technical assistance grants totalling $6.27 million. During 1998, the Bank's operations were affected by the international sanctions imposed against new lending by multilateral finance institutions in the wake of nuclear tests conducted by India in May 1998. As a result, only one loan of $250 million for the Rajasthan Urban Infrastructure Development Project was approved. The Bank also provided an amount of $4.5 million for five technical assistance grants.

In 2003, the ADB lent an amount of $1,532.0 million for nine loans for the development of rural and state roads, national highways, water supply and power development. Bank's operations in India were aimed at poverty reduction. Projects focused on reforms, public resource management, better governance, building rural and urban infrastructure and improving the environment. In infrastructure, the focus was on removing bottlenecks and on promoting the private investment in public infrastructure such as road systems, inland water transport, railways, gas-based energy, and power. In 2003, in addition to nine loans, 22 technical projects worth $14.7 million were also approved.

The ADB's loans and equity investments for private sector in 2003 stood at $102.65 million These included loan of $62.00 million for Tata-Delhi transmission, $20.00 million to Sundaram Home Finance Limited; $20.00 million to Dewan Housing Finance Corporation Limited, and equity investment of $0 65 million in Centurian Bank Limited. The cumulative private sector lending and equity investments at the end of 2003 for 27 projects amounted $374.15 million.

Approval of Lending To India by ADB in 2010

ADB approved a lending of $2.11 billion to India from OCR in 2010. The sectorwise distribution is given below:

Table 36.2   Sectorwise Distribution of Lending

	Sector	$Million
1.	Agribusiness Infrastructure Development programnie-Tranchl 1	
2.	Assam Integrated Flood and River tank Erosion Risk Management	67.5
3.	Assam Power Sector Enhancement Investment Program, Tranche 2	89.7
4.	Bihar Power Sector Improvement	132.2
5.	Bihar State Highways	300.0
6.	Himanchal Pradesh Clean Energy Development Investment Program	208.0
7.	Infrastructure Development Investment Program for Tourism.	43.4
8.	Karnataka State Highways Improvement	315.0
9.	Madhya Pradesh Power Sector Investment Program	69.0
10.	Micro, Small and Medium Enterprises Development	50.0
11.	National Capital Region, Urban Infrastructu Financing Facility	78.0
12.	North Karnataka Urban Sector Investment Program	123.0
13.	Rajasthan Urban Sector Development Investment Program	63.0
14.	Rural Roads Sector II Investment Program	222.2
15.	Second India Infrastructure Trenche Financing Facility.	250.0
16.	Sustainable Coastal Protection and Management Investment Program	51.6

It is clear from the above table that the highest-amount of loan has been approved for the State Highways particularly in Bihar and Karnataka. The second place goes to power sector including Electricity Clean Energy Development Investment Program. From the point of view of rural development it is encouraging that Rural Roads Sector has also drawn the attention of ADB along with urban development.

In addition to the above ADB continues to support India in addressing climate change issues, Work to achieve this has included the preparation of strategies to mainstream adaptation concerns in the management of the Cauvery, Chambal and Sutluj river sub-basins. ADB is also increasing support for the development of renewable energy, including solar power, to help India along a low-carbon growth path. Through the Rural Roads Sector-I project, ADB supported construction of 975 Kms of all weather roads connecting rural areas in Chhattisgarh and Madhya Pradesh. Through the Karnatka Urban Development and Coastal Environment Management Project (1999-2009) ADB helped to increase drinking water supplies, expand the drainage net-work, build the quality sewage plans in 10 coastal towns.

Evaluation

The Asian Development Bank is now 36 years old and in this short period of 36 years through its large lending and technical assistance services it has made its impact felt in the Asian and Pacific region. It has been able to provide confidence to its members and has amply justified its role as a regional development bank. It has lent over eight billion dollars to the poor members of the region for various development projects at concessional rates. It has been able to promote infrastructural projects in the region. The Bank has also made several feasibility studies in the region. The establishment of the Asian Development Bank Institute (ADBI) in 1996 is a landmark development. The ADBI identifies effective development strategies and improves the capacity for sound management of the agencies and organisations in DMCs engaged in the development work. The target clientele of ADBI's research activities includes groups and individuals engaged in the development work in DMCs. The Bank has a vital role to play in the fostering of capital markets in the region. In most developing countries in the ECAFE region capital markets are small and unsophisticated and private savings are mainly devoted to traditional pursuits. The ADB can directly and through the national development institutions assist the Asian member countries' capital markets by marketing its own bonds in these markets, by dealing in the other securities from its own portfolio, and by underwriting the issue of new industrial securities.

Some of the key activities of Asian Development Bank in 2003 comprised rebuilding the member states emerging from several years of conflict. In this, the ADB provided $1 billion of assistance to Afghanistan to support the reconstruction activities. Apart from assisting Afghanistan in providing essential infrastructure facilities, the ADB worked closely with other donor agencies in Sri Lanka to assess the rehabilitation needs of conflict-affected districts with ADB contributing $1 billion. The ADB has established a new accountability mechanism to better address the concerns of the project-affected people. The ADB is continuing reforms which will be increasingly effective in Bank's fight against poverty in the region. The Bank has strengthened senior management by creating the office of the fourth Vice-President for Knowledge Management and Sustainable Development. A new results management unit has also been established to lead implementation of ADB's action plan for managing for development results.

The Asian Development Bank's medium-term strategy has been to achieve the objectives of economic growth, poverty reduction, improving women's status, supporting human development (including population planning) and environmental protection in the developing member countries. The Asian Development Bank has also actively pursued cofinancing activities with official as well as commercial and export credit sources. It has also entered into equity investment operations.

Suggested Readings

1. Asian development Bank, *Agreement Establishing the Asian Development Bank.*
   Annual Reports.

Questions

1. Explain the main purposes and working of the Asian Development Bank.
2. Explain the role of the Asian Development Bank in the economic development of Asia's poor countries. How has India been benefited from her membership of the Asian Development Bank?

Author Index

Subject Index